International Series on Biomechanics

BIOMECHANICS IX-A

EDITORS

David A. Winter, Ph.D.	Department of Kinesiology
Robert W. Norman, Ph.D.	University of Waterloo
Richard P. Wells, Ph.D.	
Keith C. Hayes, Ph.D.	University of Western Ontario
Aftab E. Patla, Ph.D.	Department of Kinesiology
	University of Waterloo

HUMAN KINETICS PUBLISHERS
Champaign, Illinois

Proceedings of the Ninth International Congress of Biomechanics held in 1983 at Waterloo, Ontario, Canada

Production Director: Sara Chilton
Copy Editor: John Sauget
Managing Editor: Sue Wilmoth
Typesetting: Sandra Meier and Aurora Garcia
Text Layout: Janet Davenport
Cover Design and Layout: Jack Davis

Library of Congress Catalog Card Number: 82-84703

ISSN: 0360-344X
ISBN: 0-931250-52-8 (set)
ISBN: 0-931250-53-6 (IX-A)

10 9 8 7 6 5 4 3 2 1

**Human Kinetics Publishers, Inc.
Box 5076, Champaign, Illinois 61820**

CONTENTS
VOLUME IX-A

Preface

It was my special privilege to welcome all scientists and special guests to the 9th International Congress of Biomechanics. Over 400 researchers and students came from 25 countries to the first ISB Congress to be held in Canada and we at the University of Waterloo were honored to host the event. We understand that you found the accommodations and social events to your satisfaction and the scientific sessions were smoothly run and stimulating.

These Proceedings represent the net summary of the papers presented during both poster and free communication sessions. Several specially sponsored sessions were incorporated in the program. These included one on the evaluation of sports protective equipment and occupational biomechanics plus three rehabilitation and orthopaedic sessions relating to joints, ligaments, the spine, and pathological gait. Rounding out the program were a wide variety of basic and applied papers encompassing all aspects of normal, sport-related, and pathological movement.

A new feature of this 9th Congress was the New Investigators Awards generously donated by Mrs. Ursula Wartenweiler and the Canadian Society for Biomechanics. The two winning papers are noted in the program along with all invited speakers. The Wartenweiler Memorial Lecture was presented by Dr. Uros Stanic who gave a stimulating slide and movie summary of the major work of his group about the electrical stimulation of the handicapped.

Finally, I would like to thank all those many hard working people who made this Congress and these Proceedings a success. It represented far more work than we anticipated but with many willing hands the task was far less demanding and the results very rewarding.

David A. Winter

WARTEN-
WEILER
MEMORIAL
LECTURE

A Heuristic Approach to the Synthesis of Movement by Functional Electrical Stimulation Orthoses

Uroš Stanič
University "E. Kardelj" Ljubljana, Yugoslavia

Doctor, I am in a hurry,
It is life that is waiting.
Do the impossible,
For I crave for life so much!

This is a poem by the recently deceased Belgrade poet Stevan Budimirović, who was bedridden for life, and could live only by using a mechanical respiratory device. He has published several collections of poems, obtained an M.Sc. degree, and was working towards his Ph.D. when his creative life stopped.

Among the many rehabilitation methods used in the last decade in the therapy and orthotics of paretic patients, functional electrical stimulation (FES), has shown the most prominent development. By means of FES, muscles deprived of nervous control can be stimulated with the aim of evoking their contraction in order to obtain useful functional movements. One of the first orthotic aids based on the principles of FES was the electronic peroneal brace (Liberson, Holmquest, Scott, & Dow, 1961). That was in 1961; later, beginning around 1965, intensive research was started in the USA and Yugoslavia (Dimitrijević, Gracanin, Prevec, & Trontelj, 1968; Vodovnik, Crochetiere, & Reswick, 1967). Five years later, implantation technology appeared (Yergler, Wilemon, & McNeal, 1971) and soon after that, multichannel stimulation (Kralj, Trnkoczy, & Aćimović, 1974; Milner, 1972). The purpose of this talk is to describe the efforts of our group after the above-mentioned period, to establish the synthesis of gait based on biomechanical development, using the latest knowledge in medical biocybernetics and technology. Also, a temporary withdrawal from the analytical approach to the synthesis of gait and its replacement by the heuristic approach took place in order to approach the final goal, a more rapid development of new therapeutic methods and aids. Special efforts were directed towards the measurement of the kinematic and dynamic parameters of gait, together with EMG measurements, and as a consequence, towards the basic development of qualitative and quantitative methods of gait evaluation and assessment.

Here, first the positioning of the human ankle by means of a closed-loop regulator will be shown, then the methodology of multichannel stimulation will be described, together with the multichannel stimulators themselves, the measuring techniques used, and the application of the qualitative and quantitative methods in the gait evaluation process.

Controller Using FES Antagonistic Muscle Groups and Position Feedback

In the early seventies, after the first successful clinical results in the application of FES it was proposed to synthesize gait for more severely disabled paretic and paraplegic patients. The idea of a hierarchical three-level structure of orthoses seemed plausible at the time (Tomović, 1968). The hypothesis to be tested was whether it is possible to control the movements of the ankle by means of FES with the same dynamic and accuracy that a normal person can do voluntarily.

In the case of a positive result, the hierarchically lowest level, that is the actuator level, for one joint would be realized. The coordination of more complicated orthoses involving several joints would be performed by the second (processor) level, supervised by the patient. The experiment was performed on normal subjects and a hemiparetic patient (Stanic & Trnkoczy, 1974). The regulator was constructed using a hybrid computer, special surface stimulation output stages, and a flexor position transducer. Dorsal and plantar ankle muscles were stimulated. The first step towards the realization of the regulator was the linearization of some nonlinear characteristics of muscles in the saturation region. This was achieved using the results of measurements which led to the construction of a mathematical model of the natural control system of the ankle and consequent design of appropriate modulators. During the measurements, the subjects were in a half-lying position. Figure 1a shows the positioning results in condi-

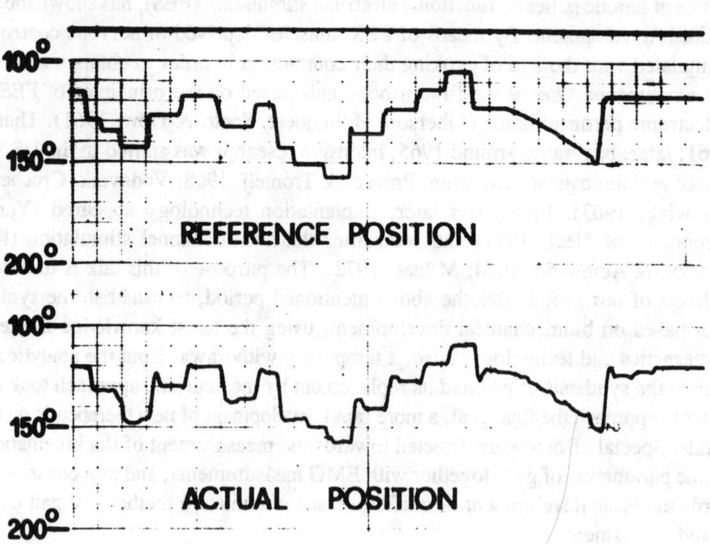

Figure 1a—Positioning of the ankle joint by stimulation of the antagonistic muscles, using PI controller and position feedback.

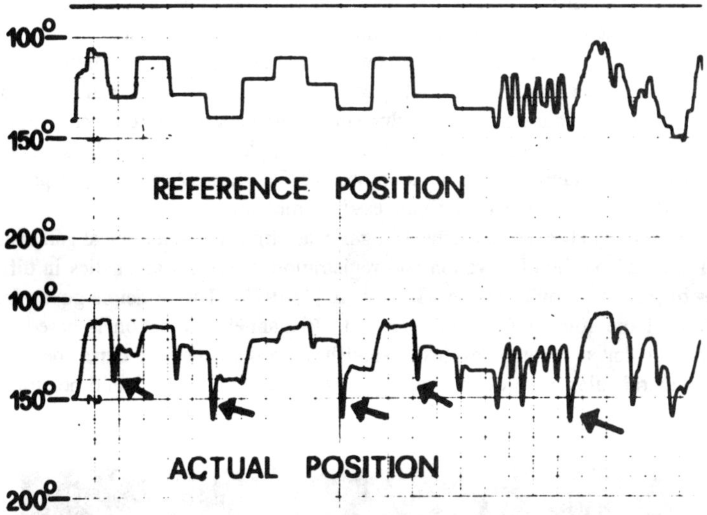

Figure 1b—Positioning of the ankle joint by stimulation of the antagonistic muscles, PI controller, and positioning feedback with an external load of 3 N•m in the direction of plantar flexion.

tions assuring that the ankle was not externally loaded. An exact analysis of the results showed that movements obtained by means of a position regulator match those with voluntary control, in terms of static and dynamic characteristics. The rise times were from 200 ms to 230 ms, and the static error was up to 2°. Figure 1b shows the results obtained after the ankle joint had been loaded by 3 N•m in the plantar flexion direction. The positioning towards dorsal flexion is not essentially worse than in the case of the unloaded joint, while overshoots of approximately 25° appeared in plantar flexion movements (as denoted with arrows). A healthy man is capable of following the changes in load and adaptively compensating for them using his sensory system. In our case there was no technical means for online force measurement. Thus the basic analytical research was abandoned and the realization of the hierarchically controlled system of multi-joint closed-loop regulators on FES principles was temporarily given up.

Correction of Paretic Gait by Multichannel Stimulation

Favorable results in the application of electrical peroneal braces used for the correction of footdrop, and positive preliminary results from three channel stimulation in the swing phase of gait of hemiparetic patients led to the following conclusion: a hemiparetic patient during standing and walking is subject to real forces (stresses) and therefore represents a real mathematical model that encompasses all of the consequences of damage to the central nervous system. Should it be possible to correct paretic gait by means of multichannel stimulation of all the muscle groups and to demonstrate the correction using qualitative evaluation, then the synthesis of gait can be achieved at the same time as determining the actuator and processor levels of hierarchically controlled multichannel FES orthosis. Several types of 6-channel stimulators have been developed. Initially, the stimulation of the flexors and the extensors of the hip, the knee and the ankle of hemiparetic patients (Stanic et al., 1978) was done, and soon after that was extended to incompletely paraplegic patients. The results led toward research of the immediate effects of multichannel stimulation in the correction of paretic patients' gait, and just recently a study has been completed dealing with its therapeutic effects (Malezič, Stanič, Klajić, & Aćimović, 1983). Figure 2 shows a paraparetic patient fitted with a 6-channel microprocessor stimulator.

The basic method in the assessment of gait was clinical kinesiological gait analysis, which is based on the observation and registration of typical anomalies in different phases of gait (Aćimović, Stanič, Gros, & Bajd, 1975). The principle governing the selection of the stimulation sequences in multichannel stimulation is based on the presumption that stimulation lessens or abolishes the anomalies and experimentally obtained general rules were established as to what muscle group should be stimulated

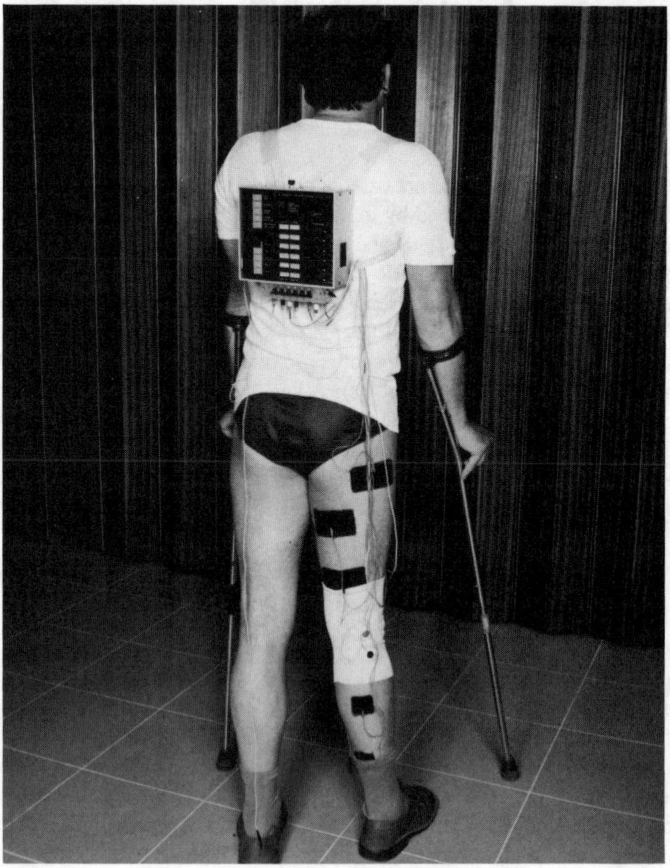

Figure 2—Paraparetic patient with a 6-channel microcomputer-controlled stimulator and electrodes positioned for stimulation of the main muscle groups.

to reduce a particular anomaly. The EMG records of leg muscle activities do not correspond to the required stimulation sequences, as was initially anticipated. Besides clinical gait analysis, the effects of therapy were evaluated by measurement of goniograms, basograms, and recording of forces produced against the shoe soles and the axial force in crutches.

Special attention was paid to computer records, which yielded meaningful, condensed information. Figure 3 shows goniograms obtained without stimulation (right side) and with stimulation (left side) for normal and handicapped subjects.

The effects of stimulations are presented in Figures 4a and 4b. Figure 4a shows the vertical component of the ground reaction force (upper part) and its distribution along the shoe sole (lower part).

The most important immediate effects of FES (Stanič et al., 1978) are reduction of the anomalies, particularly in the stance phase, and significant improvement in gait symmetry during the stance phase in at least one half of the treated patients (see Figure 4b). The results of the control study show that therapy with a multichannel stimulation lasting for several months enables faster rehabilitation in comparison with traditional methods, and results in a higher level of rehabilitation. Certain indications exist which suggest that such therapy may reactivate or essentially normalize motor activity during walking, and this has been proved by EMG measurements. The accompanying instrumentation has been purposefully developed so that all gait parameters can be measured. Diminishing its stochastic nature means a more comfortable gait and a smaller consumption of energy (Klajić & Trnkoczy, 1978). The essential result of multichannel stimulation is manifested in a more reproducible gait, and therefore its effects can be interpreted as extra information input, which contributes to reconstruction of coor-

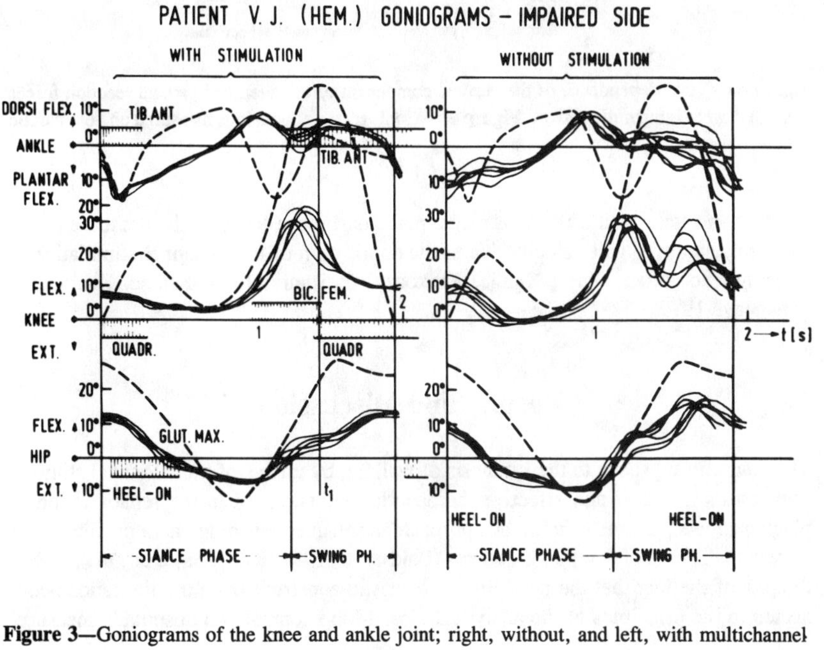

Figure 3—Goniograms of the knee and ankle joint; right, without, and left, with multichannel stimulation.

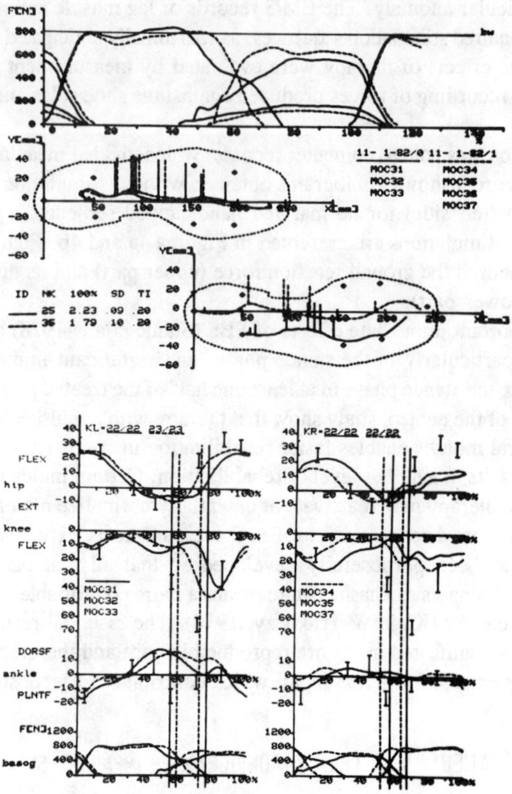

Figure 4a—Time dependence of the vertical component of the averaged ground reaction forces measured on a hemiparetic patient. **Figure 4b**—Goniograms of both legs measured on hemiparetic patient.

dination and improved activity of plegic patients. Figure 5 shows the decrease of entropy in relation to gait velocity due to the application of multichannel stimulation in a hemiplegic patient (Bajd, Kljajić, Trnkoczy, & Stanič, 1974; Kljajić, Krajnik, & Trnkoczy, 1979).

Results and Discussion

The heuristic approach to the synthesis of walking by means of multichannel stimulation results in a new and effective therapeutic method, used in the rehabilitation of plegic patients. A similar heuristic approach has enabled standing and a primitive gait pattern in the case of paraplegic patients (Bajd et al., 1981; Kralj, Bajd, & Turk, 1980). In spite of the fact that the need for the heuristic approach from the limitations connected to the principles of the analytical closed-loop control and unsuitable measure-

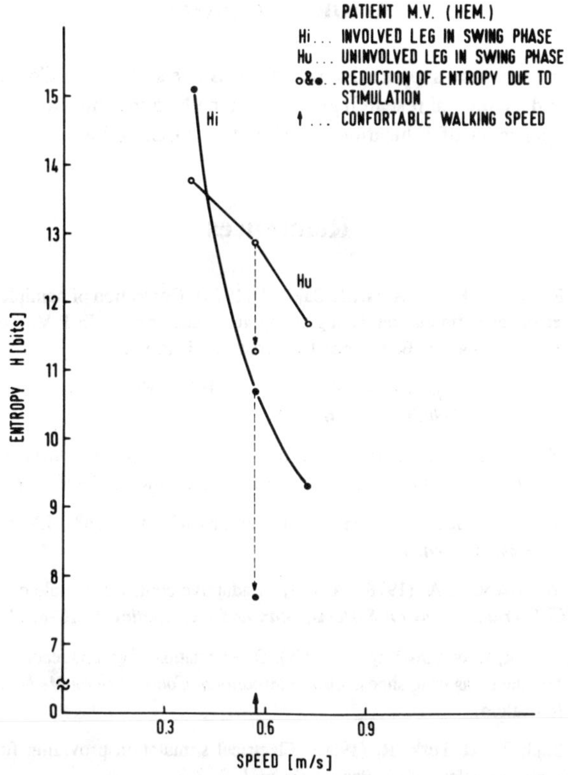

Figure 5—Relationship of gait parameter entropy as a function of speed and its considerable reduction due to multichannel stimulation.

ment technology, the analytical tools of biomechanics supported and followed the development described.

Therefore it is quite possible that the technology of implantable stimulators (Strojnik, Vavken, Aćimović, & Stanič, 1982), sensors, and microprocessors may allow the construction of a new generation of totally implanted multichannel stimulators. Such stimulators could synthesize the gait of even the most handicapped plegic patients using the principles of closed-loop adaptive regulators by means of online measurements of the kinematic and kinetic parameters of the system and their application to the mathematical model determined. Until this goal is reached, it is mandatory that patients be rehabilitated with effective methods of multichannel stimulation during the therapy phase, and to then offer them, for orthotic purposes, simple, commercially available stimulators.

The work described in this lecture is the product of the interdisciplinary group from the Rehabilitation Engineering Center of Ljubljana, Yugoslavia, performed during the last ten years. I wish to express special thanks to Dr. L. Vodovnik, who laid the foundations for this biocybernetics research in the field of FES.

Acknowledgment

These investigations were supported in part by the Slovene Research Council, Ljubljana, Yugoslavia, and by several research grants from the National Institute of Handicapped Research, Department of Education, Washington, D.C., USA.

References

AĆIMOVIĆ, R., Stanic, U., Gros, N., & Bajd, T. (1976). Correction of hemiplegic gait pattern by multichannel surface stimulation during swing and stance phase. In P.V. Komi (Ed.), *Biomechanics V-A* (pp. 444-451). Baltimore: University Park Press.

BAJD, T., Kljajić, M., Trnkoczy, A., & Stanič, U. (1974). Electrogoniometric measurement of step length. *Scand. J. Rehab. Med.*, **6**, 78-80.

BAJD, T., Kralj, A., Sega, J., Turk, R., Benko, H., & Strojnik, P. (1981). Use of the two-channel functional electrical stimulator to stand paraplegic patients. *Phys. Ther.*, **67**, 526-527.

DIMITRIJEVIĆ, M., Gračanin, R., Prevec, F., & Trontelj, T. (1968). Electronic control of paralysed extremities, *Biomed. Eng.*, **3**, 8-19.

KLAJIĆ, M., & Trnkoczy, A. (1978). A study of adaptive control principle orthoses for lower extremities. *IEEE Transactions on Systems, Man and Cybernetics*. SMC-8, 20-28.

KLAJIĆ, M., Krajnik, J., & Trnkoczy, A. (1979). Determination of ground reaction and its distribution on the foot by the measuring shoes. *XIIth International Conference on Medical and Biological Engineering*, Jerusalem.

KRALJ, A., Bajd, T., & Turk, R. (1980). Electrical stimulation providing functional use of paraplegic patients' muscles. *Med. Progr. Technol.* **7**, 3-9.

KRALJ, A., Trnkoczy, A., & Aćimović, R. (1974), Improvement of locomotion in hemiplegic patients with multichannel electrical stimulation. *Human Locomotor Engng.*, pp. 45-50.

LIBERSON, W.T., Holmquest, H.I., Scott, D., & Dow, M. (1961). Functional electrotherapy in stimulation of the peroneal nerve synchronized with the swing phase of the gait of hemiplegic patients. *Arch. Phys. Med. Rehab.* **42**, 101.

MALEZIC, M., Stanic, U., Klajić, M., & Aćimović, R. (1983). Therapeutic effects of multisite electrical stimulation of pathological gait. *Scand. J. Rehab. Med.*

MILNER, M. (1972). Human locomotion and possibilities for programmed electrical stimulation of available musculature. *South African Medical Journal*, **46**374.

STANIČ, U., & Trnkoczy, A. (1974). Closed-loop positioning of hemiplegic patient's joint by means of FES. *IEEE Transactions on Biomedical Engineering*, **21**, 365-370.

STANIČ, U., Aćimović-Janezič, R., Gros, N., Trnkoczy, A., Bajd, T., & Klajić, M. (1978). Multichannel electrical stimulation for correction of hemiplegic gait. *Scand. J. Rehab. Med.*, **10**, 75-92.

STROJNIK, P., Vavken, E., Aćimović, R., & Stanic, U. (1982). A new Ljubljana implantable peroneal stimulator. *Proc. World Congress Medical Physics and Biomedical Eng.*, pp. 12.08.

TOMOVIĆ, R. (1968). Prosthetics and orthotics of human extremities. *Mathematical Biosciences*, 151-157.

VODOVNIK, L., Crochetiere, W.J., & Reswick, J.B. (1967). Control of a skeletal joint by electrical stimulation of antagonists. *Medical and Biological Engineering,* **5**, 97-109.

YERGLER, W.G., Wilemon, W., & McNeal, D. (1971). An implantable peroneal nerve stimulator to correct equinovarus during walking. *Journal of Bone and Joint Surgery,* **53A**, 1660.

I.
MUSCLE MECHANICS

Structural and Functional Characteristics of Thigh Muscle in Athletes

N. Tsunoda
Kokushikan University, Japan

S. Ikegawa
Japan Women's University, Japan

H. Yata
Wako University, Japan

M. Kondo
Nihon University, Japan

H. Kanehisa, T. Fukunaga, and T. Asami
University of Tokyo, Japan

Several investigations have demonstrated that maximum muscular strength is closely related to the muscle cross-sectional area (Ikai & Fukunaga, 1968; Maughan, Watoson, & Weir, 1983; Nygaard, Houston, Susuki, Jorgensen, & Saltin, 1983), and that strength per unit muscle cross-sectional area is approximately the same value regardless of age (Ikai & Fukunaga, 1968) and sex in the untrained subjects (Ikai & Fukunaga, 1968; Nygaard et al., 1983; Schantz et al., 1983). On the other hand, Maughan and associates (1983) reported that although there was no significant difference in the cross-sectional area of muscle between the sprinters and distance runners, the strength of sprinters was greater than that of distance runners even in terms of unit muscle area. Furthermore, Ikai and Fukunaga (1970) demonstrated that the strength per unit cross-sectional area of muscle improved with strength training.

From these observations, it may be assumed that muscle development and/or improvement of muscle strength in the athletes will be specific to the training protocols. The purpose of this study is to clarify the morphological and functional characteristics of the thigh muscle in athletes from the view points of the cross-sectional area of the quadriceps femoris muscle and maximum voluntary strength of the knee extension.

Material and Methods

Subjects

Subjects were male Japanese elite athletes who were specialized in seven different sporting events: sumo wrestlers (15 professional and 16 amateur), oarsmen (22), volleyball players (8), speed skaters (16), soccer football (12), sprinters (10), distance runners (10), and 14 nonathletes. The means and standard errors for age, body height, and weight of subjects are presented in Table 1.

Table 1

Physical Characteristics of Subjects

	Number of Subjects	Age (yrs)	Body Height (cm)	Body Weight (kg)
Sumo wrestler				
Professional	15	17.52 (0.48)	178.44 (1.32)	99.89 (3.34)
Amateur	16	19.96 (0.33)	171.69 (0.71)	98.87 (2.30)
Oarsman	22	20.61 (0.33)	177.59 (0.89)	74.24 (1.34)
Volleyball	8	19.67 (0.24)	181.01 (2.06)	72.77 (2.14)
Speed skater	16	20.44 (0.47)	170.84 (0.84)	68.46 (0.93)
Soccer football	12	25.54 (0.80)	171.53 (0.88)	67.27 (1.07)
Sprinter	10	19.50 (0.29)	172.24 (1.54)	64.12 (1.56)
Distance runner	10	20.40 (0.32)	166.32 (1.87)	56.89 (1.28)
Nonathletes	14	20.85 (0.38)	168.00 (1.61)	62.21 (1.83)

Mean (S.E.)

Measurement of the Muscle Cross-sectional Area

The cross-sectional image of the right thigh was obtained by using the ultrasonic apparatus (ALOKA, Echo-vision SSD-120) connected with the circular compound scanner. The ultrasonic photograph of thigh was taken at the midpoint marked on the skin between the greater trochanter and the lateral condyle. The tissues such as fat, bone and muscle bundles were outlined clearly from the ultrasonic photograph (see Figure 1). The cross-sectional area of the tissues was measured by using a planimeter. In the present study, the muscle area of quadriceps femoris was analyzed as the leg extensor muscle.

Measurement of Muscle Strength

The muscle strength in maximum voluntary knee extension was measured isometrically using a specially designed strain gauge dynamometer. The subjects were requested to sit on the experimental chair (hip joint angle of 90°) and were fastened with seatbelts. The lower right leg was attached to the strain gauge sensor of the dynamometer with

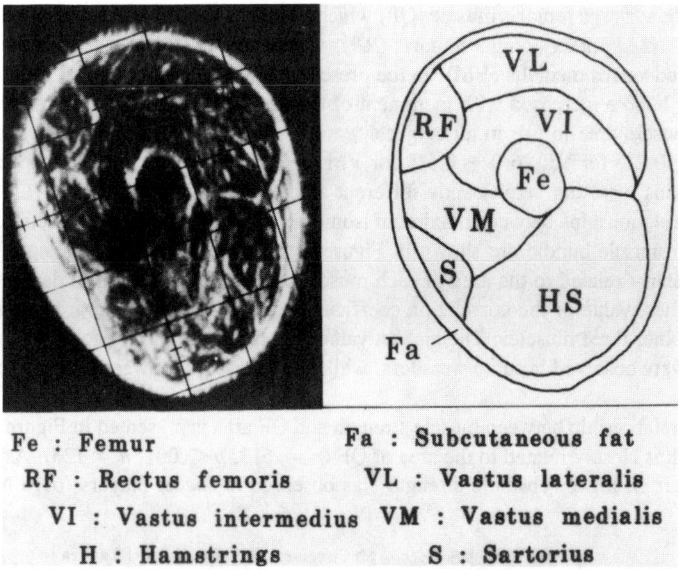

Figure 1—Cross-sectional image of the thigh using ultrasonic method.

a knee angle of 100° (180°: full extended leg). Three measurements of muscle strength were made and the highest value was used for analysis. The absolute strength (AS, strength per unit muscle area) was calculated by the following equation:

$$AS = (MS \times 8.1)/A$$

where A is the cross-sectional area of the quadriceps femoris muscle, MS is the maximum strength measured by the strain gauge dynamometer, and 8.1 is a leverage for knee extension obtained from cadavers.

Results

The cross-sectional area of the whole thigh is closely related to the fat and muscle area. High significant correlation coefficients between whole thigh area and each tissue area were obtained: $r = .897$ for muscle and $r = .884$ for fat area ($p < .001$, $n = 123$). On the other hand, the bone area was about 7.3 cm^2 and this was independent of the whole thigh area.

The sumo wrestlers demonstrated 303 cm^2 for the cross-sectional area of the whole thigh, 93 cm^2 of fat, and 203 cm^2 of muscle area, and these were significantly larger than those of other athletes ($p < .001$). The distance runners and soccer players showed smaller areas of fat than other athletes. The muscle area of all athletes except for the distance runners was significantly larger than that of the nonathletes (134 cm^2) ($p <$ .01). The distance runners indicated similar muscle area to the nonathletes.

The quadriceps femoris muscle (QF) which produces knee extension, consists of the four muscle bundles: rectus femoris (RF), vastus lateralis (VL), vastus intermedius (VI), and vastus medialis (VM). In the present study, the cross-sectional area of each muscle bundle increased with increment of the area of QF. The percentage ratios of each muscle area to QF in all subjects were 32.8 ± 0.4% (mean ± S.E.) for VL, 30.0 ± 0.2% for VI, 26.0 ± 0.5% for VM, and 11.3 ± 0.2% for RF, respectively. The ratios were not significantly different among the athletic groups.

The relationships between maximum isometric strength and the cross-sectional area of each muscle bundle are shown in Figure 2. Maximum strength was significantly and linearly related to the area of each muscle bundle. The VL muscle demonstrated the highest value of the correlation coefficient between strength and area, compared to the other three muscles. The highest values for strength and for each of the muscle areas were observed in sumo wrestlers, while the lowest values were found in distance runners.

The relationship between muscle strength and QF area is presented in Figure 3. The strength is closely related to the area of QF ($r = .613, p < .001, n = 123$). As shown in Figure 3, higher absolute strength was observed in soccer players (69.6 N/cm²),

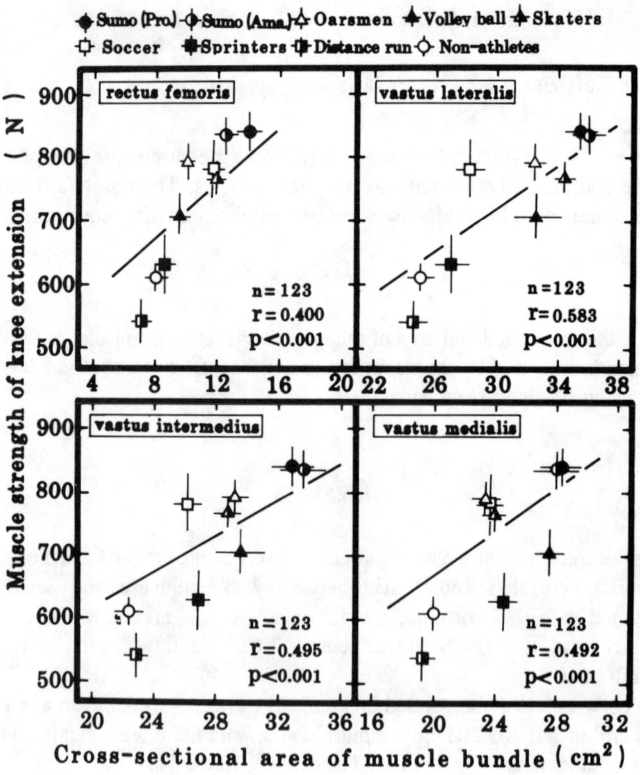

Figure 2—The relationships between maximum strength and cross-sectional area of each muscle bundle in athletes.

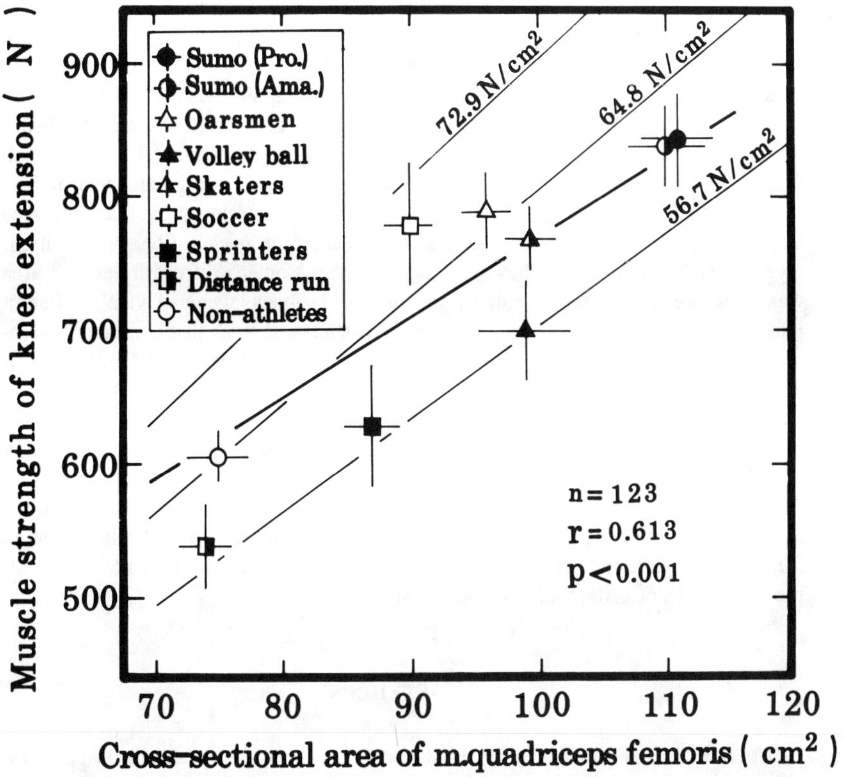

Figure 3—The relationship between maximum strength and cross-sectional area of quadriceps femoris in athletes.

and lower strength in volleyball players and sprinters (57.8 N/cm²), respectively. There were, however, no significant differences in the absolute strength among the athletes and nonathletes.

Discussion

Ikai and Fukunaga (1968), and Nygaard and associates (1983) reported that arm strength was closely related to the muscle area of the upper arm, and also Maughan and associates (1983) showed a significant linear relationship between muscle strength and muscle area for the knee extensors. These findings were obtained from untrained subjects. In the case of elite athletes who participate in vigorous training programs, a highly significant relationship between the knee extension strength and area of QF muscle was observed, as indicated in the present study.

Moreover, an interesting finding in this study was that VL muscle showed the highest correlation coefficient between strength and area, compared to the other three muscles.

Taking into consideration the large muscle area and high correlation for the VL muscle, it suggests that VL effects most of the strength of knee extension.

In the present study, the absolute strength indicated approximately the same value regardless of the sporting events. This was unexpected. Furthermore, there was no significant difference in the absolute strength between the trained and untrained subjects. These results differ from the findings of Maughan and others (1983) in which the maximum strength per unit of muscle area in sprinters was greater than that in distance runners. They suggested that the difference in muscle strength per unit area between sprinters and distance runners depended on the muscle fiber composition. Nygaard and others (1983), however, reported that isometric strength per unit area was approximately the same value among untrained subjects who had widely different muscle fiber composition. In addition, previous studies showed that maximum voluntary isometric strength was not affected by muscle fiber composition (Hulten et al., 1975; Nimmo & Maughan, 1983; Nygaard et al., 1983; Schantz et al., 1983; Thorstensson et al., 1976). Considering these findings, the quality of muscle in terms of muscle fiber composition could not be more decisive to isometric absolute strength than muscle mass.

In conclusion, the present study indicated that the size of each muscle bundle of the quadriceps femoris muscle greatly contributed to knee extension. The absolute strength for athletes in different sporting events ranged between 57.8 N/cm² and 69.6 N/cm², but no significant difference was observed.

References

HULTEN, B., Thorstensson, A., Sjodin, B., & Karlsson, J. (1975). Relationship between isometric endurance and fiber types in human leg muscles. *Acta Physiologica Scandinavica,* **93**, 135-138.

IKAI, M., & Fukunaga, T. (1968). Calculation of muscle strength per unit cross-sectional area of human muscle by means of ultrasonic measurement. *Internationale Zeitschrift für Physiologie,* **26**, 26-32.

IKAI, M., & Fukunaga, T. (1970). A study on training effect on strength per unit cross-sectional area of human muscle by means of ultrasonic measurement. *Internationale Zeitschrift für Physiologie,* **28**, 173-180.

MAUGHAN, R.J., Watoson, J.S., & Weir, J. (1983). Relationship between muscle strength and muscle cross-sectional area in male sprinter and distance runners. *European Journal of Applied Physiology,* **50**, 309-318.

NIMMO, M.A., & Maughan, R.J. (1983). Influence of variation in muscle fiber composition on the ratio of strength to cross-sectional area of m. quadriceps femoris in man. *Medicine and Science in Sports,* **15**(2), 178. (abstract)

NYGAARD, E., Houston, M., Suzuki, Y., Jorgensen, K., & Saltin, B. (1983). Morphology of the brachial biceps muscle and elbow flexion in man. *Acta Physiologica Scandinavica,* **117**, 287-292.

SCHANTZ, P., Randall-Fox, E., Hutchison, W., Tyden, A., & Åstrand, P.O. (1983). Muscle fiber type distribution, muscle cross-sectional area and maximal voluntary strength in humans. *Acta Physiologica Scandinavica,* **117**, 219-226.

THORSTENSSON, A., Grimby, G., & Karlsson, J. (1976). Force-velocity relations and fiber composition in human knee extensor muscle. *Journal of Applied Physiology,* **40**(1), 12-16.

The Functional Significance of Architecture of the Human Triceps Surae Muscle

R.D. Woittiez, R.H. Rozendal, and P.A. Huijing
Free University, Amsterdam, The Netherlands

The aim of this study is to relate the geometrical behavior of the constituting parts of the Triceps Surae (TS) muscle (m. gastrocnemius c. mediale [GM], c. laterale [GL], and m. soleus [SO] to the performance of these muscles by means of a model.

Methods

GM, GL, and SO are reconstructed based on measurements performed on material from human cadavers. The distance between the center of rotation of the knee and ankle joint (segment length) as well as the length and width of the tendon plates in the fleshy part of the muscle were measured with calipers. The muscle volume was determined by submerging the muscles and measuring the volume of the displaced water. From the heads of the gastrocnemius, fibers were teased from bundles at 35 locations and the average number of sarcomeres determined (Huijing, 1981). From this and the sarcomere optimal length (2.6 μm, Close, 1972; Huxley, 1968) the mean fiber optimal length can be calculated. For SO the fiber optimal length is estimated. Within one muscle or head the fibers are assumed to be of equal length (Huijing). Pennation of the muscles is quantified by the index of architecture (i_a) which equals the ratio of fiber length and muscle length at muscle optimal length (Woittiez, Huijing, & Rozendal, in press). The muscle length is calculated from tendon plate length, fiber length and angle of pennation, the last variable estimated directly from cadaver muscles.

The behavior of TS length during fast gait was calculated using data of Sutherland and Hagy (1972) as well as Grieve, Pheasant, and Cavanagh (1978) using regression lines relating knee and ankle joint angles to muscle length (Grieve et al.). The lever arm of TS with respect to both joints is calculated from the slope of these regression lines.

The equilibrium position of the knee joint obtained by arthography (Such, Unsworth, Wright, & Dowson, 1975) is on the average 75° (Heerkens, Woittiez, Huijing, & Rozendal, 1983) and is estimated to be 85° for the ankle joint (0° being full knee extension and plantar flexion, respectively). It is assumed that GM, GL, and SO attain their optimal length (l_o) at these joint angles.

The relation between the geometrical reconstruction of TS and its parts and their functional characteristics is calculated using a three dimensional muscle model consisting of a muscle of constant volume with fibers running between rigid kite shaped tendon plates (Woittiez, Huijing, & Rozendal, 1983). The fibers of the modelled muscle have classical length force (Close, 1972) and force velocity characteristics (concentric: Close; eccentric: Sugi, 1972). The muscles are assumed to exert a tension of 29.4 N/cm² (3 kg/cm²) (Close). The maximal velocity of contraction is taken to be 10 times the fiber resting length (Close). Peak power at velocities occurring during gait and maximal power is calculated ($P = F \cdot v$).

Results

Table 1 gives a summary of the most important variables of TS. The width of the normalized length force relations of GM, GL, and SO is 21.5, 29.5, and 14.5% of l_o, respectively (see Figure 1A). SO has the largest maximal isometric force, followed by GM and GL (see Figure 1B). The absolute length range of GL is 51.8 mm, of GM 43.4 mm, and of SO 34.8 mm (see Figure 1B). As the area under the absolute length-force relation is a measure of maximal external work performable in the direction of the line of pull of the muscle, a ranking SO, GM, GL is obtained for this variable (see Figure 1B).

Figure 2A shows the predicted length changes of GM, GL, and SO during fast walking as well as their optimal length. It is clear that for this activity SO uses approximately 60% of its physiological length range. For GM this value is approximately

Table 1

Morphological and Physiological Features of the M. Triceps Surae

	GL	GM	SO
Architectural Variables:			
Number of sarcomeres	21400	18400	—
Fiber optimal length (mm)	55.7	47.8	37.8
Muscle optimal length (mm)	175.8	202.1	239.6
Index of architecture	0.317	0.237	0.158
Volume (cm³)	104	199	250
Physiological cross section (mm²)	1867	4163	6614
Fiber angle at l_o (degrees)	20	20	25
Physiological Variables:			
Muscle length range (mm)	51.8	43.4	34.8
(% l_o)	29.5	21.5	14.5
v_{max} muscle (mm/s)	523.4	449.2	353.5
(% l_o/s)	297.5	223.3	147.5
Maximal power (watts)	36.9	69.3	69.4
v_{muscle} at p_{max} (mm/s)	263.7	242.5	143.8
(% l_o/s)	150.0	120.0	60.0
v_{fiber} at p_{max} (% l_o fiber/s)	504	539	419

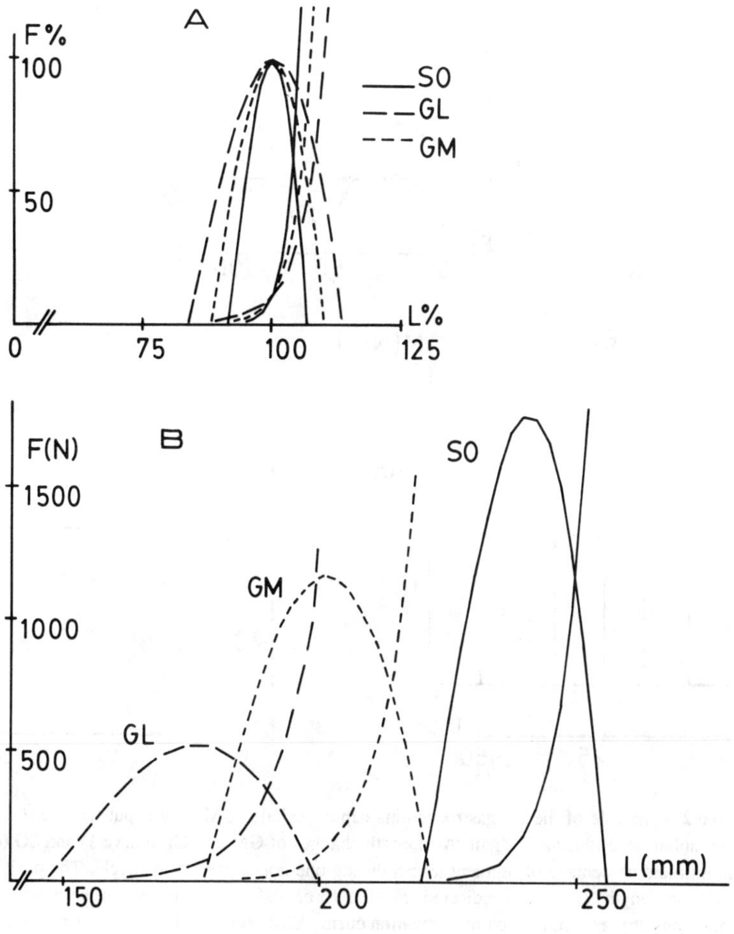

Figure 1—The normalized (A) and absolute (B) length-force relations of the m. gastrocnemius caput mediale (GM) and laterale (GL) and m. soleus (SO).

90% and for GL 75%. Figure 2B shows the production of force and torque by GM, GL, and SO if they would be maximally activated at velocities occurring during gait. The relative height of the labeled bars is indicative of the relative contribution of the three parts of TS at submaximal activity as well, if one assumes equal relative activation. The peak power deliverable at velocities occurring during fast gait are shown in Figure 2C. The relative contributions of GM, GL, and SO at maximal activation are indicated. At submaximal activations the relative contributions do not differ from the maximal situation as long as equal relative activation is the case.

At optimal length the maximal power of GM, GL, and SO is calculated to be 69.3, 36.9, and 69.4 W (TS maximal power 175.6 W). During gait, assuming maximal activation of TS peak power, production in the concentric phase of contraction would be 53.4, 29.3, and 67.6 W, respectively.

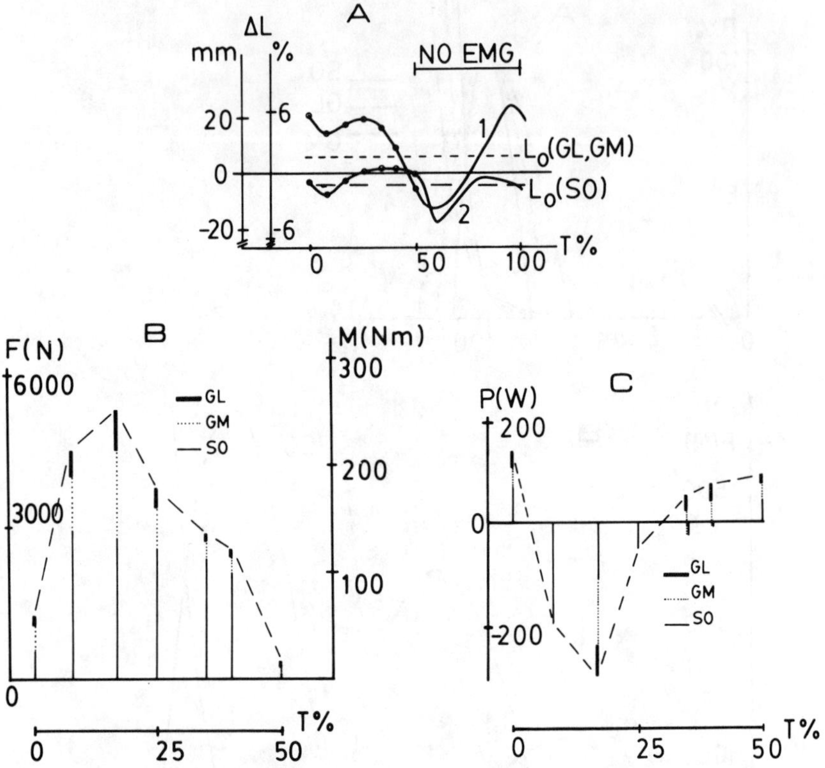

Figure 2—The role of the m. gastrocnemius caput mediale (GM) and caput laterale (GL) and the m. soleus (SO) during fast gait. **A**—Length changes of GM and GL (curve 1) and SO (curve 2) in mm and percentage of segment length during one normalized stride cycle. The position of the optimal lengths (l_o) is also indicated. **B**—The force and plantar flexion torque contributions of these muscles, assuming maximal activation during EMG activity. **C**—The power contribution of these muscles, assuming maximal activation during EMG activity.

Discussion

During fast gait all parts of TS are predicted to stay well within their physiological range (length-force relation). Differences in architecture and number of joints crossed are reflected in the percentage of length range travelled during fast gait. The force and power of SO, during the eccentric contraction in the first part of the stance phase, is considerably greater than the contributions from the other parts of TS. This is caused by the fact that GM and GL are lengthening less due to simultaneous knee flexion as well as the smaller degree of pennation of these heads. As the SO has shorter fibers the velocity of lengthening of these fibers is higher with respect to muscle lengthening than in GM and GL.

Note that it is predicted that energy taken up by TS in the stance phase is greater than the energy produced. This may be caused by a realistic effect related to the necessity to slow down segment motion during fast gait or the rather unrealistic assumption of continuous maximal activation of these muscles or to some storage and later use of elastic energy which the muscle model does not take into account.

The predicted maximal isometric torque of TS with respect to the ankle is in the same order of magnitude as that found for healthy humans (Hof & van den Berg, 1977). The maximal isometric force of SO equals the sum of GM and GL contribution. This is not the case for maximal power for which the sum of GM and GL is greater than the SO contribution. The velocity of shortening at which this maximal power occurs is considerably lower for SO than for GM and GL. The latter two have approximately equal values for this variable despite differences in architecture between them. The maximal power deliverable by TS during concentric contraction, while comparing well with the value found for isokinetic plantar flexion (Fugl-Meyer, Mild, & Hornsten, 1982), is very small compared to the peak power measured at the ankles (2000 to 3000 W) during jumping. This gives support to the idea (Vergroesen, de Boer, & van Ingen Schenau, 1982) that the contribution of power by the shortening of the calf muscles is relatively small during jumping and that there is a flow of power from the hip and knee to the ankle, mediated by the polyarticular m. rectus femoris and m. gastrocnemius (Grégoire, Veeger, Huijing, & van Ingen Schenau, 1983).

References

CLOSE, R.L. (1972). Dynamic properties of mammalian skeletal muscles. *Physiological Reviews*, **52**, 129-197.

FUGL-MEYER, A., Mild, K.H., & Hörnsten, J. (1982). Output of skeletal muscle contraction. A study of isokinetic plantar flexion in athletes. *Acta Physiologica Scandinavica*, **115**, 193-199.

GRÉGOIRE, L., Veeger, H.E., Huijing, P.A., & van Ingen Schenau, G.J. (1983). The function of mono- and biarticular muscles during the vertical jump. *Journal of Anatomy*, **137**, 404. (abstract)

GRIEVE, D.W., Pheasant, S., & Cavanagh, P.R. (1976). Prediction of gastrocnemius length from knee and ankle joint posture. In P.V. Komi (Ed.), *Biomechanics VI-A*, pp. 405-412. Baltimore: University Park Press.

HEERKENS, Y.F., Woittiez, R.D., Huijing, P.A., & Rozendal, R.H. (1983). The in vivo stiffness of the human knee joint. *Journal of Anatomy*, **137**, 434. (abstract)

HOF, A.L., & van den Berg, J.W. (1977). Linearity between the weighted sum of the EMG's of the human triceps surae and the total torque. *Journal of Biomechanics*, **10**, 529-539.

HUXLEY, H.E. (1968). Structural difference between resting and rigor muscle; evidence from intensity changes in the low angle equatorial X-ray diagram. *Journal of Molecular Biology*, **27**, 507-520.

HUIJING, P.A. (1981). Bundle length, fibre length and sarcomere number in human m. gastrocnemius. *Journal of Anatomy*, **133**, 132.

SUCH, C.H., Unsworth, A., Wright, V., & Dowson, D. (1975). Quantitative study of stiffness in the knee joint. *Annals of Rheumatic Diseases*, **34**, 286-291.

SUGI, H. (1972). Tension changes during and after stretch in frog muscle fibres. *Journal of Physiology*, **225**, 237-253.

SUTHERLAND, D.H., & Hagy, J.L. (1972). Measurement of gait movements from motion picture film. *Journal of Bone and Joint Surgery*, **54A**, 787-797.

VERGROESEN, I., de Boer, R.W., & van Ingen Schenau, G.J. (1982). Force, power and work analysis of the take-off phase of the vertical jump in three joints. *Journal of Biomechanics*, **15**, 797. (abstract)

WOITTIEZ, R.D., Huijing, P.A., & Rozendal, R.H. (1983). Influence of muscle architecture on the length force diagram: a model and its verification. *Pfluegers Archiv*, **397**, 73-74.

WOITTIEZ, R.D., Huijing, P.A., & Rozendal, R.H. (in press). A three dimensional muscle model predicting function characteristics. *Acta Morphologica Neerlando Scandinavica.*

Morphometrics and Force-Length Relations of Skeletal Muscles

E. Otten
Free University of Amsterdam, The Netherlands

Force-length relations of skeletal muscles depend upon muscle architecture (Hatze, 1981; Woittiez, Huijing, & Rozendal, 1983). In order to predict both the passive and the active force generated by a given muscle as a function of length, a model considering only morphometric information is needed. The main problem in formulating such a model is finding the necessary morphometric information to predict force-length relations within a given degree of accuracy. Force-length relations have to be measured from a large variety of muscles in order to test such a model and to obtain an insight into the variability and typical characteristics of these relations. By means of the model, functionally important differences in active and passive force-length relations can be understood in terms of morphometric differences. Since it is virtually impossible to obtain force-length relations of human muscles by direct measurement, such a model could prove to be very useful by calculating these relations from the morphometrics of human muscles.

Methods

The force-length relations of 18 skeletal muscles of the hind limb of the cat have been acquired by employing a computerized draw-bench (Poliacu Prosé & Otten, in preparation). The model was written in FORTRAN IV-PLUS and operated on a PDP computer with VERSATEC graphical output. Architectural characteristics of muscles were determined by means of a Zeiss microscope on which a drawing tube was mounted. Serial sections were made of several muscles embedded in paraffin. In addition, small pieces of muscle tissue were embedded in epon and sectioned on an ultramicrotome.

The Model

The force-length relations of the muscles of the hind limb of the cat have been normalized to the maximum active isometric force of the active component of contraction

(Fa in Figure 3A). The major difference between active force-length curves is a difference in width (S_h + S_ℓ in Figure 3A) produced by differences in relative muscle fiber length (Woittiez et al., 1983). These differences are defined in terms of the average muscle length divided by the length of the whole muscle at its normalized length ($\ell_f/\ell m$ in Figure 2).

A muscle fiber is able to contract over a certain percentage of its resting length, which in turn depends on sarcomere geometry. A muscle with relatively short fibers will contract only a small percentage of the muscle resting length, producing a narrow force-length curve. It appears that there is a good linear relation between the width of the active force-length curve and the relative muscle fiber length for the 18 muscles that have been measured. Moreover, the measurements by Gordon, Huxley, and Julian (1966) of the force-length relation of a single frog muscle fiber produce a curve width that is exactly the same as the width of the curves of the muscles with a relative muscle fiber length of 1. This is to be expected because of the conservative nature of vertebrate sarcomeres: they may vary in resting length, depending on the species, but there is always a constant ratio of the length of the actin filament to that of the myosin filament, which produces the same filament overlap functions normalized to the resting length of the sarcomere. This is an indication that the graph produced by Gordon et al. (1966) can be used as the input function for a generalized muscle model at the level of the muscle fibers, regardless of species.

If the passive characteristics of the muscle could be fully accounted for by the muscle fibers, as is almost by definition the case for the active characteristics, the point of intersection of the active and passive curve would always occur at the same normalized force F_i. According to the measurements this is not always the case. There is a tendency for muscles with relatively long fibers to have a higher F_i, or in other words, have a relatively high stiffness. This is an indication that stiffness is partly produced by an elastic element parallel to the whole muscle. After superimposing the graphs of the muscles (corrected for differencs in relative muscle fiber length) with the lowest F_i, it appears that the passive curves of these muscles are identical within the accuracy of measurements. This indicates that the residue elastic behavior of these muscles is

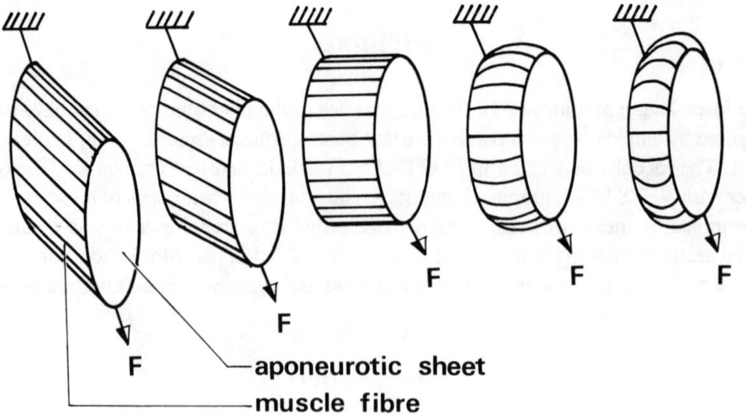

Figure 1—Geometrical events during muscular contraction as simulated by a three-dimensional model with output force F as a function of muscular length.

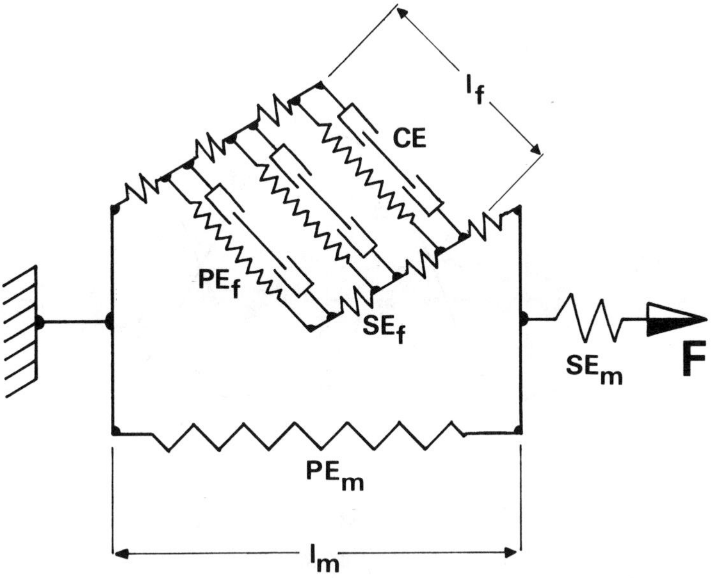

Figure 2—Series (SE_f and SE_m) and parallel (PE_f and PE_m) elastic elements together with contractile elements (CE) arranged in the model. This is only a network scheme in a two-dimensional layout. The actual three-dimensional arrangement (with about 3000 CE and PE_f elements) is as in Figure 1.

mainly produced by the muscle fibers themselves. Comparing this elastic force-length curve drawn to the normalized length scale of the muscle fiber with results of Natori, Umazume, and Yoshioka (1974) and Gordon et al. (1964), it appears that the curve falls in between these results.

In this way both the active and passive force-length relations of a single muscle fiber are given (Gordon et al.'s fiber graph of the active force component and the residue passive force component as found in this study, respectively) and these can be used as input for the muscle model to be constructed.

After a number of trial and error cycles, a three-dimensional model was produced in which the shapes and positions of the aponeurotic sheets or muscle-to-bone insertions can be chosen freely. The model simulates the change in angle between the muscle fibers and the main axis of the muscle and also the change in curvature and normalized length of all muscle fibers (see Figure 1). The elements, all having non-linear force-length relations, include two types of series elastic elements, one situated in the aponeurotic sheet (SE_f) and one situated in the tendon of the muscle (SE_m). Two types of parallel elastic elements include one parallel to each muscle fiber (PE_f) and one parallel to the whole muscle (PE_m in Figure 2). Since the characteristics of the contractile elements CE and the elastic elements PE_f are constant for every muscle as long as no velocities of contraction are introduced; and, since the characterstics of SE_m are not important when only the length of the muscle belly l_m is considered, the only muscle-dependent characteristics are those of SE_f and PE_m.

The geometrical events simulated by the model are summarized in Figure 1. Together with the active muscle fiber characteristics, these determine the active output force

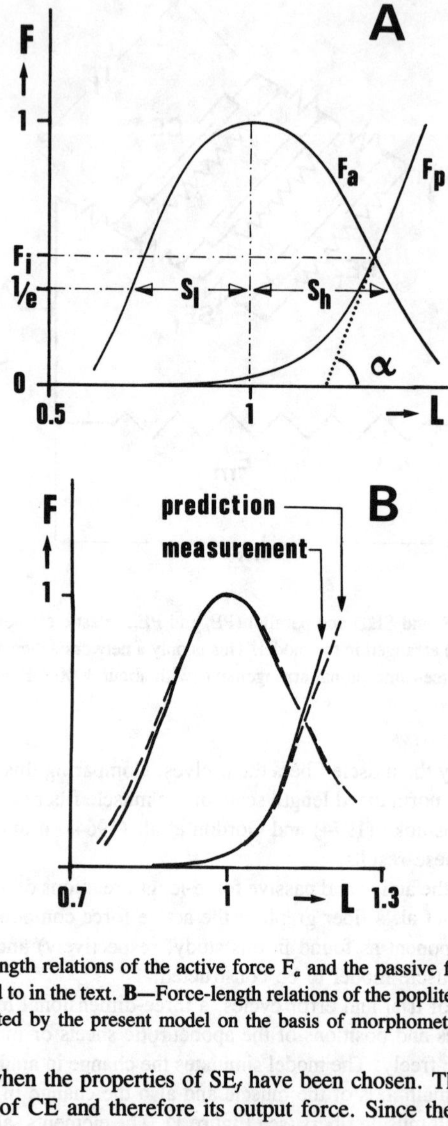

Figure 3A—Force-length relations of the active force F_a and the passive force F_p of a muscle, with symbols referred to in the text. **B**—Force-length relations of the popliteus muscle of the cat, measured and predicted by the present model on the basis of morphometrics of the muscle.

F_a of the muscle when the properties of SE_f have been chosen. These properties influence the length of CE and therefore its output force. Since the length of SE_f increases with increasing output force of CE and PE_f, it is to be expected that the influence of SE_f on the shape of the curve of the active output force of the muscle is strongest when the output force of the muscle fiber is highest. The stretch of SE_f results in a change in asymmetry (defined by s_h/s_l in Figure 3) of the active force output curve, since the highest output force of the muscle occurs at longer normalized length than sarcomere lengths indicate.

By examining the shapes of the passive force output curves of the 18 muscles and trying to fit them with exponential functions from literature (Hatze, 1981; Woittiez et al., 1983), it appears that these curves start approximately exponential but become linear when they cross the active force output curve (see Figure 3). Therefore it was decided to use a three-parameter function to describe the characteristics of PE_f and PE_m with:

$$F_p = \exp(c_1 + c_2 (L - 1)); F_p < F_c$$
$$F_p = F_c (1 - \ln F_c + c_1) + F_c c_2 (L - 1); F_p \geq F_c$$

F_c is the force where the exponential function becomes a linear one in a continuous way. Only c_2 of PE_m is a parameter in the model; constant values of F_c and c_1 for all muscles were chosen. The same was decided for SE_f. Constants for PE_m were chosen such that the properties of the muscle with the strongest passive characteristics of the hind limb of the cat (the sartorius muscle) was correctly predicted. The constants for the SE_f elements were chosen such that the aponeurotic sheet would stretch evenly over its whole surface up to a maximum of 5% of its resting length.

Results

The model presented here appears to have good predictive power. The width (s_h + s_l) and asymmetry (s_h/s_l) of the active force curve of the muscles can be fully accounted for by the geometrical events resulting from muscle architecture and the maximum stretch of SE_f. In many cases, SE_f can be omitted, indicating that a stiff aponeurosis is available or, as can be observed directly, when a muscle-to-bone insertion is present. The passive curve is well predicted when c_2 of PE_m is chosen appropriately. A relation between c_2 and the relative amount of collagen in parts of the muscle running from origin to insertion has only been shown in three muscles. A more extensive micromorphometric technique has to be developed in order to fully relate c_2 to collagen contents and collagen fiber orientation. The fact that accurate predictions can be made by using only c_2 as a parameter for PE_m indicates that PE_m is tuned to the active force-length curve. There is variation in steepness of the linear part of the passive curve relative to the steepness of the descending part of the active curve, but there is a constant way in which PE_m is brought to tension with increasing normalized length, since with only slight variation, its curve always passes through a single force value at $L = 1$.

Conclusions

It is possible to accurately predict force-length relations for both the active and the passive force components of skeletal muscles when a model is formulated in which the active and passive muscle fiber force-length functions and geometrical information on the three-dimensional architecture of the muscles are used. Estimates have to be made of the values of two parameters determining the properties of aponeurotic sheet stretch and the stiffness of the muscular fascia, based on their relative collagen contents and their collagen fiber arrangement.

References

GORDON, A.M., Huxley, A.F., & Julian, F.J. (1964). The length-tension diagram of single vertebrate striated muscle fibres. *Journal of Physiology*, **171**, 28-30.

GORDON, A.M., Huxley, A.F., & Julian, F.J. (1966). The variation in isometric tension with sarcomere length in vertebrate muscle fibres. *Journal of Physiology,* **184**, 170-192.

HATZE, H. (1981). *Myocybernetic control models of skeletal muscle.* Pretoria: University of South Africa.

NATORI, R., Umazume, Y., & Yoshioka, T. (1974). Viscous elastic properties of internal membrane of skeletal muscle fibres. *Jikeikai Medical Journal,* **21**, 135-150.

WOITTIEZ, R.D., Huijing, P.A., & Rozendal, R.H. (1983). Influence of muscle architecture on the length-force diagram. *Pflügers Archiv,* **197**, 73-74.

Length and Velocity Patterns of the Human Locomotor Muscles

M.R. Pierrynowski
University of Toronto, Toronto, Ontario, Canada

J.B. Morrison
Simon Fraser University, Burnaby, B.C., Canada

The length and velocity of a muscle has a profound effect on the magnitude of its force capacity. One of the earliest predictions of the sliding filament theory of contraction was that as a muscle is lengthened the area of overlap of the actin and myosin filaments diminishes and the tension developed decreases (Hanson & Huxley, 1953). Of greater importance to the force generating capability of a muscle is the velocity of muscle contraction. It has long been known that an inverse relationship exists between speed of contraction and muscle force (Fenn & Marsh, 1935). It is therefore of interest to examine the length and velocity changes of the human locomotor muscles during the gait cycle since their kinematics determine the physiological constraints on their force magnitudes. These data are required when individual muscle force analyses are performed.

The lines of action of the locomotor muscles have been modelled as straight lines joining origin to insertion (Crowninshield, Johnson, Andrews, & Brand, 1978; Hardt, 1978). However, many muscles deviate from a linear path due to anatomical constraints. Frigo and Pedotti (1978) improved upon this representation by using lines and arcs of circles to model the course of 11 major locomotor muscles, but they did not scale the muscle paths for different subjects and their analysis was limited to the sagittal plane. The work of Jensen and Davy (1975) strongly suggests that the muscle centroid line should be used and they present data on three of the muscles crossing the hip to support their case. Although their technique is quite accurate, a balance between accuracy and simplicity must be considered.

In this paper a male Caucasian skeleton was used to define the lines of action of 47 modelled locomotor muscles (see Table 1). Each muscle was represented by an elastic thread connected from the centroid of its area of origin to the centroid of its area of insertion. Between these two end points up to four additional points were defined from anatomical considerations, through which the muscle was constrained to pass. A muscle's line of action was obtained by joining its defined points with a combination of straight and curvilinear sections. Straight lines were used whenever a muscle ran free-

Table 1

The Lower Extremity Muscles Included in the Model

1 Psoas major	2 Iliacus
3 Gemellus superior	4 Gemellus inferior
5 Obturator externus	6 Obturator internus
7 Piriformis	8 Quadratus femoris
9 Pectineus	10 Adductor longus
11 Adductor magnus (ant)	12 Adductor magnus (mid)
13 Adductor magnus (post)	14 Adductor brevis
15 Gluteus minimus (ant)	16 Gluteus minimus (mid)
17 Gluteus minimus (post)	18 Gluteus medius (ant)
19 Gluteus medius (mid)	20 Gluteus medius (post)
21 Gluteus maximus (deep)	22 Gluteus maximus (sup)
23 Tensor fascia latae	24 Semimembranosus
25 Semitendinosus	26 Gracilus
27 Sartorius	28 Rectus femoris
29 Biceps femoris (long)	30 Biceps femoris (short)
31 Vastus lateralis	32 Vastus intermedius
33 Vastus medialis	34 Popliteus
35 Gastrocnemius (lateral)	36 Gastrocnemius (medial)
37 Plantaris	38 Soleus
39 Tibialis anterior	40 Tibialis posterior
41 Peroneus longus	42 Peroneus brevis
43 Peroneus tertius	44 Extensor digitorum longus
45 Extensor hallucis longus	46 Flexor digitorum longus
47 Flexor hallucis longus	

ly from point to point and a curve was fitted when the muscle was forced to alter its course by either bony prominences or tendon sheaths.

To scale the skeletal data to a subject a transformation technique was used (Lew & Lewis, 1977). Knowing the location of several points of prominence on a skeletal segment and the corresponding location of these same points measured in vivo on a subject, a 4 × 4 transformation matrix was calculated which transformed the data from the disarticulated skeletal segment into that segment of the subject. When applied with a group of at least four points on a body segment, a rigid body translation and rotation is performed along with a homogeneous deformation of the initial segment (skeletal bone) into the final segment (subject). These transformations were then used to locate the line of action of a muscle within the subject under study. A sagittal and frontal plane view of the lines of action of the 47 muscles, scaled to the subject (age 27 years, height 1.78 m, mass 70.8 kg), are shown in Figure 1.

To estimate the instantaneous lengths and velocities of the muscles during the walking cycle the transformation technique was again applied. This required knowing the location of markers affixed to the subject in both the anatomical position and during the movement cycle. To track the segments in 3-D space a cinematographic system was used. Two Locam cameras viewed the subject from the front and right sides and their respective images were combined using the Direct Linear Transformation (Abdel-Aziz & Karara, 1971) to estimate the 3-D locations of the markers. From a knowledge of the location of the above markers on the subject, in both the anatomical position

Figure 1—A medial and frontal view of the muscle lines of action as modelled within the subject.

and during the movement pattern, the points defining the muscles were determined using the transformation method for each frame of cine data. The points defining a given muscle were then joined by straight or curvilinear sections and the length of the muscle calculated. Finite differentiation was then used to estimate the velocity time patterns.

Sample muscle length and velocity time patterns are shown in Figure 2. All curves begin and end at right heel contact. Comparing these results to those reported by Frigo and Pedotti (1978), good agreement is obtained even though they did not consider the muscle length changes in the coronal or transverse planes. The results of the present investigation also compare quite well to those reported by Grieve, Pheasant, and Cavanagh (1978) who examined the length and velocity changes in gastrocnemius during level walking.

It is interesting to compare the muscle velocities with the muscle fiber lengths since it is known that the maximum rate of fiber shortening is proportional to the fiber length (Close, 1964). For the parallel fibered muscles maximum concentric fiber velocities were 0.43 ± 0.17 fiber lengths/s. Therefore it seems likely that during normal walking the slow-twitch muscle fibers within these muscles rarely exceed their maximum

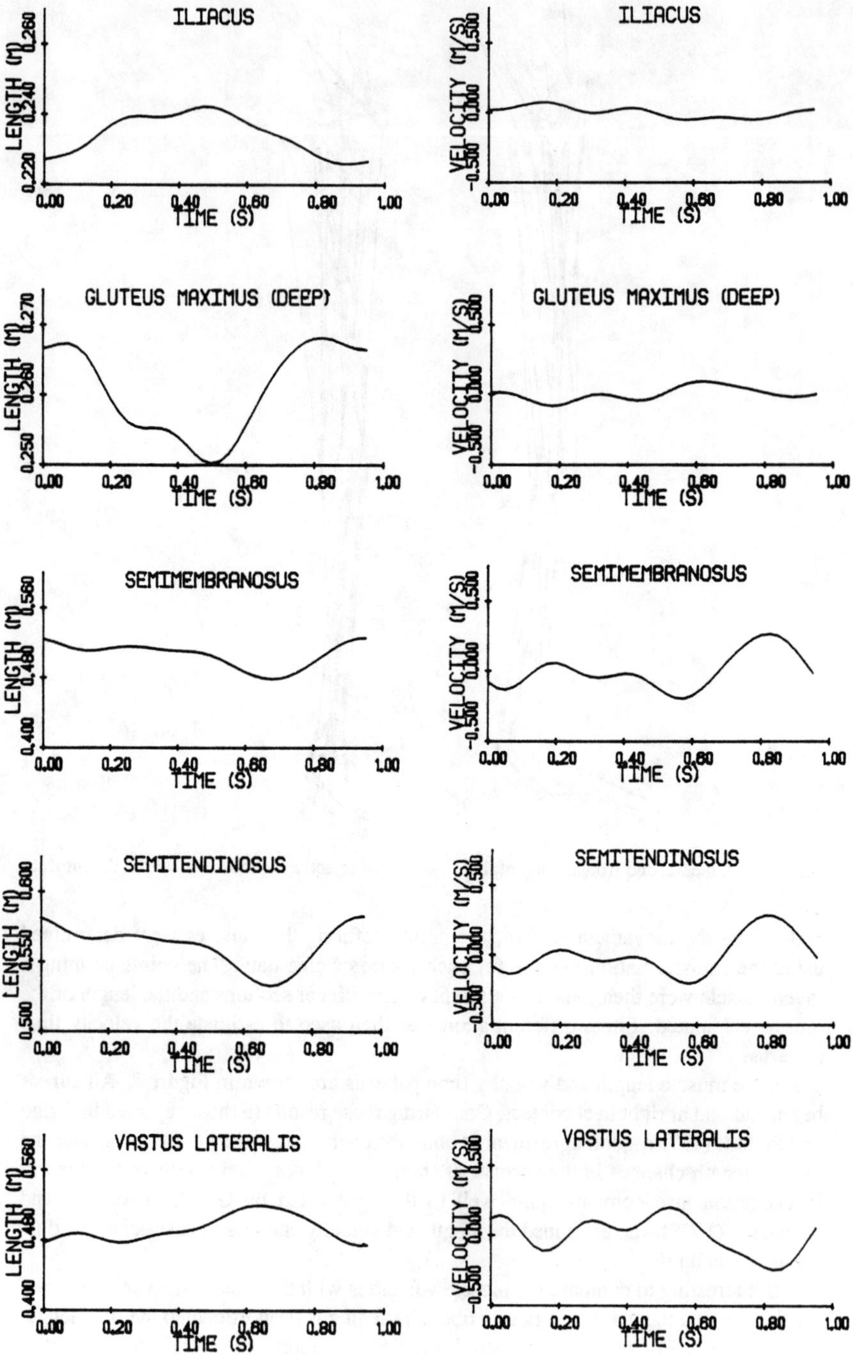

Figure 2—The muscle lengths and velocities during the walking cycle.

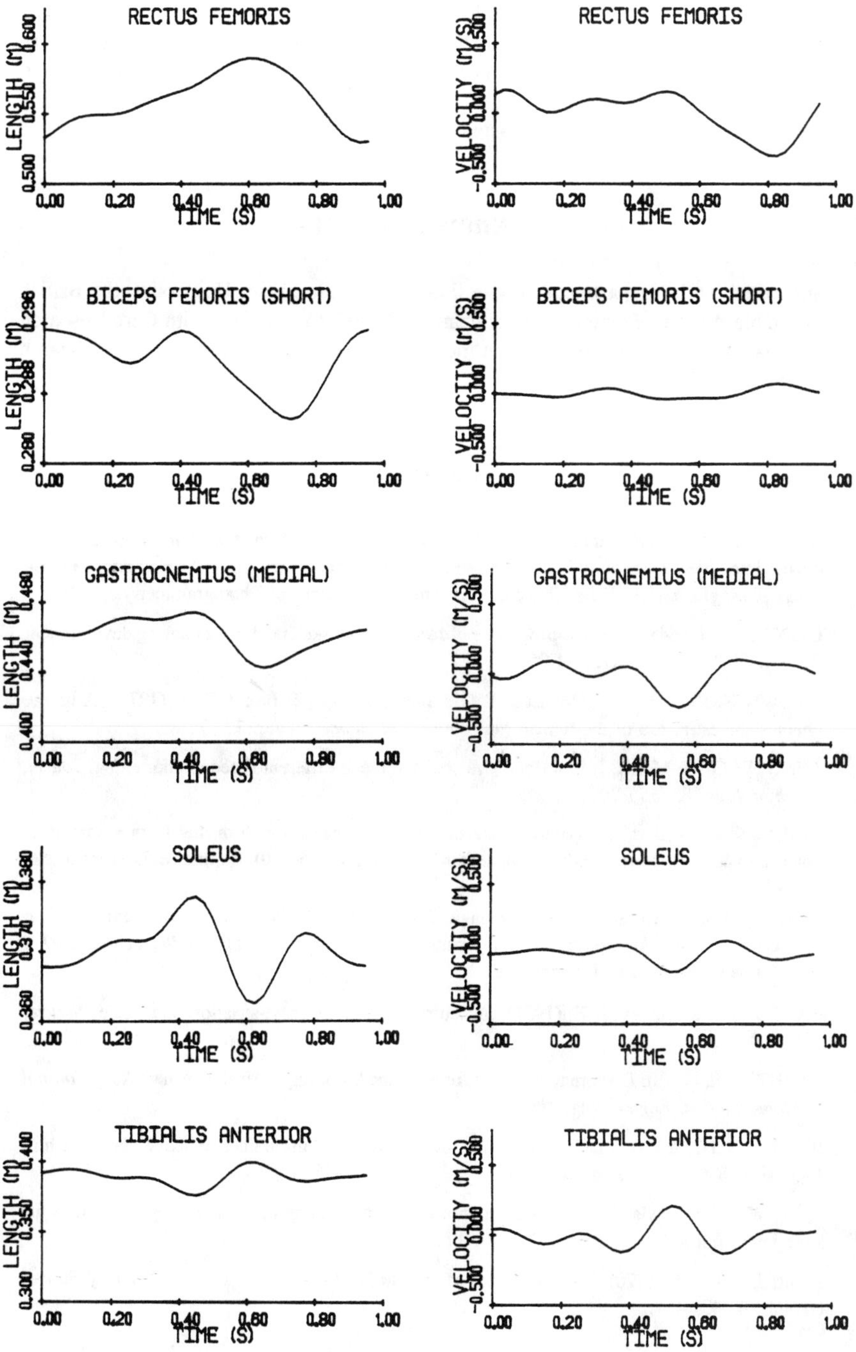

Figure 2 Cont.

rate of intrinsic shortening (1.0 fiber lengths/s). For the pennate muscles the velocities were 1.17 $\pm$ 0.55 fiber lengths/s which suggests that if their slow-twitch muscle fibers were recruited during the phases of maximal concentric velocities the slow-twitch fibers would exert no force. However, muscles tend to be inactive during conditions of maximal concentric shortening (Morrison, 1970).

Acknowledgments

Supported by the Department of Kinesiology at Simon Fraser University, the British Columbia Arthritis Society, a grant from the British Columbia Health Care Research Foundation, and a scholarship from the Natural Sciences and Engineering Research Council (CANADA).

References

ABDEL-AZIZ, Y.I., & Karara, H.M. (1971). Direct linear transformation from comparator coordinates into object space coordinates in close-range photogrammetry. In *ASP symposium on close-range photogrammetry.* Falls Church, VA: American Society of Photogrammetry.

CLOSE, R.I. (1964). Dynamic properties of fast and slow skeletal muscles during development. *Journal of Physiology.* **173**, 74-95.

CROWNINSHIELD, R.D., Johnson, R.C., Andrews, J.G., & Brand, R.A. (1978). A biomechanical investigation of the human hip. *Journal of Biomechanics.* **11**, 75-85.

FENN, W.O., & Marsh, B.O. (1953). Muscular force at different speed of shortening. *Journal of Physiology,* **85**, 277-297.

FRIGO, C., & Pedotti, A. (1978). Determination of muscle length during locomotion. In E. Asmussen & K. Jorgensen (Eds.), *Biomechanics VI-A* (pp. 355-360). Baltimore: University Park Press.

GRIEVE, D.W., Pheasant, S., & Cavanagh, P.R. (1978). Prediction of gastrocnemius length from knee and ankle joint posture. In E. Asmussen & K. Jorgensen (Eds.), *Biomechanics VI-A* (pp. 405-411). Baltimore: University Park Press.

HANSON, J., & Huxley, H.E. (1953). Structural basis of the cross-striations in muscle. *Nature,* **172**, 530-532.

HARDT, D.E. (1978). Determining muscle forces in the leg during normal human walking. *Journal of Biomechanical Engineering,* **100**, 72-78.

JENSEN, R.H., & Davy, D.T. (1975). An investigation of muscle lines of action about the hip. *Journal of Biomechanics,* **8**, 103-110.

LEW, W.D., & Lewis, J.L. (1977). An anthropometric scaling method with application to the knee joint. *Journal of Biomechanics,* **10**, 171-181.

MORRISON, J.B. (1970). The mechanics of muscle function in locomotion. *Journal of Biomechanics,* **3**, 431-451.

The Effect of Joint Angle on Cross-Sectional Area and Muscle Strength of Human Elbow Flexors

S. Ikegawa
Japan Women's University, Japan

N. Tsunoda
Kokushikan University, Japan

H. Yata
Wako University, Japan

A. Matsuo, T. Fukunaga, and T. Asami
Tokyo University, Japan

In many previous studies, it has been reported that muscle strength is closely related to muscle area and that the absolute strength (strength per unit cross-sectional area) can be calculated. The absolute strength and muscle cross-sectional area are a function of the joint angle. The purpose of this study was to clarify the effect of joint angle on the muscle area and muscle strength in human elbow flexors.

Method

Six healthy volunteers (1 female, 5 male) participated in the experiment. The cross-sectional area of three different elbow flexor muscles (biceps brachii, brachialis, and brachioradialis) were measured by means of ultrasonic apparatus at four different elbow angles of 90°, 120°, 150°, and 170°. The 5 MHz frequency of ultrasonic wave was chosen to get a clear view of each tissue (muscle, fat, and bone). The cross-sectional area of each muscle bundle was estimated from ultrasonic picture by using a planimeter.

The maximum isometric voluntary strength for elbow flexion (measured strength) was measured at every 10° from 90° to 170° of joint angle by strain gauge dynamometer.

In order to obtain the actual tension exerted by elbow flexor muscles, the leverage for each muscle in the joint was estimated from a cadaver. All obstructing structures of the cadaver's arm were cleared off until free movement was possible with minimum friction. The humerus was held in clamps and strain gauge load cells were attached to the tendon of each muscle and to the carpus. The output of these two load cells were connected to an X-Y recorder, and the ratio of tension, which was recorded at tendon to that of wrist belt, was calculated.

The muscles contributing to elbow flexion are biceps brachii, brachialis, brachioradialis, pronator teres, and extensor carpi radialis. In the present study, pronator teres and extensor carpi radialis were neglected because of the small force arm of these muscles.

The absolute strength (AS, strength per unit cross-sectional area) was calculated from the following equation:

$$AS = MS \times \frac{1}{\frac{Ma1}{L1} + \frac{Ma2}{L2} + \frac{Ma3}{L3}}$$

where MS = Measured strength
Ma1 = Cross-sectional area of biceps brachii
Ma2 = Cross-sectional area of brachialis
Ma3 = Cross-sectional area of brachioradialis
L1 = Leverage of biceps brachii
L2 = Leverage of brachialis
L3 = Leverage of brachioradialis.

The tension developed by each muscle was estimated from the product of muscle cross-sectional area and the absolute strength.

Results and Discussion

As shown in Figure 1, the cross-sectional area of muscle (biceps brachii and brachialis) decreased linearly with increments of joint angle up to 170° (180° was full extension of elbow joint). The rate of decrease of the muscle area for biceps brachii and brachialis were about 0.43 cm^2 and 0.10 cm^2 per every 10° of extension of elbow joint, respectively. The lengthening of the muscle due to extension of elbow joint causes the reduction in cross-sectional area of muscle because the muscle volume is constant.

The ratio of the tension measured at wrist joint to that at tendon of each muscle (leverage) is indicated in Figure 2. At 90° of elbow joint, the leverage was 4.3 for biceps brachii, 6.2 for brachialis, and 3.0 for brachioradialis and it increased slightly with an increase in joint angle. A previous study by Ikai and Fukunaga, (1968) reported the leverage for elbow flexor muscle to be 4.9. They estimated the attachment of elbow flexor muscle on the bone between tuberositus radii and tuberositus ulnae on an X-ray photograph. The value obtained by Ikai and Fukunaga falls between values for biceps brachii and brachialis obtained in the present study.

In Figure 3, measured strength (MS) and absolute strength (AS) were plotted against the angle of the elbow joint. The MS indicated peak value of 240 N at 100° and it decreased with increasing joint angle. This relation between strength and joint angle is in good agreement with the results of Singh and Karpovich (1966). The AS was about 38 N/cm^2 at 90° and increased to 48 N/cm^2 at 170°. Ikai and Fukunaga (1968)

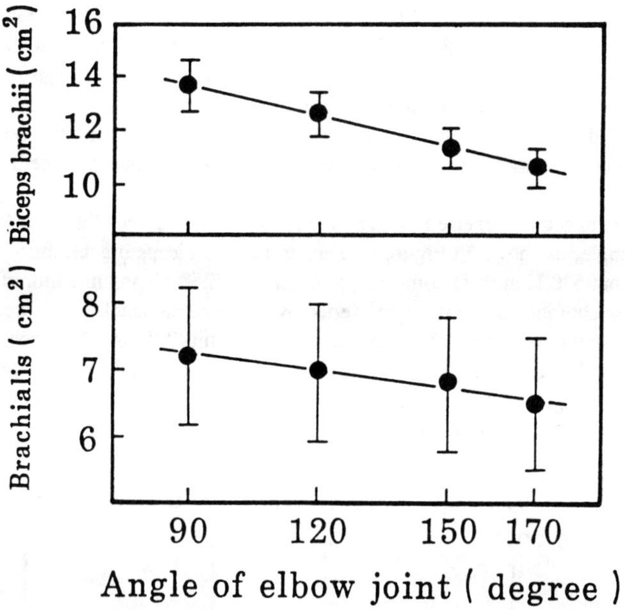

Figure 1—The relationship between joint angle and cross-sectional areas for the biceps brachii and brachialis (mean ± S.E.).

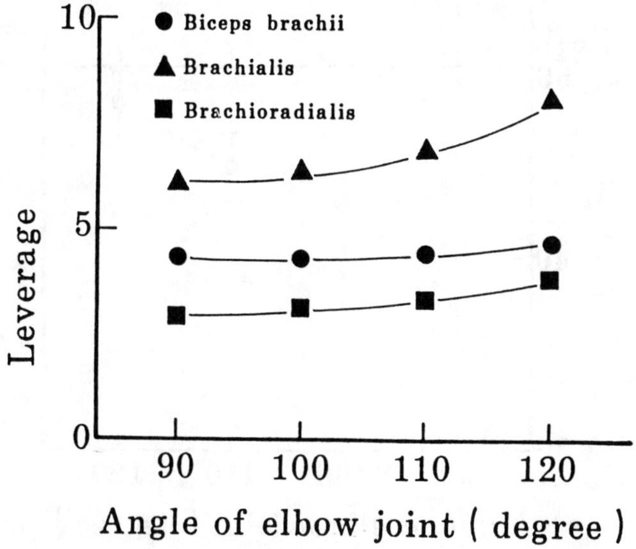

Figure 2—Changes in leverage with joint angle.

found AS to be 6.7 kg/cm² (65.7 N/cm²) when the MS was measured at 90° and muscle area at full extension (180°) of the elbow joint. They reported that the AS was 5.0 kg/cm² (47.0 N/cm²) when both muscle area and strength were measured at 90°. The AS of 38 N/cm² at 90° in the present study was lower than Ikai's value. Ikai and Fukunaga did not measure the area of brachioradialis muscle. The difference in the AS value between Ikai and this study therefore may be attributed to methodological differences.

The strength actually exerted by each muscle bundle increased slightly with increasing joint angle, as shown in Figure 4. The strength of biceps brachii muscle was the higher (about 530 N at 90°) compared to brachialis (250 N) and brachioradialis (150 N). The percent ratios of the strength exerted by each muscle bundle to the total strength for elbow flexion were about 57% for the biceps brachii, 27% for the brachioradialis, and 20% for the brachialis. From these results it may be considered that the increase of muscle strength with increasing elbow joint angle is caused by the functional characteristics of the muscle which is indicated by length-tension curve.

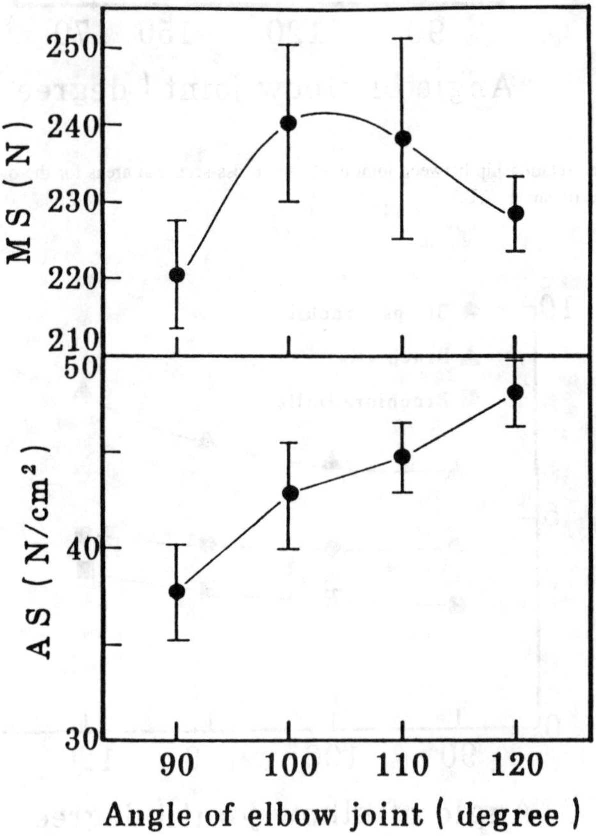

Figure 3—Changes in MS (measured strength) and AS (absolute strength) with joint angle (mean ± S.E.).

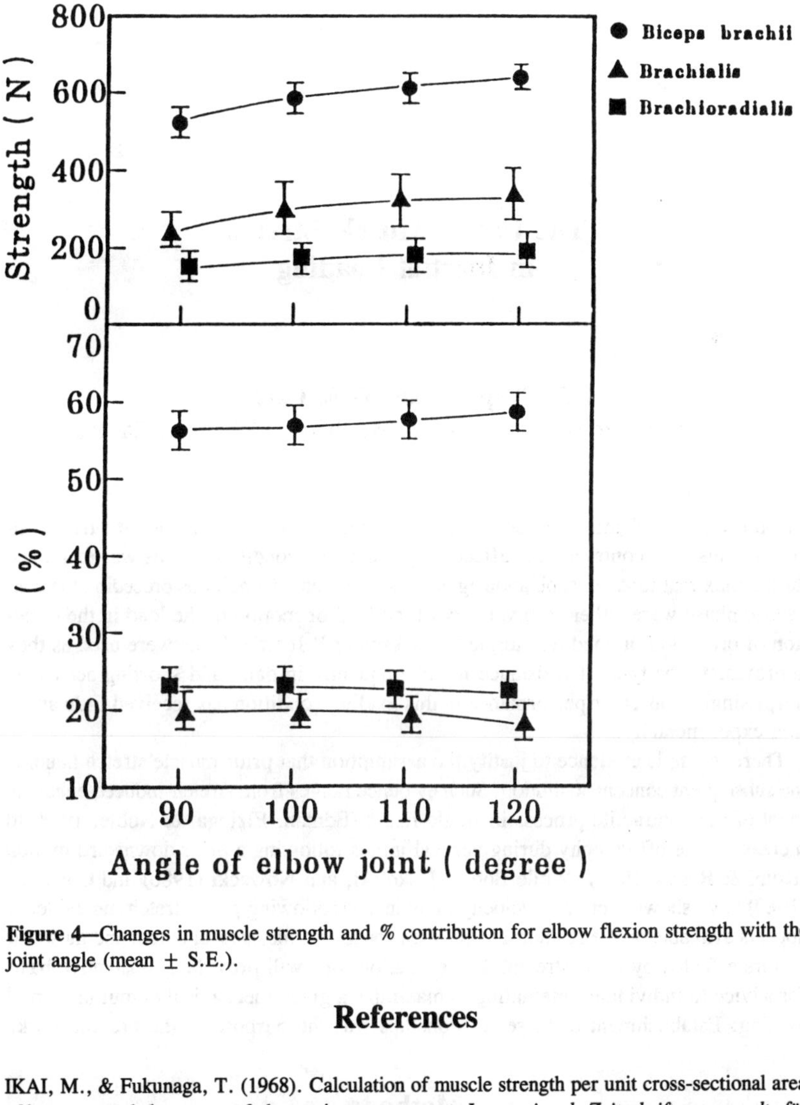

Figure 4—Changes in muscle strength and % contribution for elbow flexion strength with the joint angle (mean ± S.E.).

References

IKAI, M., & Fukunaga, T. (1968). Calculation of muscle strength per unit cross-sectional area of human muscle by means of ultrasonic measurement. *Internationale Zeitschrift angewandte für Physiologie*, **26**, 26-32.

SINGH, M., & Karpovich, P.V. (1966). Isotonic and isometric force of forearm flexor and extensor. *Journal of Applied Physiology*, **21**(4), 1435-1437.

The Use of Muscle Stretch in Inertial Loading

A.E. Chapman and G.E. Caldwell
Simon Fraser University, Burnaby, British Columbia, Canada

The present investigation was designed to examine how mechanical output during concentric muscular contraction is affected by preceding conditions. This was examined during maximal forearm supination against inertial loads. Conditions preceding the concentric phase were either zero velocity of the load or motion of the load in the direction of pronation, a condition termed "backswing." Inertial loads were used as they approximate the type of resistance found frequently in daily and sporting activities. Surprisingly, muscular performance in this loading condition has received little attention experimentally.

There is ample evidence to justify the assumption that prior muscle stretch benefits the subsequent concentric motion. Such evidence ranges from stretch-induced enhancement of the contractile process in single fibers (Edman, Elzinga, & Noble, 1978) to increased take-off velocity during vertical jumps following a prior downward motion (Komi & Bosco, 1978). While Bober, Jaskolski, and Nowacki (1980) and Chapman (1980) have shown increased velocity of an inertia following prior stretch, no evidence appears available on the relationship between concentric mechanical output and muscular factors effected by prior stretch. These relationships will provide a substantive basis for advice to individuals attempting to maximize a given mechanical output in inertial loading. Establishment of these relationships was the purpose of the present work.

Methods

Three subjects performed supination of the right (dominant) forearm in a standing posture with the upper arm vertical, the forearm horizontal, and trunk movement restricted. Elbow position was monitored by means of a horseshoe shaped cup surrounding the joint. Supination was performed with maximal voluntary effort under both isometric and dynamic conditions against various inertias.

Apparatus

Subjects grasped a handle, instrumented to record torque and tangential acceleration, fixed to an axle colinear with the forearm. A Polarized Light Goniometer (Grieve,

1969) measured handle angle. The distal end of the axle carried a metal bar mounted transversely along which masses of 4 kg were fixed. The angle and acceleration transducers were given calibration inputs of 90° and 9.81 m/s², respectively. A torque calibration input of 4.905 N•m was supplied by a 1 kg mass hung 0.5 m from the axle. Outputs from each transducer were amplified and sampled at 200 Hertz by means of a 12-bit A/D converter interfaced with an Apple II+ computer.

Procedures

During an experimental session subjects performed a variety of isometric and dynamic contractions totalling 10 in a random sequence. Isometric contractions were performed at different fixed handle angles. Dynamic contractions from rest involved setting the handle at one of a variety of starting angles with the subject relaxed and requiring maximal supination effort on a given command. Dynamic contractions with a backswing required the experimenter to initiate motion of the load in the direction of pronation with the subject relaxed. On a given command maximal supinator torque was required of the subject. This was performed with several initial angular velocities of the load from a variety of positions. The moment of inertia (MI) of the load was changed in each session.

Data Treatment

Angle of the handle was computed in radians (ref. zero = palm downwards). The product of angular acceleration and forearm MI was added to measured torque to yield absolute torque of supination. Angle and torque were smoothed with a dual pass 3-point moving average and angular velocity was obtained by finite difference differentiation over two time intervals.

Time and angle were recorded at the onset of torque and at zero angular velocity. The muscular work performed during the backswing was computed by integrating torque with respect to displacement. The angle at zero angular velocity (starting angle SA), amplitude of stretch (AS), muscular work (EW), and time (ET) of backswing represented the chosen mechanical conditions preceding the concentric phase.

During the concentric phase the angular velocity and positive work (CW) achieved after 0.1, 0.25, and 1.5 rad and 0.1 and 0.2 s after zero velocity were noted. These variables represented mechanical output.

Analysis

For the purpose of assessing the effect of prior conditions, multiple linear regression was employed using output as the dependent variable and prior conditions and load MI as independent variables. Only independent variables with significant T values ($p < .05$) were used. Regressions were developed for both dependent variables as follows:

$$D(a) = B_o + B_i I_i + B_{i+1} I_{i+1} \ldots B_n I_n$$

where D, the dependent variable = either angular velocity or positive work
a = either 0.1, 0.25, or 1.5 rad or 0.1 or 0.2 s of concentric motion at which D was obtained

B_o = the coefficient predicting D when all independent variables were zero
B_{i-n} = coefficients of independent variables I_{i-n}
I_{i-n} = selected combinations of independent variables SA, AS, MI, EW, and ET.

Results

Figure 1 shows torque and angular velocity with respect to time during contractions from rest and with a backswing. Zero angular velocity occurred at the same forearm angle in each case. The effect of the backswing on torque can be seen clearly as a higher, earlier peak in comparison with the rest contraction. In fact the peak torque was greater than the isometric value at that forearm angle even though it occurred at a concentric angular velocity. The large early torque with a backswing and rest velocities achieved after 0.1, 0.25, and 1.5 rad of concentric motion which were greater than from rest (see Figure 1). However, the difference between backswing and rest velocities decreased as the amplitude of the motion increased, with velocities achieved after 1.5 rad differing little. Clearly the angular velocities following the backswing are also greater

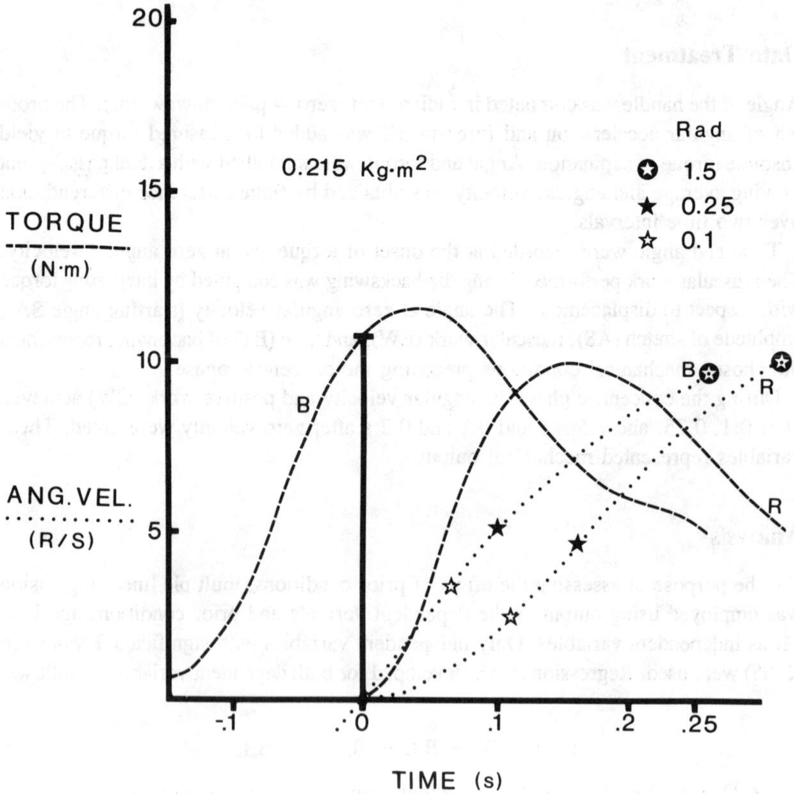

Figure 1—Torque and angular velocity over time for rest and backswing contractions.

at any given time following zero velocity. A further salient feature is that the torque from a backswing decays faster than from rest, indicating that as time progresses the mechanical impulses from each contraction become closer in value.

Figure 2 (one subject) illustrates how angular velocity achieved after various amplitudes of concentric motion varied with respect to MI. The solid and open symbols represent, respectively, the middle of the range of values achieved from rest and with a backswing. Figure 2 shows the expected result of reduced angular velocity with increased MI. The benefits of backswing in achieving angular velocity are clearly seen, although the absolute effect tended to decrease with increased MI and also with increased concentric angular displacement.

Since both amount of backswing and angle at zero angular velocity were variable during backswing trials, multiple linear regression analysis was used to account for their potential effects (see Table 1). This combined data from all subjects reflects the individual results. Due to the inertial nature of the loading AS, EW and ET were highly correlated, so that any one could serve as an independent variable in regression. Amplitude of backswing (AS) was chosen as that most easily observed in daily activity.

In contrast to trends of angular velocity, concentric work over a given amplitude of motion increased with increased MI. Backswing was of less importance in achieving angular velocity as amplitude and time progressed, as decreasing coefficients of AS indicate, while the amount of backswing showed increasing importance for concentric work. As both amplitude and time of contraction progressed, the angular velocity benefited from greater muscle stretch as the negative values of SA indicate. The ability of the linear regression to account for concentric work was less than the ability to account for angular velocity as the values of R^2 show.

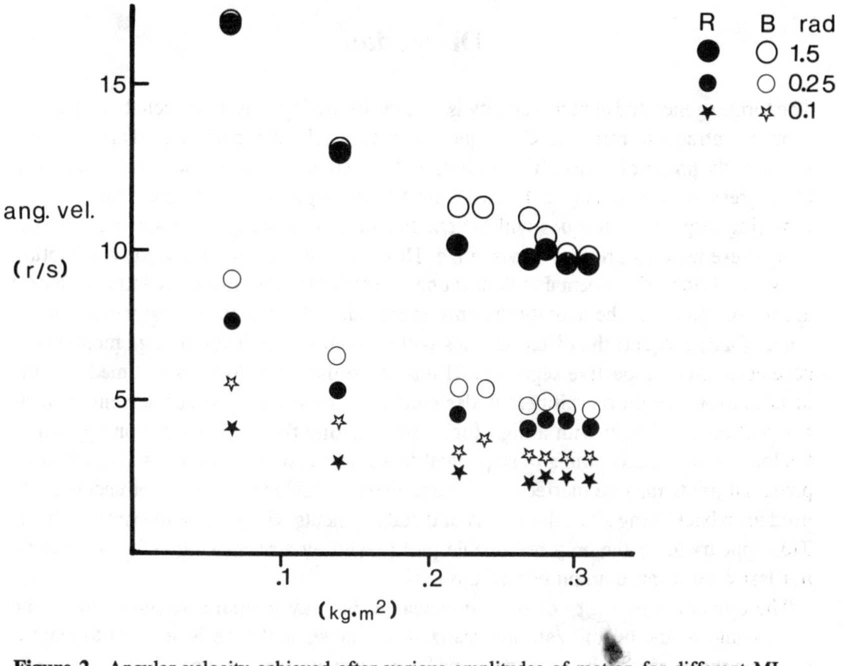

Figure 2—Angular velocity achieved after various amplitudes of motion for different MI.

Table 1
Results of Regression Analysis

	B	SA	AS	MI	R^2	N
AV						
0.1 rad	5.11	—	1.08	−9.92	0.76	156
0.25 rad	8.59	—	1.02	−16.59	0.79	156
1.5 rad	19.39	−1.55	0.63	−35.28	0.82	149
0.1 s	11.29	—	3.91	−35.52	0.73	156
0.2 s	18.42	−1.20	2.32	−44.58	0.79	141
CW						
0.1 rad	0.55	—	0.69	1.35	0.59	156
0.25 rad	1.88	—	0.99	2.58	0.47	156
1.5 rad	11.82	—	1.03	6.39	0.09	149
0.1 s	4.22	—	3.58	−13.64	0.60	156
0.2 s	13.59	—	5.14	−13.41	0.47	141

Coefficients of independent variables which accounted for the variability (R^2) of the dependent variables are shown in the left hand column. Dependent variables are angular velocity (AV in rad/s) and concentric work (CW in joules) at 0.1, 0.25 and 1.5 rad and 0.1 and 0.2 s of concentric motion. Independent variables are angle at zero angular velocity (SA in rad), amplitude of backswing (AS in rad) and moment of inertia (MI in kg•m²). B_o is the value of the dependent variable when all independent variables equal zero. Data are from all 3 subjects combined (N = # of contractions).

Discussion

The torque generated at zero velocity is greater following muscle stretch than in a concentric contraction from rest. Consequently large accelerations of the load are achieved in the early phase of concentric motion, and the significance of backswing factors in the regression equations (see Table 1) attest to its importance. Therefore any activity requiring large early torques will benefit from the backswing. In single-segment activity these benefits are easy to visualize. However, in normal multisegmental motion the general flow of segmental motion is one in which the more massive central segments are moved first. As the activity then progresses distally to optimize the motion of the object (ball, racquet) the distal muscles will be required to develop large moments to accelerate their respective segments. If this transmission of force is so timed that the distal muscles begin recruitment under conditions of stretch, the subsequent concentric motion will begin with a high force, maximizing the net acceleration applied to the load. Without this timing of segmental recruitment, successive backswings at more proximal joints may be ineffective. Antagonistic muscular activity is not necessary to produce a backswing since the inertia of distal segments will provide the stretch effect. This appears to be the only reasonable part played by a backswing as its benefits do not last over a large amplitude of motion.

The dynamic properties of muscle appear to be only temporarily modified by the backswing, as the benefits are not realized if a substantial time is allowed to elapse, as in the movement of large inertias and/or movement over long periods of time. The major benefits derived from a backswing are to allow activation time to develop and to gain advantage from high eccentric forces, rather than to substantially change the dynamic properties of muscle.

The regression equations provide empirical evidence on which to base advice for the use of muscle in single-segment activity. Obviously the MI of the load moved has a dominant effect on the angular velocity achieved. As the MI is decreased, the relative effects of the SA and AS increase. This results from the fact that with low inertias the time required to achieve a given concentric velocity is reduced. Consequently, the transitory decay in backswing benefits (probably Edman's "force-enhancement") assumes greater importance.

References

BOBER, T., Jaskolski, E., & Nowacki, Z. (1980). Study on eccentric-concentric contraction of the upper extremity muscles. *Journal of Biomechanics*, **13**, 135-138.

CHAPMAN, A.E. (1980). The effect of "Wind-up" on forearm rotational velocity. *Canadian Journal of Applied Sport Sciences*, **5**, 215-219.

EDMAN, K.A.P., Elzinga, G., & Noble, M.I.M. (1978). Enhancement of mechanical performance by stretch during tetanic contractions of vertebrate skeletal muscle fibres. *Journal of Physiology*, (Lond.), **281**, 139-155.

GRIEVE, D.W. (1969). A device called POLGON for the measurement of the orientation of parts of the body relative to a fixed external axis. *Journal of Physiology*, (Lond.), **201**, 70P.

KOMI, P.V., & Bosco, C. (1978). Utilization of stored elastic energy in leg extensor muscles by men and women. *Medicine and Science in Sports and Exercise*, **10**, 261-265.

Performance of Females with Respect to Males: The Use of Stored Elastic Energy

J.L. Hudson
Rice University, Houston, Texas, USA

M.G. Owen
Temple University, Philadelphia, Pennsylvania, USA

In the research literature on biomechanical assessment of athletic performance, there are few studies designed to examine the performance of females with respect to males. However, a precedent-setting strategy for sex-dependent comparisons has been established in the physiological assessment of performance. According to Wilmore (1981), comparisons between the sexes should be made only if the two groups have comparable duration and intensity of training, coaching, and competition.

A theoretical aspect of the biomechanics of performance which has received recent attention is the use of stored elastic energy. Two studies (Asmussen & Bonde-Petersen, 1974; Komi & Bosco, 1978) include data for females as well as males. Since the combined results of these studies are inconclusive, the present study was designed to use the paradigm of Wilmore to investigate the performance of females with respect to males in the use of stored elastic energy.

Review of Literature

Cavagna, Dusman, and Margaria (1968) have shown that skeletal muscles can perform greater positive work if the immediately preceding condition is lengthening contraction rather than isometric contraction. The explanation of this finding is that during a lengthening contraction elastic energy is stored which can be reutilized in a subsequent shortening contraction. The conditions necesary for the use of stored elastic energy are existent in vertical jumps performed with a countermovement but not in jumps taken from a static squat position.

The protocol used by Asmussen and Bonde-Petersen (1974) and Komi and Bosco (1978) to investigate the use of stored elastic energy involved the collection of force-time data as subjects performed countermovement and static jumps. The young adult subjects of Asmussen and Bonde-Petersen included 5 females and 14 males. The respective mean values for females and males were: weight, 58.0 and 75.6 kg and countermovement jump height, 30.4 and 41.5 cm. The female-to-male ratio for countermovement jump height was 73% and the mean use of stored elastic energy for all subjects was 23%. The physical education students who served as subjects for Komi and Bosco included 25 females and 16 males. The respective means for females and males were: age, 20.6 and 24.0 years; height, 165.6 and 176.7 cm; weight, 58.2 and 75.4 kg; countermovement jump height, 23.3 and 40.3 cm; and use of stored elastic energy, 92 and 49%. The female-to-male ratio for countermovement jump height was 58%.

Methods

Subjects for this study were selected with the intent of obtaining female and male groups with comparable jumping experience. Of the eight members of each group, four were college varsity track and field athletes who were training or competing in jumping activities and four were young adults who were training or competing in distance running events but were not participating in jumping activities. After anthropometric data were collected, each subject performed maximum vertical jumps in the countermovement (CMJ) and static jump (SJ) styles. Force records were used to confirm that no unweighting phase occurred in the SJ trials. Congruency between the starting position of the SJ and the lowest position of the CMJ was achieved through recordings from an electrogoniometer which was attached to the left knee. The right side of the subject was filmed at 100 frames/s. The digitized segmental end points were smoothed with cubic spline subroutines (ICSMOU and ICSSCU of the International Mathematical and Statistical Library). Body segment data were generated in accordance with the model of Hanavan (1964).

For the energy analysis, the velocity of the center of mass of the body was used to establish phases of the jump: the eccentric phase started at peak downward velocity and ended when velocity reached zero; the concentric phase began with the initiation of upward velocity and terminated with peak upward velocity. The linked segment method of Winter, Quanbury, and Reimer (1976) was used to compute the instantaneous energy for each of the four segments: head-arms-trunk, thighs, shanks, and feet. This method sums potential, translational kinetic, and rotational kinetic energy as follows:

$$E_i = m_i \cdot g \cdot h_i + \tfrac{1}{2} m_i \cdot v_i^2 + \tfrac{1}{2} I_i \cdot \omega_i^2 \qquad (1)$$

where E_i = instantaneous energy of ith segment
m_i = mass of ith segment
g = gravitational constant
h_i = height of the center of mass of ith segment
v_i = linear velocity of the center of mass of ith segment
I_i = moment of inertia about the center of mass of ith segment
ω_i = angular velocity of ith segment.

The instantaneous energy of the body was calculated by summing the segmental instantaneous energies:

$$E_B = \sum_{i=1}^{4} E_i \qquad (2)$$

where E_B = instantaneous energy of the body.

The net energy of the body was determined by summing the absolute change in the instantaneous energy of the body from frame to frame:

$$E_{NB} = \sum_{j=2}^{f} \left| E_{B_j} - E_{B_{j-1}} \right| \quad (3)$$

where E_{NB} = net energy of the body
f = number of frames in given period.

The equation of Asmussen and Bonde-Petersen was employed to compute the use of stored elastic energy from the net energy of the body in the concentric and eccentric phases of the CMJ and the eccentric phase of the SJ:

$$\text{USEE} = (E_{NB_{CMJ_c}} - E_{NB_{SJ_c}}) \cdot E_{NB_{CMJ_e}}^{-1} \times 10^2 \quad (4)$$

where USEE = use of stored elastic energy
$E_{NB_{CMJ_c}}$ = net energy of the body during concentric phase of CMJ
$E_{NB_{SJ_c}}$ = net energy of the body during concentric phase of SJ
$E_{NB_{CMJ_e}}$ = net energy of the body during eccentric phase of CMJ.

Analysis of variance techniques were used to test the differences between female and male groups in the use of stored elastic energy and the following six performance variables: The height of CMJ was defined to be the displacement of the center of mass above the standing reference position; the depth of CMJ was determined to be the displacement of the center of mass below the standing reference position; the knee angle was established to be the angle of greatest flexion at the knee during the CMJ; the concentric time was defined as the duration of the concentric phase of the CMJ; the depth ratio was computed to be the depth of CMJ divided by the standing height; and the velocity ratio was calculated to be the peak upward velocity in the CMJ relative to the peak upward velocity in the SJ.

Results

The means and standard deviations for each group are given in Table 1 for the three physical and seven performance variables.

Discussion

In testing the differences between the means of the females and males, two of the three physical variables (height and weight) were significant at the .05 level. Also, the subjects of this study stood taller and weighed less than their counterparts in the studies of Asmussen and Bonde-Petersen and Komi and Bosco.

Table 1

Group Characteristics (Mean ± S.D.)

Variables	Females	Males
Age (years)	20.9 ± 2.1	22.1 ± 3.6
Height (cm)*	168.8 ± 9.6	180.8 ± 4.6
Weight (kg)*	57.3 ± 8.1	71.0 ± 9.6
USEE (%)	37.4 ± 38.1	51.0 ± 22.7
Height of CMJ (cm)	36.4 ± 8.3	44.3 ± 8.8
Depth of CMJ (cm)	29.3 ± 7.5	32.4 ± 4.0
Knee angle (°)	85.2 ± 13.9	88.7 ± 5.0
Concentric time (s)	.245 ± .055	.235 ± .027
Depth ratio	17.4 ± 4.3	17.9 ± 2.1
Velocity ratio	6.2 ± 6.4	5.7 ± 2.1

*Group difference significant at the .05 level.

Although there was an apparent difference between females (37%) and males (51%) in the use of stored elastic energy, the large within-group variance prevented statistical significance. There was similarity between the male subjects of Komi and Bosco, who used 49% stored elastic energy, and the male subjects of this study; however, the female subjects of Komi and Bosco, who used 92% stored elastic energy, were divergent from the female subjects of this study. The subjects of Asmussen and Bonde-Petersen used the least stored elastic energy (23%).

One explanation for the range of values in the use of stored elastic energy lies in the variation of jumping positions employed in these studies. Asmussen and Bonde-Petersen did not include information about the position of the body during the squat which was assumed in preparation for the thrust of the CMJ and SJ. Komi and Bosco controlled the knee angle at 90° in the squat preceding the SJ trials but did not report the maximum knee flexion for the squat of the CMJ trials. From Equation 1 it can be seen that the height from which a thrust is taken has a relationship to the energy which is computed. Thus, if the SJ starts from a position which is higher than the corresponding CMJ, the energy computed for the SJ will be too low. If an underestimation of the net energy of the body during the concentric phase of the SJ is used in Equation 4, an overestimation for the use of stored elastic energy will result. Conversely, using a greater depth of descent in the SJ than CMJ is associated with an underestimation for the use of stored elastic energy.

Since the males of this study used a knee angle of 89° in the CMJ and SJ, it is possible that the male subjects of Komi and Bosco used a knee angle of approximately 90° in the CMJ as well as in the SJ. Therefore, the similarity in use of stored elastic energy in these two groups could be explained by the apparent similarity in squat position. If the female subjects of Komi and Bosco were similar to the female subjects of this study, a knee angle of about 85° was used in the CMJ. Because a knee angle of 85° represents greater flexion in the CMJ compared to the SJ with a knee angle of 90°, it is possible that the females of Komi and Bosco had greater depth of descent in the CMJ than SJ. Thus, an overestimation in the use of stored elastic energy could be inferred. Similarly, the low value for use of stored elastic energy reported by Asmussen and Bonde-Petersen could result from SJ trials involving greater depth of descent than CMJ trials.

There was a measurable difference in height of CMJ for females (36 cm) and males (44 cm) but the difference between groups was not statistically significant. After expressing height of CMJ relative to standing height, the females (22%) were similar to the males (24%). Since the females of this study jumped 82% of the height of the male subjects, these groups are closer in ability than the groups with 73% difference (Asmussen & Bonde-Petersen) and 58% difference (Komi & Bosco).

The greater depth of CMJ shown by the male subjects was in proportion to their greater standing height. Small differences between the groups may have occurred in posture since the females had more flexion at the knee than the males. There was a difference of 10 ms between the groups in the amount of time taken in concentric contraction. Both groups had a 6% improvement in velocity of projection when the eccentric/concentric CMJ was compared to the concentric SJ. An interesting divergence in the data was the greater variability in the female group, particularly in the use of stored elastic energy, depth of CMJ, knee angle, and concentric time.

The typical female performance related well with the typical male performance in jumping as indicated by variables of position (depth of CMJ, knee angle) and time (concentric time). Moreover, both groups showed an equivalent ability to coordinate the eccentric aspect of jumping with the concentric aspect by increasing the velocity in CMJ compared to SJ.

Conclusion

From the results of this study, it appears that there are no differences in the use of stored elastic energy and related biomechanical variables between groups of comparably skilled females and males.

References

ASMUSSEN, E., & Bonde-Petersen, F. (1974). Storage of elastic energy in skeletal muscles of man. *Acta Physiologica Scandinavica*, **91**, 385-392.

CAVAGNA, C.A., Dusman, B., & Margaria, R. (1968). Positive work done by a previously stretched muscle. *Journal of Applied Physiology*, **24**, 21-32.

HANAVAN, E.P. (1964). *A mathematical model of the human body.* AMRL Tech. Rept. 64-102, Wright-Patterson Air Force Base, Ohio.

KOMI, P.V., & Bosco, C. (1978). Utilization of stored elastic energy in leg extensor muscles by men and women. *Medicine and Science in Sports*, **10**, 261-265.

WILMORE, J.H. (1981). Women and sport: An introduction to the physiological aspects. In J. Borms, M. Hebbelenck, & A. Venerando (Eds.), *Women and sport: An historical, biological, physiological and sportsmedical approach* (pp. 109-111). Basel: S. Karger.

WINTER, D.A., Quanbury, A.O., & Reimer, G.D. (1976). Instantaneous energy and power flow in normal human gait. In P. Komi (Ed.), *Biomechanics V-A* (pp. 334-340). Baltimore: University Park Press.

Elastic Response of Isotonically Contracting Skeletal Muscle

M.D. Grabiner
University of Southern California,
Los Angeles, California, USA

The elastic properties of actively contracting skeletal muscle have long been of interest to biomechanists. Investigation of isometric and eccentric conditions has utilized single muscle fibers as well as in vivo animal/human muscle (Bosco, Komi, & Ito, 1981; Cavanagh & Komi, 1979). Under sustained isometric conditions increasing tension levels are associated with non-Hookian increases in the length of the elastic elements functionally in series with the contractile mechanism (Hill, 1950; Wells, 1965; Wilkie, 1956). Typically when an isometrically tetanized muscle is allowed to shorten under partially or fully-decreased load a biphasic time-displacement function is produced. The initial, more rapid phase of shortening is associated with the recoil of the lengthened elastic elements. The second, slower phase is credited to the contractile element in accordance with its force velocity properties (Hill, 1950). Observations of this type have resulted in the conclusion that the recovery of stored elastic energy allows muscle to achieve a shortening velocity not otherwise possible if solely dependent upon its contractile attributes.

It has been observed that the majority of human activities rarely begin from a resting static condition. That is, most movement is preceded by an eccentric lengthening of the agonists, resulting in an extension of the series elastic elements and subsequent recovery of a portion of the stored elastic energy. The contribution of elastic recoil to these types of contractions has been reported as occurring in the initial positive work phase (Thys, Cavagna, & Margaria, 1975) and has been associated with increased physiological efficiency (Thys, Faraggiana, & Margaria, 1972). The mechanical contribution to performance has been reported for activities such as the vertical jump (Cavagna, Dusman, & Margaria, 1968).

There have been no data reported regarding the contribution of the elastic response in concentric contractions in human movement initiated from a resting state. This is, perhaps, indicative of the difficulty of measuring in vivo preacceleration muscular force. An elastic recoil was suggested, however, by Amis, Dowson, and Wright (1980) in their analysis of unrestricted maximal effort elbow flexion and extension. The purpose of this investigation was to noninvasively determine the contribution of the estimated stored elastic energy to an unrestricted concentric contraction associated with a mono-articulated, fixed-distance task not preceded by a stretch cycle and/or a period of sustained isometric contraction.

Every concentric contraction is preceded by an isometric phase referred to as the motor reaction time, latent period, or electromechanical delay (EMD). The EMD is predominantly associated with the time necessary for the lengthening of the elastic elements (Viitasalo & Komi, 1981). Previous research has studied the effects of a number of independent variables on the duration of this period. However, none has attempted an estimation of the mechanical/elastic behavior that is manifested as a function of muscular tension, perhaps because of a lack of a definitive measurement tool for the preacceleration muscular tension. In this investigation, based upon the isometric characteristics of the EMD, the integrated electromyogram (IEMG) during the EMD was used as an index of generated isometric tension.

Methods

Fifteen male subjects (x = 27.4 years, 75.43 kg, 180.28 cm) performed ten isotonic elbow extensions in a horizontal plane from three initial elbow positions (120, 100, 80° from full extension) and four movement velocities over a range of 45° (.785 rad). The normalized movement velocities for each joint position included the subjects' maximal effort (minimal movement time), twice, three times, and four times the minimal movement time (submax1, submax2, and submax3). The subjects' forearms were attached to and supported by a specially built dynamometer which provided electrogoniometric and accelerometric signals to an on-line PDP/1103 computer system. Surface EMG was used to monitor the bioelectric activity of the triceps brachii (medial head) which 1) served to denote the onset of the EMD and, 2) following integration and normalization provided a measure of the isometric tension generated during EMD.

The estimated elastic lengthening during EMD was calculated independently of the value obtained for IEMG and consisted of the product of the duration of the EMD (ms) and the value representing the mean rate of the elastic element lengthening during EMD (arbitrary units per second). The latter was based upon the estimated contractile element shortening velocity during movement (arbitrary units per second) obtained from the acceleration data. This procedure was consistent with previous research that has, for instance, used the elapsed time to peak tension as representative of contraction velocity (Petrofsky & Phillips, 1979). This value was assumed to be consistently representative, albeit an underestimation, of the elastic element lengthening because of the intimate theoretical structural relationship between the elastic and contractile components of muscle. The data was initially treated as a 4 × 3 repeated measures MANOVA and was subsequently followed by a path analysis. The latter served as a means of identifying the contribution of the derived value of elastic response to the initial positive work phase, represented by the ratio of the magnitude of the peak acceleration to the elapsed time to the peak acceleration (average jerk).

Results

As expected, linear increases in movement time for all initial joint positions were significantly different from one another. These changes, however, were not reflected in the various mechanical and bioelectrical variables of interest. The changes in the duration of the EMD (see Figure 1) are inconsistent across the various joint positions. These trends could be reflective of the viscoelastic nature of muscle and tendon. That is, as internal muscular force increased the internal resistance to deformation did as well, accounting for, in part, the shorter EMD at increased levels of effort. However, the IEMG and, subsequently, tension of submax1 through submax3 were not significantly different from one another although significantly less than the maximal effort condition. This intimates a more complex relation between the level of activation, initial muscle length, and EMD.

The derived value for the elastic response during EMD was found to vary in a fashion more consistent with hypothesized expectations than EMD. Encouragingly, it was statistically consistent with the independently measured value of IEMG (see Figure 1). The elastic response for the maximal condition was significantly greater than the submaximal efforts which were not different from one another. There are notable fac-

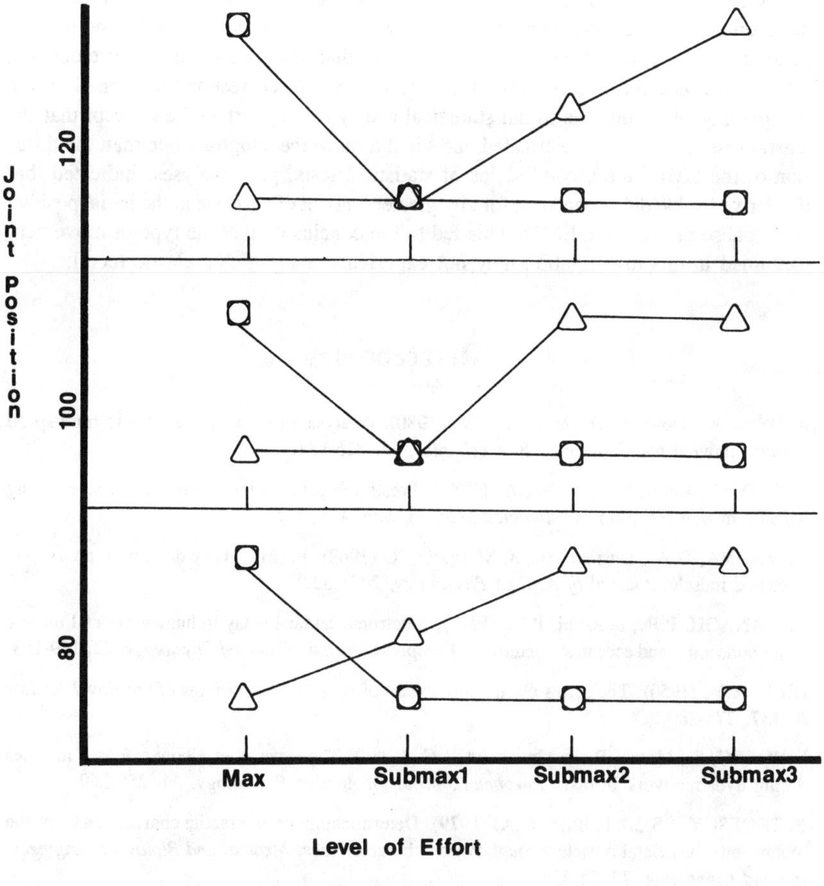

Figure 1—Elastic component response (O), EMD (Δ), and EMG (□) as a function of joint position and level of effort. Significant differences ($p < .05$) are indicated as a function of vertical level on the arbitrary Y-scale.

tors which may have influenced the data and need to be considered with respect to the assessment. For instance, the triceps brachii and anconeus muscles have been shown to switch roles as agonist when movement velocity decreases (LeBozec, Maton, & Cnockaert, 1980); therefore, all three heads of the triceps and the anconeus should be monitored to observe changes in EMD and the levels of bioelectric activity during EMD. With respect to the path analysis, despite the initial findings of the MANOVA, it was found that neither the EMD nor the derived elastic response differed in the proportion of variance accounted for in the initial work phase. It is possible that the ratio used as a measure of the initial positive work phase was not a sensitive enough measure to detect the contribution of an elastic recoil. The instantaneous jerk may well provide a more valuable variable. Indeed, the elastic recoil for a movement such as the one investigated may occur sometime after the contribution of the contractile element has peaked, serving to (a) help maintain the duration of the peak acceleration and/or (b) decrease the magnitude of the rate of deceleration. If this is so then the entire positive phase of the acceleration curve needs to be studied for possible elastic contributions.

The lack of a contribution by a recoil may be, in part, due to the relatively brief duration and low-tension levels associated with the unrestricted EMD. The tension generated during the EMD is much lower than that associated with either elongated isometric or eccentric conditions and, therefore, would correspond to a small elastic lengthening. Although the initial statistical tests gave support to the concept that the elastic response could be estimated and yield a more meaningful value than the duration of the EMD, the second series of statistical tests, path analyses, indicated that the derived value did not account for any greater amount of variance in the initial positive work phase than did the EMD. This led to the conclusion that the type of movement examined in this investigation may not experience a significant elastic recoil.

References

AMIS, A.A., Dowson, D., & Wright, V. (1980). Analysis of elbow forces due to high-speed forearm movements. *Journal of Biomechanics,* **13**, 825-831.

BOSCO, C., Komi, P.V., & Ito, A. (1981). Prestretch potentiation of skeletal muscle during ballistic movement. *Acta Physiologica Scandinavica,* **111**, 135-140.

CAVAGNA, G.A., Dusman, B., & Margaria, R. (1968). Positive work done by a previously stretched muscle. *Journal of Applied Physiology,* **24**21-32.

CAVANAGH, P.R., & Komi, P.V. (1979). Electromechanical delay in human skeletal muscle under concentric and eccentric conditions. *European Journal of Applied Physiology,* **42,** 159-163.

HILL, A.V. (1950). The series elastic component of muscle. *Proceedings of the Royal Society B,* **137**, 273-280.

LeBOZEC, S., Maton, B., & Cnockaert, J.C. (1980). The synergy of elbow extensor muscles during dynamic work in man. *European Journal of Applied Physiology,* **44**, 255-269.

PETROFSKY, J.S., & Phillips, C.A. (1979). Determination of contractile characteristics of the motor units in skeletal muscle through twitch characteristics. *Medical and Biological Engineering and Computing,* **17**525-533.

THYS, H., Faraggiana, T., & Margaria, R. (1972). Utilization of muscle elasticity in exercise. *Journal of Applied Physiology,* **32**, 491-494.

THYS, H., Cavagna, G., & Margaria, R. (1975). The role played by elasticity in exercise involving movements of small amplitude. *Pfluegers Archiv,* **354**, 281-286.

VIITASALO, J.T., & Komi, P.V. (1981). Interrelationships between electromyographic, mechanical, muscle structure, and reflex time measurements in man. *Acta Physiologica Scandinavica,* **11**, 97-103.

WELLS, J.B. (1965). Comparison of mechanical properties between slow and fast mammalian muscles. *Journal of Physiology,* **178**, 252-269.

WILKIE, D.R. (1956). Measurement of the series elastic component at various times during a single muscle twitch. *Journal of Physiology,* **134**, 527-530.

Electromyographic Analysis on Utilization of Elastic Energy in Human Leg Muscles

K. Funato, H. Ohmichi, and M. Miyashita
University of Tokyo, Tokyo, Japan

The possibility that mechanical energy may be temporarily stored in the series elastic components of active muscles for reuse in a following contraction was presented by Asmussen and Bonde-Petersen (1974). Later Bosco, Tarkka, and Komi (1982) reported that the increase in performance was attributed to a combination of utilization of elastic energy and myoelecric potentiation of muscle activation. Such an improvement of performance would often appear as a higher mechanical efficiency.

The present study was to investigate the effect of the velocity of prestretching on the amount of the utilization of stored elastic energy from the viewpoints of mechanical efficiency and myoelectrical activity of knee extensors during stretch-shortening cycle exercise.

Methods

Subjects

Six healthy male students from our laboratory: 24.0 ± 1.5 years of age, 171.7 ± 2.9 cm body height, and 65.8 ± 5.9 kg body weight participated in this study. All subjects were fully informed of the purpose and procedures associated with this study, and consented to participate.

Experimental Procedures

Each subject continuously performed stretch-shortening cycle exercise for three minutes on a force platform (Kistler). The subject bent his knees to a given squatting position and immediately extended up to the upright position. The frequency of stretch-shortening cycles was controlled by an auditory metronome at a speed of 20, 35, 50, 65, and 80 cycles/min.

Angular amplitude of knee joint was recorded with an electrogoniometer attached on the right knee joint. It was at zero degrees when subject was standing erect. The knee goniogram was displayed on an oscilloscope set in front of the subject so that he could control the given amplitude of knee joint motion by himself.

Two squatting amplitudes were used in this study, i.e., half (60°) and deep (90°) knee joint angles. During exercise, the subject was instructed to keep his hands at his waist, his body in an upright position, and place both his toes and heels on the force platform throughout the exercise.

In the last 5 min of 20-min rest, 3-min exercise, and 25-min recovery, expired air was collected in the Douglas bags to determine oxygen requirements. Expired air volume was measured by a dry gasometer. The O_2 and CO_2 gas fractions were determined with a gas analyzer (model 1H06, SANEI Inc., Tokyo), which was periodically calibrated using the micro-Scholander technique. Mean metabolic power was calculated as oxygen requirement per second.

Potential external energy was determined by the double electrical integration of the vertical acceleration of the body center of gravity recorded by the force platform. The external energy divided by the duration of one cycle of stretch-shortening exercise was represented as mean mechanical power. Mechanical efficiency was determined as a ratio of mean mechanical power to mean metabolic power.

Electromyograms (EMGs) were continuously recorded from the muscles of rectus femoris and vastus medialis during exercise. Bipolar surface electrodes made of a silver plate 10 mm in diameter were attached 4 cm apart on the belly of the muscles. EMG signals obtained during the last 30 s of each 3-min exercise period were amplified, rectified, and then integrated using an analogue integrator (model EI-600G, Nihonkoden Co., Ltd., Tokyo).

As shown in Figure 1, the integrated EMG for one cycle of stretch-shortening exercise (Total iEMG) contains two phases; during knee flexion-negative work phase

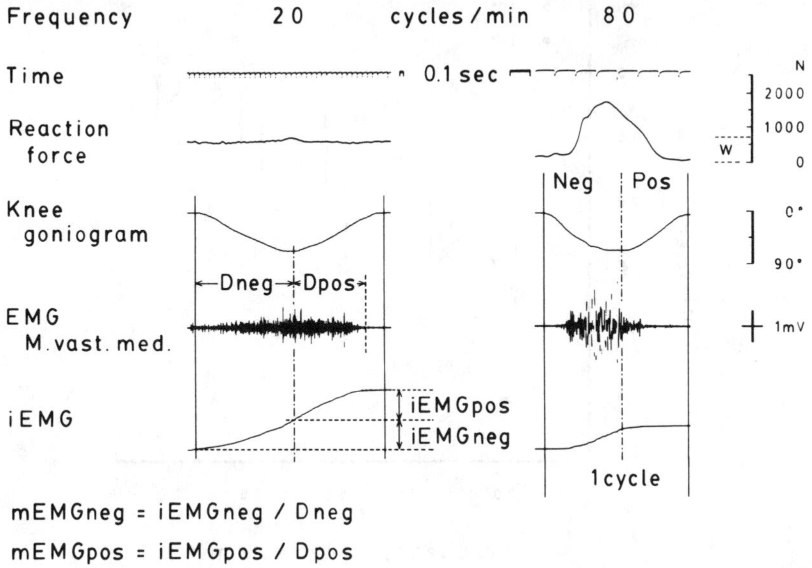

Figure 1—Electrogoniogram of knee joint, EMG from m. vastus medialis, and integrated EMG at the frequency of 20 cycles/min (left) and at 80 cycles/min (right).

(iEMGneg) and during knee extension-positive work phase (iEMGpos). Mean amplitude of EMG in negative and positive phases was calculated as iEMGneg and iEMGpos per the corresponding duration of muscle activation, respectively (mEMGneg and mEMGpos).

Results

The exercise oxygen requirement increased with the frequency of the stretch-shortening cycle exercise, from 20.6 ± 6.8 to 55.1 ± 20.2 ml/kg in the 60° condition and from 36.0 ± 5.1 to 87.9 ± 14.4 ml/kg in the 90° condition. Mean mechanical power was also proportional to frequency in a range of 10.0 to 128.7 W and 33.1 to 209.8 W for the 60° and 90° conditions, respectively.

Accordingly, mechanical efficiency for both conditions, plotted against the different frequencies, fell on a straight line with high r value ($Y = 11.23 + 0.22X$, $r = 0.78$, $p < 0.001$). As seen in Figure 2, the mean mechanical efficiency of six subjects improved with the increase of frequency from about 13% to 28%. However, there was no statistically significant difference in efficiency between 60° and 90° conditions.

Figure 3 shows the iEMG and mEMG in relation to frequency. Total iEMG decreased with frequency while iEMGneg remained around 50% of total iEMG at 20 cycles/min. iEMGpos decreased with frequency to 10% (60°) and to 6% (90°) at the frequency of 80 cycles/min. Therefore, the reduction of total iEMG was a result of that of iEMG during positive phase.

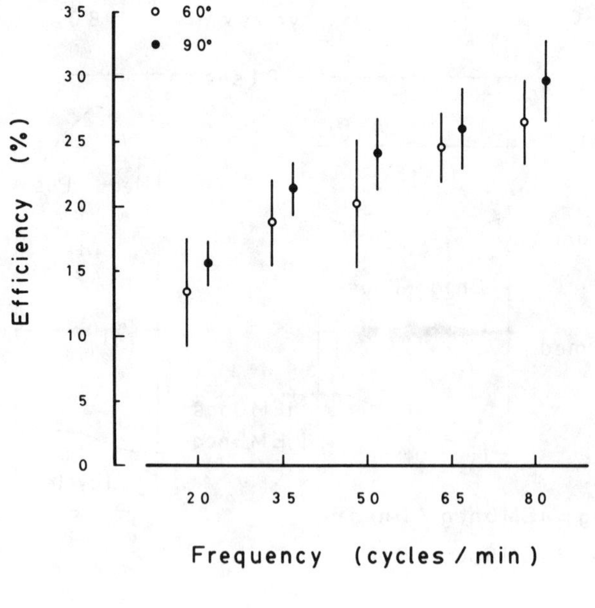

Figure 2—Mechanical efficiency against frequency of stretch-shortening cycle in the 60°- and 90°-squatting positions.

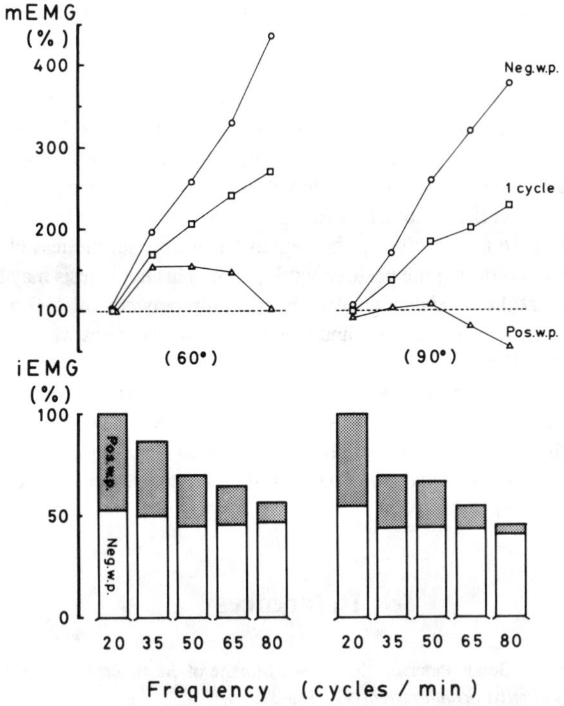

Figure 3—iEMG and mEMG in relation to frequency of stretch-shortening cycle. The iEMG and mEMG from vastus medialis are represented as percent of total iEMG and total mEMG at 20 cycles/min as 100%, respectively. Each value indicates the mean for all subjects.

On the other hand, mEMGneg proportionally increased with frequency, that is, the value obtained at 80 cycles/min was 3.7 to 4.4 times greater than that at a frequency of 20 cycles/min but that of positive phase did not change so notably with the increase of frequency.

A linear relation was observed between mean metabolic power and mEMG. The correlation coefficients for all subjects ranged from 0.78 to 0.99.

Discussion

In the present study, the mechanical efficiency ranged from 9.55% to 34.95%, and there was a strong positive correlation between efficiency and frequency of stretch-shortening cycle exercise (see Figure 2). On the other hand, at the same frequency of movement, there was no statistically significant difference in mechanical efficiency between the two amplitudes of movement (60° and 90°). These results suggest that mechanical efficiency must be definitely affected by the frequency of repetitive exercise, but not by the amount of work done. Therefore, it may be said that the higher the prestretching velocity of knee extensor muscles, the greater the amount of elastic energy stored in the prestretched muscle and reused in the following concentric contraction. These results coincided well with the previous reports (Bosco & Komi, 1979; Bosco, Komi, & Ito, 1981) that the amount of elastic energy was greater when the prestretching speed was higher.

As was seen in Figure 3, iEMG during one cycle of exercise decreased with exercise frequency, which was mainly due to the decrease in iEMGpos. This result indicates that although the quadriceps are the primary muscle groups used for knee extension, they are less activated during the positive work phase at higher frequencies of stretch-shortening cycle exercise. In other words, it may be hypothesized that the more elastic energy stored in the muscle during the negative work phase, the less electrically activated the muscle is during the positive work phase. This hypothesis may be supported by the results that there exists a positive relationship between mEMG and frequency of stretch-shortening cycle exercise, and that mEMGneg increases with frequency more rapidly than mEMGpos.

Taking into account the results as described above and the linear relationship between mean metabolic power and mEMG, it is assumed that physiological energy consumption during negative work might increase with the frequency of the stretch-shortening cycle exercise, and some portion of this increased energy could be reused during positive work.

References

ASMUSSEN, E., & Bonde-Petersen, F. (1974). Storage of elastic energy in skeletal muscle in man. *Acta Physiologica Scandinavica,* **91**, 385-392.

BOSCO, C., & Komi, P.V. (1971). Potentiation of the mechanical behavior of the human skeletal muscle through prestretching. *Acta Physiologica Scandinavica,* **106**, 467-472.

BOSCO, C., Komi, P.V., & Ito, A. (1981). Prestretching potentiation of human skeletal muscle during ballistic movement. *Acta Physiologica Scandinavica,* **111**, 135-140.

BOSCO, C., Tarkka, I., & Komi, P.V. (1982). Effect of elastic energy and myoelectrical potentiation of triceps surae during stretch-shortening cycle exercise. *International Journal of Sports Medicine,* **3**, 137-140.

Storage and Utilization of Elastic Energy in Musculature

J. Denoth
Swiss Federal Institute of Technology (ETH),
Zürich, Switzerland

Intended movements are usually preceded by a countermovement which may have a positive influence on performance. In vertical jumping the influence of the height at takeoff on the resultant height is well known (Asmussen & Bonde-Petersen, 1974; Komi & Bosco, 1978). This effect on performance is explained by the property of musculature to store energy; this mechanism has been described as follows: "When an active skeletal muscle is forced to stretch, it stores elastic energy which can be—at least partially—recovered during the subsequent shortening phase" (Komi, 1979).

From the point of view of mechanics, however, the important factor is not the energy, but the momentum. This will be explained in the following example. When "describing" the vertical jump as a free fall of a rigid body onto a spring, the height before the jump and height reached after the jump are equal if the spring is perfectly elastic. This result can be explained by energy considerations too, because the mass-spring-system has only one degree of freedom and no energy is transformed into heat. But since muscle activity produces heat, energy considerations, as in this example, are only meaningful in thermally closed systems.

The purpose of this paper is to illustrate theoretically some properties of muscular mechanics. The influence of countermovements on performance is studied in a two-link model. The model describes the vertical throw of a shot. The resultant height, which is considered as the performance, as well as the work and heat production of the musculature are discussed.

Methods

When modeling a movement, the following procedure is commonly used: Reality is simplified into a mechanical model, the differential equations are formulated and then solved by numerical integration.

The two-link model which is chosen for the following computation (see Figure 1) represents the upper arm and the forearm (mass ml). The upper arm is fixed for the time period from beginning of the countermovement until the release of the shot. The forearm, the hand, and the shot (mass m) are combined into one rigid body which will be referred to as the forearm. The upper arm and the forearm are connected via a muscle. This muscle is considered to act as a massless force generator, comprising a contractile element (CE) and a series elastic element (SE). The mechanical properties of the series elastic element are given as follows (Hatze, 1981; Yamada, 1973):

$$FSE = f1 \cdot y + f2 \cdot y^2 + r \cdot \dot{y} \tag{1}$$

where FSE = force on the series element
y = "length" of the series element
f1 = spring constant
f2 = constant describing nonlinearity
r = damping coefficient

The contractile element is that proposed by Hill (1970), in which the muscle characteristic parameters a, b, and Po depend upon the level of muscle activity (Philips & Petrofsky, 1980). The following dependencies were used (Denoth, 1982):

$$Po(Z) = p1 \cdot Z$$
$$a(Z) = a1 \cdot Z$$
$$b(Z) = b1 \cdot Z - b2 \cdot Z^2$$

with Z being the "intrinsic state" (Hill, 1970) or "active state" of the contractile element. The intrinsic or active state Z is not regulated during the movement by the nervous system. The time history of Z is chosen as:

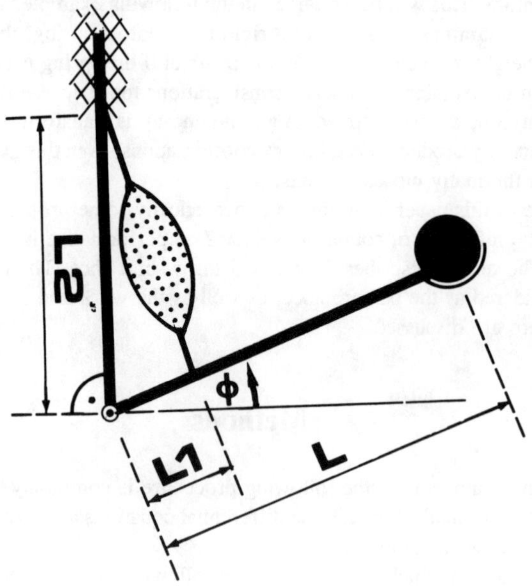

Figure 1—The two-link model of the upper arm, the forearm, the hand, and the shot.

$$Z(t) = Zo \cdot (1 - \exp[-t/T]) \tag{2}$$

The range of the contractile element is narrowed to $(\bar{X} - \Delta\bar{X}, \bar{X} + \Delta\bar{X})$; $\bar{X}$ is the "average" length of the CE.

During a movement, transfer of energy takes place. That part of the energy which is transferred into heat is given by (Denoth, 1982; Hatze, 1981; Hill, 1970):

$$Q = \int a \cdot b \, dt - \int a \cdot \dot{x} \, dt + r \cdot \int \dot{y}^2 \, dt \tag{3}$$

The elastic energy stored in the series element or utilized by the series element is given by:

$$\Delta ASE = [\tfrac{1}{2} f1 \cdot y^2 + \tfrac{1}{3} f2 \cdot y^3]_{y1}^{y2} \tag{4}$$

and the work done by the contractile element is:

$$\Delta ACE = \int KCE \cdot \dot{x} \, dt \tag{5}$$

When simulating the vertical throw of a shot, several parameters have to be set at certain values. For the computations presented here these parameters were chosen as follows:

m1 = 1 kg, m = 2 kg; L = 0.4 m, L1 = 0.05 m, L2 = 0.40 m; p1 = 1000 N, a1 = 250 N, b1 = 0.4 m/s, b2 = 0.2 m/s; f2 = 0, r = 0; T = 0.03 s; Δx = 0.02 m; f1 = $0.25 \cdot 10^5$ N/m, $0.5 \cdot 10^5$ N/m, $1 \cdot 10^5$ N/m, $2 \cdot 10^5$ N/m, $4 \cdot 10^5$ N/m.

The initial conditions were:

case (i): $\dot{\phi}(0) = 0$; x (0) = −0.018 m, −0.014 m,..0.018m
case (ii): $\phi(0) = 0$; $\dot{\phi}(0)$ = 0 °/s, −100 °/s, −200 °/s,

The critical load of the muscle was set at 1200 N (= 120% of the maximum isometric force).

Results

With increasing angular velocity of the forearm, the resultant height is increased (see Figure 2). At an angular velocity of −400 °/s, the critical load for the muscle is in this case exceeded. Theoretically, the muscle is torn apart, which is possible in this model because the active state of the muscle is not regulated during the movement. During the countermovement the CE can be stretched. With this stretching the range in which the muscle can perform work is enlarged, hence the resultant height becomes increased. When comparing the resultant height with and without countermovement, the influence of the countermovement is relatively large.

Phenomena of resonance are frequently observed in mechanics when an oscillator is coupled onto a system. This is also the case in this model, as can be demonstrated

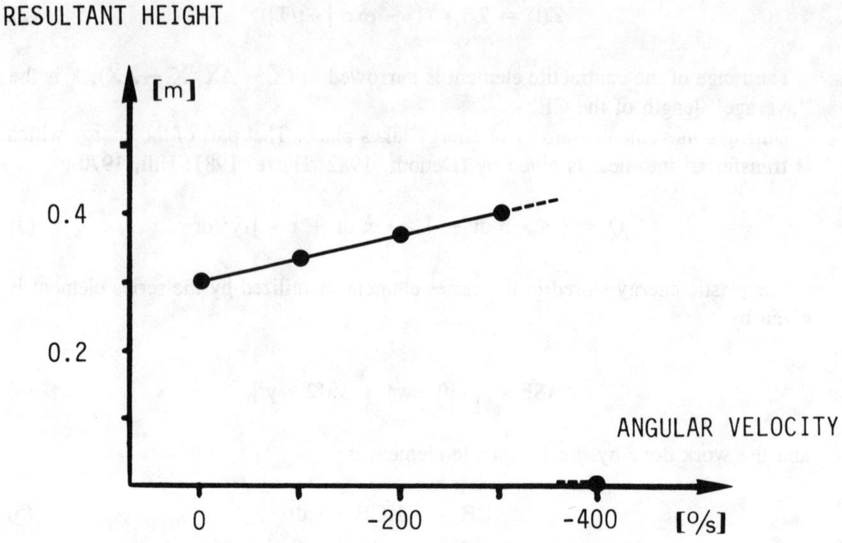

Figure 2—Resultant height versus angular velocity of the forearm (fl = 1 • 10^5 N/m).

with the resultant height by varying only the stiffness of the SE (see Figure 3).

The resultant height reaches a maximum if the force-time characteristics of the CE are "equal" to the frequency of the "series elastic-mass-system." When the system is in resonance, and the force amplitude and the time period of the throw are large,

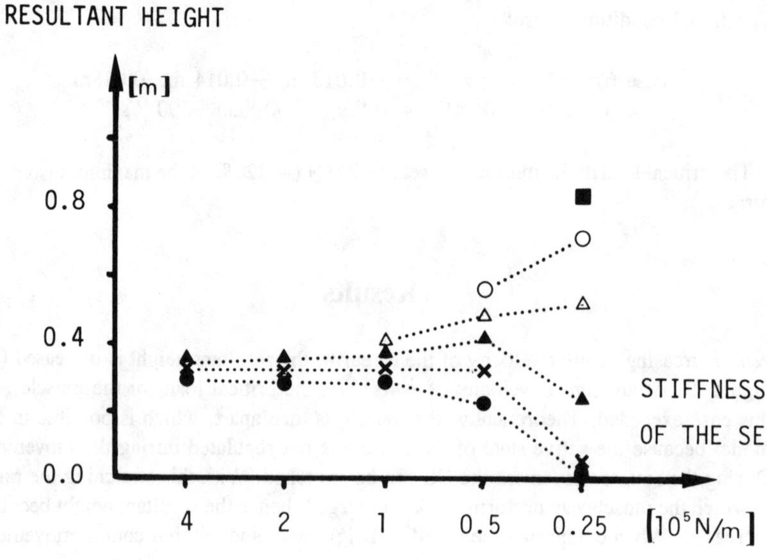

Figure 3—Resultant height versus stiffness of SE at different angular velocities (●: 0 °/s; x : −100 °/s; ▲: −200 °/s; △: −300 °/s; ○: −400 °/s).

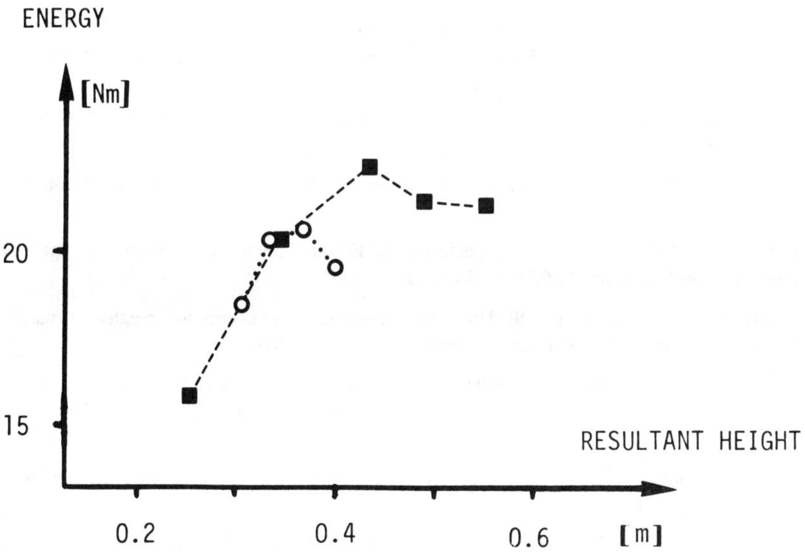

Figure 4—Energy versus performance at different angular velocities and for two different SE (fl = $1 \cdot 10^5$ N/m, $0.5 \cdot 10^5$ N/m).

the resultant momentum is large, too. The resultant heights demonstrated in Figure 3 are not the maximum heights, since not all of the parameters were varied.

Increased (sporting) performance generally demands an increased energy expenditure but it may be that the muscle spends less energy at maximum performance (see Figure 4) than at lower performance levels.

Conclusions

The resultant height—and similarly the performance in vertical jumping—depends largely on the countermovement. This relation is based on the muscle momentum, which is determined by the "resonance phenomenon." The stored elastic energy is given by the maximum strain of the SE and consequently is a result of the phenomenon of resonance as well. Therefore, the phenomenon of resonance, and not the stored elastic energy, is the primary reason for the resultant height.

Acknowledgment

This work was supported by the Swiss National Foundation.

References

ASMUSSEN, E., & Bonde-Petersen, F. (1974). Storage of elastic energy in skeletal muscle in man. *Acta Physiologica Scandinavica*, **91**, 385-392.

DENOTH, J. (1982). Biomechanische Probleme der muskulären Leistung. In W. Groher & W. Noack (Eds.), *Sportliche Belastungsfähigkeit des Haltungs—und Bewegungsapparates.* Stuttgart: Thieme.

HATZE, H. (1981). *Myocybernetic control models of skeletal muscle.* Pretoria: University of South Africa.

HILL, A.V. (1970). *First and last experiments in muscle mechanics.* Cambridge: Cambridge University Press.

KOMI, P.V. (1979). Neuromuscular performance: Factors influencing force and speed production. *Scandinavica Journal of Sports Sciences,* **1**, 2-15.

KOMI, P.V., & Bosco, C. (1978). Utilization of stored elastic energy in leg extensor muscles by men and women. *Medicine and Science in Sports,* **10**, 261-265.

PHILIPS, C.A., & Petrofsky, J.S. (1980). Velocity of contraction of skeletal muscles as a function of activation and fiber composition. A mathematical model. *Journal of Biomechanics,* **13**, 549-554.

YAMADA, M.D. (1973). *Strength of biological materials.* Huntington, NY: Robert E. Krieger Publishing Company.

Bone-on-Bone Forces at the Ankle Joint During a Rapid Dynamic Movement

V. Galea and R.W. Norman
University of Waterloo,
Waterloo, Ontario, Canada

Knowledge of the magnitude of bone-on-bone forces at joints is of interest to people working in the areas of arthritis, prosthetics, ergonomics, sports, and dance. However, most of the available literature is on relatively slow movement such as walking (Proctor & Paul, 1982) and running (Burdett, 1982).

Calculation of the resultant moment and joint reaction forces has been the subject of considerable literature. These calculations, using link segment mechanics, may be performed easily enough, but they do not take into account the force of contraction of the muscles crossing the joint. Bone-on-bone forces are a result of the joint reaction force and the muscle and ligament forces. Use of the joint reaction force alone significantly underestimates the load experienced by the joint surface.

Various models have been developed which seek to calculate the bone-on-bone force through non-invasive means (Crowninshield & Brand, 1981). The challenge lies in the calculation of the instantaneous muscle forces during the movement in question. Estimation of muscle force utilizing knowledge of muscle geometry, isometric force output, neural activation, force/length and force/velocity relationships has been attempted by some authors (Hof & Van den Berg, 1981; Norman, 1977). These muscle forces were then implemented in a series of dynamic equations using linked segment mechanics in the calculation of bone-on-bone forces or moments at a joint.

The purpose of this study was to estimate the bone-on-bone forces at the ankle joint during a rapid ballet movement, a spring from flat feet onto the toes. Muscle forces were calculated using an expansion of the muscle force output model proposd by Norman (1977), by adding the *linear* muscle force/length and force/velocity relationships to the static force component and the neural activation. The force values from muscles gastrocnemius (GAST), soleus (SOL), tibialis anterior (TA), peroneus longus (PL), flexor hallucis longus (FHL), and extensor hallucis longus (EHL) were then incorporated into dynamic equations which modelled the resultant bone-on-bone forces at the ankle joint.

Methods and Procedures

The calculation of the ankle joint bone-on-bone force components was performed using the following equations generated from the free body diagram in Figure 1:

$$F_{ax} = G_x + EHL \times Cos(\theta_f + 180) + TA \times Cos(\theta_f + 180) + PL \times Cos(\theta_s + 180) + FHL \times Cos(\theta_s + 180) + GAST/SOL \times Cos(\theta_s + 180) \quad (1)$$

$$F_{ay} = G_y + EHL \times Sin(\theta_f + 180) + TA \times Sin(\theta_f + 180) + PL \times Sin(\theta_s + 180) + FHL \times Sin(\theta_s + 180) + GAST/SOL \times Sin(\theta_s + 180) \quad (2)$$

where G_x and G_y are the measured horizontal and vertical ground reaction forces.

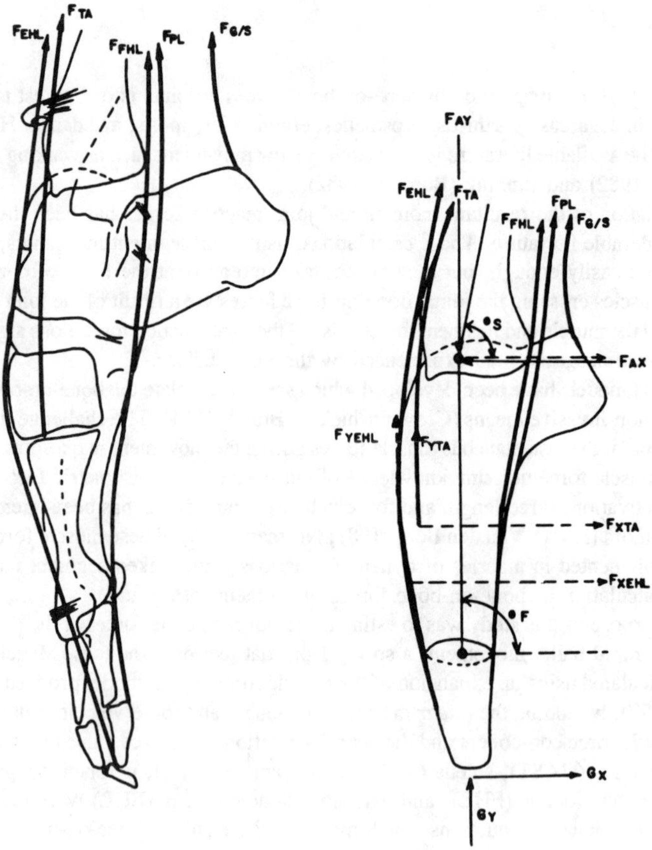

Figure 1—Schematic of the foot on full pointe showing the lines of action of the muscles. See text for force equations.

The individual muscle forces were calculated using the following model:

$$F_m = F_I \times \frac{EMG}{EMG_I} \times \frac{F(V)}{F_I(V=0)} \times \frac{F(L)}{F(L_o)} \tag{3}$$

where F_m = the instantaneous muscle force (e.g., EHL, TA, etc., in Equations 1 and 2
F_I = the isometric force output (N) during a calibration contraction
EMG_I = the EMG produced during F_I
EMG = the instantaneous value of the processed EMG activity of F_m
$\frac{F(V)}{F_I(V=0)}$ = the ratio of muscle force at velocity ± V as a proportion of its force output when V equals zero
$\frac{F(L)}{F(L_o)}$ = the ratio of muscle force at length (L) as a proportion of its force output at its resting length (L_o).

Muscle length and velocity changes were calculated using a modification of the trigonometric method of Frigo and Pedotti (1978). Centroids of the origin and insertion areas of each muscle were approximated and an idealized mean fiber joining the two centroids represented the muscle length. Velocities were calculated using a finite differences technique. Muscle force modulation due to velocity was accomplished by utilizing the force/velocity curve proposed by Jorgensen (1976). A modified version of this curve appears in Figure 2. The curve was linearized in four sections and extrapolated to the extreme ranges of velocities in the movement studied. The ratio $F(V)/F_I(V = 0)$ in Equation 3 was thus obtained. Force modulation due to length was similarly acquired. The length/tension curve for fusiform rabbit muscle (Fidelius, 1967) was chosen and modified to approximate the length/tension relationship for the purposes

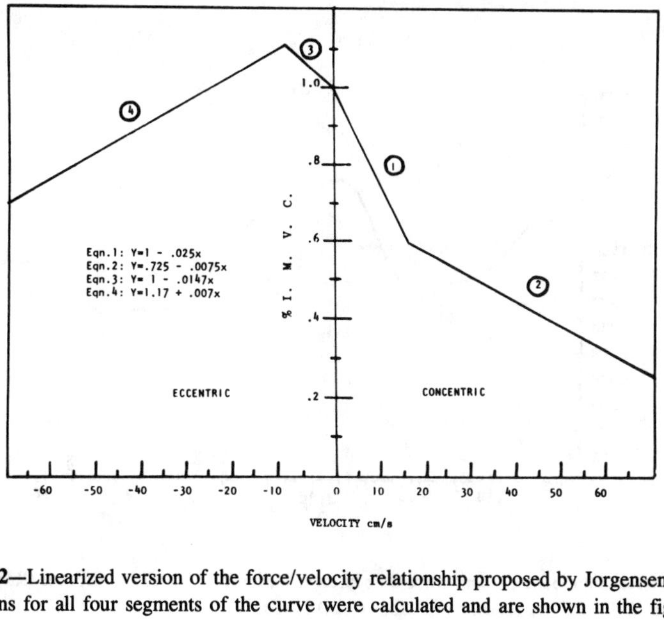

Figure 2—Linearized version of the force/velocity relationship proposed by Jorgensen (1976). Equations for all four segments of the curve were calculated and are shown in the figure.

of this study. The portion of the curve attributed to the human physiological range of the movement was linearized. Once the equation of the line was known, $F(L)/F(L_o)$ was found and was also used in Equation 3. L_o was considered to be the lengths of the muscles at the joint angles (knee, ankle, m-p) to which the limb moved when the muscles were relaxed and there were no gravitational effects, i.e., the leg was suspended horizontally.

Three female ballet dancers performed four trials of relevés on pointe. The movement consisted of a spring from two feet to two feet and was repeated four times during one trial. After undergoing calibration procedures to determine isometric force and EMG outputs from gastrocnemius, soleus, tibialis anterior and peroneus longus, subjects performed the movements on a force plate. Trials were filmed at a rate of 60 frames/s and the film, force plate and EMG data were synchronized. FHL and EHL were assigned an isometric force output value of 40 N/cm² for purposes of this study.

Results and Discussion

Peak bone-on-bone forces at the ankle ranged from 5255 N to 7030 N and almost always occurred when the dancer achieved full pointe. An example of typical resultant bone-on-bone (BOB) force time histories appears in Figure 3. This curve was observed in almost all cycles analyzed. The segment of the cycle where the dancer achieves full pointe begins at RPT (right pointe) and terminates at the second RTO (right toe off).

Individual muscle forces were sometimes as high as 2677 N (soleus) and 2466 N (tibialis anterior). The instantaneous velocities of the muscle seemed to demonstrate a considerable effect on the respective muscle force outputs, sometimes reducing it to 38% of its isometric value. Results from the length modulation factors failed to reveal a significant effect. A certain amount of error was inherent in the muscle model used.

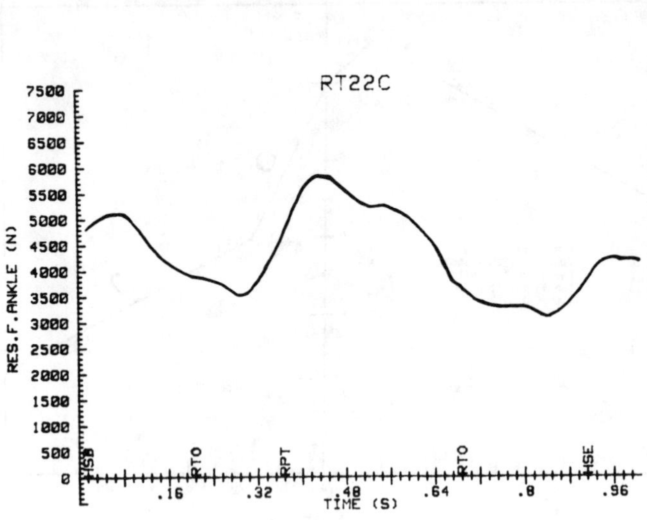

Figure 3—Resultant bone-on-bone force time history. Push off (HSB-RTO); airborne (RTO-RPT); full pointe (RPT-RTO); return to flat foot (RTO-HSE).

Table 1

Peak Resultant Bone-on-Bone (BOB) Forces and Joint Reaction Forces (JRF) Observed Over the Course of the Movement Cycles Monitored for Subject RT

Trial	JRF(N)	Phase	BOB(N)	Phase
RT21A	607	RPT	6436	RPT
RT21B	777	RTO	6224	RPT
RT21C	756	RTO	6083	RPT
RT22A	565	RTO	6131	RPT
RT22B	829	RTO	5698	RPT
RT22C	859	RTO	5836	RPT

Forces were overestimated by as much as 40% in some cycles, yet even allowing for error in the model, values remained surprisingly high.

Examination of the values in Table 1 and observation of the curve (see Figure 3) reveal that the BOB forces are as high as 10 times body weight. Comparison with peak joint reaction forces in Table 1 indicates that muscle cocontraction considerably escalates the load at the joint surface. In a movement such as this it is unrealistic to assume no cocontraction. Peak bone-on-bone forces for walking have been reported at 3.9 times body weight (Proctor & Paul, 1982), and for running, up to 13 times body weight (Burdett, 1982). The values reported in this study, occurring at the extremes of the ankle joint range of motion, are comparable with those observed during running. One must question the possible consequences of repetitively imposing forces of this nature on the joint surface.

Ultimate (to failure) compression stress values range between 13.8 to 28 KN/cm^2. The forces reported in this study are well below that value. The problem here is not one of high stress leading to fracture; rather it is the repetitive loading at the joint surface. An unhealthy response to repetitive loading results when the frequency of loading precludes the remodeling necessary to prevent fracture. A common cause of this unhealthy response is performing when the body is fatigued. Further, joint degeneration is thought to be accelerated by impulsive repetitive loading (Radin, Parker, Pugh, Sternberg, Paul, & Rose, 1973). Osteoarthritic changes may result from habitual performance of movements with improper technique or performing with fatigued muscles. The ankle joint is particularly susceptible to this form of joint degeneration leading to considerable problems later in life.

References

BURDETT, R.G. (1982). Forces predicted at the ankle joint during running. *Medicine and Science in Sports and Exercise,* **14**(4), 308-316.

CROWNINSHIELD, R.D., & Brand, R.A. (1981). The prediction of forces in joint structures: Distribution of intersegmental resultants. In D.I. Miller (Ed.), *Exercise and Sports Sciences Reviews,* **9**, 159-181.

FIDELUS, K. (1968). Some biomechanical principles of muscle cooperation in the upper extremities. *Proceedings from the First International Biomechanics Seminar (Zurich, 1967)* (pp. 172-177). Basel: Karger.

FRIGO, C., & Pedotti, A. (1978). Determination of muscle length during locomotion. In E. Asmussen & K. Jorgensen (Eds.), *Biomechanics VI-A* (pp. 355-360). Baltimore: University Park Press.

HOF, A.L., & Van den Berg, J. (1981). EMG to force processing. Articles I-IV. *Journal of Biomechanics*, **14**, 747-792.

JORGENSEN, K. (1976). Force-velocity relationships in human elbow flexors and extensors. In P.V. Komi (Ed.), *Biomechanics V-A* (pp. 145-151). Baltimore: University Park Press.

NORMAN, R.W.K. (1977). The use of electromyography in the calculation of dynamic joint torque. Unpublished doctoral dissertation, Pennsylvania State University at University Park.

PROCTOR, P., & Paul, J.P. (1982). Ankle joint biomechanics. *Journal of Biomechanics*, **15**(9), 627-634.

RADIN, E.L., Parker, H.G., Pugh, J.W., Sternberg, R.S., Paul, I.L., & Rose, R.M. (1973). Response of joints to impact loading. III: Relationship between trabecular microfractures and cartilage degeneration. *Journal of Biomechanics*, **6**, 51-59.

The Influence of Training With Concentric and Eccentric Work on the Force and Velocity Characteristics of Muscle

C. Urbanik and O. Ubukata
Academy of Physical Education, Warsaw, Poland

Experimental studies of muscle force-velocity characteristics indicate that the evaluation of different training methods is a difficult and complex problem. In various methods the isometric, concentric and eccentric nature of muscle work can be isolated. According to opinions expressed by some authors (Berger, 1962; Denishin & Kuzniecow, 1972; Dobrowolskij, 1972; Hettinger, 1972; Iwanow, 1968), the training effectiveness is closely linked with the type of work the muscles perform. Opinions of research specialists in this matter differ considerably. Different values of training load have been applied ranging from 70% of maximum load for concentric work to 140% for eccentric work (Dobrowolskij, 1972; Hakkinen & Komi, 1982; Iwanow, 1968). A doubt arises whether the largest increase in the force-velocity characteristics results from the character of the muscle work or from the size of the training load.

In this paper an attempt is made to assess the changes in the force-velocity characteristics of lower human limbs under the influence of training with eccentric and concentric muscle work having similar values of training load. A hypothesis has been formed that the force and velocity increase is not dependent on the characteristics of the muscle work but only on the amount of training load.

The measurement of training load has been an unresolved problem in many sports (Wazny, 1978). In our experiment the proposition of Fidelus (1974) has been accepted in order to equalize the training loads in the experimental groups, on the basis of external work, measured by the amount of kinetic energy.

Material and Methods

Eighteen students from the AWF Warsaw aged from 20 to 23 years and who did not practice sport in clubs took part in the experiment.

They were divided into two groups who trained with concentric (10 subjects) and eccentric (8 subjects) work. The basic idea was to train both these groups with equal training loads as measured by the kinetic energy absorbed or produced. This was accomplished using the combined training/measuring device seen in Figure 1.

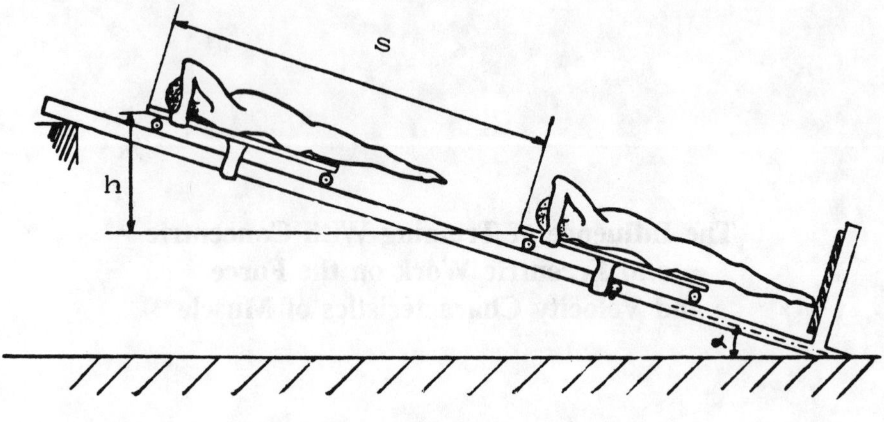

Figure 1—Training cart and inclined plane.

The subject lay on the cart and either extended against the platform (concentric contractions) or was released from a predetermined height and stopped the cart (eccentric contractions). In both cases the speed of the cart was measured when the lower limb was just out of contact with the platform. On the basis of this speed the amount of kinetic energy was calculated and was found to be similar for all the subjects. The training lasted for 6 weeks (3 times per week) and was characterized by a training load variation in a weekly cycle as practiced in sports training schemes. In order to assess the increase in the force-velocity characteristics of the lower limbs in response to a defined training program, measurements were carried out of moment of force in static conditions for the knee and hip joints extensors and for the ankle joint plantar flexors. For the knee joint extensor muscles, measurement of movement velocity was also carried out as a function of a variable external load.

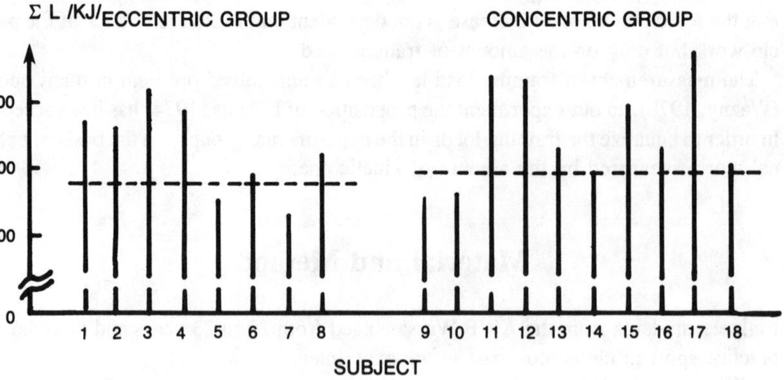

Figure 2—The total amount of work performed in training for all subjects.

Results

The amount of work performed in all the training for the particular subjects is presented in Figure 2. The broken line indicates the average value of training work for all the cases examined, which differ in both the groups by 3.1%.

On the bais of the experiment, percentage differences for both groups, of muscle moment in a static condition for the hip and knee extensors, as well as for ankle joint plantar flexors, have been presented (see Figure 3).

The results indicate that a maximum increase of force took place after the training was ended. In the case of the hip joint and knee joint, the maximum value was achieved

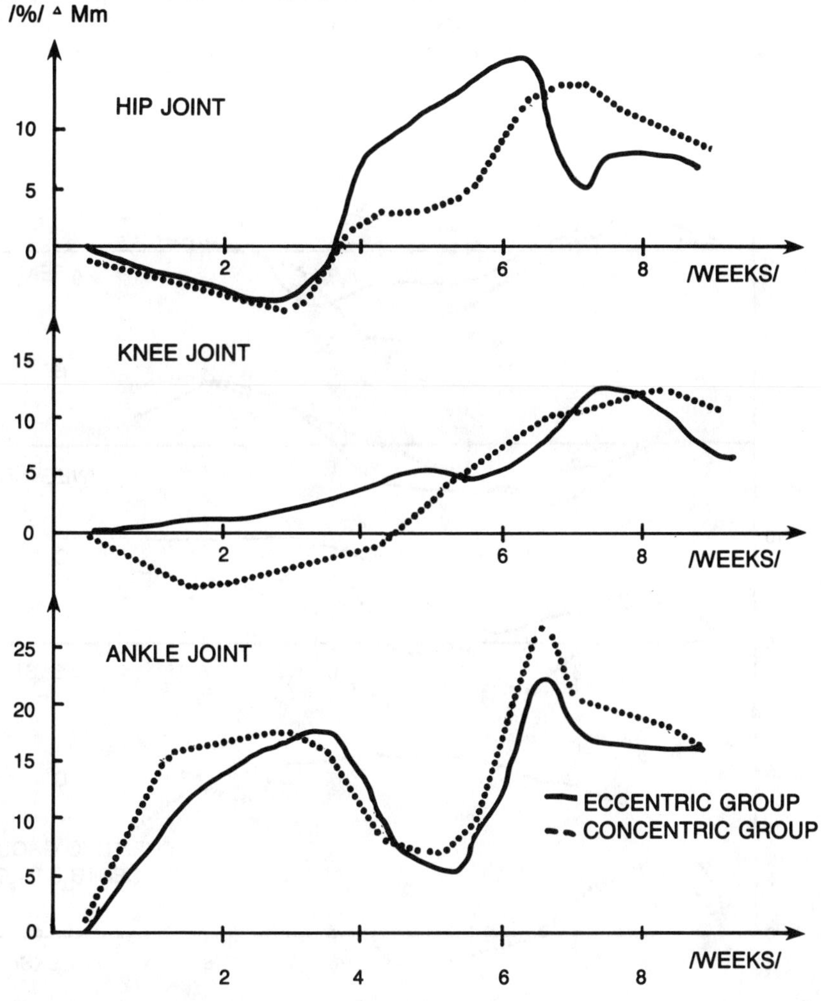

Figure 3—The percent change of increase of the moment of force in static condition, occurred during the training of the concentric (broken line) and eccentric (solid line) group.

earlier in the eccentric group. In the concentric group, at first a drop of force was observed and then a slow increase took place with the peak value being reached about a week later. The increase rate did not, in fact, differ statistically in both groups.

The extent and character of the force increase in the case of the ankle joint plantar flexors did not differ between groups, and has two peak values, one after about 3 weeks of training, and other higher one a few days later after the training.

Figure 4 presents percentage variations of leg extension velocity at the knee joint with external load having values of 20, 60, 100, and 140 Nm. At first a decrease was evident and then later a velocity increase was noticed in both the groups having the load of 20 Nm and 60 Nm. In case of higher loads of 100 Nm and 140 Nm the increase curve is different. In the case of the eccentric group the movement velocity increase followed immediately. In the concentric group, however, at first a drop took place and later the velocity increased. Maximum movement velocity increase in both the groups is similar and was evident after the training ended. Differences in the increase rate

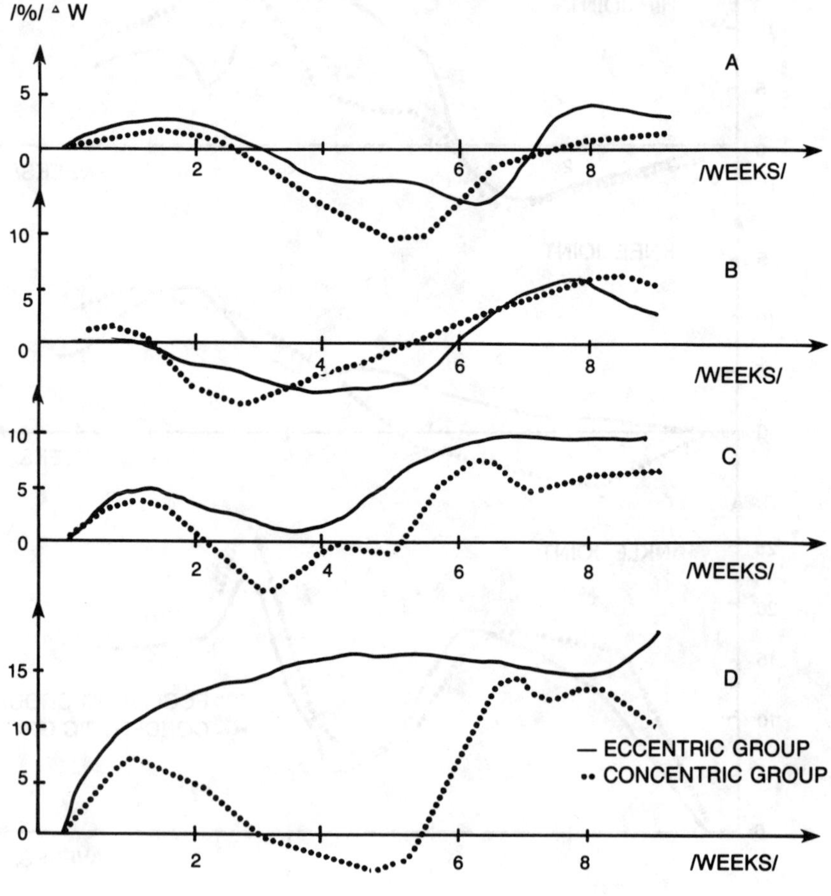

Figure 4—The percent change of the velocity increase of the knee joints move during the training with different external load (A—20 Nm, B—60 Nm, C—100 Nm, D—140 Nm).

in case of small and large loads were clearly visible. It can be assumed that the training used by us influences, to a great extent, the velocity increase with the larger external resistance.

The correctness of our hypothesis has been evaluated on the basis of maximum increase of muscle force in static and dynamic conditions. The differences obtained are statistically of no importance. However, the time and character of reaching the maximum varies.

Acknowledgment

Supported in part by Grant No. 10/7 from Central Committee of Physical Culture and Sport.

References

BERGER, R. (1962). Comparison of static and dynamic strength increases. *Research Quarterly*, **33**(3), 329-333.

DENISHIN, D.N., & Kuzniecow, W.W. (1972). Development of velocity and force qualities with use of special training equipment in young athletes. *TiPFK*, **4**, 8-12.

DOBROWOLSKIJ, I.M. (1972). Development of velocity and force qualities by using exercises of various character of muscle work. *TiPFK*, **7**, 23-27.

FIDELUS, K. (1974). Propositions of uniformed measurements of the training load. *Ach. Sp.*, **9**, 3-10.

HETTINGER, T. (1972). *Isometric muscle training*. Stuttgart: Georg Thieme Verlag.

HAKKINEN, K., & Komi, P.V. (1982). Specificity of training-induced changes integrative functions of the neuro-muscular system. *World Weightlifting*, **3**, 44-46.

IWANOW, J.I. (1968). Practical use of various aspects of muscular work in connection with rationalisation of velocity-force training of weight lifting athletes. *TWNIFK*, Moscow, 105-113.

WAZNY, Z. (1978). Evolution of record and analysis forms of training load. *Ach. Sp.*, **12**, 3-16.

A New Method for Determining the Force-Velocity Relationship in Human Quadriceps Muscle

S. Jarić, D. Ristanović, P. Gavrilović, V. Ivančević
Faculty for Physical Education, Belgrade, Yugoslavia

The relationship between the velocity of muscle shortening, v_m, and maximum contraction force, F_m, is given by Hill's characteristic equation:

$$(F_m + a)(v_m + b) = b(F_o + a), \tag{1}$$

where F_o is the maximum isometric force at zero speed and a and b are the dynamic constants to be fitted (Hill, 1938). Contrary to measurements on isolated animal muscles, the evaluation of Equation 1 in humans is complex and cumbersome (Tihanyi, Apor, & Fekete, 1982). The main aim of the present study was to evaluate the Hill equation parameters by studying in vivo the voluntary isometric contraction of the quadriceps femoris muscle in humans under various loading conditions. Preliminary reports of this work have appeared (Gavrilović, Ristanović, & Jarić, 1981; Jarić, Ristanović, & Gavrilović, 1981).

Methods

The mechanics of muscle shortening were described by means of two-dimensional analysis and force platform technique. Twenty young males, ranging in age from 20 to 34 years, acted as subjects in this study. Ten were national class athletes, sprint runners, high jumpers and shot-putters (group A), and the others were graduate students at the School for Physical Education in Belgrade (group B).

Details of the procedure used have been published previously (Gavrilović et al., 1981; Jarić et al., 1981). In short, the subject stood on a force platform (Kistler) used to measure the maximum values of the vertical ground reaction force exerted during contact with it. The movement of the hip joint center was registered two-dimensionally with the use of a light-emitting diode (LED) attached to the trochanter and projected onto the sagittal plane. Another diode was attached to the lateral malleolus. The diode positions were registered with a Selspot optoelectronic system camera (Selcom AB).

The signals from the camera were amplified and graphed on a Honeywell multichannel recorder, along with the signals from the platform. The force and both coordinates were shown simultaneously against time for the feet contact period.

The subject's feet were in a plantar flexed position. From this starting position, with the knee joint angles ranging from 65° to 80° and from an approximately vertical trunk position (monitored by diode positions), he made a vertical jump exerting maximum effort. In order to measure different velocities of muscle shortening at various knee-angles, four jumps were made on the platform: one was analyzed when the quadriceps muscles were contracted against no load, and the others, when three additional loads of 20, 40, and 60 kg were applied.

From the plot, the vertical components of the muscle contraction velocities and those for the forces exerted by the legs on the platform were read out for knee angles of 100°, 120°, 140°, 150°, and 160°. The subject was required to keep the upper body vertical; in that case the lever arm of the ground reaction with respect to the hip joint center is approximately zero. This was also monitored from plots of diode positions.

Biomechanical Analysis

We have already shown (Jarić et al., 1981) that the measured values of the quadriceps muscle force (F_m), acting through the patella, and its velocity of shortening (v_m), can be expressed as:

$$\frac{F}{F_m} = \frac{v_m}{v} = \frac{mx + n}{\cos\frac{x}{2}}, \quad F = Ae^{Bx}, \qquad (2)$$

where v is the takeoff velocity of the diode attached to the trochanter, F is the measured force, x (which ranged from 70° to 160°) is knee joint (tibio-femoral) angle, $m = 5.2 \cdot 10^{-4} \text{deg}^{-1}$, $n = 0.049$, $A = 261$ N, and $B = 0.0178 \text{ deg}^{-1}$ (Gavrilović et al., 1981). Equation 2 enables the calculations of the values v_m. Since it is known that the maximum quadriceps muscle force exerting on the patellar tendon attachment was found to be at the knee angle of about 112° (Jarić et al., 1981), the corrected knee extensor force F_{mc} was done by means of the following formula (Jarić et al., 1981):

$$F_{mc} = \frac{F \cdot F_m\,(112°)}{Ae^{Bx}}, \qquad (3)$$

where $F_m(112°) = 10.1$ kN was calculated using the conventional belt method of testing leg force (Gavrilović et al., 1981). Prior to calculation, all the measured forces F were reduced by 16% of the body weight. This reduction approximately corresponds to the reaction force measured by the platform when the muscle contraction control stopped at any moment of the contact phase of the takeoff. The obtained value of the force was then divided by two in order to obtain the force per one quadriceps muscle. The values of the parameters a and b of Equation 1 were calculated using the least-squares fitting procedure.

Results

The mean values (X), standard errors (SE) for the mean features of the quadriceps muscle shortening, and coefficients of variations (k) of the data points are summarized in Table 1. The mean heights (H) for both groups are similar, but the mean weights (G) are different. By means of the mean values of the parameters a, b, and F_o, the Hill curves drawn from Equation 1 for both groups of the subjects are shown in Figure 1. Using the method proposed by Kaneko (1978), the maximal mechanical powers of

Table 1

Parameter Estimates

		G (N)	H (m)	a (kN)	b $(\frac{m}{s})$	F (kN)	k (%)	a/F 10^{-2}	F/G
A	X	970	1.85	.32	.30	9.05	9.2	3.4	94
	SE	64	.02	.05	.01	.40	.9	5	4
B	X	794	1.83	.76	.31	7.60	14.5	10.6	88
	SE	14	.03	.14	.02	.55	1.5	1.6	6
	$p <$	.05	—	.05	—	.05	.01	.01	—

See text for definition of symbols.

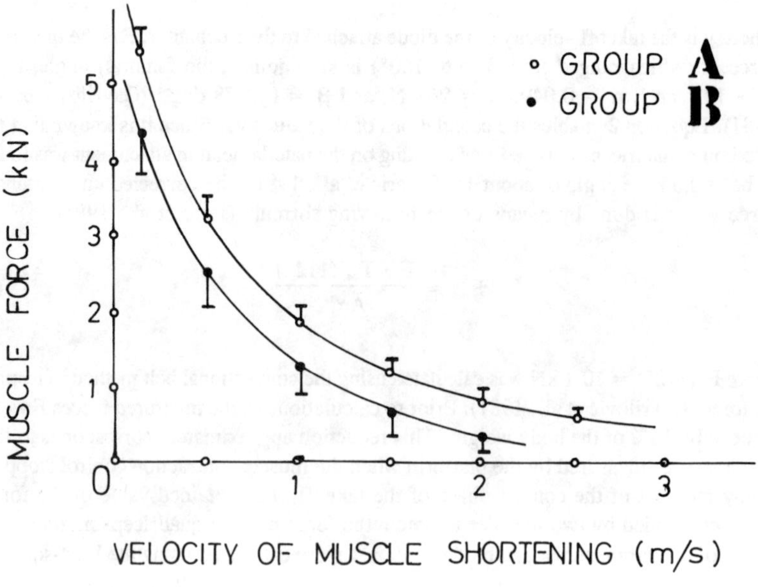

Figure 1—Graphs of the force-velocity relation calculated from Hill's Equation 1.

the quadriceps muscle were $P_A = 1.87$ kW and $P_B = 1.26$ kW, and the corresponding velocities of the muscle shortening, $v_A = 1.32$ m/s and $v_B = 0.72$ m/s.

Discussion

The obtained estimates of a/F_o and $v_o = bF_o/a$ are proved approximately consistent with data from in vitro and in vivo experiments (Hill, 1938; Spector, Gardiner, Zernicke, Roy, & Wodgatte, 1980; Thorstensson, Grimby, & Karlsson, 1976; Tihanyi et al., 1982). Incomplete agreements between our and other authors' results are probably due to the influence of the leg inertial forces and insufficient time to reach maximum quadriceps force which are not considered in the experiments of other authors. Increased influence of the rectus femoris muscle force in the last phase of the takeoff produced by an increased angular velocity in the hip joint could also be the reason for these discrepancies.

The influence of a great number of factors cannot be generally considered in this type of experiment, such as: other muscle group forces, inertial forces, ankle compliances, reflex activities, psychological factors, etc. In any case, the investigated means of absolute and relative parameter values, expected differences between groups, and agreement of the experimental results with the curves fitted showed that the applied methodology seems to be correct.

References

GAVRILOVIĆ, P., Ristanović, D., & Jarić, S. (1981). In vivo study on the effect of muscle length on its maximum force of contraction. *Period. Biol.* **83**, 135-137.

HILL, A.V. (1938). The heat of shortening and the dynamic constants of muscle. *Proceedings of the Royal Society B*, **126**, 136-195.

JARIĆ, S., Ristanović, D., & Gavrilović, P. (1981). The force-velocity relation in quadriceps muscle shortening. *Period. Biol.*, **83**, 153-155.

KANEKO, M. (1978). The effect of previous state of shortening on the load-velocity relationship in human muscle. *Physiological Society of Japan*, **40**, 12-14.

SPECTOR, S.A., Gardiner, P.F., Zernicke, R.F., Roy, R.R., & Edgerton, V.R. (1980). Muscle architecture and fiber composition in human knee extensor muscles. *Journal of Neurophysiology*, **44**, 951-960.

THORSTENSSON, A., Grimby, G., & Karlsson, J. (1976). Force-velocity relations and fiber composition in human knee extensor muscles. *Journal of Applied Physiology*, **40**, 12-16.

TIHANYI, J., Apor, P., & Fekete, G. (1982). Force-velocity-power characteristics and fiber composition in human knee extensor muscles. *European Journal of Applied Physiology*, **48**, 331-343.

The Effects on Electromechanical Delay of Muscle Stretch of the Human Triceps Surae

M. Muro
Tokyo College of Pharmacy, Hachioji, Japan

A. Nagata
Niigata University, Niigata, Japan

Electromechanical delay (EMD) is defined as the time from the onset of the myoelectric signal to the onset of the development of tension in the muscle. EMD has been reported to be the time required to stretch the series elastic component in a muscle (Cavanagh & Komi, 1979). Elasticity and viscosity as well as the release of Ca^{++} from the sarcoplasmic reticulum could be expected to affect the delay time. Furthermore, EMD changes could be estimated from the latency caused by a brief relaxation in a muscle just prior to its twitch. Since this latency coincides with the sequence of events which couple mental excitation and muscle contraction, it is suggested that latency relaxation might be directly related to coupled excitation and contraction (Gilai & Kirsch, 1978). If EMD and latency relaxation are related in this way, EMD would be a useful measure to study excitation-contraction coupling; however, the underlying mechanisms are still unknown.

Recently, Norman and Komi (1979) have reported that shorter EMD is seen in muscle which is predominantly fast twitch (biceps brachii). Viitasalo and Komi (1981) also reported an inverse relationship between EMD time and percent of fast twitch fiber in whole muscle. Moreover, Norman and Komi (1979) and Cavanagh and Komi (1979) have shown EMD times to be shorter in eccentric than in concentric contractions. The latter authors have suggested that EMD is likely to be related to the rate of tension development. It appears, then, that the time required to stretch the series elastic component represents the major portion of EMD times and it is speculated that the time is affected by motor unit recruitment pattern and by different characteristics of slow and fast twitch muscle fibers to trigger E-C coupling (Viitasalo & Komi). However, these findings are based on only a moderate muscle stretch length.

The aim of the present investigation was to study EMD changes in human triceps surae muscles as a function of muscle stretch and to determine whether, on a heavily stretched muscle, myoelectric signals would be changed or not. In order to test the notion that EMD is directly related to muscle stretch length, we attempted to alter evoked EMG responses at various muscle lengths and observed both the EMD and M wave.

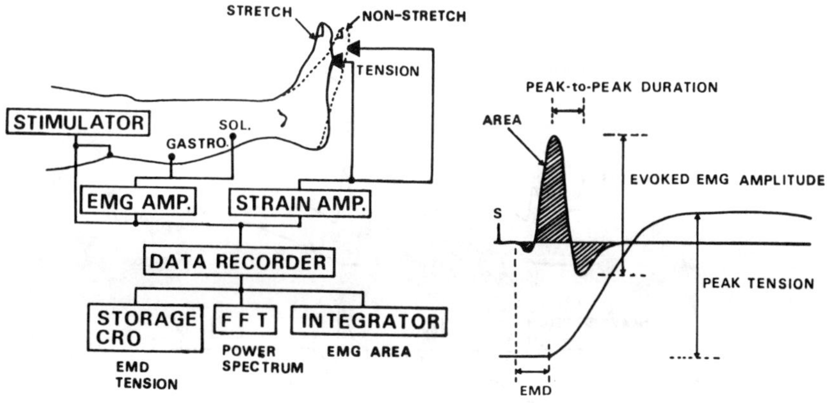

Figure 1—Schematic diagram of experimental methods and evoked myoelectric signal and tension curve.

Methods

The investigation was carried out on ten healthy subjects, aged 19 to 25 years, all of whom gave informed consent to undergo the experimental procedures. Each experiment was repeated at least five times on each subject. Subject responses were similar. Subjects were seated in an arm chair. The knee was fully extended and the foot was tied with a leather strap to a rigid foot plate. A pressure sensor (Kistler 9251A) was located on the underside of the foot to measure muscle tension and the ankle angle could be adjusted to 70, 80 or 90° to vary muscle length (see Figure 1).

EMG measurements were made on the lateral head of the gastrocnemius (MG) and on the soleus (SOL) via two pairs of surface electrodes with a ground over the lateral malleolus. The muscles were stimulated by means of an electrode placed behind the knee over the posterior tibial nerve which received a single rectangular pulse of 0.7 ms duration. The subjects were instructed to keep the muscles of the experimental leg relaxed. This condition was monitored via the EMG.

Results

Figure 2 illustrates single sweep records of evoked EMG (M and H wave) and twitch tension. Following each stimulation a short delay (EMD) of approximately 10.9 ms was observed before the onset of tension. Amplitudes of the evoked EMG were 14.8 mV higher in the heavily stretched compared with the unstretched muscle. Amplitudes of the H and M waves in the heavily stretched muscle were, respectively, 87 and 114% of those in the unstretched muscles (see Figure 3). The differences were statistically significant. The EMD was 11.7 $\pm$ 1.6 ms in the unstretched muscles and 7 $\pm$ 1.2 ms in the heavily stretched muscle.

The literature has suggested that EMD differences between stretched and unstretched muscles should be related to structural changes of the series elastic component (SEC) of the muscle. However, the present results show that stretching produces changes not

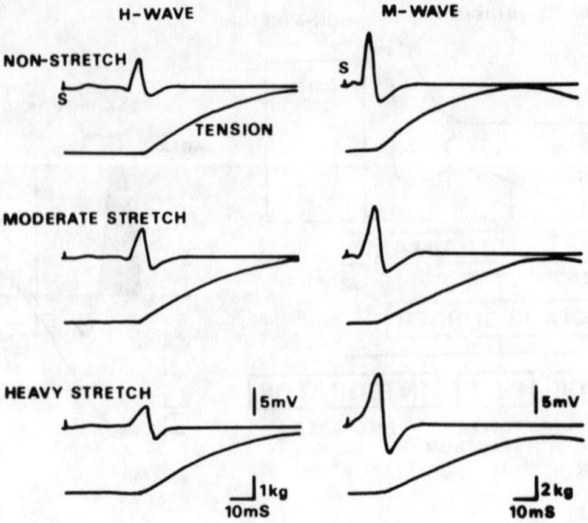

Figure 2—Typical evoked M and H waves and tension curves during three stages of muscle stretching (nonstretch, moderate and heavy stretch).

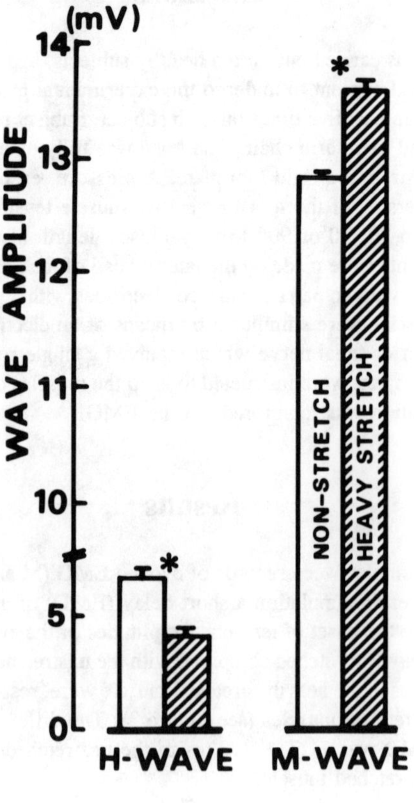

Figure 3—Statistical histogram of M and H wave amplitude by stimulation of the nerve.

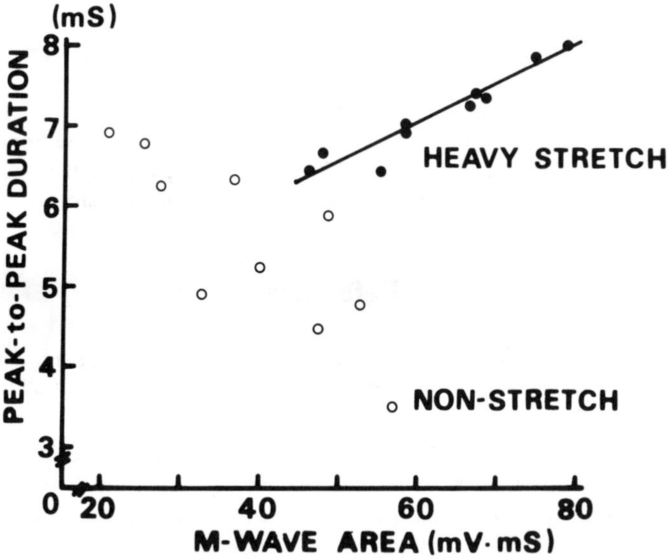

Figure 4—Relationship between the area and the duration (peak to peak) induced the M wave calculation. Note that in the heavy stretched muscle relationship between both parameters is shown significant.

only in the SEC but also action potential (AP) changes evoked by stimulation. In addition to peak-to-peak amplitude changes induced by stretch (see Figure 3), changes in peak-to-peak duration and M wave area were observed. Figure 4 shows that M wave area and peak-to-peak duration are directly related in the heavily stretched muscle (correlation significant at $p < .01$) and inversely related but with considerable scatter in the nonstretched muscle.

Discussion

The aim of these experiments was to test the hypothesis that EMD is changed due to stretching of the SEC and conduction velocity of the action potential along the muscle fiber (Cavanagh & Komi, 1979; Norman & Komi, 1979; Viitasalo & Komi, 1981). The present results also show that EMD can be altered by muscle stretching which should have an effect on the length of the SEC; but in addition, there is an inverse relationship between EMD and evoked EMG amplitudes and a prolonged peak-to-peak duration with increasing stretch. These action potential changes produced by stretch could be because of changes in Ca^{++} release related to changes in sarcomere length.

In isolated muscle, the latency relaxation has been reported to be reduced with stretched muscle. The present data coincide with the classical results which were attributed to the release of Ca^{++} from the sarcoplasmic reticulum and to sarcomere length (Bartels, Skydsgaard, & Sten-Knudsen, 1979; Gilai & Kirsch, 1978; Haugen, 1982a, 1982b). Cavanagh and Komi (1979) speculated that EMD depended upon many variables: conduction of the AP along the T tubule system; release of Ca^{++} by the sarcoplasmic reticulum; cross bridge formation between action and myosin filaments and the subsequent tension development in the contractile element; and stretching of the SEC by the contractile element.

The present results show that AP and EMD changes occur with heavy muscle stretch and support the notion that with extended sarcomere length Ca^{++} release from the sarcoplasmic reticulum would be facilitated. This factor, among others noted by Cavanagh and Komi (1979) is important in explaining electromechanical delay.

References

BARTELS, E.M., Skydsgaard, J.M., & Sten-Knudsen, O. (1979). The time course of the latency relaxation as a function of the sarcomere length in frog and mammalian muscle. *Acta Physiologica Scandinavica,* **106**, 129-137.

CAVANAGH, P.R., & Komi, P.V. (1979). Electromechanical delay in human skeletal muscle under concentric and eccentric contractions. *European Journal of Applied Physiology,* **42**, 159-163.

GILAI, A., & Kirsch, G.E. (1978). Latency-relaxation in single muscle fibers. *Journal of Physiology,* **282**, 197-205.

HAUGEN, P. (1982a). Increase of resistance to stretch during the latent period in single muscle fibers of the frog. *Acta Physiologica Scandinavica,* **114**, 187-192.

HAUGEN, P. (1982b). Short-range elasticity after tetanic stimulation in single muscle fibers of the frog. *Acta Physiologica Scandinavica,* **114**, 487-495.

NORMAN, R.W., & Komi, P.V. (1979). Electromechanical delay in skeletal muscle under normal movement conditions. *Acta Physiologica Scandinavica,* , **106**, 241-248.

VIITASALO, J.T., & Komi, P.V. (1981). Interrelationships between electromyographic, mechanical, muscle structure and reflex time measurements in man. *Acta Physiologica Scandinavica,* **111**, 97-103.

Effects of Training on Force-Velocity Characteristics

J.T. Viitasalo
University of Jyväskylä, Jyväskylä, Finland

The force-velocity characteristics described as the so-called force-velocity (f-v) diagram of a muscle or muscle group have been shown to be responsive to training (Ikai, 1970). The effects of training on the f-v curve are thought to be connected to training stimuli so that exercises with high movement velocities change the "velocity end" of the curve, while high resistance, slow velocity strength training mainly affects the "force end" of the curve (Ikai, 1970; Koning, Binkhorst, Vissers, & Vos, 1982; Lesmes, Costill, Coyle, & Fink, 1978; Thorstensson, 1977). Training programs for sportsmen are often planned in such a way that different parts of the f-v curve are trained at different phases of a training year. Testing of the f-v curve for the leg extensor musculature easily and quickly in field conditions, however, has been problematic. Our previous articles (Viitasalo, 1982, 1984) present a simple method for measurement of the f-v curve; in the test the subject jumps with and without extra loads on a force platform or on a platform (Bosco, 1980) connected to a digital timer. In this method the height of rise of the body center of gravity (h.C.G.) is used as an estimate of vertical velocity and the extra loads are used to build the force-axis of the f-v curve. The method has been proved to give reproducible results (Viitasalo, 1984). Although the method gives only an estimate of the "true" f-v curve of the leg extensor musculature the curve constructed by this method is called the force-velocity curve in this paper from here on.

The present article deals with (a) short term (fatiguing) effects of training on the f-v characteristics measured one and twenty-four hours after the training session (Study I), (b) the sensitivity of the f-v curve to high jump training over a seven month period (Study II) and (c) the sensitivity of the f-v curve to volleyball training over a six month period (Study III).

Methods

Subjects

In the first study 9 male students studying physical education volunteered as subjects. A detailed description of these subjects is presented in another article in this book

(Viitasalo, 1984). In the second study 5 high jumpers at top Finnish national level were tested. Their mean (±S.D.) age, weight, height and high jumping record were 22.6 ± 1.6 years, 74.7 ± 2.9 kg, 186.6 ± 3.9 cm and 211.9 ± 3.3 cm, respectively. In the third study the men's Finnish national volleyball team (12 players) served as subjects. They were 23.0 ± 2.8 years old, 192.2 ± 5.8 cm tall and their average weight was 87.1 ± 6.8 kg.

Measurements and Data Analysis

In each study after warming-up the subjects performed vertical jumps on a force platform (Study I and II) or on a platform (Bosco, 1980), which was connected to a digital timer (Study III). In both cases the flight time of the jump was used for calculation of the height of the rise of the body center of gravity (see Viitasalo, 1984). The jumps were performed from a static semisquatting position without a countermovement (SJ, pure concentric contraction) and with a preliminary countermovement (CMJ, eccentric stretch followed by a concentric contraction) as described previously by Komi and Bosco (1978). In the first study extra barbell loads of 0, 20, 40, 60, and 80 kg on the shoulder were used, in the second study a 100 kg load was also used, while in the third study the loads were 0, 20, 30, 40, and 60 kg.

Training

In the first study the reproducibility and short term effects of training on the f-v curve were tested by eight different measurements. The first type of training was heavy strength training with high resistance and slow movement velocities, while the second type of training was an explosive type of jumping training including plyometric jumps. The subjects chose the two types of training from their own regular training programs. The only criterion was that both of the training types had to be "heavy" and fatiguing. The pre-tests (tests 2 and 6) were done 1 hr before and the post-tests 1 hr (tests 3 and 7) and 24 hr (tests 4 and 8) after the training sessions. The two training sessions were separated by a time interval of 2 to 5 days (see Figure 1, Viitasalo, 1984).

In the second study the high jumpers were tested 17 times over a two year period. For this article the test values of March, May, August, and October 1982 are included. The training programs of the jumpers were documented and analyzed and will be published elsewhere. In the third study the Finnish men's volleyball team was tested in April, June, August, and twice in September 1982 before the World Championships in early October. Their physical training for the explosive power of leg muscles included strength training using a barbell and jumping, especially plyometric training.

Statistical Analysis

Changes in the f-v curves during the follow-up periods were tested for each load level and for both types of jumps (SJ and CMJ) using the dependent t-test, and the coefficient of variation (C.V.).

Results

In the analysis of short term effects of the strength and jump training on jumping height no statistically significant differences between the pre- and post-training values of the jumping height in SJ and CMJ conditions were found. The coefficients of variation between the pre- and post-exercise tests and between the two post-exercise tests in connection with the strength and jump training (see Table 1) were at the same level (5.1 to 6.5%) showing no systematically different effects of the two training types on the h.C.G.

The effects of sports training in high jumping and in volleyball on the f-v characteristics were studied over several months. Since the changes in the squatting and countermovement jumps were very similar, the following results are shown only for the CMJs. Among both the groups the changes in jumps with small extra loads tended to be greater than in jumps with heavy loads. The h.C.G. in CMJ with a 20 kg load was found to increase statistically significantly among the high jumpers from March to August (9.8%, $p < .05$), while after this up to October a significant decrease, as also in CMJ with 0 load, was found (7.2% and 8.9%, respectively, $p < .05$). Among the volleyball players the changes in the f-v characteristics during the five month period (see Figure 1) were rather similar to those among the high jumpers, the changes for CMJ with 0 and 20 kg loads being statistically significant (9.1% and 6.2%, respectively, $p < .05$).

Table 1

Coefficients of Variation (%) Between the h.C.G. of the Successive Tests

Jumps and Extra Loads		Comparison Between Tests 2-3	3-4	6-7	7-8
0 kg	SJ	6.3	5.7	5.2	3.1
	CMJ	3.0	3.7	2.7	4.8
20 kg	SJ	3.6	3.4	5.2	4.4
	CMJ	6.0	5.8	5.1	4.3
40 kg	SJ	5.2	7.2	4.1	4.1
	CMJ	3.4	4.7	6.3	6.6
60 kg	SJ	9.2	3.5	11.7	8.5
	CMJ	8.2	11.4	9.0	7.6
80 kg	SJ	6.2	7.4	7.6	10.7
	CMJ	13.7	12.3	7.9	8.1
$\bar{x}$	SJ	5.1	6.5	6.5	6.2
	CMJ	4.0	3.2	2.6	2.5

Jump training was performed between the tests 2 and 3 and a strength type of training between the tests 6 and 7. For other symbols see text.

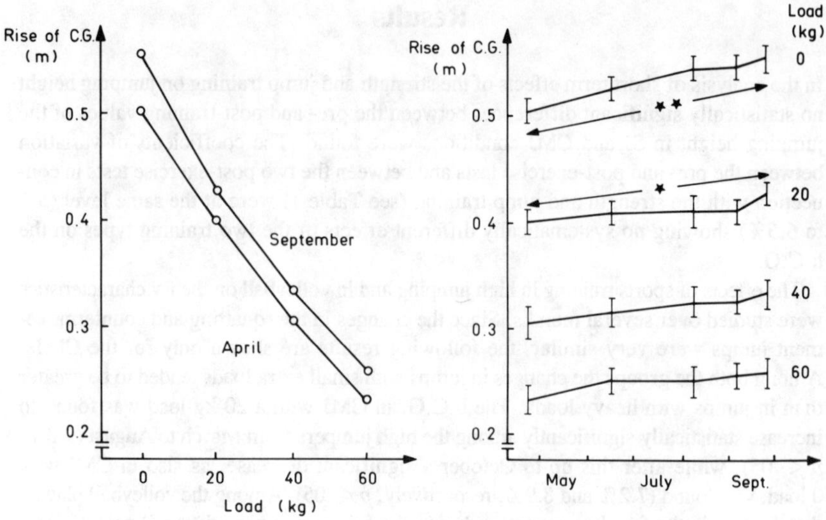

Figure 1—The average f-v curve of the volleyball team in April and September 1982 (left) and the average (±S.D.) vertical jumping height with the different extra loads (right) between April and September 1982 (*$p < .05$, **$p < .01$).

Discussion

The effects of training one hour before taking the vertical jumping height measurements were found to be minimal; the changes were not systematic or statistically significant and the coefficients of variation between the pre- and post-exercise tests were in the order of 5.1 to 6.5%, which are at the same level as found for duplicate same-day and between-days measurements (5.6 to 6.7%, Viitasalo, 1984). The fast recovery from both the strength and jumping training is in line with some previous results, where the maximal isometric force, and the rate of force production and relaxation have been found to recover to 90-100% in ten minutes (Viitasalo & Komi, 1981).

The average h.C.G. of the high jumpers reached its peak value during the competition season (June to August) and decreased rapidly during the off-season (September to October). The best values of the h.C.G. for the volleyball players were measured one week before the World Championships. Thus the changes in the f-v characteristics were very predictable. The training of both the high jumpers and the volleyball players was aimed at improving the explosive strength of the leg extensor musculature. This is why the tendency for greater changes in the h.C.G. of jumps with light weights than in jumps with heavy weights is in line with the idea of specificity of training and with the results of some previous studies (Ikai, 1970; Koning et al., 1982; Lesmes et al., 1978; Thorstensson, 1977).

To summarize, the effects of a single strength and jump training session on the vertical jumping height measured one and twenty-four hours after the training session were found to be minimal, while training in high jumping and volleyball at top Finnish national level was found to produce significant changes in the force-velocity characteristics. The way of measuring force-velocity characteristics of the leg extensor muscles, using a platform with a digital timer and extra loads on the shoulder, proved to be quick and easy for the testing of sportsmen in field conditions.

Acknowledgment

Supported in part by the Finnish Olympic Committee. Ossi Aura, Vesa Martikkala, Kari Miettunen, and Otto Pajala are acknowledged for assistance in collecting the data.

References

BOSCO, C. (1980). Pallavolo, 5

DeKONING, F.L., Binkhorst, R.A., Vissers, A.C.A., & Vos, J.A. (1982). Influence of static strength training on the force-velocity relationship of the arm flexors. *International Journal of Sports Medicine,* **3**, 25-28.

IKAI, M. (1970). Training of muscle strength and power in athletes. *FIMS,* Oxford.

KOMI, P.V., & Bosco, C. (1978). Utilization of stored elastic energy in leg extensor muscles by men and women. *Medicine and Science in Sports,* **10**(4), 261-265.

LESMES, G.R., Costill, D.L., Coyle, E.F., & Fink, W.J. (1978). Muscle strength and power changes during maximal isokinetic training. *Medicine and Science in Sports,* **10**, 266-269.

THORSTENSSON, A. (1977). Observations on strength training and detraining. *Acta Physiologica Scandinavica,* **180**, 491-493.

VIITASALO, J.T. (1982). Jalkalihasten voima/nopeus-ominaisuuksien mittaaminen kenttäolosuhteissa. *Valmennus-lehti,* **7**, 44-45.

VIITASALO, J.T. (1985). Measurement of force-velocity characteristics for sportsmen in field conditions. In D. Winter, R. Norman, R. Wells, K. Hayes, & A. Patla (Eds.), *Biomechanics IX-A* (pp. 96-101). Champaign, IL: Human Kinetics Publishers.

VIITASALO, J.T., & Komi, P.V. (1981). Effects of fatigue on isometric force- and relaxation-time characteristics in human muscle. *Acta Physiologica Scandinavica,* **111**, 87-95.

Measurement of Force-Velocity Characteristics for Sportsmen in Field Conditions

J.T. Viitasalo
University of Jyväskylä, Jyväskylä, Finland

The force-velocity diagram describes the momentary condition of the neuromuscular system for the production of force at different contraction velocities. An inverted relationship between these variables has been obtained with isolated muscles (Lewin & Wyman, 1928), as well as with intact human skeletal muscle (e.g., Asmussen, Hansen, & Lammert, 1965). Force-velocity relationships in in vivo human muscles have been studied most often with different dynamometers, using constant linear or angular velocities (e.g., Thorstensson, 1976) which by the nature of the isokinetic apparatus, have usually been only 30% or less of the maximal voluntary speed. Thus, these conditions are not similar in all cases to the natural movements of a human being, which include accelerations, decelerations, and eccentric stretching phases prior to the concentric or shortening phases.

In some recent investigations (Bosco & Komi, 1979; Bosco, Viitasalo, Komi, & Luhtanen, 1982) force-velocity curves have been constructed for humans in vertical jumping, using the vertical ground reaction force measured with different extra loads and the respective knee angular velocities as an estimate of force and velocity. This experimental design also makes it possible to use jumps with and without a preliminary countermovement or drop jumps from different heights in order to study the effects of the prestretching of the active muscles on the force-velocity curve.

The present study was designed to compare two methods for constructing the force-velocity curve of the leg extensor musculature: Bosco and Komi's (1979) method and a simplified method (Viitasalo, 1982), where the average knee angular velocity was substituted by the height of the rise of the body center of gravity (h.C.G.) and where the average concentric ground reaction force was substituted by the weight of the subject plus the extra barbell load on his shoulder. Although in either of these methods the curves constructed are just an estimate of the true force-velocity curve of the leg extensor musculature, they are called force-velocity curves from here on.

Methods

Subjects

Nine male students studying physical education volunteered as subjects. Their average (±S.D.) age, weight and height was 22.5 ± 1.4 years, 75.9 ± 6.3 kg and 183.5 ±

7.5 cm, respectively. Six of the subjects were sportsmen in track and field events at top Finnish national level and the other 3 subjects were habitually physically active and in good shape.

Measurements

During the testing of sportsmen in field conditions (see Viitasalo, 1984) the h.C.G. was calculated using the flight time, which was measured with a digital timer connected by a resistive (or capacitative) platform (Bosco, 1980). For this study a force platform was used. After warming up, the subjects performed a test battery of twenty maximal vertical jumps on the force platform with one minute interpauses. The test battery consisted of jumps performed with and without a preliminary countermovement (CMJ and SJ, respectively, see Komi & Bosco, 1978) with different amounts of extra barbell load (0, 20, 40, 60, and 80 kg) on their shoulders. The CMJ and SJ with the same extra load were performed successively and the loads were changed in the order of 0, 20, 40, 60, 80, 80, 60, 40, 20, and 0. Thus each jump was performed twice during the same test occasion. The test battery was performed eight times according to the experimental scheme shown in Figure 1.

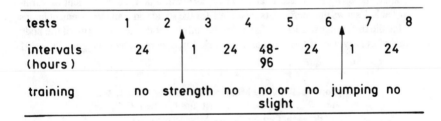

tests	1		2		3		4		5		6		7		8
intervals (hours)		24		1		24		48-96		24		1		24	
training		no		strength		no		no or slight		no		jumping		no	

Figure 1—Experimental scheme for the eight tests, time intervals between the tests and physical activity between the tests.

Analysis

Simultaneously with the ground reaction force, the angular displacement of the right knee joint (an electrical goniometer) was stored on magnetic tape (Recal Store-7). The force-time signal from the force platform was analyzed with a HP-1000 laboratory computer for the average concentric force ($\bar{F}_{conc}$), concentric contact time and for the height of rise of the body center of gravity (h.C.G.) using the vertical take-off velocity (V_v) as follows:

$$\text{h.C.G.} = \frac{V_v^2}{2^*g}$$

where $g = 9.81\ m/s^2$.

V_v was calculated using the flight time (t_{air}) as follows:

$$V_v = 1/2^*t_{air}^*g$$

The concentric contact time and the angular displacement during the concentric phase of contraction were used to calculate the average knee angular velocity ($\bar{\omega}_{knee}$).

Statistical Calculations

The reproducibility of the measurements of the h.C.G., $\bar{F}_{conc}$ and $\bar{\omega}_{knee}$ during the same test was analyzed by calculating the Pearson correlation coefficient and the coefficient of variation (C.V.) for each type of jump by using the duplicate measurements of each test occasion. The correlation coefficients from the eight tests were averaged using the z-transformation for each type of jump. Similarly the reproducibility between days was calculated for each type of jump by comparing tests 1 and 2, and 5 and 6. The interdependence between h.C.G. and $\bar{\omega}_{knee}$, and between $\bar{F}_{conc}$ and weight + extra loads was calculated using the Pearson correlation coefficient and for averaging using the z-transformation.

Results

The reproducibility for the measurements of the h.C.G., $\bar{\omega}_{knee}$ and $\bar{F}_{conc}$ within the test was analyzed using the linear coefficient of correlation and the coefficient of variation. Table 1 shows the coefficients separately for the static and countermovement jumps and for the different extra loads. The reproducibilities of the height of rise of the body center of gravity and of the average concentric force were on average good (r = .91

Table 1

Reproducibility of the Measurements of the h.C.G., $\bar{\omega}_{knee}$ and $\bar{F}_{conc}$ With the Same Test and for the h.C.G. Between Two Successive Measurement Days

Jumps and Extra Loads		Within Test						Between Days	
		h.C.G.		$\bar{\omega}_{knee}$		$\bar{F}_{conc}$		h.C.G.	
		r	C.V.	r	C.V.	r	C.V.	r	C.V.
0 kg	SJ	.93	5.0	.85	7.6	.96	5.5	.84	3.6
	CMJ	.95	4.3	.87	4.6	.90	7.6	.99	3.4
20 kg	SJ	.96	5.8	.70	13.1	.85	4.3	.93	5.7
	CMJ	.89	4.8	.85	9.0	.96	5.1	.75	3.6
40 kg	SJ	.93	6.3	.77	10.7	.87	6.3	.97	2.6
	CMJ	.86	6.8	.59	7.9	.90	5.0	.92	7.2
60 kg	SJ	.91	7.9	.92	10.0	.89	5.5	.89	8.7
	CMJ	.91	7.1	.85	9.9	.96	3.5	.89	7.1
80 kg	SJ	.87	8.4	.90	8.7	.92	4.3	.79	7.2
	CMJ	.89	9.5	.61	11.8	.92	5.7	.75	9.8
x	SJ	.92	6.7	.85	10.0	.91	5.2	.90	5.6
	CMJ	.91	6.5	.78	8.6	.93	5.4	.91	6.2

r = Pearson coefficient of correlation, C.V. = coefficient of variation.

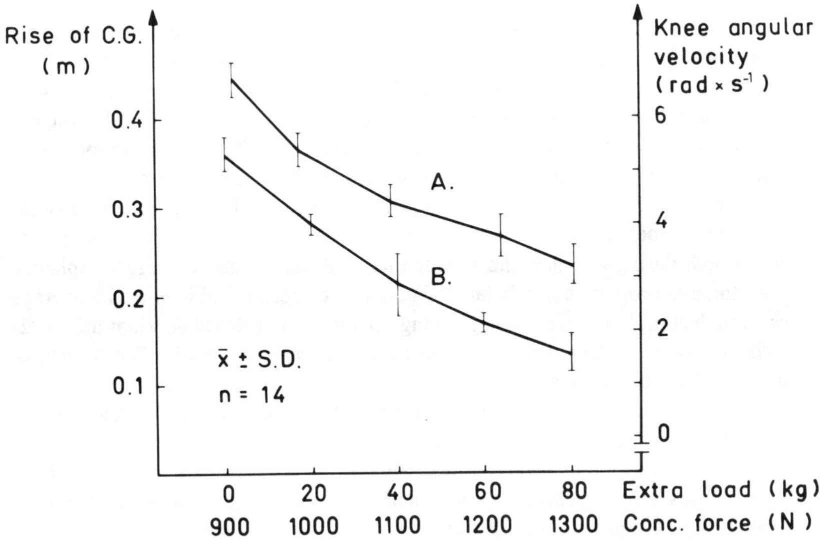

Figure 2—F-v curves measured in CMJ condition and expressed with averaged knee angular velocity and concentric force (curve A) and with h.C.G. and amount of extra load (curve B).

to .93, C.V. = 5.2 to 6.7%) while the respective values for the average knee angular velocity were lower (r = .78 to .86, C.V. = 8.6 to 10.0%). The between-days reproducibility was tested only for the h.C.G. On average the coefficients of correlation and coefficients of variation between two successive measurement days were .90 to .91 and 5.6 to 6.2%, respectively.

Figure 2 depicts the average force-velocity curves of fourteen different tests calculated by Bosco and Komi's (1979) and the current method.

The interrelationships between the two methods were measured by using the linear correlation coefficient. The h.C.G. and the average angular velocity during the concentric phase of contact showed a high correlation ($r = .90, p < .001$), while the respective correlation coefficient between the average concentric force and the different extra loads was a little lower ($r = .78, p < .001$).

Discussion

The shapes of the average force-velocity curves calculated by Bosco and Komi's (1979) method and by the method described in this article were very similar as shown in Figure 2. Also when the h.C.G. and $\bar{\omega}_{knee}$, and $\bar{F}_{conc}$ and amount of extra load were compared using the linear correlation coefficient, the similarities were confirmed. When these two methods were compared in respect to the reproducibility of measurements, it was found out that the reproducibility for the height of rise of body center of gravity was better than that of the average knee angular velocity. Since the h.C.G. is determined by the vertical take-off velocity, it can be suspected that the way to produce V_v, that is, coordination and timing of activity of the different extensor muscles of the lower extremities during the contraction, differed within the subjects. This idea is supported by the results of our previous study (Bosco & Viitasalo, 1982) where myoelectrical activities of the leg extensor muscles were shown to vary considerably between successive vertical jumps. However, in spite of these obvious differences at muscular level the results of contraction, that is, the vertical jumping height, among athletes (this study) as well as among untrained subjects (Bosco & Viitasalo, 1982) are quite reproducible.

When the reproducibilities of the vertical jumping heights are compared between jumps with different amounts of extra load, a tendency for higher coefficients of variation and lower coefficients of correlation as a function of load increase is seen (see Table 1) both during one day and between days. Probably this is largely explained by coordinative properties and balance, because the reproducibilities of the average force as such at high force levels in jumping (this study and Bosco & Viitasalo, 1982) as well as maximal strength measured on a dynamometer (Viitasalo, Saukkonen, & Komi, 1980) have been shown to be good.

In conclusion, the method for measurement of the force-velocity characteristics in field conditions using the vertical jumping height with extra loads was shown to give rather similar results to the method of Bosco and Komi (1979) where the average knee angular velocity and concentric ground reaction force were used. The present method was found to give reproducible values for the h.C.G., both in same-day and between-days comparisons, in jumps with extra loads from 0 to 80 kg.

Acknowledgment

Supported in part by the Finnish Olympic Committee. Vesa Martikkala is acknowledged for assistance in collecting the data.

References

ASMUSSEN, E., Hansen, O., & Lammert, O. (1965). *The relation between isometric and dynamic muscle strength in man.* Communications from the testing and observation institute of the Danish National Association for Infantile Paralysis. No. 20.

BOSCO, C. (1980). Pallavolo, 5.

BOSCO, C., & Komi, P.V. (1979). Potentiation of mechanical behaviour of the human skeletal muscle through prestretching. *Acta Physiologica Scandinavica,* **106**, 467-472.

BOSCO, C., & Viitasalo, J.T. (1982). Potentiation of myoelectrical activity of human muscles in vertical jumps. *Electromyographics and Clinical Neurophysiology,* **22**, 549-562.

BOSCO, C., Viitasalo, J.T., Komi, P.V., & Luhtanen, P. (1982). Combined effect of elastic energy and myoelectrical potentiation during stretch-shortening cycle exercise. *Acta Physiologica Scandinavica,* **114**, 557-565.

KOMI, P.V., & Bosco, C. (1978). Utilization of stored elastic energy in leg extensor muscles by men and women. *Medicine and Science in Sports,* **10**(4), 261-265.

LEWIN, A., & Wyman, J. (1928). The viscous elastic properties of muscle. *Proceedings of the Royal Society,* **B100**, 218-243.

THORSTENSSON, A. (1976). Muscle strength, fibre types and enzyme activities in man. *Acta Physiologica Scandinavica,* Suppl. 443.

VIITASALO, J.T. (1982). Jalkalihasten voima/nopeus-ominaisuuksien mittaaminen kenttäolosuhteissa. *Valmennus-lehti*, **7**, 44-45.

VIITASALO, J.T. (1984). Effects of training on force-velocity characteristics. In D. Winter, R. Norman, R. Wells, K. Hayes, & A. Patla (Eds.), *Biomechanics IX-A* (pp. 91-95). Champaign, IL: Human Kinetics Publishers.

VIITASALO, J.T., Saukkonen, S., & Komi, P.V. (1980). Reproducibility of measurements of selected neuromuscular performance variables in man. *Electromyographics and Clinical Neurophysiology*, **20**, 487-501.

The Dynamic Behavior of a Three Link Model of the Human Body During Impact With the Ground

J. Denoth
Swiss Federal Institute of Technology (ETH),
Zürich, Switzerland

The loading of the human body in comparison with its load capacity is an important factor in biomechanics, sports, and orthopedics. It is essential, for example, to have an objective evalutation of the effectiveness of orthopedic shoe corrections. Load analysis is also an interesting matter from the theoretical point of view. It is the aim of theoretical investigations to develop more realistic models in order to obtain a better understanding of loading of the human body.

The central problem in load analysis, where geometry is defined, is the function of muscles as internal force generators. The non-rigidness of the segments of the body is also a problem but it is not discussed here. Where both "initial" and "boundary" conditions are defined and the functional behavior of musculature is known, the movement is also defined. Conversely, to infer loading only from movement is generally not possible. Different muscles may produce the same angular momentum. In order to find definite solutions, cost functions can be introduced and minimized (Pedotti, Krishman, & Stark, 1978; Seireg & Arvikar, 1975) or the muscles can be "regulated" by controlling mechanisms (Hatze, 1981; Pierrynowsky, 1982).

The purpose of this paper is to illustrate the muscular influence of loading on the human body. A three link model of the body will be simulated during impact with the ground. The ground reaction, joint and muscle forces for different geometries, and levels of muscle activity are discussed.

Methods

"Reality" was simplified to a mechanical model, the differential equations were formulated and solved by numerical integration.

The three link model (see Figure 1) which was chosen for the following computations represents the upper part of the body, the thigh and the leg (Nigg & Denoth, 1980). The muscles are considered to act as massless force generators, comprising a

Figure 1—The three link model. The four muscles are: vasti femoralis, biceps femoris, iliacus, and glutei.

contractile element (CE) and a series elastic element (SE). The mechanical properties of the series elastic element are as follows (Hatze, 1980; Yamada, 1973):

$$FSE = f1 \cdot y + f2 \cdot y^2 + r \cdot \dot{y}$$

where FSE = force on SE, y = "length" of SE, f1 = spring constant, f2 = constant describing the nonlinearity, and r = damping coefficient.

The contractile element is that proposed by Hill (1970), in which the muscle characteristic parameters a, b, and Po depend upon the level of muscle activity (Philips & Petrofsky, 1980). The following dependencies were used:

$$Po\ (Z) = p1 \cdot Z;\ a\ (Z) = a1 \cdot Z;\ b\ (Z) = b1 \cdot Z - b2 \cdot Z^2$$

with Z being the "instrinsic state" (Hill, 1970) or the "active state" of the contractile element. The "intrinsic" or "active state" as a function of time is regulated by the nervous system.

At the moment of impact the movement can be divided into active and passive parts (Denoth, 1977; Nigg, 1980). The passive part is characterized by the fact that the active state Z of the musculature—and not its tension—will not be changed by the influence of the external force. This will always be true if the time of impact is smaller than the latent period of the total muscle innervation system. This latent period is in the order of 30 ms. The time history of Z before impact and during the passive phase for a muscle is chosen as:

$$Z(t) = Zo\ (1 - \exp\ [-t/T])$$

The ground reaction force is defined by the mechanical properties of the ground and its surface (Nigg & Denoth, 1980).

Results

During the passive phase, the ground reaction force (vertical component) is only insignificantly influenced by the muscle activity (see Figure 2). This seems reasonable since the muscular force is too small during this phase to stop the downward motion of the thigh, that is to increase the "rotational inertia" of the thigh. The muscular force influences the ground reaction force at the boundary of the passive phase and the active phase. When the activity is increased, the unloading after the peak is relatively small. For smaller activations, the unloading is relatively large.

At the knee joint the muscle activity is already "observed" during the impulse. In other words, the force at the knee joint (F^{KNEE}) during the passive phase can be divided into two parts (Nigg, Denoth, & Neukomm, 1979):

$$F^{KNEE} = F_J^{KNEE} + F_M^{KNEE}$$

where F_J^{KNEE} = "force of impulse" and F_M^{KNEE} = "resultant force of muscle activities"

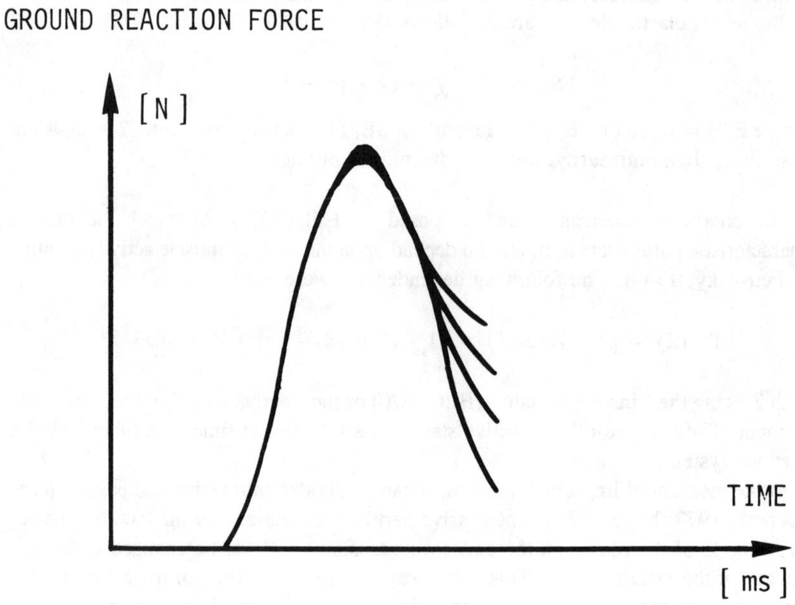

Figure 2—Ground reaction force vs. time for different activities of the vasti femoris (vertical component).

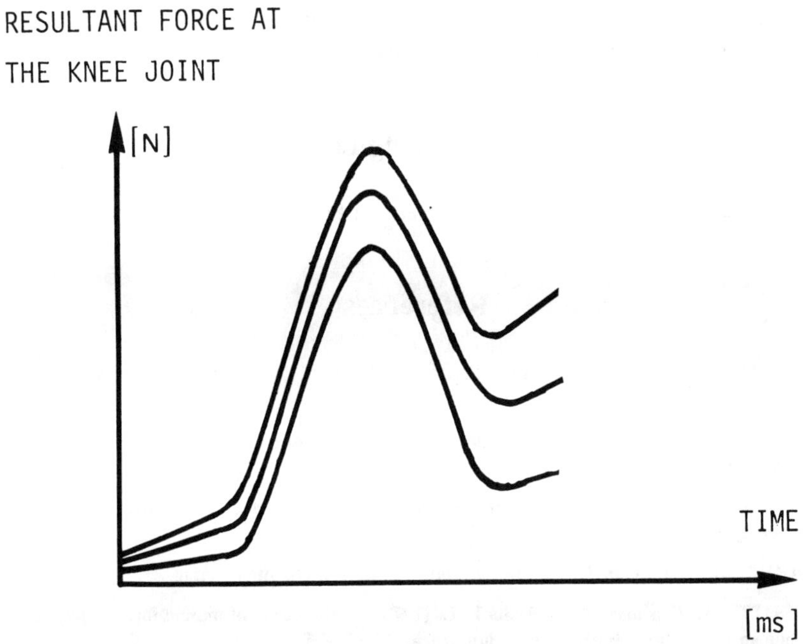

Figure 3—Resultant force at the knee joint vs. time for different active states of the vasti femori (vertical component).

Given a certain muscle, the force produced by the muscle depends on the knee angle, the angular velocity of the knee, and the active state of the muscle. Experimental analysis of the influence of the musculature on the resultant force at the knee joint would be complicated because the impulse is short in time and the motions of the thigh and the leg are difficult to detect as a result of skin/fat motion.

When several muscles are active in flexion or extension during the passive phase, the following can be stated:

The force which each muscle can produce is given by its characteristic properties, its geometry and dimensions, the joint angle, and the angular velocity of the joint and its active state. This is contradictory to the idea of introducing cost functions. For example, in slow movements, that part of the resultant joint force produced by muscle A may be larger than that of muscle B. In fast movements, however, this may not be true.

The horizontal component of the ground reaction force and of the force at the knee joint depend largely on the static and dynamic friction coefficients. In this approach depending on the initial and boundary conditions an ''unstable'' state is possible. This can be observed in the horizontal component of the ground reaction force which changes its sign several times during the passive phase.

Conclusion

During the passive phase that part of the resultant joint reaction force produced by a muscle is given by the physiological and anatomical parameters of the muscle, the movements of the body segments, and by the active state of the muscle.

Acknowledgment

This work was supported by the Swiss National Foundation.

References

DENOTH, J. (1977). Der Einfluss des Sportplatzbelages auf den menschlichen Bewegungsapparat. *Medita*, **9**, 164-167.

HATZE, H. (1981). *Myocybernetic control models of skeletal muscle*. Pretoria: University of South Africa.

HILL, A.V. (1970). *First and last experiments in muscle mechanics*. Cambridge: Cambridge University Press.

NIGG, B.M., & Denoth, J. (1980). *Sportplatzbeläge*. Zürich: Juris Verlag.

PEDOTTI, A., Krishnan, V.V., & Stark, L. (1978). Optimization of muscle force sequencing in human locomotion. *Mathematical Biosciences*, **38**, 57-76.

PHILIPS, C.A., & Petrofsky, J.S. (1980). Velocity of contraction of skeletal muscles as a function of activation and fiber composition. A mathematical model. *Journal of Biomechanics*, **13**, 549-554.

PIERRYNOWSKY, M.R. (1982). A physiological model for the solution of individual muscle forces during normal human walking. Unpublished doctoral dissertation, Simon Fraser University.

SEIREG, A., & Arvikar, R.J. (1975). The prediction of muscular load sharing and joint forces in the lower extremities during walking. *Journal of Biomechanics*, **8**, 89-102.

YAMADA, M.D. (1973). *Strength of biological materials*. Huntington, NY: Robert E. Krieger Publishing Company.

Alterations in Electrical and Mechanical Components of Muscle Contraction Following Isometric and Rhythmic Exercise

G. Kamen
Indiana University, Bloomington, Indiana, USA

Since its introduction by Weiss (1965), the fractionated reaction time (RT) model has become an important tool for the study of sensorimotor functioning. The central components of the fractionated reaction time model have been used to study areas such as the aging process, cortical control mechanisms underlying rapid responses (cf. Kroll & Clarkson, 1977), and more recently, the effects of task complexity. Although the fractionated RT paradigm has been used to study muscular fatigue, little effort has been made to fractionate the motor time component, and thus determine what changes in the periphery accompany the onset of muscular fatigue. This investigation sought to partition the motor time component of RT into two components and determine the response of these variables following fatiguing exercise. Motor time was partitioned into two component latencies—a mechanical latency component (delay between the onset of muscle electrical activity and the development of force), and a tension development time component (time needed to develop a 15% MVC following onset of force production). These criterion measures were assessed following fatiguing exercise at each of two intensities (low-intensity and high-intensity) and with each of two types of muscle contraction (isometric and rhythmic).

Methods

Subjects consisted of 12 college-age males who were tested on six different days. The first two days were used to stabilize baseline measures. Four different exercise conditions were administered on each of the last four days.

For testing knee extensor reaction time, the subject was seated on a bench with an inextensible band around the ankle. A metal plate connected to the ankle band contacted an electromagnet, which was energized to produce a force equal to 15% of the subject's MVC. At the onset of a visual stimulus, the subject responded with a rapid knee extension contraction. Thus, the subject had to exert a force of 15% MVC before movement could begin. A strain gauge connected in series with the electromagnet recorded the force produced on a rapidly moving pen recorder (250 mm/s). Separate strain gauges were used for MVC and RT measures in order to record the mechanical response with a sensitivity of less than 2 N. Surface electrodes placed over the belly of the rectus femoris recorded the onset of electromyographic activity on a storage oscilloscope. Preliminary measurements with this system indicated that attempts to produce small twitches of the knee extensors were accompanied by noticeable force recordings whenever EMG activity was present.

Two of the exercise tasks involved a submaximal isometric contraction of the knee extensors using either a low intensity (25% MVC) or a high intensity (50% MVC) load. The subject was required to hold the weight for as long as possible. The other two tasks involved rhythmic exercise during which the subject was required to lift a load of either 25% MVC or 50% MVC repeatedly at a rate of 24 repetitions per minute. Thus, the four exercise conditions were: low-intensity isometric, high-intensity isometric, low-intensity rhythmic and high-intensity rhythmic. Four bouts of each exercise task were administered with the criterion measures assessed following each bout. Besides the mechanical latency and tension development time measures described above, maximum dF/dt was assessed during each trial, as well as the time needed to reach the point at which max. dF/dt was attained.

Results

Mechanical Latency (ML). Pooled across the four exercise conditions, no significant change occurred in mechanical latency across exercise bouts, as seen in Figure 1. The greatest change in ML occurred during the high-intensity isometric exercise condition which effected a 30% increase in mechanical latency (32.9 ms pre-exercise vs. 42.7 ms post-bout 4). However, analysis of simple main effects for the 50% MVC isometric exercise condition alone showed that no significant change occurred over exercise bouts.

Tension Development Time (TDT). Averaged across the four exercise conditions, TDT showed a statistically significant increase from 52.1 ms prior to exercise to 58.3 ms following the last exercise bout ($p < .01$), indicating that the time needed to reach 15% MVC (and begin the ballistic movement) increased as a result of fatiguing exercise. No significant interactions were noted for each exercise intensity (25% or 50% MVC) or exercise type (isometric vs. rhythmic). Analysis of simple main effects showed that all four of the exercise conditions produced significant increases in TDT over exercise bouts *except* for the 50% MVC isometric condition which seemed to cause no change in tension development time.

Max. dF/dt. Pooled across exercise conditions, maximum rate of tension development (dF/dt) decreased only slightly from 7590 N/s prior to exercise to 7129 N/s following bout 4. Analysis of simple main effects showed that none of the exercise conditions produced significant alterations in max dF/dt.

Time Needed to Reach dF/dt. Although max dF/dt varied only slightly, a significant lengthening occurred in the time needed to reach peak rate of tension development (*p*

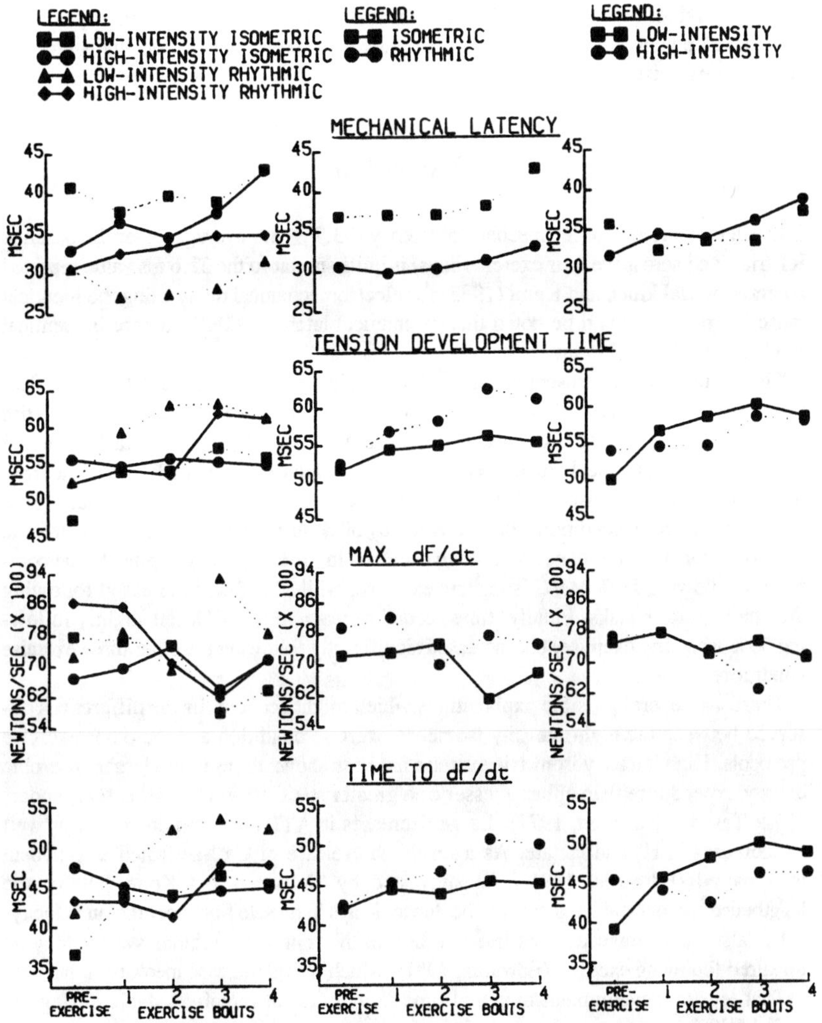

Figure 1—Effect of the four exercise tasks on mechanical latency, tension development time, max. dF/dt, and time to max. dF/dt. Means of the 12 subjects are shown in the left column. The middle column shows the comparison of the two isometric conditions vs. the two rhythmic conditions, while the graphs on the right demonstrate the comparison between the low-intensity conditions and the high-intensity conditions.

< .01). Pooled across exercise conditions, the time needed to reach max. dF/dt increased from 42.3 ms prior to exercise to 47.5 ms following bout 4. Moreover, a significant bout × exercise intensity level interaction was observed as well ($p < .01$). Examination of Figure 1 would seem to indicate that time to max. dF/dt increased during the low-intensity exercise tasks but changed little following high-intensity exercise. Analysis of simple main effects showed that only the low-intensity rhythmic exercise resulted in statistically significant changes in the time needed to reach dF/dt. This lengthening approached but did not reach statistical significance in the low-intensity

isometric and the high-intensity rhythmic exercise. The time needed to reach peak tension development actually decreased slightly over bouts during the high-intensity isometric exercise.

Discussion

The value reported here for mechanical latency (33.7 $\pm$ 8.8 averaged over the baseline RT trials and across the four exercise days) is quite similar to the 32.6 ms value reported recently by Hakkinen and Komi (1983) for electromechanical delay using the identical muscle group. It should be noted that mechanical latency as defined here is identical to electromechanical delay.

The results of the present investigation would appear to indicate that the electromechanical response of fatigued muscle may vary depending upon the characteristics of the exercise task. The high-intensity (50% MVC) isometric exercise task appeared to exhibit an electromechanical response profile which was distinctly different from that of the other exercise conditions. Mechanical latency remained stable during the other three exercise conditions, but increased by 30% during the high-intensity isometric condition. On the other hand, no change occurred in the tension development time component following 50% MVC isometric exercise, while TDT was increased following the other exercise tasks. Finally, time needed to reach max. dF/dt fell slightly following 50% isometric exercise, but increased significantly following the other three exercise conditions.

There are several possible explanations which might account for the differences observed between the high-intensity isometric exercise condition and the other exercise protocols. High-intensity isometric exercise has been shown to result in a larger anaerobic energy expenditure than either a lesser or a greater static effort (Karlsson & Ollander, 1972; Tesch & Karlsson, 1977). Large decreases in ATP concentration occur as well as increases in H^+ and lactate. As a result, the release of Ca^{++}-activated actomyosin is impaired (Edwards, 1981). As suggested by Hakkinen and Komi (1983), the lengthened mechanical latency could be due to delays in muscle fiber conduction velocity. It has also been suggested that transmission in the transverse tubular system may be impaired following exercise (Edwards, 1981), which would increase mechanical latency.

That tension development time and time to max. dF/dt remained stable during the 50% MVC isometric exercise condition would seem to indicate that one or more factors may be exerting a facilitating effect to counteract the failure of the muscle contractile component. This finding corroborates the results of a recent study of muscular fatigue influences on the Achilles tendon reflex, in which contraction time and reflex force remained stable following 50% MVC isometric exercise, but diminished following 25% isometric exercise (Kamen, submitted for publication). One factor which could enhance muscle contraction following the high-intensity isometric exercise is the phenomenon of post-contraction, which involves enhanced muscular force as a result of a prior contraction (Hick, 1953). However, there is no overwhelming evidence to support the idea that the postcontraction effect would be enhanced at greater submaximal tension levels.

Tension development time reflects the ability to produce a low-force (in this case, 15% MVC) contraction as rapidly as possible. The ST fibers could be expected to be most important at these low force levels. The 50% MVC isometric exercise task presumably involves the greatest accumulation of lactate, which might be reutilized by ST fibers (Hulten, Thorstensson, Sjodin, & Karlsson, 1975). This could also account for the stability of TDT and tim to max. dF/dt following high-intensity exercise.

References

EDWARDS, R.H.T. (1981). Human muscle function and fatigue. In R. Porter & J. Whelan (Eds.), *Human muscle fatigue: Physiological mechanisms* (pp. 1-18). London: Pitman Medical.

HAKKINEN, K., & Komi, P.V. (1983). Electromyographic and mechanical characteristics of human skeletal muscle during fatigue under voluntary and reflex conditions. *Electroencephalography and Clinical Neurophysiology,* **55**, 436-444.

HANSON, C.J. (1983). Influences of task complexity on reaction time component latencies. *Abstracts—Annual meeting of the North American Society for the Psychology of Sport and Physical Activity*, p. 9.

HICK, W.E. (1953). Some features of the after-contraction phenomenon. *Quarterly Journal of Experimental Psychology,* **5**, 166-170.

HULTEN, B., Thorstensson, A., Sjodin, B., & Karlsson, J. (1975). Relationship between isometric endurance and fibre types in human leg muscles. *Acta Physiologica Scandinavica,* **93**, 135-138.

KARLSSON, J., & Ollander, B. (1972). Muscle metabolites with exhaustive static exercise of different duration. *Acta Physiologica Scandinavica,* **86**, 309-314.

KROLL, W., & Clarkson, P.M. (1977). Fractionated reflex time, resisted and unresisted fractionated reaction time under normal and fatigued conditions. In D.M. Landers & R.W. Christina (Eds.), *Psychology of Motor Behavior and Sport—1977* (pp. 106-129). Champaign, IL: Human Kinetics Publishers.

TESCH, P., & Karlsson, J. (1977). Lactate in fast and slow twitch skeletal muscle fibres of man during isometric contraction. *Acta Physiologica Scandinavica,* **99**, 230-236.

WEISS, A. (1965). The locus of reaction time change with set, motivation, and age. *Journal of Gerontology,* **20**, 60-64.

Relationship Between Fatigability and Stretching Levels of Lower Leg Muscles During Maximal Voluntary and Nerve Stimulated Contractions in Humans

K. Shimoshikiryo, A. Nagata, M. Muro, and H. Sunamoto
Toho University School of Medicine, Tokyo, Japan

The relationship between tension and EMG exerted during muscle contractions has been analyzed using surface and needle electrodes in vivo and vitro. The amount of tension decrease is different in trained muscles, and trained subjects exert greater tension than untrained subjects during maximal voluntary contractions (MVC), because they are able to maintain more excitability than untrained subjects at the motor level of the central cortex. Fatigability has been measured with static strength changes of muscles during isometric contractions. However, there are no data on muscle fatigue when measured according to variable muscle length (stretching).

The purpose of this study was to investigate the mechanism of fatigue systems in peripheral muscles, and nerve stimulated contractions with stretched lower leg muscles, comparing trained and untrained subjects.

Methods

Subjects used in this study were 6 healthy male students (ages 19 to 21 years). All subjects gave their informed consent about this experiment. Three subjects who exercised regularly were called the "trained group," and the other subjects were called the "untrained group." MVC and nerve stimulated contractions with stretched lower leg muscles (ankle degree: 90°) during the plantar flexion exercise (sustained contraction time: 60 s) were practiced as shown in Figure 1.

Surface EMGs were recorded with two adhesive cup electrodes (10 mm) filled with electrode paste on the skin over the gastrocnemius (GASTRO.) and the soleus (SOL.). The amplified EMG signals were rectified and filtered, and furthermore, IEMG was measured as the area of myographical signals under the rectified and filtered conditions. The mean power frequency (MPF) with power spectral frequency and energy density were calculated, sampling at 10,000 samples/s for a sampling period of 100

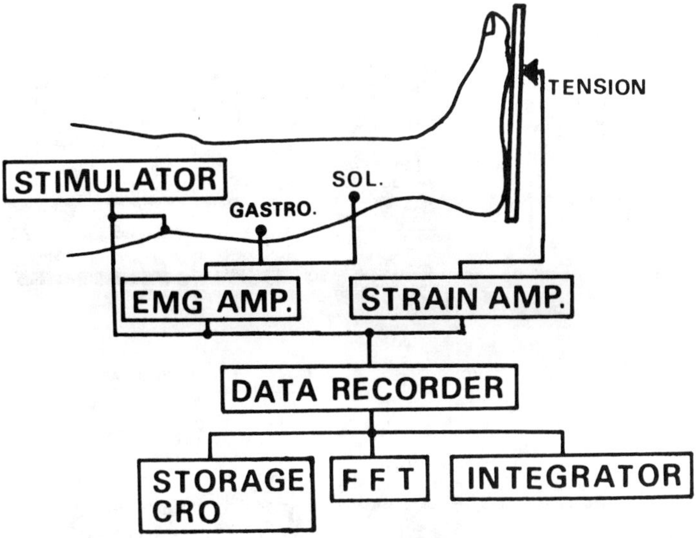

Figure 1—Experimental diagram of sustained contraction exercise and electric nerve stimulation.

ms by the method of Uhlich's algorithm Fourier transform (KANOMAX). Muscle tension was measured with a piezoelectric pressure sensor (KISTLER 9251 A). The tibial nerve was stimulated at the fossa poplitea with frequency of 50 Hz, an amplitude of 55 V using rectangular pulses of 500 ms. The evoked EMG (M wave) was recorded by bipolar surface electrodes on the external side of the gastrocnemius.

Results

Muscle tensions and EMGs of the gastrocnemius (G.) and the soleus (SOL.) during MVC or nerve stimulated contractions are shown in Figure 2. The leg muscles were gradually stretched over a one minute period. MVC tension of both subjects remained almost constant while being slowly stretched over the contraction duration. However, the tension of trained subjects did not change at all during the nerve stimulated contractions, while that of untrained subjects decreased significantly, immediately after the muscle contraction (MVC) began. Mean power frequency changes of the gastrocnemius in the time course during MVC for both trained and untrained groups are shown in Figure 3. There were no significant differences between groups, or before and after sustained MVC. The IEMG of the trained group was shown to be higher than that of the untrained group but there were no significant differences before and after sustained MVC for either group.

The experimental muscle was maximally stimulated via the tibial nerve for 60 s at a constant frequency (50 Hz). At 2 s, 10 s, and at subsequent 10 s intervals, evoked surface action potentials were generated as shown in Figure 4. There was no change in the time course during MVC or evoked surface action potentials for either trained or untrained subjects. However, the action potentials of untrained subjects decreased more rapidly than that of trained subjects.

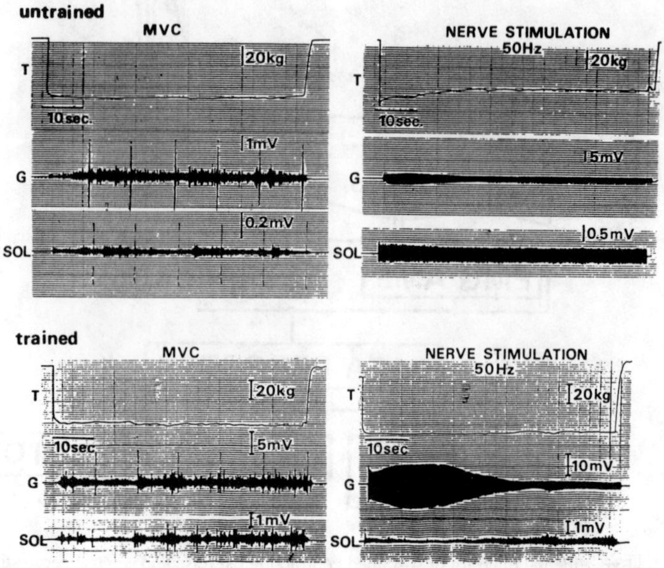

Figure 2—Tension (T), EMG signals of the gastrocnemius (G) and the soleus (SOL) during maximal voluntary and nerve stimulated contractions of trained and untrained subjects.

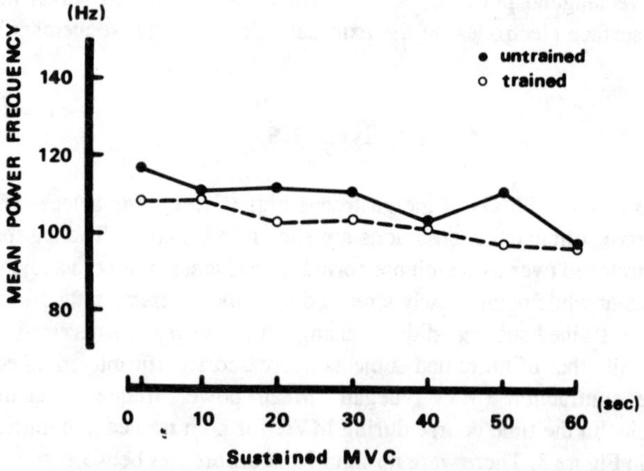

Figure 3—Mean power frequency in the time course during sustained maximal voluntary contractions of untrained and trained subjects (gastrocnemius).

Evoked EMG (M wave) amplitudes during MVC were measured by the technique of nerve stimulation for stretched muscles and nonstretched muscles and for trained and untrained groups quantitatively. These amplitude changes were presented in Figure 5. Though M wave amplitudes increased slightly in the time course of muscle contraction from before to after MVC, there was not a significant difference before and after

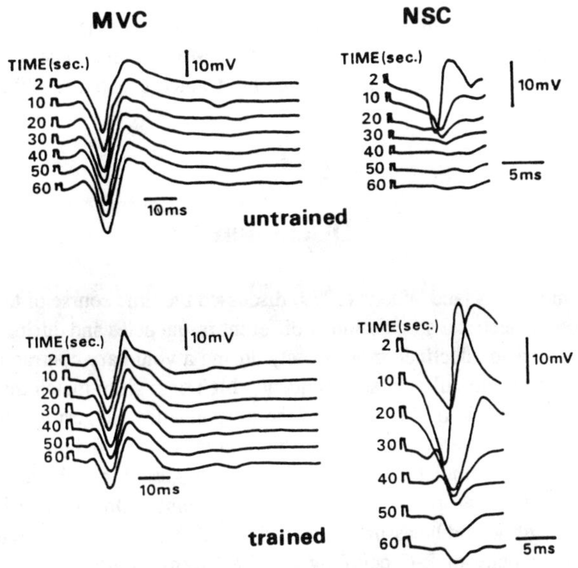

Figure 4—Action potentials recorded at 10 s intervals during maximal voluntary and nerve stimulated contractions of trained and untrained subjects.

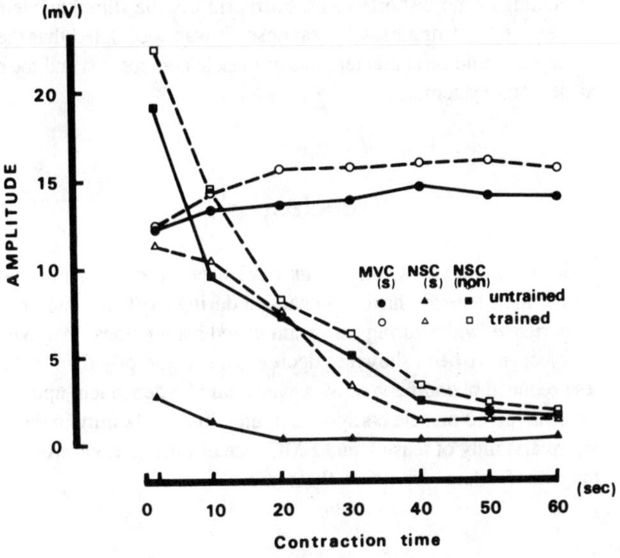

Figure 5—M wave amplitude in the contraction time course during maximal voluntary and nerve stimulated contractions with stretched and nonstretched muscles of untrained and trained subjects.

stimulation. Also, there were no differences between trained and untrained groups. When nerve stimulated contractions were maintained in the state of non-stretched muscles, the M wave amplitudes decreased rapidly 40 s after the onset of muscle contractions. When nerve stimulations were added the M wave amplitudes of the untrained group decreased more rapidly than those of the trained group after 30 s time course in both groups.

Discussion

Bigland-Ritchie, Jones, and Woods (1979) discussed the time course of tension decline during repetitive electrical stimulation at different frequencies and during MVC. They found that this tension declined more slowly during a voluntary contraction than during prolonged nerve stimulation at a frequency high enough to match the initial force of a voluntary contraction. The results of the present study suggest that the stretching muscle should be recovered shortly from peripheral muscle fatigue during maximal voluntary and nerve stimulated contractions. It is suggested that chemical material in the muscle would be transported from muscles efficiently. On the other hand, muscle sarcolemma length would be assumed to be changed at the sustained stretching conditions, and the velocity of E-C coupling and Ca^{++} transportation were maintained to generate high level of the muscle tension. Moreover, the decreasing rate of conduction velocity (gradient slope of M wave) of the untrained subjects was observed to fall significantly more than those of the trained subjects. These phenomena should be assumed to depend upon special muscle length characteristics and muscle fiber movement (more stretching or more shortening). Furthermore, the stimulus intensity would increase until it evoked a strong muscle response. It was speculated that these tension and M wave changes would be characteristics of muscle components and the excitability in the central nervous system.

Conclusion

From our experiments the following results have been observed: subjects have kept fully activated and their tensions have not changed during MVC in spite of their trained or untrained experience, while during nerve stimulated contractions, M wave amplitudes of untrained subjects have been shown to decrease more abruptly than those of trained subjects. These reduced phenomena of M wave should be dependent upon inactivation of muscle fibers, involved muscle components, and non-excitability in the central cortex. However, invariability of tension and EMG signals during MVC would result from more stretching than natural length of these muscles.

Acknowledgment

Supported in part by Grant No. 57780113 from the Japanese Ministry of Education in 1982.

References

BIGLAND-RITCHIE, B., Jones, D.A., & Woods, J.J. (1979). Excitation frequency and muscle fatigue: Electrical response during human voluntary and stimulated contractions. *Experimental Neurology,* **64**, 414-427.

Biomechanical Implication of the Artificial and Natural Strengthening of Wrist Joint Structures

S. Kornecki, K. Kulig, and J. Zawadzki
Wroclaw, Poland

The transmission of muscular force to the environment requires that some joints be stabilized to enable others to perform simultaneously. The stabilization of muscles is necessary because the degrees of freedom (DOF) of the motor system are ten times the DOF of the joints when performing a complex motor task (Fidelus, 1971; Kornecki, Zawadzki, & Bober, 1983; Morecki, 1976). The active stabilization of joints is a dynamic and innate property of muscles and an important attribute of the central nervous system.

Kornecki, Zawadzki, and Bober (1981) and Bober, Kornecki, Lehr, and Zawadzki (1982) found that the maximum force applied by subjects' upper extremities to a handle (as designed by Kornecki, 1982) with one DOF constituted only 75% of the total force exerted to a stationary handle. This means that under conditions which are typical for professional sports, active constraints absorb a substantial percentage (about 25%) of the human potential force.

The purpose of this study is to determine the effects of artificial and natural wrist joint reinforcement on the force transmitted by the upper extremity. It is worthy to note that the wrist requires more reinforcement than any other part of the upper extremity (Kornecki et al., 1983).

Two experiments were conducted in this study. The first experiment (I) used a plaster cast applied to the wrist joint as artificial reinforcement. The second (II) employed isometric training of forearm flexors and extensors as natural reinforcement.

Methods

The subjects of the first experiment were 10 male students of physical education having a mean age of 22, a mean height of 176 cm and a mean weight of 69 kg. The subjects horizontally pushed a stationary handle and a nonstationary handle with one DOF both with and without a plaster splint. The highest force obtained during three trials was used for analysis. In the second experiment the subjects were 12 male students of physical education having a mean age of 23, a mean height of 181 cm and a mean weight of 81 kg. This experiment involved isometric training of forearm muscles by maximum contraction of forearm flexors and extensors.

Each contraction was for 5 s, and the interval between the three sequences of five contractions each was 3 to 5 min. The intervals and the reaction force recorded by a strain gauge were controlled by the subjects. The training was conducted on 10 consecutive days, and was preceded and followed by the experimental runs.

The following maximum muscle torques were recorded: 1) flexors and extensors of the wrist joint (sum = T_W); 2) elbow extensors and arm flexors (sum = T_{EX}). The same experimental set-up was used to record handle push force as described earlier except that no plaster reinforcement was used.

The measurements of the maximum push force on stationary and nonstationary handles were taken on a special stand (see Figure 1). The basic components of the stand were two handles with 0 or 1 DOF (Kornecki et al., 1981). The reaction forces F(t) and angular displacements ϕ(t) of the handle were recorded simultaneously (see Figures 1 and 2). The measurement tolerance was about 5% in both cases; all lengths were measured with a tolerance of 0.5 cm.

The subjects were positioned on the stand and grasped a spherical handle with the palm and forearm pronated 90° from their anatomical positions. The angles of 110° in the elbow joint and 20° in the shoulder sagittal plane were set before each run. The lever arms of wrist flexors and extensors were measured from the knuckle III to the line joining the two epicondylus radialis and ulnaris. The measurements were taken with the hand creating a natural prolongation of the forearm, with both body segments located 90° from their anatomical positions. For all other measurements the positioning of body segments was identical to those described by Hettinger (1972).

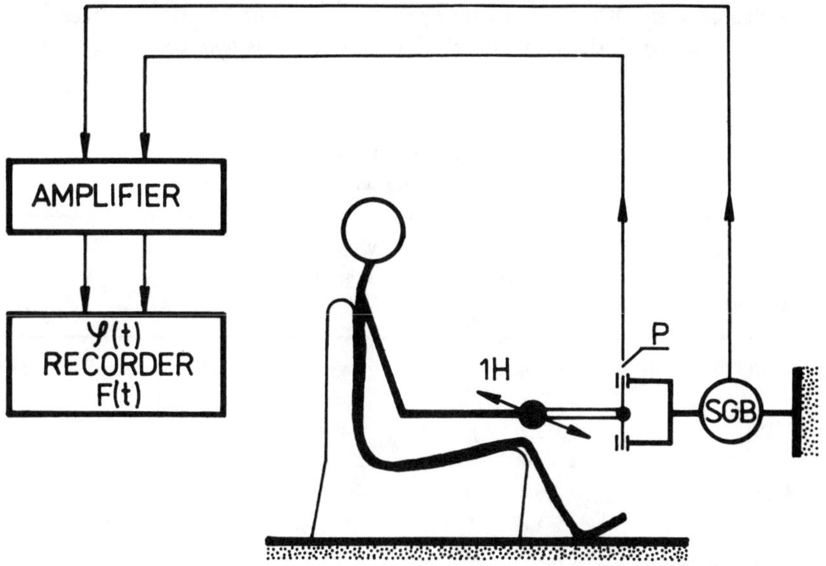

Figure 1—Block diagram for measurements used in experiments I and II. P = Potentiometer, SGB = Strain Gauge Bridge, 1H = Degree of Freedom in Horizontal Plane.

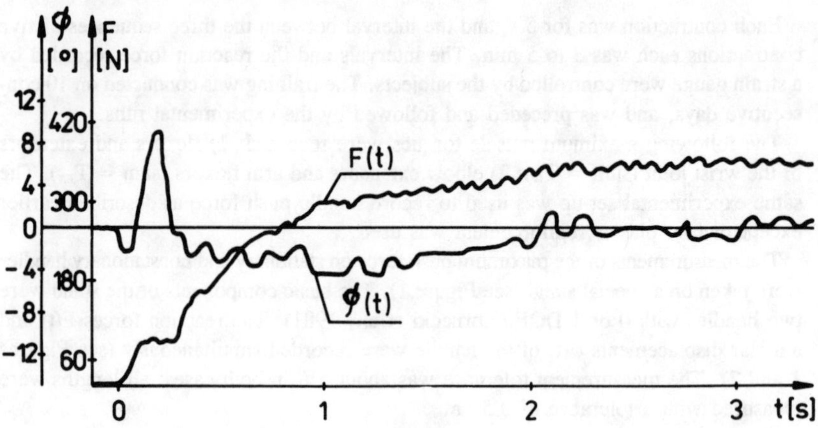

Figure 2—A sample chart of nonstationary handle push force F(t) and displacements $\phi(t)$, (positive values indicate movement to the right and negative values indicate movement to the left). Oscillograph paper speed was 50 mm/s.

Results

The results of the first experiment are shown in Table 1. The cast did not interfere with the results because the magnitude of force exerted on the stationary handle (F_o) did not change when plaster casts were applied to the wrist. In the case of the nonstationary handle, the maximum force increased 16% when the cast was used. The increase was significant ($p = .002$) and the relative differences between F_o and F_l (force exerted on a one DOF handle) were significant ($p = 0.02$). These differences were found to be only 1.9% when artificial reinforcement was used compared to 16.4%

Table 1

Selected Set of Parameters Obtained in Experiment I (N = 10)

Parameter	Without Cast on the Wrist $\bar{x} \pm$ S.D.	With Cast on the Wrist $\bar{x} \pm$ S.D.	%Δ	p
F_o [N]	424.4 ± 76.2	430.5 ± 96.3	1	.68
F_1 [N]	353.2 ± 59.9	409.5 ± 57.3	16	.002
%ΔF [%]	16.4	1.9	—	.02
ϕ [°]	3.6 ± 2.3	1.6 ± 0.8	—	.02
f [Hz]	4.9 ± 1.4	4.3 ± 1.6	—	.28

Table 2

Selected Set of Parameters
Obtained in Experiment II (N = 12)

	Before Training		After Training			
Parameter	$\bar{x}$	S.D.	$\bar{x}$	S.D.	%Δ	p
T_W [Nm]	28.3	4.9	32.3	5.0	14	0.005
T_{EX} [Nm]	143.4	28.4	164.8	33.8	15	0.20
F_o [N]	598.5	141.2	679.5	92.0	14	0.02
F_1 [N]	454.3	58.2	524.6	51.9	15	0.0002
%ΔF [%]	21.8	—	22.3	—	—	0.84
ϕ [°]	4.0	3.1	4.7	2.9	—	0.55
f [Hz]	5.2	0.8	5.5	0.8	—	0.38

without reinforcement. Under the above conditions muscular stabilization appears to absorb only a negligible percentage of the potential force. An increase in F_1 is accompanied by significant decreases in the average amplitude ϕ of a nonstationary handle ($p = 0.02$) and insignificant changes in frequencies (f).

The results of the second experiment are shown in Table 2. The torque of the wrist flexors and extensors [T_W] increased 14% as a result of training. The increase was significant with $p = .005$. Consequently, a 15% increase in maximal force applied to the nonstationary handle was recorded as a result of the training. The difference was again highly significant with $p = .0002$. An increase in T_{EX} resulted mainly from a very high (37%) and significant ($p = .001$) increase in the force of forearm extensors. This was because the force of the arm flexors remained unchanged. There was no significant difference in percent ΔF because of proportional increases in F_o and F_1. The change was 21.8% and 22.3% before and after training, respectively. During the course of the experiment we found no significant change in ϕ or f.

Summary

Muscular stabilization provides restrictions on the free DOF. Due to the excess DOF the efficiency of performing various motor tasks may decrease considerably. About 25% of the potential force was absorbed by muscular stabilization. It is assumed that muscular stabilization was activated by the negative feedback in the neuromuscular control system. The loop was activated only when the wrist joint muscles were stretched. This process was balanced by the critical tension related to the proprioreceptor-generated signals that decreased the extensor's tension, which decreased the force transmitted by the wrist joint to the unstable handle.

The results of the first experiment support the hypothesis because the plaster constraint replaced active constraints of the joint thus dampening the effects of the negative feedback. The subjects were found to apply equal force to both stationary and nonstationary handles when passive reinforcement was applied. The above results supported the research by Walsh (1966) who concluded that the force of a muscular contraction could be increased if the muscle ligaments and the Golgi receptors associated with it were anesthetized.

The natural reinforcement of the wrist joint through isometric training yielded the same effect as passive plaster constraints to increase the magnitude of the force applied to a nonstationary handle. Because of the inability to train selected groups of multijoint muscles separately, natural reinforcement did not result in increased efficiency of the stabilizing process.

Finally, increasing the force applied by the upper extremity to a nonstationary object could be accomplished by either natural or artificial reinforcement of the wrist joint. The choice of reinforcement depends on the nature and conditions of the task, and possibly game rules and regulations.

References

BOBER, T., Kornecki, S., Lehr, R.P., & Zawadzki, J. (1982). Biomechanical analysis of human arm stabilization during force production. *Journal of Biomechanics,* **15**, 825-830.

FIDELUS, K. (1971). *Biomechaniczne parametry konczyn gornych czlowieka (Biomechanical parameters of the motor system of man's upper limbs).* Warsaw: PWN.

HETTINGER, T. (1972). *Isometrisches muskeltraining (Isometrical muscle-training).* Stuttgart: Georg Thieme Verlag.

KORNECKI, S., Zawadzki, J., & Bober, T. (1981). Effect of forced stabilization on magnitude of revealed muscle force. In A. Morecki, K. Fidelus, K. Kedzior, & A. Wit (Eds.), *Biomechanics VII-A* (pp. 107-113). Baltimore: University Park Press.

KORNECKI, S., Zawadzki, J., & Bober, T. (1983). The efficiency of muscular stabilization in the wrist joint. In H. Matsui & K. Kobayashi (Eds.), *Biomechanics VIII-A* (pp. 244-250). Champaign, IL: Human Kinetics Publishers.

MORECKI, A. (1976). *Manipulatory biomiczne (Bionical manipulators).* Warsaw: PWN.

WALSH, E.G. (1966). *Fizjologia ukladu nerwowego (Physiology of the nervous system).* Warsaw: PZWL.

Effect of Arm Elevation on Muscle Circulation and Exercise Performance

H. Yata
Wako University, Japan

T. Fukunaga, A. Matsuo, K. Hyodo, T. Ryushi, and T. Asami
University of Tokyo, Japan

When the position of an arm or leg is altered in a vertical plane, the blood circulation in it may be affected by the change of hydrostatic pressure. Holling and Verel (1957) observed that the reduction of brachial arterial pressure caused by elevation of the arm corresponds to the hydrostatic pressure difference between heart level and the level of elevation, and also that forearm blood flow was decreased as the arm was lifted. Hartling, Noer, and Trap Jensen (1976) also reported that the mean leg muscle blood flow increased from 48 ml/100 ml/min in a body position with the legs elevated 65 cm above heart level to 101 ml/100 ml/min in a sitting position with dependent legs 70 cm below heart level. These studies clearly show that as the arm or leg is elevated from heart level, muscle blood flow is decreased by the reduction of hydrostatic pressure. This hydrostatic pressure change in tissue may exert influence on the muscle work performance and energy metabolism. The purpose of the present study was to observe the effect of arm position on work performance and exercise metabolism in human muscle.

Method

Six healthy adult men aged 22-28 years volunteered to act as experimental subjects. The subject sat on the specially designed apparatus and held two different arm positions. In one the arm was held at heart level, and in the other the arm was elevated vertically. The exercise was performed at these two different arm positions. The subjects performed dynamic hand grip contractions with intensity of ⅔, ½, and ⅓ of maximum voluntary strength till exhaustion. Forearm blood flow was determined by venous occlusion plethysmography using Whitney's strain gauge, which was attached to the forearm. Venous blood of 5 ml was taken by a teflon catheter inserted into deep antecubital vein. The content of O_2 and CO_2 as measured by Van Slyke method. Partial pressure of O_2 and CO_2 and pH in blood was determined by IL meter Type 213. Blood lactate, pyruvate, and glucose were analyzed by enzymatic method. The measurement was done during resting condition and immediately, 1 min, 2 min and 5 min after the exercise. The room temperature was kept at 25 to 27°C during the experiment.

Results and Discussion

The partial pressure of O_2 and CO_2, blood flow, and forearm oxygen uptake for the two arm positions are shown in Figure 1. When the arm was held in an elevated position, lower blood flow and pO_2 were observed but oxygen uptake was not affected by arm elevation. Holling and Verel (1957) reported that the reduction of brachial arterial pressure caused by the elevation of the arm corresponded to the hydrostatic pressure difference between heart level and the level of elevated arm, and also that forearm blood flow was decreased as the arm was lifted. Hartling et al. (1976) observed that the mean leg muscle blood flow increased from 48 ml/100 ml/min in the body position with the legs elevated 65 cm above heart level to 101 ml/100 ml/min in a sitting position with dependent leg 70 cm below heart level. They suggested that when the limb was lifted from heart level, the vascular transmural pressure in the elevated limb was decreased. This decrement of the pressure may cause a diminution of the lumen of vessels through passive elastic recoil and decreased blood flow.

The effect of arm elevation on maximum number of hand grip contractions is shown in Figure 2. At an intensity of ⅓ of maximum strength, a decreased number of contractions was observed in the elevated arm position. However, at intensities of ½ and

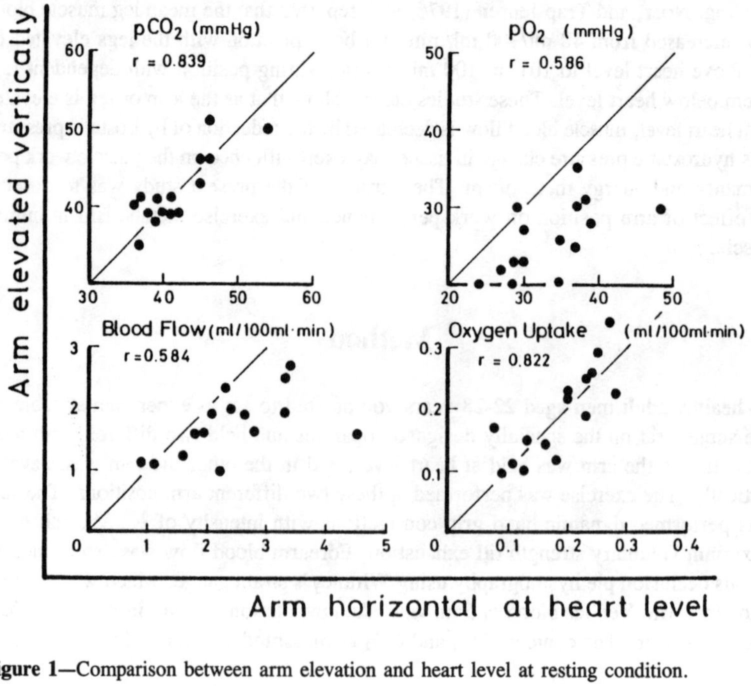

Figure 1—Comparison between arm elevation and heart level at resting condition.

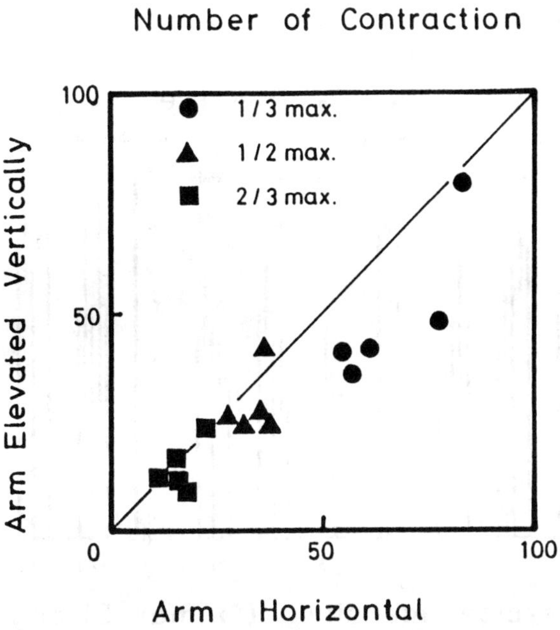

Figure 2—Effect of arm elevation on maximum number of hand grip contraction.

⅔ of maximum strength there was no difference in performance between the arm positions.

In Figure 3 the blood flow, a-v O_2 difference, and oxygen uptake in the forearm are indicated at each exercise intensity. In the arm elevated position, the muscle blood flow indicated significantly lower values than that when the arm was held at heart level. On the other hand, higher a-v O_2 difference was observed in the elevated arm position. The oxygen uptake in muscle tissue, which was obtained from the product of muscle blood flow and a-v O_2 difference, did not show any significant change due to arm elevation. Holling and Verel (1957) indicated that in resting condition the forearm oxygen consumption was not altered in spite of decrement of the flow, when the forearm was elevated from heart level. As shown in the present study, in the maximal exercise condition the decrement of blood flow in the elevated arm was compensated by the increment of a-v O_2 difference and the oxygen uptake in muscle tissue was not affected.

Figure 4 shows the relation between lactate concentration and blood flow after exercise. Lactate concentration is expressed as the difference between venous and arterial levels. The product of lactate concentration and blood flow therefore was considered as the magnitude for the lactate to be washed out from muscle by venous blood. As shown in Figure 4, lactate concentration increased with decrease of blood flow, while lactate production was almost the same, independent of blood flow.

From these results it is considered that when the arm is lifted from heart level, the vascular transmural pressure in the elevated arm is decreased, and the decreased muscle blood flow due to arm elevation induces the decrement in muscle performance.

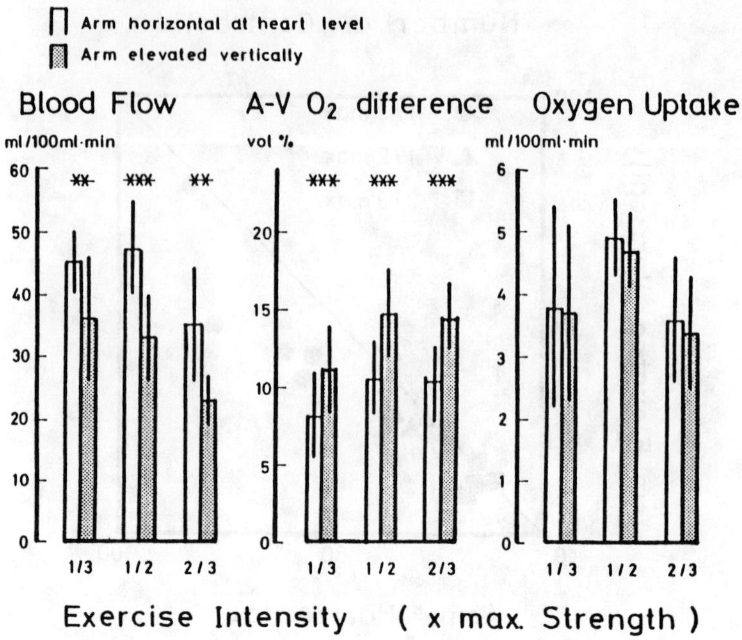

Figure 3—Values of blood gas at each exercise intensity in both arm positions.

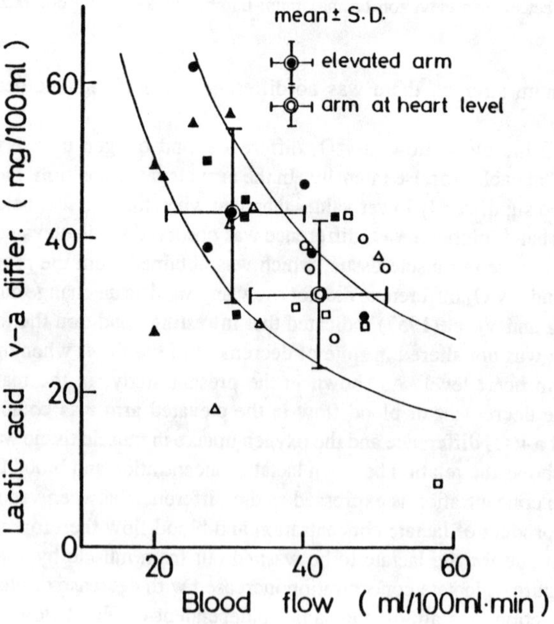

Figure 4—Relation between lactate concentration and blood flow.

References

HARTLING, O., Noer, I., & Trap Jensen, J. (1976). Leg muscle blood flow during reactive hyperemia. *Pflügers Archiv,* **336**, 131-135.

HOLLING, H.E., & Verel, D. (1957). Circulation in the elevated forearm. *Clinical Science,* **16**, 197-213.

Force Variability in Isometric Tasks

L.G. Carlton and K.M. Newell
University of Illinois, Urbana-Champaign, Illinois, USA

The relationship between the level of force produced in an isometric task and the consistency with which that force can be produced has been the focus of a number of studies seeking to describe the force-force variability function (i.e., Fullerton & Cattell, 1892; Schmidt, Zelaznik, Hawkins, Frank, & Quinn, 1979). Previous descriptions have yielded equivocal results suggesting that the force-force variability relationship is an increasing square root function (Fullerton & Cattell, 1892); a linear and proportional function (Schmidt et al., 1979); and an inverted U-shaped function (Sherwood & Schmidt, 1980).

There are a number of factors which could account for the various estimates of the peak force-force variability relationship based upon experimental procedures. Examples of these are insufficient force levels to adequately describe the function and insufficient data points at each force level to obtain veridical estimates of variability. However, a more fundamental reason for the discrepancies may lie in the manner in which peak force is produced. Previous studies have controlled only for peak force level although time to peak force and the rate of force production could contribute to the relationship. Maximum peak force increases as time to peak force increases, possibly up to 2 s (Kamen, 1983). Thus, modulating time to peak force leads to changes in the percentage of maximum that the subject is operating at. This in turn may have an effect on peak force variability.

The present experiment examines the relationship between peak force and peak force variability at various percentages of maximum force (5 to 90%) with 3 criterion times to peak force. By examining peak force variability at a consistent time to peak force, it was anticipated that estimates of the force-force variability relationship could be examined independent of shifts in time to peak force across the various percentages of maximum conditions. Previous work from our laboratory indicates that subjects systematically shift time to peak force as a function of the percentage of peak force being produced. This is in contrast to pulse-step models of force production (Ghez, 1979; Ghez & Vicario, 1978) and the impulse variability model (Schmidt et al., 1979), where peak force is scaled with a constant time to peak force.

Method

Three subjects attempted to exert a specific peak force at a criterion time to peak force, using elbow extension. The force production apparatus consisted of a rigid bar connected to a piezo-electric force transducer. Forces were produced isometrically with the upper arm and forearm in the horizontal plane with the elbow at 90°.

There were two phases in the experiment. In the initial phase, Day 1, 25 trials were taken to obtain maximum peak force for three times to peak force (100, 200, 400 ms). Based upon these trials, the average maximum force obtained with a time to peak force of 100, 200, and 400 ms was calculated. The second phase consisted of conditions where subjects attempted to match a criterion peak force and time to peak force. The experimental protocol was identical, regardless of the experimental condition. The subject was presented with a warning light which was followed by a stimulus light with a random foreperiod. Approximately 8 seconds after the completion of the response, the template and the just produced force-time curve were shown on a graphics screen. A few seconds later, the warning light was presented to start a new trial. Each subject completed 8 conditions varying in the percentage of maximum force at each of 3 times to peak force. Testing was completed over 13 days.

Experimental Design

The maximal peak force values for the 100, 200, and 400 ms conditions were set as 100% performance and values representing 5, 10, 15, 25, 50, 75, 85, and 90% of maximum peak force were calculated. The order of presentation was random for each subject. Time to peak force and trials within a percentage of maximum condition were both blocked. There were 80 trials at each percentage of maximum condition with an inter-trial interval of 20 s. Verbal feedback as to peak force produced and time to peak force was given after each trial in addition to the visual feedback.

Results

The results presented focus on peak torque, peak torque variability, and temporal characteristics of the response. The results from the maximum force trials obtained on Day 1 indicated that as time to peak force increases, maximum peak force increases. The mean maximum forces for the three time to peak force conditions ranged between 47 to 68 N (100 ms), 53 to 100 N (200 ms), and 60 to 120 N (400 ms) for the three subjects.

The remaining analyses were conducted over the last 50 trials in each condition. The data for time to peak force indicated that subjects were successful in generating peak torque values in times close to the criterion. Generally, mean times to peak force were within 10% of the criterion time. There were no systematic shifts across percentage of maximum conditions. This control was desirable in that it allowed for comparisons of response variability as a function of force level without having fluctuations in variability due to modulations in time to peak force.

The standard deviation of peak force, as a function of peak force level, is shown in Figure 1. As force level increased, the variability in peak force also increased. The function is distinctly curvilinear with proportionally greater variability at lower peak forces. There did not appear to be different variability functions for the time to peak force conditions. The variability was more strongly related to peak force. However, when the standard deviation is plotted as a function of the percentage of maximum (5 to 90%), three separate curves emerge with greater variability associated with the longer time to peak force.

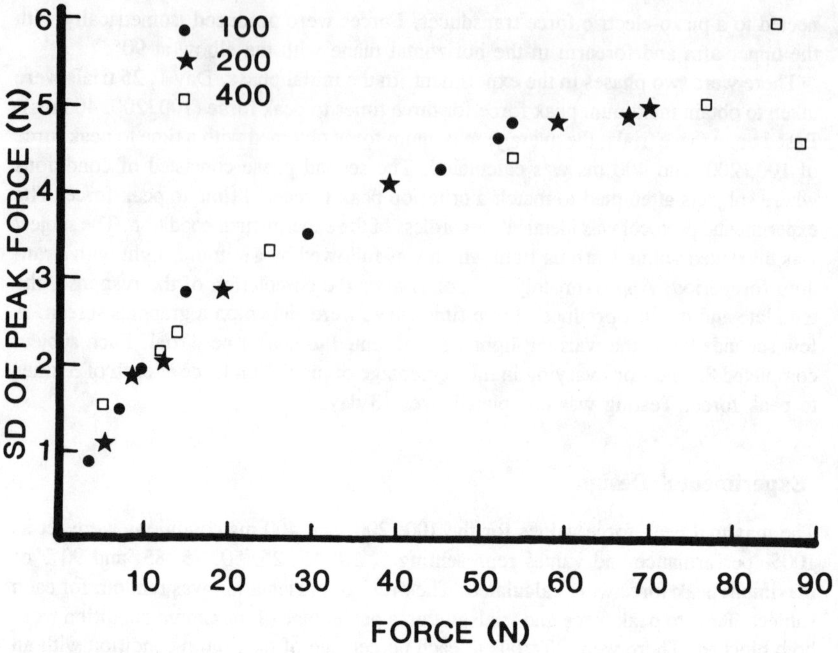

Figure 1—The peak force-peak force variability function.

While the time to peak force was maintained at the criterion values across percentage of maximum conditions, there were systematic changes in the time to peak rate of change of force. Figure 2 displays the shifts which occurred for each of the time to peak force conditions. As the percentage of maximum peak torque increased from 5 to 15%, peak rate of change of force increased for each of the time to peak force conditions. At the higher percentages of maximum the time to peak rate of change decreased. The changes seem quite large given that the time to peak torque remained constant.

Discussion

Analysis of peak force variability indicated that the peak force-peak force variability function is negatively accelerating, similar to that reported by Fullerton and Cattell (1892). This is contrary to predictions of the motor output variability model put forth by Schmidt et al. (1979) and a number of other suggested force-force variability relationships (e.g., Sherwood & Schmidt, 1980). The present study investigated the range of force levels which can be produced within the time to peak force constraints imposed, thus maximizing the potential to detect deviations from linearity. Impulse characteristics may be important for understanding the force-force variability function. Modifications in rate of change of force as well as temporal characteristics may influence the consistency with which forces can be produced. The present results do not

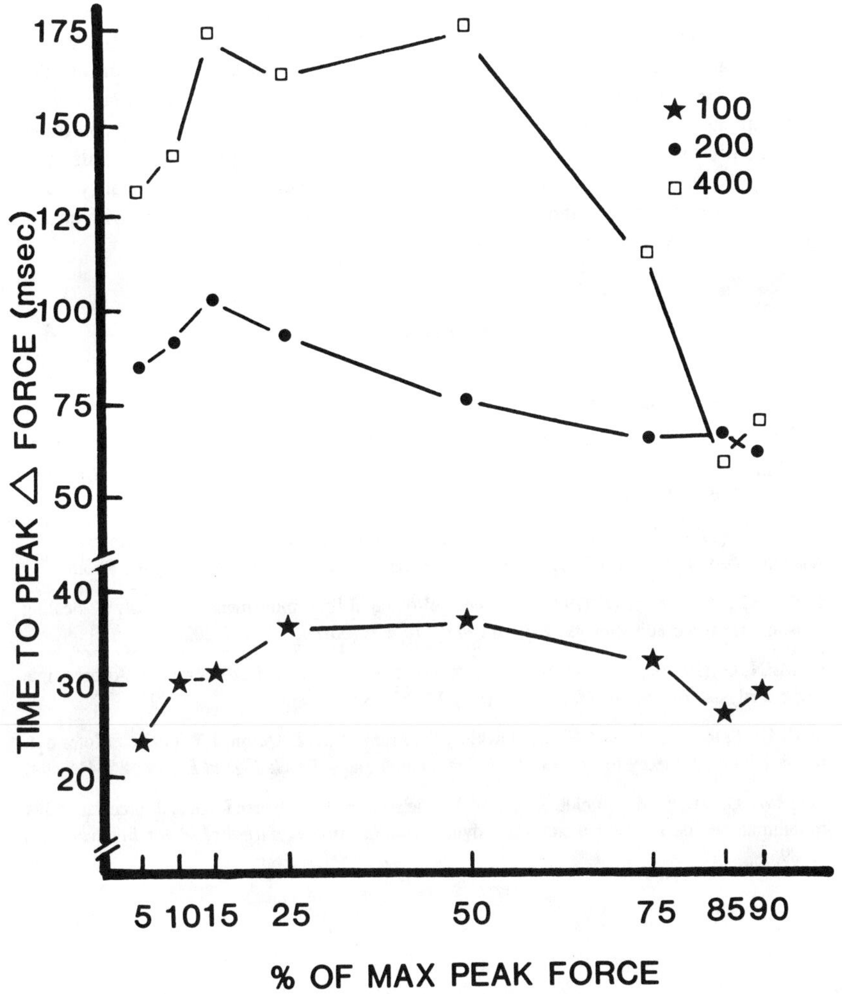

Figure 2—Time to peak rate of change of force across percentages of maximum peak force.

provide a strong test of whether changes in impulse characteristics are related to peak force variability. Since time to peak force was not systematically varied across time to peak force conditions, a direct test was not conducted.

The results do indicate, however, that there are systematic modulations of time to peak rate of change of force across percentage of maximum conditions. This is inconsistent with the pulse-step model of force production (Ghez, 1979; Ghez & Vicario, 1978). This model is based on isometric force production experiments demonstrating that increments of force are characterized by linear scaling properties. As force level increased, the temporal characteristics of the impulse remain unchanged. Similar findings are reported by Freund and Budingen (1978) but close examination of their data

(1978, Figure 1) suggests that systematic lengthening of time to peak force occurred with increments of peak force. Also, recent work from our laboratory indicates that subjects vary the rate of force production for any given force level. This indicates that there may be an optimal rate of force production for any peak force and that subjects systematically vary time to peak force when this parameter is not artificially controlled as in the present experiment. It appears that the linear scaling properties of pulse-step and impulse variability models, based on kinematic principles from physical systems, do not accurately reflect modifications in force production characteristics with changing force demands.

References

FREUND, H.J., & Budingen, H.J. (1978). The relationship between speed and amplitude of the fastest voluntary contractions of human arm muscles. *Experimental Brain Research,* **31**, 1-12.

FULLERTON, G.S., & Cattell, H. McK. (1892). On the perception of small differences. *University of Pennsylvania Philosophical Series.* No. 2.

GHEZ, C. (1979). Contributions of central programs to rapid limb movement in the cat. In H. Asanuma & V.J. Wilson (Eds.), *Integration in the nervous system.* Tokyo: Igaku Shoin.

GHEZ, C., & Vicario, D. (1978). The control of rapid limb movement in the cat. II. Scaling of isometric force adjustments. *Experimental Brain Research,* **33**, 191-202.

KAMEN, G. (1983). The acquisition of maximal isometric plantar flexor strength. A force-time curve analysis. *Journal of Motor Behavior,* **15**, 63-73.

SCHMIDT, R.A., Zelaznik, H.N., Hawkins, B., Frank, J.S., & Quinn, J.T. (1979). Motor output variability: A theory for the accuracy of rapid motor acts. *Psychological Review,* **86**, 415-441.

SHERWOOD, D.E., & Schmidt, R.A. (1980). The relationship between force and force variability in minimal and near-maximal states and dynamic contractions. *Journal of Motor Behavior,* **12**, 75-89.

II.
BIOMECHANICS OF BONES, JOINTS, AND LIGAMENTS AND AND THEIR REPLACEMENT

Mechanical Aspects of Osteoarthritis

J. Pugh
Hospital for Joint Diseases Orthopaedic Institute,
New York, New York, USA

Osteoarthritis is generally accepted to be a mechanically mediated disease process that has well-defined biochemical, metabolic, and histologic manifestations. The mechanical changes occurring in the disease process associated with the bone, cartilage, and synovial fluid are outlined and related to current experimental observations involving lubrication, subchondral cancellous bone architecture (and associated structure-controlled mechanical properties), and the symbiotic mechanical relationship between the subchondral cancellous bone and the adjacent articular cartilage. The possible roles of dynamic loads in the activities of daily living are delineated.

There is considerable controversy over what really is the cause of osteoarthritis. One contention (Freeman, 1973) is that a breakdown occurs in the lubrication mechanism of the joint. This could be due to chemical changes in the fluid or the cartilage. The central feature here may be an inability of the synovial fluid to provide proper nourishment to the cartilage. It has long been recognized (Trueta, 1968) that cartilage itself is avascular and must be nourished indirectly by the synovial membrane. Another contention (Radin, Paul, & Rose, 1972) is that subchondral bone changes are responsible for the disease and that these changes cause the cartilage and fluid to change their role of nutrition and lubrication slightly. The nature of the bone changes was thought to be either chemical or structural, but it is now conceded (Lanyon, 1974; Prager, 1970) that they are structurally induced by a Wolff's law mechanism that governs the functional adaptation of bone.

Numerous studies have focused on the later stages of osteoarthritis. The gross cartilage fibrillations, bone remodeling, and osteophyte formations occurring in these advanced stages render the joint in a state much different from that of its original prearthritic condition. Any cartilage and bone studies done on these grossly affected joints really offer only hindsight into the disease process. In contrast, studies performed on normal joints offer insight into the delineation of signs that may ultimately allow early diagnosis and identification of prearthritic joints before any gross changes occur.

The effect of the subchondral bone on the functioning of joints has often been described as passive and minimal. Although numerous authors have suggested the possible effects of the subchondral bone on the biomechanics of the joint, the diagrams shown often do not recognize the essential structural inhomogeneity of the subchondral cancellous architecture. In fact, this structural heterogeneity has the same dimensions as the structural features of the overlying articular cartilage (Pugh, 1975). These features of the articular cartilage play a major role in joint lubrication. Disruption of the underlying inhomogeneity of the cancellous bone has been shown to produce clinical osteoarthritis in dogs (Ewald, Poss, Pugh, & Schiller, 1982). A definite structural and mechanical interaction exists between the articular cartilage and the subchondral bone.

An increase in the elastic stiffness of the subchondral bone in a series model for the mechanics of a joint is shown to increase the peak dynamic forces in the articular cartilage, in the same manner that a decrease in the stiffness of the articular cartilage decreases the peak dynamic forces in the subchondral bone during the activities of daily living. Thus a model based on either bone or cartilage remodeling in response to changing dynamic loads can explain the onset and progression of osteoarthritis—cartilage and bone respond to changes in stress through an alteration in structure and mechanical properties. The subchondral bone associated with articular cartilage undergoing the earliest stages of osteoarthritis has been shown to possess significantly altered structure and mechanical properties relative to that of normal joints (Pugh, Rose, & Radin, 1974).

Chondromalacia

It is generally agreed that chondromalacia patellae (softening of the articular cartilage on the underside of the patella) is a form of osteoarthritis. This is reasonable since the patellofemoral joint is subjected to the highest dynamic and static loads of any joint in the human body during the activities of daily living. Studies of the patella during the process of chondrodegeneration have resulted in significant insights into the osteoarthritic process (Raux, Townsend, Miegel, & Raux, 1975; Townsend, Rose, Radin, & Raux, 1977).

The internal cancellous bone stucture of the patella has been analyzed and characterized. It consists primarily of sheets and rods interconnected. Density was found to be dependent on intertrabecular distance rather than thickness. A significant variation in the structure was found to exist over the articular surface of the patella, and the underlying bone was shown to also vary significantly in structure according to the articular changes. The mechanical stiffness of the cancellous bone in a direction normal to the articular surface was also shown to vary considerably across the patella. The focal lesions seen in clinical chondromalacia were found to be strongly associated with the areas of maximum stiffness *gradient* rather than stiffness itself (Townsend, Rose, Radin, & Raux, 1977).

Thus the concept that it is stiffness *gradient* rather than the value of the stiffness per se that controls cartilage degeneration received considerable support from the studies on the patella. This is also consistent with the clinically observed degeneration associated with subchondral cysts, implants placed in proximity to the articular surface, and other focal lesions in the subchondral cancellous region.

Experimentally-Induced Osteoarthritis

Rabbits

A set of white rabbits was placed on a table (Radin, Parker, Pugh, Steinberg, Paul, & Rose, 1973), and one hind limb was subjected to an impulsive load by means of a motor-driven cam. The opposite leg served as the control for each rabbit. Cartilage and bone changes were induced in the experimental limb as early as four days into the experiment. Subchondral cancellous bone remodeling was quantified by means of tetracycline labeling. This remodeling was observed to be microfracture-induced.

Cartilage degeneration consistent with osteoarthritis was thus shown to be created by daily intervals of physiologically reasonable impact loads. The onset of these loads was such that the propensity for remodeling in the subchondral region of the joint was exceeded, and cartilage degeneration occurred before the joint could alter its structure to accommodate the increased level of mechanical function.

Sheep

One group of sheep was subjected to prolonged activity on hard walking surfaces and housed on asphalt (Radin, Orr, Kelman, Paul, & Rose, 1982). The control group was walked on compliant wood chip surfaces and pastured. Significant differences were found in the knee joints of the two groups after two and one-half years. The remodeling in the experimental group knees was much more extensive than that in the control group. Focal changes in articular cartilage were associated with the changes observed in the underlying subchondral cancellous bone. Prolonged, repetitive impulsive loading was thus shown to have significant effects on both articular cartilage and its underlying bone.

In contrast to the studies on the rabbits, the sheep work indicated that alterations imposed on activities of daily living not severe enough to overwhelm the tissues' ability to modulate sufficiently to handle the increased stresses involved does not lead to rapid tissue destruction. It was conjectured that these chronic low-level insults would lead to frank osteoarthritis over time.

Dogs

Methacrylate arthroplasty was performed in dogs in an effort to evaluate this as a clinical procedure in humans that preserves the biological running surface (Ewald, Poss, Pugh, Schiller, and Sledge, 1982). After resection of the femoral head at the cartilaginous-osseous junction, the cancellous bone was removed down to the subchondral bony plate and the void packed with methacrylate bone cement. The femoral head was then anatomically reduced on the femoral neck prior to polymerization of the methacrylate.

The femoral heads of the dogs all showed loss of articular cartilage over the weight-bearing dome. The histologic features of osteoarthritis were evident in all dogs with methacrylate arthroplasty, while the contralateral controls showed well preserved articular cartilage and subchondral bony plate.

The literature showed no statistically significant differences in the elastic properties of polymethylmethacrylate and cancellous bone. Mechanical testing of a sample of femoral heads with methacrylate arthroplasty compared with the contralateral controls revealed no statistically significant differences in elasticity. The only completely consistent explanation for the observed onset of osteoarthritis lies in the change in subchondral homogeneity of structure and properties.

The normal femoral head of the dog is supported by cancellous bone having a spatial variation in structure and properties controlled by the locations of the individual trabeculae that comprise the composite. In contrast, the subchondral areas of the dogs with the methacrylate arthroplasties had the same global properties in elastic response as that of the preexisting cancellous bone substrate, but significant differences in the microscopic distribution of the values of the elasticity. Since the articular cartilage prior to surgery had existed in harmony with this heterogeneous substrate, the replacement of the substrate with a material having homogeneous properties resulted in the articular cartilage being subjected to significantly increased shear stress. It is common knowledge that cartilage is remarkably prone to failure when loaded in shear. It is constructed to resist compressive stresses that place the superficial tangential zone in tension.

Thus, articular cartilage viability was shown to be dependent on the architecture as well as the material properties of the underlying subchondral cancellous bone. In the normal hip joint, the microtopography of the articular cartilage is fashioned by the supporting subchondral bone during ossification and chondrogenesis. By replacing the cancellous bone with a material of markedly different architecture, the force transmission across the hip was altered and the loading capacity of the articular cartilage was exceeded.

Summary

It is clear that the articular cartilage and subjacent cancellous bone exist in mechanical symbiosis in a typical joint. Wolff's law, when interpreted in light of the observation that the bone and cartilage are already overloaded when compared to the loads subjected to engineered structures made by man and designed to resist failure, shows that the mechanical symbiosis is a tenuous one. A range of loads results in supportive mechanical interactions between the articular cartilage and the underlying bone that ensures their mutual viability. Loads too small or too large are known to result in disruption of the metastable mechanical symbiosis.

In addition, it is clear that osteoarthritis can be induced by any procedure that results in a disruption of the Wolff's law mediated mechanical symbiosis that has been established for a particular joint. These procedures, whether excessive impact, disuse, or change in architecture of the bone supporting the articular cartilage, ultimately result in the production of excessive shear stress in the articular cartilage.

Stiffness gradients in the subchondral cancellous bony bed are therefore critical to the development of osteoarthritis. The focus for articular degeneration is associated with the region of maximum stiffness gradient, whether macroscopic or microscopic. Stiffness gradients that exist normally on a microscopic scale in the subchondral cancellous bone, interestingly, result in minimal shear stress in the articular cartilage because the cartilage has formed in such a manner to interact with the underlying bony substrate to reduce these potential shear stresses. Consequently, removal of those stiffness gradients by replacement of the cancellous bone with a homogeneous material such as methylmethacrylate places the cartilage in shear and renders it prone to failure. In addition, production of extreme gradients in stiffness relative to the preexisting gradients already present will result in the production of cartilage failure.

Osteoarthritis as a mechanical phenomenon is thus a relatively complex interaction of materials of varying mechanical properties. The metastability of joints with regard to mechanical breakdown remains the central feature of this thesis.

References

EWALD, F., Poss, R., Pugh, J., Schiller, A., & Sledge, C. (1982). Hip cartilage supported by methacrylate in canine arthroplasty. *Clinical Orthopedics,* **171**, 273-279.

FREEMAN, M. (1973). *Adult articular cartilage.* New York: Grune and Stratton.

LANYON, L. (1974). Experimental support for the trajectorial theory of bone structure. *Journal of Bone and Joint Surgery,* **56B**, 160-166.

PRAGER, W. (1970). Optimization of structural design. *Journal of Optimization Theory and Applications,* **6**, 1-21.

PUGH, J., Rose, R., & Radin, E. (1974). Quantitative studies of human subchondral cancellous bone. *Journal of Bone and Joint Surgery,* **56A**, 313-321.

PUGH, J. (1975). The role of cancellous bone in joint function. *Journal of Bone and Joint Surgery,* **57A**, 575-576.

RADIN, E., Paul, I., & Rose, R. (1972). Hypothesis: Role of mechanical factors in the pathogenesis of primary osteoarthritis. *Lancet,* **1**, 519-522.

RADIN, E., Parker, H., Pugh, J., Steinberg, R., Paul, I., & Rose, R. (1973). Response of joints to impact loading. III. Relationship between trabecular microfractures and cartilage degeneration. *Journal of Biomechanics,* **6**, 51-57.

RADIN, E., Orr, R., Kelman, J., Paul, I., & Rose, R. (1982). Effect of prolonged walking on concrete on the knees of sheep. *Journal of Biomechanics,* **15**, 487-492.

RAUX, P., Townsend, P., Miegel, R., Rose, R., & Radin, E. (1975). Trabecular architecture of the human patella. *Journal of Biomechanics,* **8**, 1-7.

TOWNSEND, P., Rose, R., Radin, E., & Raux, P. (1977). The biomechanics of the human patella and its implications for chondromalacia. *Journal of Biomechanics,* **10**, 403-407.

TRUETA, J. (1968). *Studies of the development and decay of the human frame.* Philadelphia: W.B. Saunders.

Chondrocyte Volume and Elastic Properties of the Cartilage

U. Rehder, L. Weh, G. Binzus, and B. Ebens
University of Hamburg,
Hamburg, Federal Republic of Germany

The histologic patterns of arthrosis are the subject of many investigations. Little is known about the correlation of pathomorphologic changes and the mechanical behavior of articular cartilage. In this study we investigated morphometrical parameters as chondrocyte volume and number, and referred it to the elastic-plastic properties of cartilage surface.

Methods

One-hundred and twenty cartilage cylinders from 12 patients were taken from arthrotic tibia plateaus during endoprosthetic procedure. The diameter of these specimens was 7 mm. Indentation tests were performed using a plane indenter with a diameter of 3 mm. The indentation velocity was 1.2 mm/min. Five cycles of successive loading and unloading were applied under constant perfusion with Ringer's saline solution. Indentation/time and force/time curves from 0 to 10 N were recorded. From these curves the Young modulus/indentation curves (see Figure 2) were calculated. A "Deformation Resistance Factor (DR-Factor)" was defined as the slope of the Young modulus/indentation curve using the mean of the slopes of the first and the second unloading and the second loading.

Cut sections of the cartilage were fixed in formol 10%, embedded in paraffin and stained by HE. The area of the chondrocytes was measured by an image analyzing system ("Videoplan," Kontron). The cell number and the cell volume was calculated from the counted and planimetric measured cells using the method of Weibel (1979). The cartilage volume per single chondrocyte was calculated.

Results

The first loading curve (see Figure 1, curve 1) showed a typical additional peak of the Deformation Resistance Factor. The following unloading (see Figure 1, curve 2,

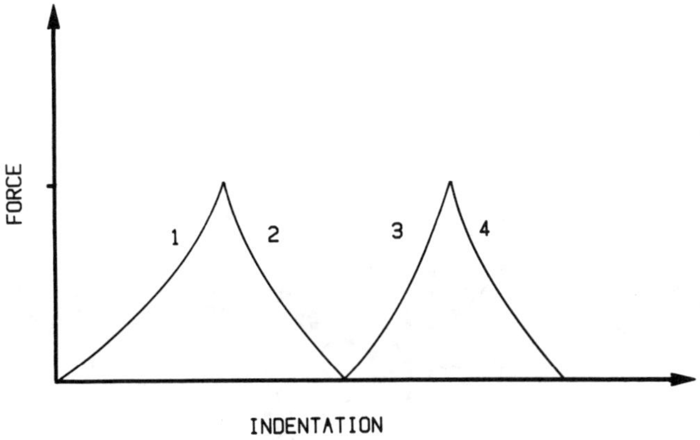

Figure 1—Force/indentation curve with two loading and unloading cycles.

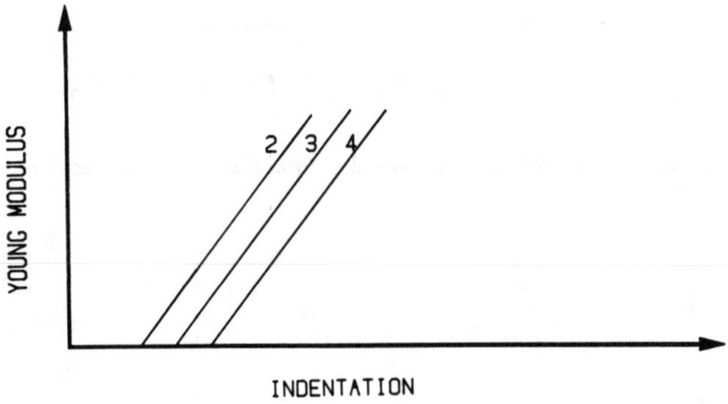

Figure 2—Young modulus/indentation curves, showing the constant slope of the first and second unloading and the second loading curve. This slope represents the "Deformation Resistance Factor."

4) and loading (see Figure 1, curve 3) revealed linear changes with indentation. The slope of these curves was nearly equal (see Figure 2). Correlation of histomorphometric and biomechanic parameters was determined by curve fitting nonlinear interactions of these items; regressions were made using 3rd order polynomials. The DR-Factor decreased with increasing cartilage thickness (see Figure 3).

The increase of relative chondrocyte volume of the total cartilage substance was accompanied by a slight increase of the DR-Factor (see Figure 4). The increase of cartilage volume per single chondrocyte was correlated to a decrease of the DR-Factor (see Figure 5). The increase of cell number per volume went together with an increase of the DR-Factor (see Figure 6).

Discussion

The fixation of the cartilage specimens in formol leads to a shrinking of cartilage and

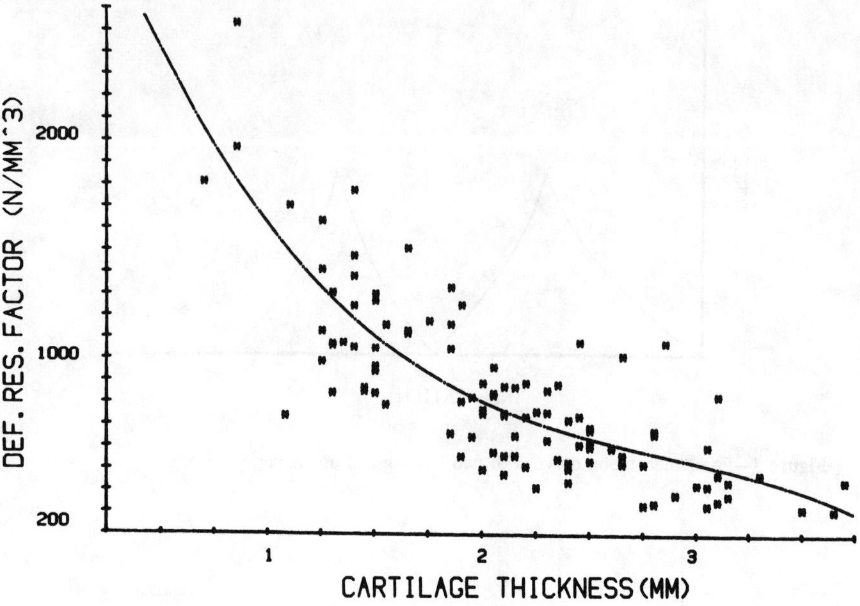

Figure 3—Correlation of cartilage thickness and Deformation Resistance Factor (n = 106, r = .815, p < .00001).

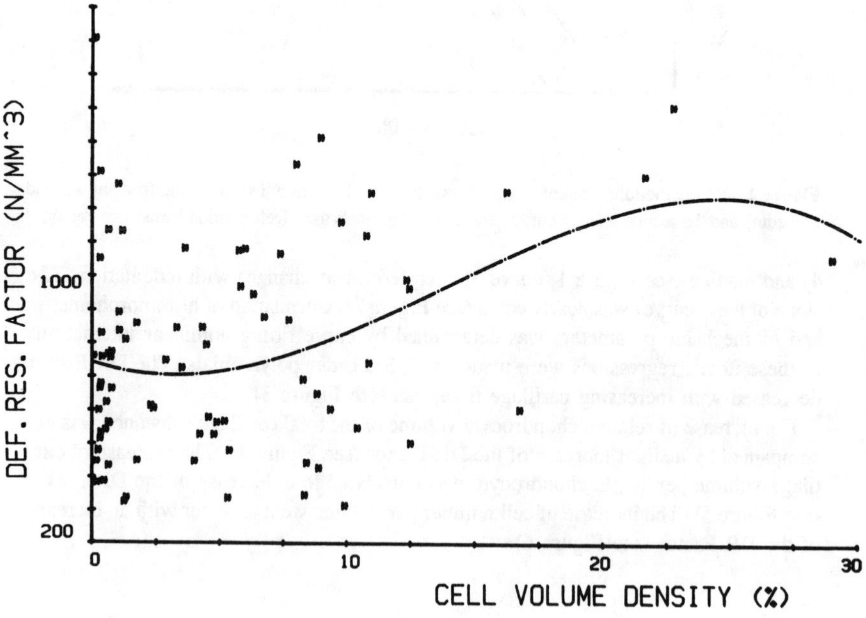

Figure 4—Cell volume density and Deformation Resistance Factor (n = 80, r = .398, p < .00026).

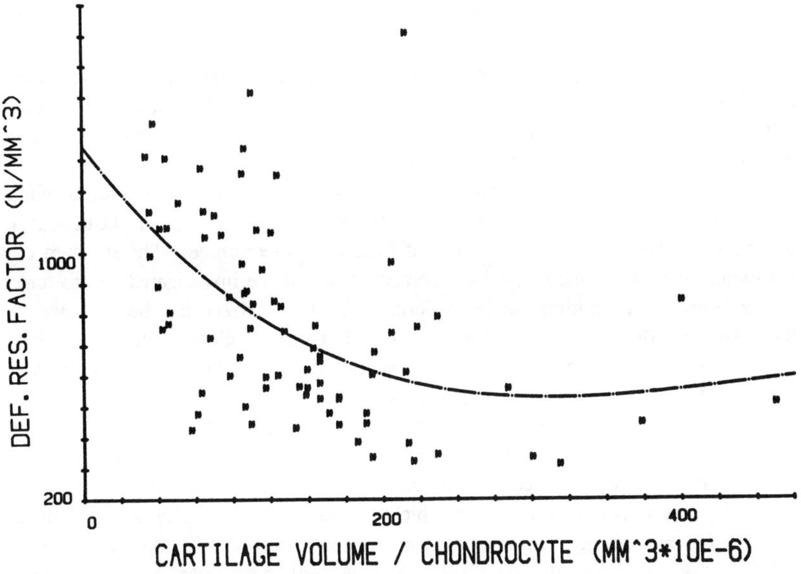

Figure 5—Correlation of cartilage volume per single chondrocyte and the Deformation Resistance Factor ($n = 80$, $r = .521$, $p < .00001$).

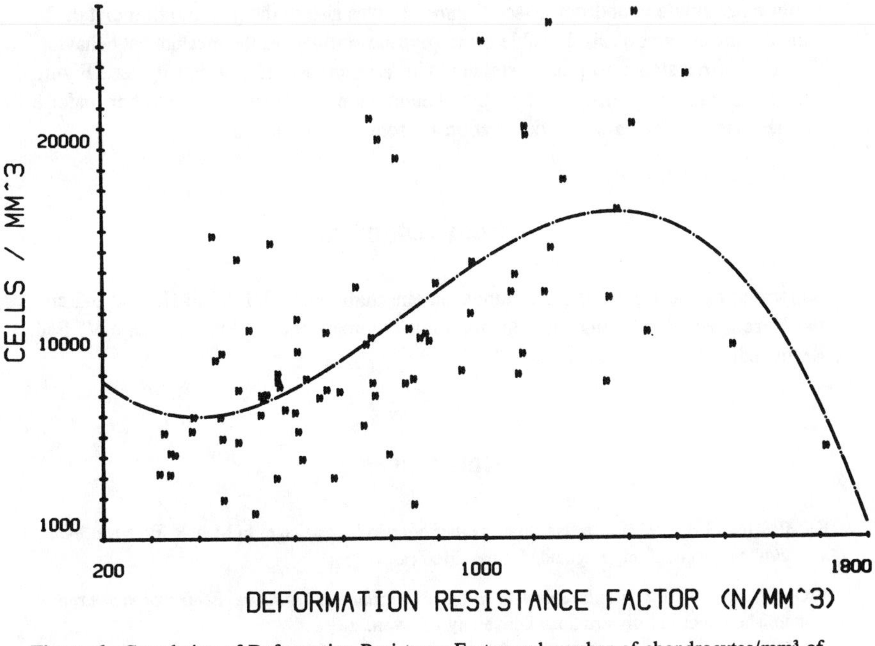

Figure 6—Correlation of Deformation Resistance Factor and number of chondrocytes/mm³ of cartilage tissue ($n = 80$, $r = .535$, $p < .00001$.)

chondrocytes. The selection of relative morphometrical parameters largely eliminated this systematic error.

In former experiments plane and hemispherically-ended indenters had been used (Kempson, 1979). We chose the plane indenter to keep the contact stress independent of the depth. The applied force of 10 N yielded a nominal compressive contact stress of 1.42 N/mm2, well within the physiological stress range.

When measuring cartilage cylinders the known problems must be taken into account. The bulging of the lateral cylinder wall influences the values of the plastic-elastic parameters. There is an increased loss of fluid and glucosamines. The standard experimental conditions maintained the comparability of the investigated parameters.

The constant cell number under a defined cartilage surface has been shown by Stockwell and Meachim (1979). The constant volume of chondrocytes under a defined surface area in normal cartilage of patella with different thickness can be calculated (Weh, Rehder, & Eggers-Stroeder, in press).

The cartilage thickness seems to be a function of the mechanical challenge. Vignon (1973) showed that the thickest cartilage in the hip joint was found in the area with the highest calculated load bearing. In our experiments the resistance to deformation of cartilage decreased with increasing thickness (see Figure 3). This seems to be an important aspect of cartilage adaptation and decompensation in arthrosis.

There is only a slight influence of cell volume density on the Deformation Resistance Factor (see Figure 4). On account of the small volume percentage the cells primarily do not cause a correlation by its mechanical properties. The metabolic chondrocyte activities are more likely to determine this cartilage behavior.

The inverse correlation of resistance of cartilage to deformation and the cartilage volume per single chondrocyte (see Figure 5) gives rise to the presumption that diffusion and the amount of displaceable proteoglycanes influences the mechanical behavior. The rise of resistance to load correlates with an increase of cell density (see Figure 6). This seems to be influenced by the phenomenon of constant cell number under a defined surface area and the correlation presented in Figure 3.

Acknowledgment

Supported by the Deutsche Forschungsgemeinschaft We 943/1-1 and Ho 388/5-2 and the Verein zur Förderung und Bekämpfung rheumatischer Erkrankungen e.V. Bad Bramstedt.

References

KEMPSON, G.E. (1979). Mechanical properties of articular cartilage. In M.A.R. Freeman, *Adult articular cartilage.* Kent, England: Pitman Medical.

VIGNON, E. (1973). *Le vieillissement du cartilage articulaire et l'arthrose. Etude morphometrique.* Unpublished doctoral dissertation, University of Lyon.

WEH, L., Rehder, U., & Eggers-Stroeder, G. (in press). Viskoelastizität und Zellvolumendichte des Patellaknorpels. *Zeitschrift für Orthopädie.*

WEIBEL, E.R. (1979). *Stereological methods* (Vol. 1). London: Academic Press.

The Influence of Intraarticular Drugs on the Mechanical Properties of Articular Cartilage

B. Ebens, L. Weh, and G. Binzus
Orthopaedische Klinik der Universitaet Hamburg,
Federal Republic of Germany

Biochemical and clinical analyses of the influence of intraarticular drugs on cartilage properties are lacking in the literature. Variations in the mechanical properties of cartilage are scarce. As cartilage functions as a bearing surface, the effect of intraarticular drugs on cartilage mechanical properties would seem to be of importance. In a previous experiment we analyzed the changes in the surface hardness of arthrotic cartilage after incubating it in vitro (Weh, Fröschle, & Damen, 1981). The surface hardness does not, however, characterize the mechanical properties sufficiently. In the present study we used indentation tests to describe the mechanical properties of arthrotic articular cartilage. The cartilage samples were soaked in concentrated solutions of different drugs. The effects of the drugs were determined from measurements of the mechanical hysteresis and indentation. The human arthrotic cartilage samples used were obtained from eleven hip and six knee joints. The patients' ages ranged from 53 to 81 years. The withdrawal was done under sterile condition during operations by means of a punch. Cylinders of cartilage with subchondral bone were taken perpendicular to the surface of the hip and tibeal plateau. The spongiosa was separated up to a subchondral cortical lamella. The samples were 7 mm in diameter. The thickness was measured with a micrometer-screw and gave values from 2 to 4.5 mm. Only cartilage cylinders with an approximately constant height over the whole area were used for the evaluation. The macroscopic analysis of the samples, used in this investigation, showed little to moderate fibrous variations at the cartilage surface. Under addition of a nutritive solution the cartilage samples were subsequently sealed in foil of 5 μm thickness. Thus a drying of the cartilage was avoided. About 3 to 4 hours after withdrawal, the biomechanical analyses were performed.

For the present analysis a specially modified testing machine (Fa. Zwick, Ulm, Germany) was used. The load was applied through a 3 mm diameter plain indenter. The applied stress was 1.42 MN/m^2 and the indenter speed was 2×10^{-5} m/s.

The range of indentations expected was of the order 1 mm. A displacement measuring system consisting of a 5 kHz measuring amplifier (KWS3072, Hottinger-Baldwin Messtechnik) and an extensometer (W100) gave displacement to within $\pm$ 17 μm.

The quantitative analysis of the mechanical properties of human cartilage requires the investigation of the mechanical hysteresis as a character of the damping property of the cartilage and the indentation under load. The cartilage cylinders were cyclically loaded up to 10 N. The load was applied vertically to the cartilage surface. The loading and unloading phase followed immediately after one another. The time-dependent course of the force and the indentation were simultaneously recorded on an x-t recorder (ZSK 2, Fa. Rhode & Schwarz). At the end of the fifth load-cycle, the residual deformation was measured. After the first mechanical analysis, the samples were incubated in 25 ml solution, consisting of Hank's solution, calf-serum, antibiotics, and the applied drugs in different concentrations. The drugs were added in the following concentrations:

0. Empty series (as comparison)
1. Dilution 1/1024 ampulla
2. Dilution 1/256 ampulla
3. Dilution 1/64 ampulla
4. Dilution 1/16 ampulla
5. Dilution 1/4 ampulla
6. Dilution 1/1, equivalent 1 ampulla

Applied Drugs

(1). ARTEPARON (R) FORTE (Luitpold-Werke)
Compound: 1 amp equivalent 2 ml consists of 125 mg Mucopolysaccharide-polyschwefelsäureester

(2). DONA (R) 200 S (Opfermann Arzneimittel)
Compound: 1 amp equivalent 2 ml consists of 400 mg D-Glucosaminsulfate, Lidocain HCL 10 mg

(3). NEYARTHROS (R) III (Vitorgan Arzneimittel)
Compound: consists of high molecular cell-extracts, 2 10^{-6}g, Natriumlauryl-sulfate 20 μg

(4). DIPROSONE (R) DEPOT (Byk-Essex)
Compound: 1 amp equivalent 1 ml consists of Beta-methasone-17, 21-dipropinat 6.43 mg, Betamethasone 21-dinatriumphosphat 2.63 mg

(5). PEROXYNORM (R) (Grünenthal GmbH)
Compound: 1 amp equivalent 4 mg consists of 12 mg dry substance, 4 mg Orgoteine and 8 mg Saccharose

After 12 days, the mechanical tests were repeated.

Results

A nonlinear regression analysis was performed on the measures of hysteresis, indentation, and residual deformation. The statistical analysis was performed by a micro-computer (HP 9835A). The results show the percentage changes in the cartilage mechanical properties as a result of the treatments in the different drugs and drug concentrations.

With ARTEPARON (R) FORTE a considerable increase of the plastic component was found at high concentrations. The percentage variation of the indentation and residual deformation revealed a similar pattern.

DIPROSONE (R) DEPOT caused a distinct, significant change of the hysteresis. At high concentrations an increase of the plasticity was calculated, and at lower con-

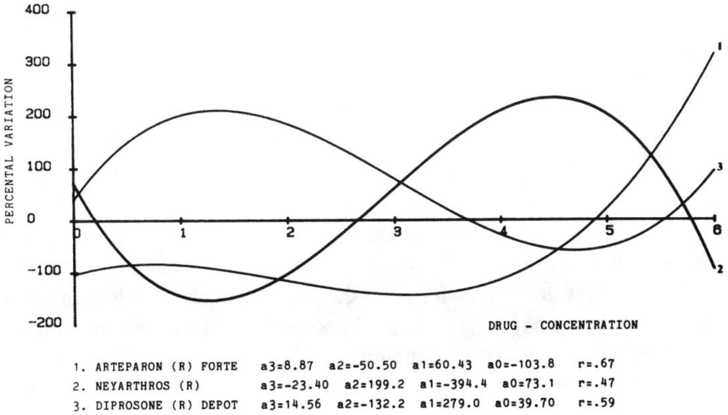

Figure 1—Concentration-dependent percentage variation of the mechanical hysteresis.

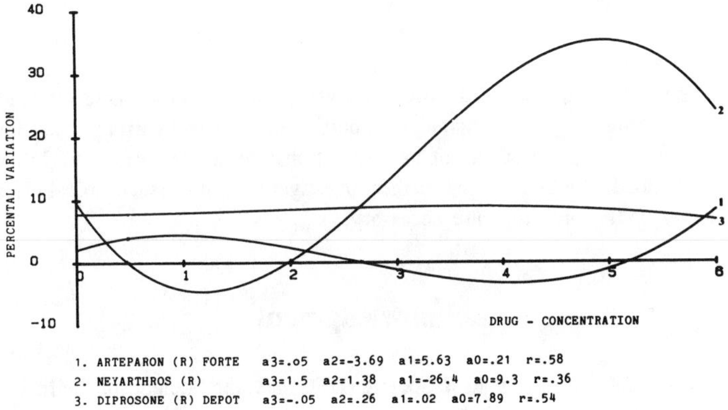

Figure 2—Concentration-dependent percentage variation of the indentation.

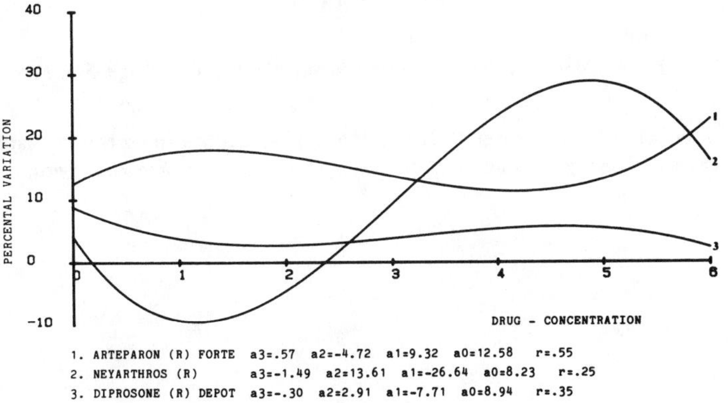

Figure 3—Concentration-dependent percentage variation of the residual-deformation.

centrations an increase of the elastic component was calculated. The variation of the indentation and residual deformation are classified as small.

NEYARTHROS (R) generally increased the plastic properties at high concentrations and increased the elasticity at low concentrations. The same phenomena were found from the indentation and residual deformation. The incubations with PEROXYNORM (R) and DONA (R) 200 S did not cause a significant change of the mechanical properties and thus no relationship is shown in the figures.

In the present study, we tried to investigate alterations of the mechanical properties of arthrotic articular cartilage under the influence of intraarticular drugs in vitro. The question is, can the results be transferred to the physiological conditions in vivo? The conditions during the 12-day incubation are important. In the present study, it was not possible to simulate the alternating pressure, which is necessary for the articular cartilage, while retaining sterile conditions. It has been found that the hydrostatic pressure induces the differentiation of the hyaline cartilage. The cartilage metabolism takes place under hypoxic conditions. Marcus (according to Sokoloff & Malemud, 1973) tested the influence of the partial oxygen pressure in chondrocyte cultures and found no influence on the proliferation of chondrocytes with a 7% lowering of the oxygen concentration. The fibrocyte growth, however, decreased about 20%. In our investigation, atmospheric oxygen pressure existed over the drug-solution. This means an unphysiologic oxygen supply to the cartilage.

It can be summarized that in the present investigation a considerable influence of intraarticular drugs on the mechanical properties of articular cartilage was found. Although the methods, especially of the incubation, are problematic, drug influence was demonstrated. We believe that further investigations on the mechanical effect of intraarticular applicable drugs are necessary.

Acknowledgment

Supported by the German Scientific Society (DFG) and the Society for the Investigation of Rheumatic Diseases, Bad Bramstedt, Germany.

References

SOKOLOFF, L., & Malemud, C. (1973). In vitro culture of articular cartilage. *Fed. Proc.*, **32**, 1502.

WEH, L., Fröschle, G., & Dahmen, G. (1981). Einfluss intraarticular applizierbarer Pharmaka aud die mechanischen Eigenschaften des Gelenknkorpels in-vitro. *Acta Rheumatologica*, No. 5.

Response of the Articular Facet Joint to Axial Loads

K.H. Yang, C.R. Tzeng, and A.I. King
Wayne State University, Detroit, Michigan, USA

The load-bearing role of the articular facets was first quantified by Prasad and King (1974). Lumbar and thoracic vertebrae of embalmed cadavers were instrumented with strain gauges and an intervertebral body load cell (IVLC) was inserted at the L4 level to measure the load borne by the disc at that level during whole-body caudo-cephalad acceleration. It was found that a significant portion of the load was borne by the facets and that spinal curvature affected the magnitude of the facet load. This load was computed from the difference of the total load at the IVLC level and the disc load measured by the IVLC. However, an assumption was needed to compute the total load. It was taken to be equal to the product of the mass of the torso above the IVLC and the applied acceleration. Subsequently, Hakim and King (1976) performed tests on isolated lumbar segments and demonstrated the existence of facet loads up to 30% during the application of static as well as dynamic loads. Adams and Hutton (1980) found that they carried approximately 16% of the applied axial compression in the erect standing posture.

In addition to these bioengineering studies, there have been many anatomical and clinical studies which provide indirect evidence that low back pain (LBP) can be attributed, in part, to the facet joint. Pedersen, Blunck, and Gardner (1956) found that the joint capsule of the facets was richly innervated with nerve endings. Wyke (1980, 1982) reviewed the neurology of LBP and proposed that one of the causes of LBP was the stimulation of receptor endings in the joint capsule of the facets. Mooney and Robertson (1976) found that the facet joint can be a persistent contributor to the chronic pain complaints of individuals with low back and leg pain. He also cited a review of 2500 consecutive lumbar disc operations. There was complete relief of both sciatica and low back pain in only 60% of the patients.

The aim of this paper is to make direct measurements on the stiffness of the facet joint as a first step towards the understanding of the mechanism of facet load transmission. A series of experiments was conducted on isolated facet joints subjected to compressive and tensile loads at different rates of loading.

Methods

Unembalmed cadaveric specimens of the lumbar spine were dissected into segments consisting of two vertebrae and one disc—a functional spinal unit (FSU). The posterior structure of the FSU was dissected free from the vertebral bodies at the pedicles. The facet joints were left intact. The superior and inferior ends of the specimen were imbedded in Ostalloy 117, a low melting-point alloy (m.p. = 47°C) to facilitate the application of load on the facet joints. To ensure that the joint was aligned in its natural position, the superior end of the specimen was imbedded in a cup of Ostalloy first. After the metal had set, it was suspended from the cross-bar of an Instron testing machine with the inferior end hanging freely. This end was then placed in an empty cup which was attached to the loading ram of the Instron. Molten alloy was then poured into the cup and the specimen was aligned in the desired position before the metal cooled. A Steinmann pin was placed through both ends of the specimen and the cup walls to ensure that the applied load was transmitted through the facet joints. The joints were maintained in a moist condition during testing by spraying normal saline solution on the exterior surfaces of the facet capsules.

Three different types of experiments were performed:

(1) axial loading
(2) axial loading with the posterior ligaments cut
(3) loading to failure.

A total of ten different facet joints was tested. In Experiments 1 and 2, the specimens were tested several times in tension and compression at three different rates of loading, 0.0125, 0.125, and 1.25 mm/s. In general, the displacements were small for these two experiments. In the third experiment, large displacements were encountered before tensile or compressive failure occurred.

Results

The results are presented in the form of load-deflection curves. Figure 1 shows typical curves obtained in the first and second experiment during compressive loading of the

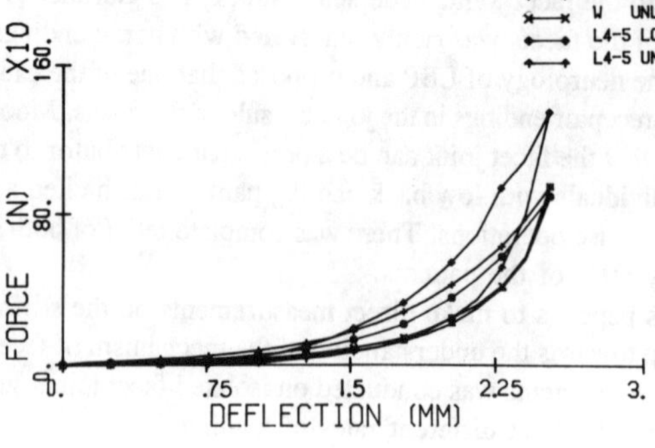

Figure 1—Load-deflection curve of L1-L2 (with and without posterior ligaments) and L4-L5 facet joints.

facets. The non-linear behavior and the hysteresis loops are of interest. It should be noted that the L4-L5 facet joint is stiffer than that of the L1-L2. Also the soft tissue between adjacent laminae contribute nothing to joint stiffness in compression.

Figure 2 reflects a typical response of the facet joint in tension. Its tensile strength without ligamentous support is considerably lower than that with all ligaments intact. The results were not sensitive to the loading rates used.

When the facet joint was loaded to failure in compression, it was due to capsular rupture without bony failure. As shown in Figure 3, the stiffness reached a maximum value of about 800 N/mm at a failure load of approximately 6 kN. Capsular failure also occurred in tension with a major portion of the load borne by the ligaments and soft tissue between adjacent laminae.

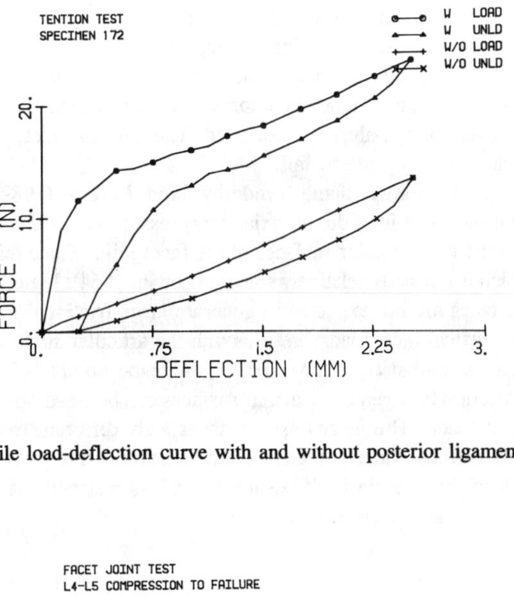

Figure 2—Tensile load-deflection curve with and without posterior ligaments.

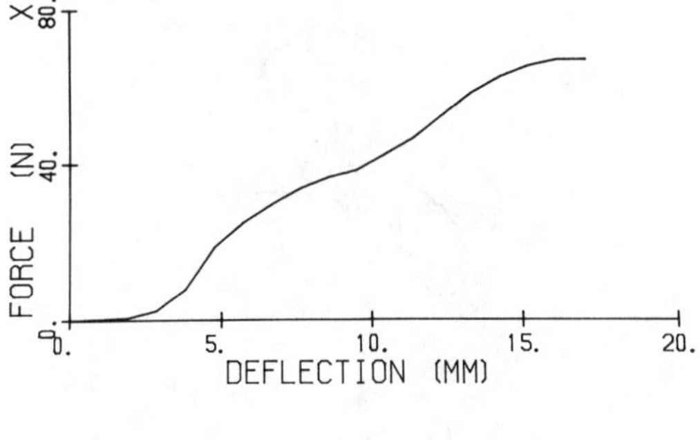

Figure 3—Load-deflection curve of an L4-L5 facet joint loaded to failure in compression.

Discussion

In the first 2 experiments, the facet joint deformation was 2 to 3 mm. The relative rotation of the articulating elements could not be measured optically. Hence, a compression test can represent spinal extension and a tension test is representative of spinal flexion. Figure 4 shows a cut-away view of the inferior facet in contact with the pars interarticularis. Although it is not possible to establish the precise regions of contact and the magnitude of the contact pressure, it can be assumed that the inferior tip of the inferior facet bottomed out on the pars of the vertebra below it, creating stiffness values in excess of 700 N/mm. It can also be postulated that the area of mutual contact is rather small and the unit pressures are very high. Adams and Hutton (1980) reported finding joints which were denuded of articular cartilage and had developed osteophytes.

Stiffness data in compression show large variations from subject to subject. This supports the hypothesis that the generation of facet load is highly independent on the geometry of the pars and of the facets. In general, for all specimens tested, the initial stiffness is relatively low and is primarily a compressive response of the soft tissue between the facet and the pars. Additional compression brings about a dramatic increase in stiffness, indicating that there is now 'bony' contact via a thin intervening layer of soft tissue. The failure mode in compression is also instructive. The facet joint is torn open before any bony failure has occurred. The inferior facets pivot about their inferior tips to cause the capsule to fail.

The recent study by Patwardhan, Vanderby, and Lorenz (1982) is somewhat misleading in that the facet load due to axial compression is not the same as contact pressure between the two articular surfaces of the facet joint. Since this joint is a true synovial joint which is normally relatively smooth (Lewin, 1964), high contact pressures normal to the surfaces are not expected to generate large frictional forces which can be interpreted as vertical facet loads, even though the articular surfaces tend to lie in a vertical plane in the lumbar region. Moreover, the response in tension is completely different. Thus, friction between the articular surfaces can be ruled out as a mechanism of generating facet loads. The mechanism is necessarily different from that in compression. The low stiffness values in tension can be attributed to soft tissue response; that is the stretch of the capsule itself, which is far less resistant than bony contact, in the case of compressive facet loads.

Figure 4—A cut-away view of a facet joint bottoming out against the pars interarticularis.

Conclusions

Facet joint stiffness in compression is much higher than that in tension.

The force deflection curves for the facet joint in tension and compression are both non-linear.

Failure of the facet joints occur primarily in the capsular ligaments for both tensile and compressive loads. Subluxation appears to be a tensile mode of failure.

The stiffness value of L4-L5 is slightly higher than that of L1-L2 for the same subject.

Acknowledgment

This research was supported in part by NIH Grant No. GM 20201 and in part by a Center of Excellence Award from Wayne State University.

References

ADAMS, M.A., & Hutton, W.C. (1980). The effect of posture on the role of the apophyseal joints in resisting intervertebral compressive force. *Journal of Bone and Joint Surgery.* **62B**, 358-362.

HAKIM, N.S., & King, A.I. (1976). Static and dynamic articular facet loads. *Proceedings of The 20th Stapp Car Crash Conference* (pp. 607-639).

LEWIN, T. (1964). Osteoarthritis in lumbar synovial joints. *Acta Orthopedica Scandinavica,* Suppl. 73.

MOONEY, V., & Robertson, J. (1976). The facet syndrome. *Clinical Orthopedics and Related Research,* pp. 149-156.

PATWARDHAN, A., Vanderby, Jr., R., & Lorenz, M. (1982). Load bearing characteristics of lumbar facets in axial compression. *1982 Advances in Bioengineering, ASME,* pp. 155-160.

PEDERSEN, H.S., Blunck, C.F.J., & Gardner, E.D. (1956). The anatomy of the lumbosacral posterior rami and meningeal branches of spinal nerves (sinuvertebral nerves): with an experimental study of their functions. *Journal of Bone and Joint Surgery,* **38A**, 377.

PRASAD, P., King, A.I., & Ewing, C.L. (1974). The role of articular facets during $+G_z$ acceleration. *Journal of Applied Mechanics,* **41**, 321-326.

WYKE, B. (1980). The neurology of low back pain. In M.I.V. Jayson (Ed.), *The Lumbar Spine and Back Pain* (2nd ed.) (pp. 265-339). Pitman Medical, Turnbridge Wells, Kent.

WYKE, B. (1982). Receptor systems in lumbosacral tissues in relation to the production of low back pain. In A.A. White, III, & S.L. Gordon (Eds.), *Symposium on Idiopathic Low Back Pain, AAOS* (pp. 97-107). C.V. Mosby Company, St. Louis.

Kinematics of the Elbow

K.N. An, B.F. Morrey, and E.Y. Chao
Mayo Clinic/Mayo Foundation, Rochester, Minnesota, USA

Kinematic data of human joints under normal conditions is essential for prosthesis design. Kinematics of the elbow joint have been studied experimentally by many investigators. Unfortunately, due to the limitations in experimental techniques and ambiguities in the definitions of measurements, some controversies have been created (Amis, Dowson, Unsworth, Miller, & Wright, 1977; London, 1981; Morrey & Chao, 1976). The variation of carrying angle during elbow joint flexion has been described as either following an oscillatory pattern (Dempster, 1955) or simply varying linearly from valgus to varus positions (Morrey & Chao). More recently, the carrying angle has even been reported to be continuously in the valgus position throughout flexion motion (London). In addition, the axial rotatory motion of the ulna during elbow flexion was observed to be significant by one group of investigators and to be negligible by others. In this study, these experimental results are reexamined based on a rigorous analytic approach. Hopefully, these discrepancies can be resolved.

Methods

In the analysis, it is assumed that flexion-extension of the elbow joint occurs about a single axis (Z_2 or Z_3) which passes through the center of the concentric arcs formed by the trochlear sulcus and the capitellum (London, 1981). Proximally, a line perpendicular to this axis of rotation makes a lateral opening angle, α, with the long humeral axis in the frontal XZ plane. Distally, this perpendicular line also makes an oblique angle, β, with the long ulnar axis in the frontal XZ plane (see Figure 1). Assume θ to be the flexion angle of the elbow joint which takes place at the trochlear-proximal ulnar joint. The relationship between the coordinate systems defining the proximal humerus (X_1, Y_1, Z_1) and distal ulna (X_4, Y_4, Z_4) could be obtained by a series of consecutive coordinate transformations,

$$\begin{vmatrix} X_4 \\ Y_4 \\ Z_4 \end{vmatrix} = \begin{vmatrix} C\beta C\theta C\alpha - S\beta S\alpha & C\beta S\theta & -C\beta C\theta S\alpha - S\beta C\alpha \\ -S\theta C\alpha & C\theta & S\theta S\alpha \\ S\beta C\theta C\alpha + C\beta S\alpha & S\beta S\theta & -S\beta C\theta S\alpha + C\beta C\alpha \end{vmatrix} \begin{vmatrix} X_1 \\ Y_1 \\ Z_1 \end{vmatrix} \quad (1)$$

where C = Cos and S = Sin.

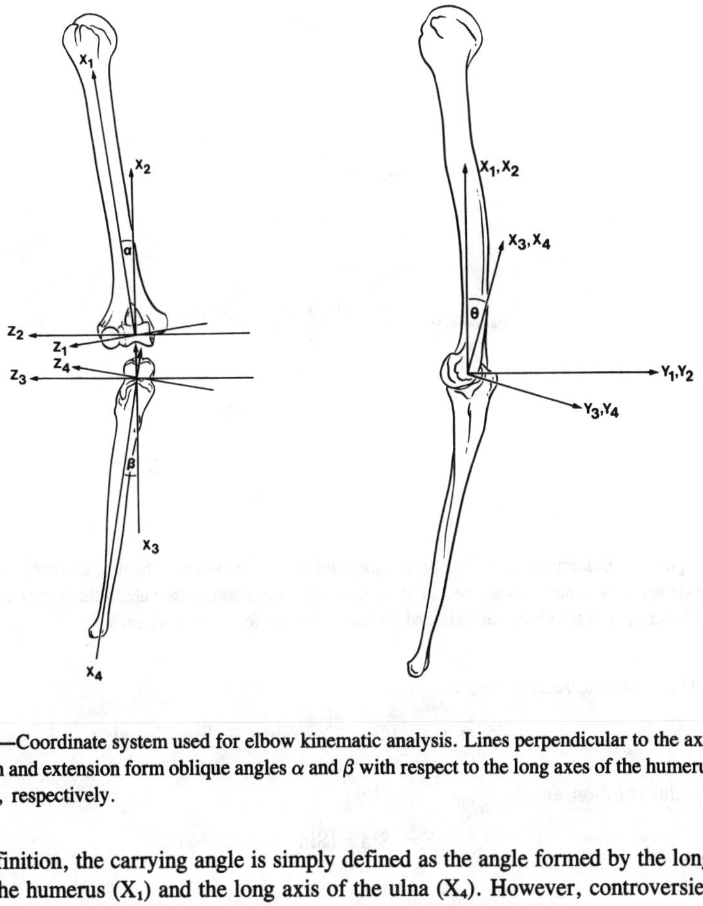

Figure 1—Coordinate system used for elbow kinematic analysis. Lines perpendicular to the axis of flexion and extension form oblique angles α and β with respect to the long axes of the humerus and ulna, respectively.

By definition, the carrying angle is simply defined as the angle formed by the long axis of the humerus (X_1) and the long axis of the ulna (X_4). However, controversies have arisen in the past due to inconsistencies in measurement methods.

First the carrying angle can be defined as the angle formed by the long axis of the humerus and the long axis of the ulna measured in the plane containing the axes of joint flexion and the humerus. In other words, the angle is measured by constantly viewing it in a direction perpendicular to the humerus (i.e., toward Y_1). Mathematically, this angle, CA_1, can be obtained by projecting X_4 on the X_1Z_1 plane and measuring the angle formed with the X_1 axis.

$$CA_1 = \tan^{-1} \frac{\text{Cos}\beta\ \text{Cos}\theta\ \text{Sin}\alpha + \text{Sin}\beta\ \text{Cos}\alpha}{\text{Cos}\beta\ \text{Cos}\theta\ \text{Cos}\alpha - \text{Sin}\beta\ \text{Sin}\alpha} \quad (2)$$

Secondly, the carrying angle has been defined as the abduction-adduction angle of the long axis of the ulna with respect to the long axis of the humerus when using the Eulerian angle approach (see Figure 2). For finite three-dimensional angular displacement of the ulna about the humerus, the three Eulerian angles are defined as flexion-extension about Z_1; forearm axial rotation about X_4; and abduction-adduction rotation about the axis orthogonal to both the Z_1 and X_4 axes. Following this definition, the transformation matrix relating systems 1 and 4 can be obtained in terms of these three Eulerian angles. The individual Eulerian angles can thus be obtained,

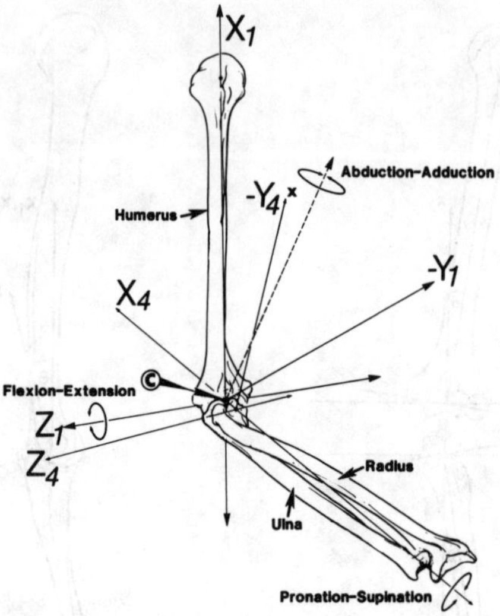

Figure 2—Eulerian angle description of the relative motion between the ulna and humerus. Flexion-extension takes place about the Z_1 axis and axial rotation takes place about the X_4 axis. Abduction-adduction rotates about an axis orthogonal to both the Z_1 and X_4 axes.

Abduction-adduction angle,

$$\phi = \sin^{-1} [\text{Cos}\beta \text{Cos}\theta \text{Sin}\alpha + \text{Sin}\beta \text{Cos}\alpha] \tag{3}$$

Axial rotation angle,

$$\psi = \text{Sin}^{-1} [\text{Sin}\theta \text{Sin}\alpha / \text{Cos}\phi]. \tag{4}$$

Results

For demonstration of the variation of carrying angle during elbow joint flexion, the anthropometric data of these two oblique angles, α and β, from Lanz (Lanz & Wachsmuth, 1959) were used.

Based on the first definition of measurement, the carrying angles were calculated and displayed as function of the joint flexion angle (see Figure 3). For the sake of easier interpretation, the reference line for the angle measurement in Figure 3 was shifted when the angle, CA_1, reached values beyond 90°. Within ranges of both small flexion angles (up to 50°) and large flexion angles (130° and above), the carrying angles are relatively constant with changing flexion angles. There is a transition region near 90° of elbow flexion where CA_1 changes rapidly with the joint flexion angle. With certain combinations of the α and β angles (e.g., α = 7.8°, β = 8.5°), the carrying angles do not reach either 0° or 180°. In other words, the ulna will stay on the lateral side of the humerus throughout the range of elbow flexion.

The carrying angle, measured by the abduction-adduction angle, decreases progressively as the elbow joint is flexed (see Figure 4). When α is large, this carrying

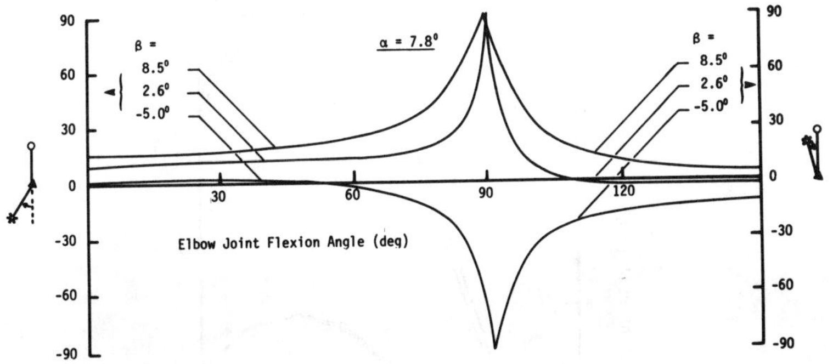

Figure 3—Carrying angle measured by projecting the ulna into the plane of the humerus and plotted as a function of the joint flexion angle. The relationship changes with changing values of α and β.

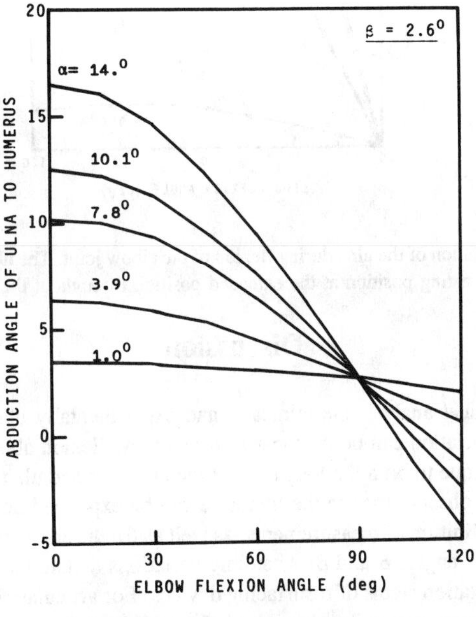

Figure 4—Carrying angle measured by the Eulerian abduction-adduction angle of the ulna with respect to the humerus and plotted as a function of the joint flexion angle.

angle could change from 15° valgus in the extended position to 5° varus in the fully flexed position. However, when α is small, the carrying angle is almost constant throughout the range of flexion.

Axial rotation of the ulna during flexion of the elbow joint was calculated by Equation 4 and is displayed in Figure 5. The ulna rotates internally with respect to its resting position at the extended position through the range of flexion. Internal rotation of the ulna progressively increases up to 90° of elbow flexion and decreases thereafter.

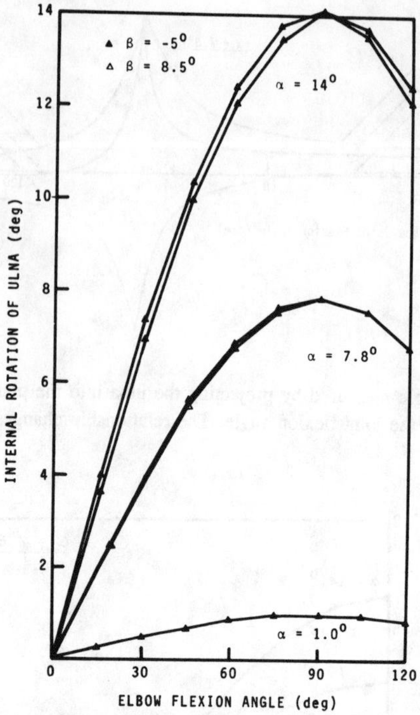

Figure 5—Axial rotation of the ulna during flexion of the elbow joint. The ulna rotates internally with respect to its resting position at the extended position throughout the range of flexion.

Discussion

Based on theoretical analysis,the clinically and experimentally observed patterns of carrying angle variation can be demonstrated by elbow flexion about a single hinged axis which is oblique to both the long axes of the humerus and ulna. From this study, the controversial observations in the literature can be explained due to (a) the inconsistency in the definition of measurements, as well as (b) the anthropometric variations of the two oblique angles α and β. Anatomic variations dictate the orientation of the intramedullary fixation stems of resurfacing devices. For articulating semiconstrained prostheses, the humeral stem is not usually influenced by these considerations, but the ulnar device must follow the contour of the bone. Therefore, in conclusion, we believe that for both types of elbow prosthesis designs, a uniaxial hinged consideration is adequate. However, the oblique alignment of the axis with both stems for proximal humeral and distal ulnar fixations should be properly incorporated, especially for resurfacing devices.

Acknowledgment

This study is supported by NIH Grant AM 26287.

References

AMIS, A.A., Dowson, D., Unsworth, A., Miller, J.H., & Wright, V. (1977). An examination of the elbow articulation with particular reference to variation of the carrying angle. *Engineering in Medicine,* **6**, 76-80.

DEMPSTER, W.T. (1955). *Space requirements of the seated operator: Geometrical, kinematic and mechanical aspects of the body with special reference to the limbs.* WADC Technical Report.

LANZ, T. von, & Wachsmuth, W. (1959). *Praktische Anatomie.* Berlin, Gottingen, Heidelberg: Springer-Verlag.

LONDON, J.T. (1981). Kinematics of the elbow. *Journal of Bone and Joint Surgery,* **63A**, 529-535.

MORREY, B.F., & Chao, E.Y. (1976). Passive motion of the elbow joint. *Journal of Bone and Joint Surgery,* **58A**, 501-508.

The Identification of Force Constraint Functions for Human Joints Under Dynamic Conditions

A.J. Komor
Institute of Sport, Warsaw, Poland

The proper definition of force constraint functions affecting human joints (the maximum available muscular torques) is an important problem in simulation and optimization of motion in various sports disciplines and in some rehabilitation cases. From the point of view of optimum control theory the optimal solution is entirely dependent on proper formulation of constraint functions. In biomechanics many examples exist of models of human mechanical systems for optimization purposes with no constraints (Morawski & Wiklik, 1978; Remizov, 1980) or with constraints in very simplified forms, for example, "step" or static (Gawronski, 1975; Ghosh & Boykin, 1976). Usually the results obtained from these simulations are presented without any discussion of the adequacy of the assumed constraints.

This work presents results of experiments and simulations concerning the proper choice of force constraint functions which approximate the natural properties of muscle groups acting under dynamic conditions.

Methods

The investigations consisted of an experimental and a simulation part. Extension of the knee joint was taken as an example of motion.

Experiment

The experimental part concerned the measurement of the kinematic parameters of a natural motion; changes in knee angle α and angular velocity ω and integrated EMG activity of the main muscles of both the extensors (RF, VM, VL) and the flexors (SM, ST). The subjects under investigation were asked to perform the motion using a maximum velocity-minimum time criterion. The measurements of the dynamics were repeated for various additional loads. The maximum muscular torque limits under static conditions were also measured. The measurements were performed by computer-aided test equipment (see Figure 1).

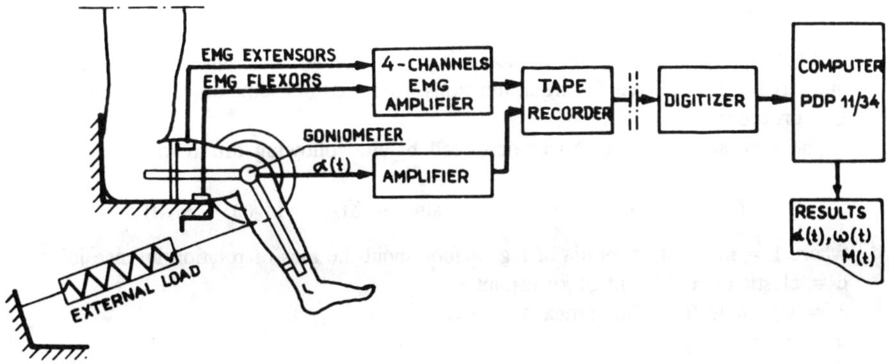

Figure 1—Test equipment. The results obtained from all the experiments enabled the formulation of three dimensional characteristics of maximum muscular torque limits of a subject under investigation.

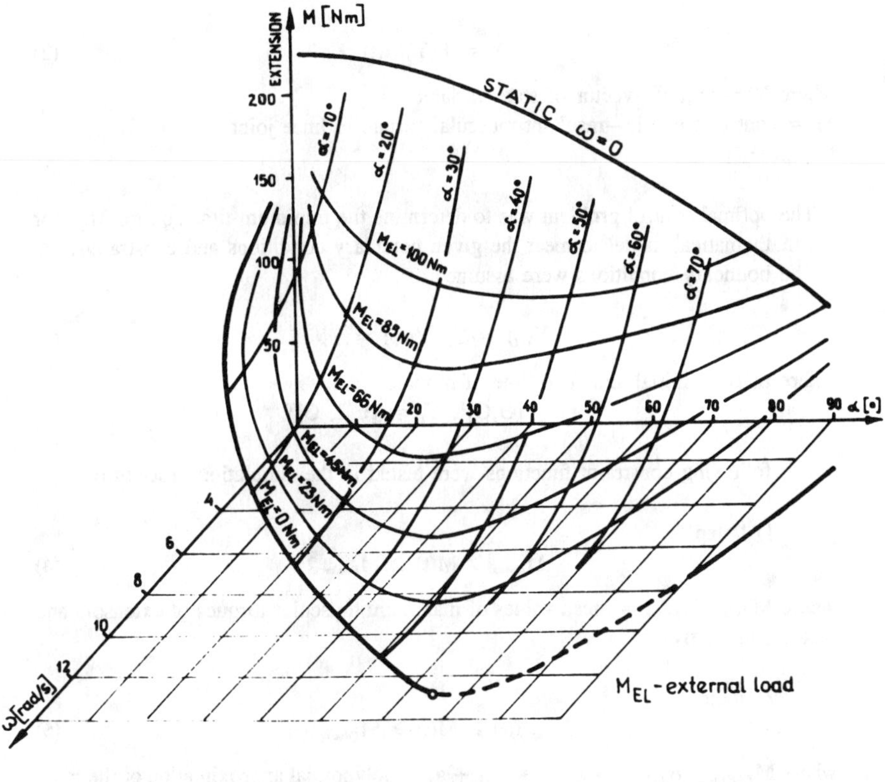

Figure 2—Characteristics of maximum muscular torque limits.

Simulation

The simple mathematical model of motion in the knee joint, the minimum time criterion, and three forms of force constraint functions were used in a simulation and optimization procedure.

The motion in knee joint was described by the following differential equation:

$$I \bullet \ddot{\alpha} + c \bullet \alpha + m \bullet g \bullet s \bullet \sin\alpha + M_{EL} = M_E(t) - M_F(t) \qquad (1)$$

where I = moment of inertia of leg + foot about the axis of rotation in knee joint
c = elasticity coefficient of knee joint
α = leg angle from the vertical
m = mass of leg + foot
g = acceleration due to gravity
s = location of center of gravity of leg + foot
M_{EL} = additional external load (torque)
M_E = torque of extensors
M_F = torque of flexors

Equation 1 was transformed to the standard state vector form:

$$\dot{Y} = F(Y,M,t) \qquad (2)$$

where Y = $[\alpha,\omega]^T$—vector of state variables
M = control variable—resultant muscular torque in knee joint
t = time

The optimal control problem was to determine the minimum-time control M(t) for the mathematical model to meet the given boundary conditions and constraints.

The boundary conditions were assumed as:

$$Y(t_I) = Y_I;\ Y(t_F) = Y_F \qquad (3)$$

where t_I, t_F = initial and final time of motion

$$Y_I = [O,O]^T;\ Y_F = [\pi/2,\ O]^T$$

The following constraint functions were tested in the simulation procedure:

1) "step"

$$M_{Emax} \geqslant M(t) \geqslant M_{Fmax} \qquad (4)$$

where M_{Emax}, M_{Fmax} = mean values of maximum muscular torques of extensors and flexors, respectively;

2) static

$$M_{Emax}\ (\alpha) \geqslant M(t) \geqslant M_{Fmax}\ (\alpha) \qquad (5)$$

where $M_{E(F)max}\ (\alpha) = a_1 \bullet \alpha^2 + a_2 \bullet \alpha + a_3$ = polynomial approximation of the maximum torque limits of the extensors and flexors; measured under static conditions

3) dynamic (Komor, 1981)

$$M_{Emax}\ (\alpha,\ \omega) \geqslant M(t) \geqslant M_{Fmax}\ (\alpha,\ \omega) \qquad (6)$$

where $M_{E(F)max}\ (\alpha,\ \omega) = a_1 \bullet \alpha^2 + a_2 \bullet \alpha + a_3)\ (1 - \frac{\omega}{\omega_{max}})$ = approximation of the maximum muscular torque limits of the extensors and flexors measured under dynamic conditions

ω_{max} = maximum angular knee joint velocity without any external load

The simulation was performed on a PDP 11/34 computer.

Two indices for comparing the experimental and simulation results were formulated:

$$I_1 = \int_{t_I}^{t_F} [Y_S(t) - Y_E(t)]^2\ dt \rightarrow \min \qquad (7)$$

where Y_S = vector of state variables from simulation

Y_E = vector of state variables from experiment

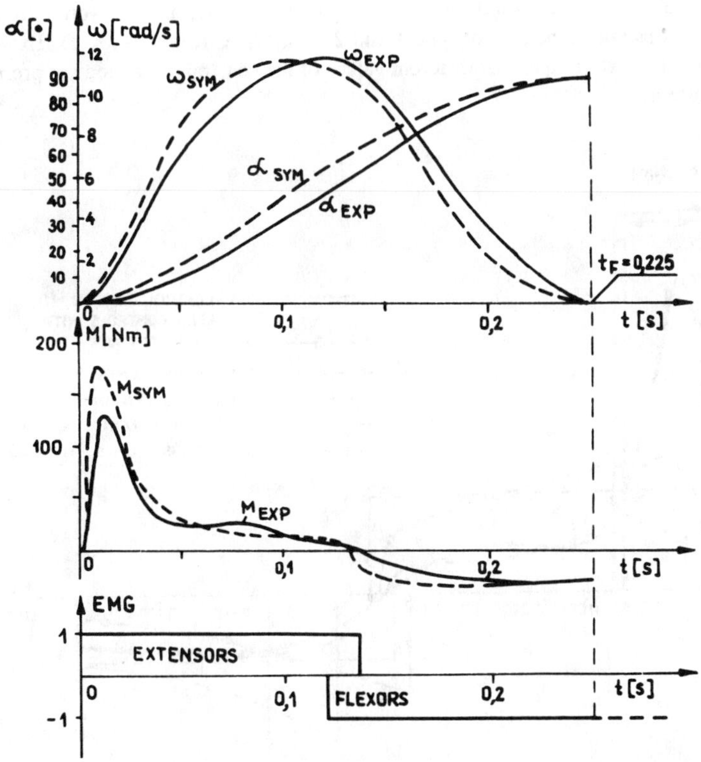

Figure 3—Example of simulation results in comparison with experimental results, M_{EL} = 0 Nm.

$$I_2 = \sum_{i=1}^{n} |t_{iE} - t_{iS}| \rightarrow \min \quad (8)$$

where t_i = switching (on or off) times from (E) experiment and (S) simulation
n = sequential number of switching action.

The minimum values of both indices in the simulation with given constraints were equivalent to the best fit of these constraints to the natural properties of the muscle groups.

Results

The simulation results showed that the final position was reached regardless of the type of constraint. The lowest values of the indices I_1 and I_2 were obtained for type 3 constraints. The simulation results for this case, in comparison with the experimental results, are presented in Figure 3.

Discrepancies in the switching action between the simulation and the experiment were observed with the use of the first and second types of constraint. In order to obtain the natural switching the additional weighting multipliers W_E and W_F were introduced into the constraint functions of type 1 and 2 ($0 < W_E < 1$, $0 < W_F < 1$). Examples of optimal control torques for different values of the constraint function are presented in Figure 4.

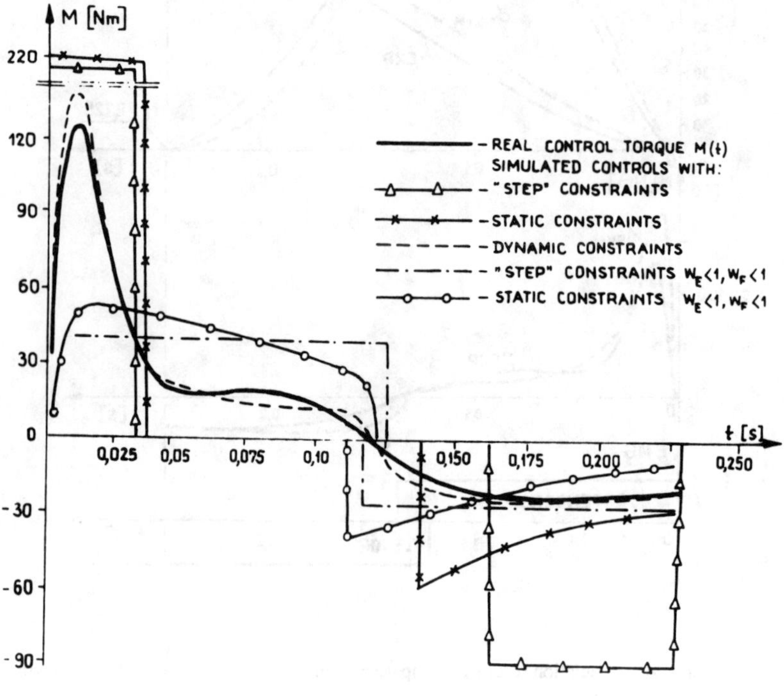

Figure 4—Example of optimal control torques for different variants of constraints, $M_{EL} = 0$ Nm.

Conclusions

The solution of any motion obtained in simulation (optimization procedure) is highly sensitive to the forms of force constraint functions. Dynamic constraints (type 3) seem to be the best, especially for the problems of optimization of motion in various sport disciplines. These applications require particularly precise determination of the model and constraint parameters. The forms 1 and 2, due to their simplicity and easy technical realization, may be used as characteristics of control actuators in prosthesis or manipulator design. Formula 2 with weighting multipliers is especially recommended when simplified but "natural" switching of control torques is required.

References

GAWRONSKI, R. (1975). *Problemy bioniki w systemach wielkich (Bionics of large systems).* Warsaw: MON.

GHOSH, T., & Boykin, W. (1976). Analytic determination of an optimal human motion. *Journal of Optimization Theory and Applications,* **19**, 327-346.

KOMOR, A., et al. (1981). Modelowanie technik ruchu (Simulation of motion techniques). *Report of Institute of Sport,* No. 105-06-13. Warsaw.

MORAWSKI, J., & Wiklik, K. (1978). Application of analog and hybrid simulation in sport. In *Proceedings of the IMACS Symposium* (pp. 251-254). North-Holland Publ.

REMIZOV, L. (1980). Optimal running on ski in downhill. *Journal of Biomechanics,* **13**, 941-945.

The Temperature Dependence of the Elastic Properties of Soft Biological Materials

M. Maes, W.F. Decraemer, and E.R. Raman
Laboratory for Experimental Physics, R.U.C.A.,
Antwerpen, Belgium

Although soft biological materials show a viscoelastic behavior, it is quite usual to talk about their elastic properties. We then mean the properties deduced from the so-called quasistatic force-length relation: this is the relation obtained by submitting the test specimen to relatively large, but slow, length changes.

We examined the influence of temperature on such force-length relations, obtained in uniaxial tension and after preconditioning the specimen.

Theoretical Model

A model based on the fiber structure of soft biological tissue is fitted to the experimental results (Decraemer, Maes, & Vanhuyse, 1980). Essentially it assumes that the tissue contains a large number (N) of fibers which bear the load. These fibers have equal cross-section (a) and Young's modulus (k) but different initial lengths (l_i) distributed normally with mean (μ) and standard deviation (s). It gives the following expression for the relation between force (K) and length (l):

$$K(l) = \frac{b}{\sqrt{2\pi}s} \int_0^l \frac{l - l_i}{l_i} \exp\left[- \frac{(l_i - \mu)^2}{2s^2}\right] dl_i \qquad (1)$$

Equation 1 contains 3 parameters: b, μ, and s, whose values are obtained from a nonlinear least squares fit.

Experimental Results

The quasistatic force-length relations for a specimen of human skin at different temperatures are shown in Figure 1. The temperature dependence of the corresponding parameters is shown in Figure 2. For other materials, such as tympanic membrane and tendon, the results are similar.

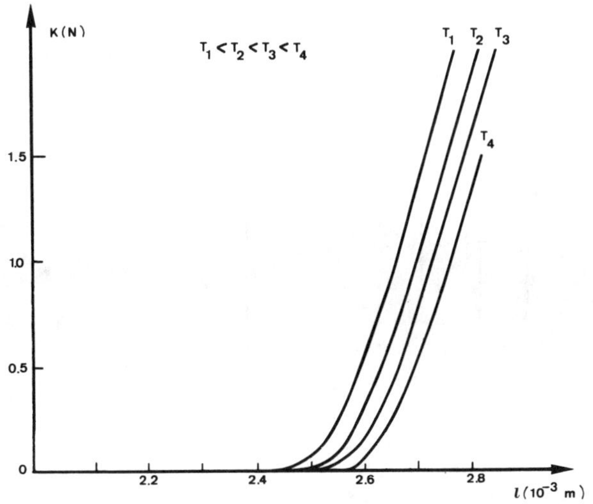

Figure 1—Quasistatic force-length relations at different temperatures for human skin.

Discussion

The experimental results fit our model provided we make the following suppositions:

—within the experimental temperature range, the Young's modulus and the cross-section of each fiber are constant.

—the length of each fiber increases linearly with temperature: $l_i = L_i\,(1 + \alpha\theta)$, with $\theta = T - T_{ref}$.

The first supposition implies that b (=N•a•k) is independent of temperature, which is in good agreement with the experimental values (see Figure 2A).

From the second supposition and from the definition of μ and s as the mean and standard deviation of the normal distribution of l_i values, it follows that

$$\mu = M\,(1 + \alpha\theta) \tag{2a}$$

$$s = S\,(1 + \alpha\theta) \tag{2b}$$

The μ values show a very good linear dependence on temperature (see Figure 2B), while for the s values, it is very difficult to draw any conclusion on their temperature dependence because of the uncertainty on their values (see Figure 2C).

Equation 1 can now be rewritten as:

$$K(\lambda,T) = \frac{b}{\sqrt{2\pi}S} \int_0^{\lambda L_o} \frac{\lambda L_o - L_i}{L_i} \exp\left[-\frac{(L_i - M)^2}{2S^2}\right] dL_i \tag{3}$$

where $\lambda = \frac{1}{l_o}$ (l_o: rest length of the strip at temperature T)

$l_o = L_o\,(1 + \alpha\theta)$

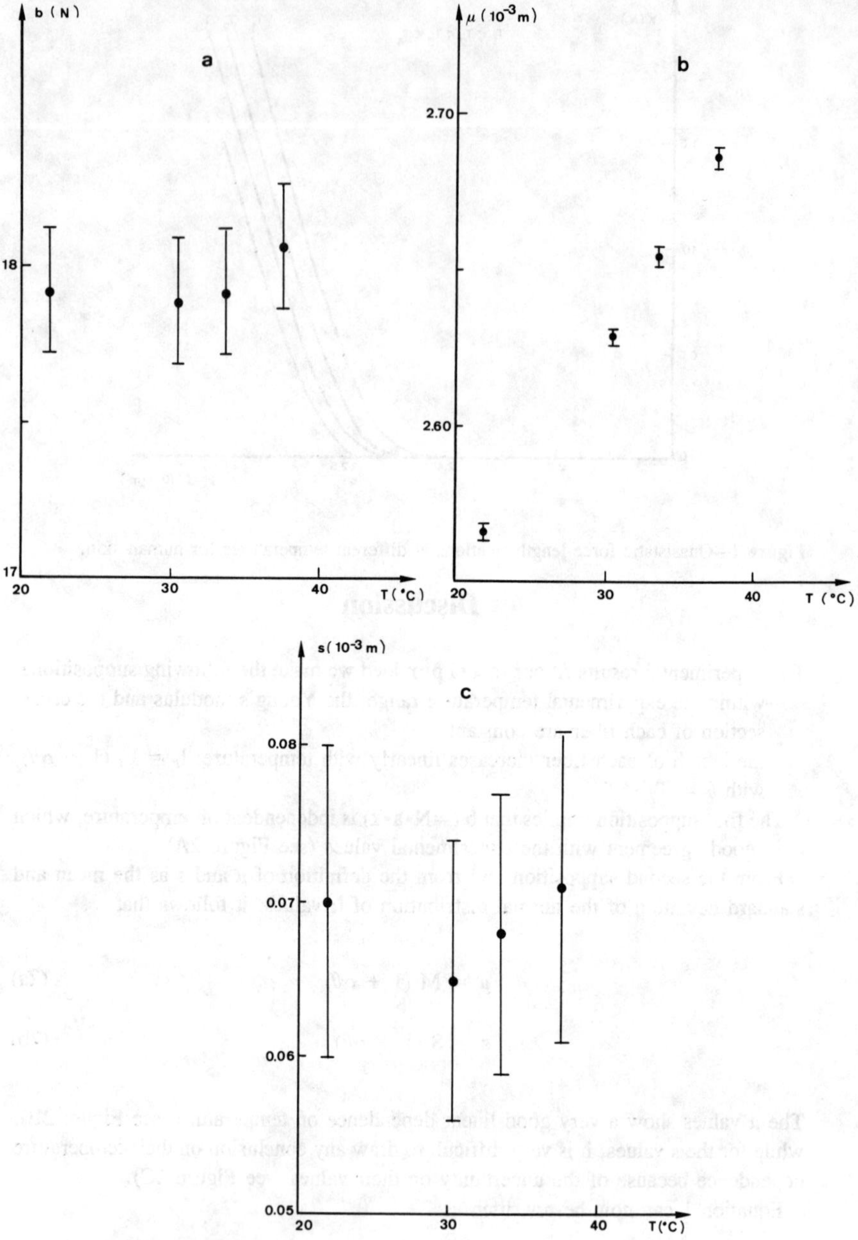

Figure 2—Temperature dependence of the parameters b, μ, and s.

Equation 3 shows that

$$K(\lambda,T)\Big|_{\lambda\,=\,cte} = K^*(\lambda) \qquad (4)$$

or the force at a given strain λ is independent of the temperature. This is in good agreement with the experimental results (see Figure 3).

Equation 4 provides us a simple method of maintaining a constant strain level at different temperatures, without any calculations and without knowing the rest length: by stretching the specimen until the force reaches the value K*, we have the same value for λ at l_1, l_2, l_3, and l_4 (see Figure 4). This is a very important result for dynamic measurements (small sinusoidally varying length changes superimposed on a larger initial strain). The dynamic properties of the examined biological tissues are strongly dependent on the initial strain level, so, in order to obtain comparable dynamic measurements at different temperatures, they should be performed at equal strain levels.

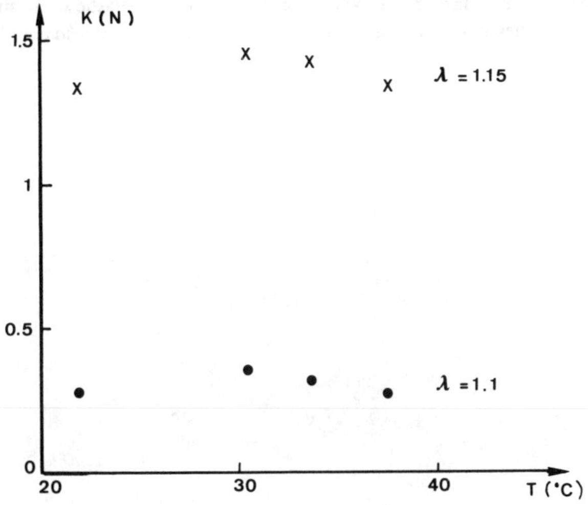

Figure 3—The force at constant strain in function of temperature.

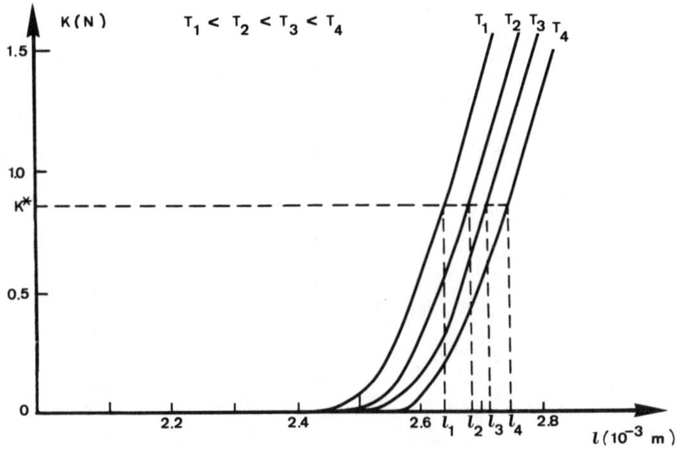

Figure 4—A method for determining the lengths l_1, l_2, l_3, and l_4, for which we have the same value for λ.

Conclusion

The force-length relation for soft biological tissues based on the structural model can be used at different temperatures provided that the length of each fiber is assumed to increase linearly with temperature. A method is proposed to maintain a constant strain level at different temperatures, without calculating the rest length of the specimen.

Reference

DECRAEMER, W.F., Maes, M., & Vanhuyse, V.J. (1980). An elastic stress-strain relation for soft biological tissues based on a structural model. *Journal of Biomechanics,* **13**, 463-468.

Foot Ligament Length and Elongation Estimation Using a Two Step Radiographic Reconstruction Technique

P. Allard, P.S. Thiry, and M. Duhaime
McGill University, Ecole Polytechnique,
and Sainte-Justine Hospital, Montreal, Canada

The human foot has a complex structural arrangement consisting of 30 intrinsic and extrinsic muscles and more than 100 ligaments connected to 26 irregularly shaped bones. These components interact together to maintain the stability of the foot. The ease with which the foot can adapt itself to different situations is well illustrated by its flexibility during heel-strike and its rigidity during heel-off (Sethi, 1977). Sprains, foot ailments, and disorders complicate the biomechanical interaction between muscles, ligaments, and bones. To better understand the normal foot mechanics, a biomechanical model was developed using data obtained from a standardized biplanar radiographic technique.

The location of the ligaments' insertion sites as well as the spatial position of the bones in the foot were computed from the radiographs. To estimate the length and elongation of the ligaments, the data collected were introduced in a model simulating various movements imposed around fixed articular axes of rotation defined after Hicks (1953).

Material and Method

The data collection technique and the model summarized in Figure 1 were inspired by that which Youm and Yoon (1979) developed for the hand. Their technique consists of arbitrarily encrusting three metallic spherical markers into each bone of the hand to uniquely identify its spatial location. The markers were also utilized in defining a coordinate reference system for each bone. Such technique eliminates the errors associated with the selection of a coordinated reference system defined by means of anatomical landmarks.

This tridimensional reconstruction technique can be readily applied to determine precisely the position of the insertion sites of the ligaments located in the foot's periphery. However, to locate the anchoring points of all the interosseous ligaments without disturbing the geometrical spatial relationship of the bone system, this technique had to be modified and adapted in an original manner.

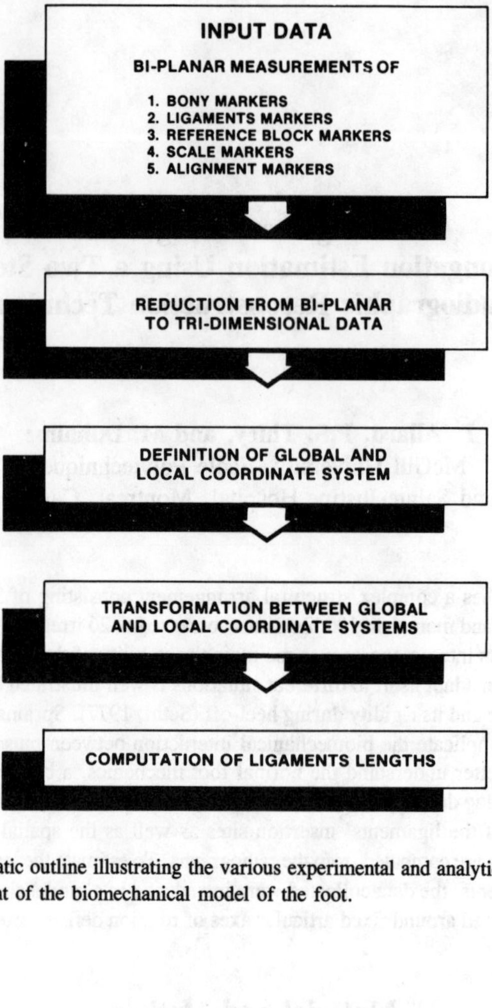

Figure 1—Schematic outline illustrating the various experimental and analytical steps involved in the development of the biomechanical model of the foot.

Data Collection

Three lower limbs amputated at mid-shank were utilized in this study. The specimens were kept frozen for preservation purposes. After thawing at room temperature for twelve hours, each foot was initially superficially dissected to explore its spatial location. The whole (attached) foot and shank were then positioned on a rotating platform allowing for anterior-posterior and lateral radiographs (Allard, Thiry, Duhaime, & Geoffroy, 1982). Afterwards, the foot was systematically dissected (detached) to locate the ligaments' insertions. The procedure consisted for each ligament in locating its insertions by inserting in the site a spherical marker identified by a color-coded pin. The detached bones were then returned to the rotating platform for another set of orthogonal radiographs. The locations of the markers' image on the radiographs were carefully recorded with an X-Y digitizer.

Model

In its current state, the model consists of fourteen osseous elements, namely, the tibia, fibula, seven bones of the tarsus and five metatarsals. The phalanges were not initially considered since they have already been exhaustively studied by Stokes, Hutton, and Scott (1979). Sixty ligaments and the plantar fascia are modeled as elastic elements. The deltoid, the long plantar ligament, the bifurcated ligament, the transverse metatarsal ligaments as well as the plantar fascia have several distal fibers. They have been included in the model as to have 61 proximal fibers and 74 distal fibers.

As shown in Figure 1, the mathematical manipulation required for the reduction of the experimental anthropometric data consists initially of a transformation from biplanar to tridimensional data by means of equations relevant to plane geometry. The coordinates determined by this method are in the global reference systems of the *detached bones* and of the whole foot (*attached bones*). Figure 2 illustrates the position of the bony markers, A, B, and C, on a bone. The attached global coordinates system (AGCS) of the foot is arbitrarily defined and located on the biplanar radiographs. The attached local coordinates system (ALCS) is located at the centroid g, of the triangle formed by markers ABC. The axes have been defined after Youm and Yoon (1979). The detached global and local coordinates systems (DGCS, DLCS) were respectively determined in a similar manner as for the attached systems.

This reduction of data is then followed by a transformation between the local and global coordinates reference systems. Five transformation equations related to the attached and detached coordinates systems allow for the reconstruction of the spatial position of each ligament, bone and Hicks' axes or rotation. These are: a translation of the DGCS to the centroid g of the DLCS, a rotation of the DGCS with respect to the DLCS, an identity equation between the DLCS and ALCS, an inverse rotation between the ALCS and the AGCS, and an inverse translation between the ALCS and the AGCS.

The computer program calculates the ligaments' length with the foot positioned at 90°. As an example of the potential application of such a model, a simulation was carried out for a full plantarflexion of 50° about the talocrural joint.

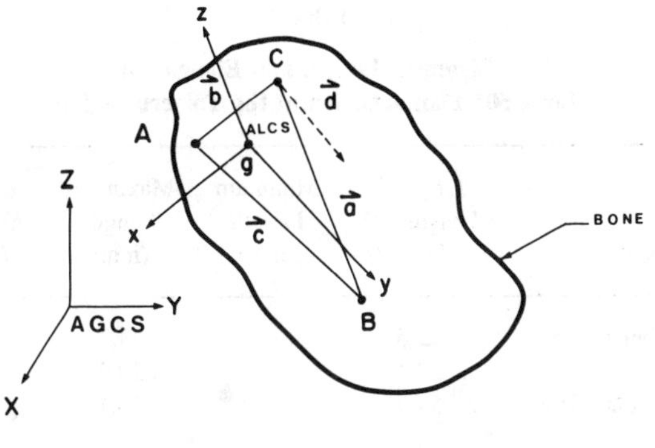

Figure 2—Schematic representation of the global (AGCS) and local (ALCS) attached coordinates systems.

Results and Discussion

Table 1 presents the computed length values as well as those measured or reported in the literature for the same ligaments. The computed values compare well to those measured experimentally.

Before presenting the results of the simulation, the problem associated with estimation of the ligament's initial length must be discussed. Each ligament has a length corresponding to its initial unstressed state (zero applied force). This length, called the initial length, needs to be known to calculate the ligament's elongation. This measurement of the initial length is a most complex problem. In this study, it has been defined as the length of the ligament corresponding to the foot's neutral position.

Table 2 presents the results obtained after a simulation of a single movement about the talocrural joint representing a 50° plantarflexion of the foot. It seems that the ligaments in the lateral side are the most important for maintaining the foot's stability since they are equally solicited, as shown by an elongation of approximately 2% to

Table 1

Calculated and Measured Length Values of Ligaments

Ligament Length	Calculated Length (mm)	Measured Length (mm)
Tibionavicular fibers	28	22
Anterior talofibular	23	21
Posterior talofibular	53	47*
Plantar fascia	147	150**

*Sosna & Sosna (1977)
**Lapidus (1943)

Table 2

Ligament Length and Elongation for a 50° Plantarflexion at the Talocrural Joint

Ligaments	Original Length 90° (mm)	Minimum Length (mm)	Maximum Length (mm)	Elongation Absolute (mm)	Elongation Relative (%)
Anterior talofibular fibers	22.6	—	23.2	0.6	2.6
Calcaneofibula	20.4	—	20.9	0.5	2.4
Posterior talofibular fibers	53.0	—	54.1	1.1	2.0
Deltoid					
Posterior talotibial fibers	65.8	63.0	—	−2.8	−4.3
Anterior talotibial fibers	33.6	27.1	—	−6.5	−19.4
Tibionavicular fibers	28.2	—	35.6	7.4	26.2
Calcaneotibial fibers	20.4	14.4	—	−6.0	−29.4

3%. On the medial side, most fibers of the deltoid ligament are inactive as illustrated in some cases by an important retraction. Only the tibionavicular fibers present a substantial elongation of 7.4 mm. These results may be explained by Hicks' (1953) observations that the plantarflexion movement is not purely occurring in the sagittal plane. There is also a contribution from a slight supination of the foot.

In this paper, a two-step radiographic reconstruction technique was presented and its application to a simple movement about the ankle joint was carried out to illustrate the potential of a ligamentous biokinematical model of the foot.

Acknowledgments

Financial support from the Friedreich's Ataxia Association of Canada and from Fonds de la recherche en santé du Québec, is gratefully acknowledged.

References

ALLARD, P., Thiry, P.S., Duhaime, M., & Geoffroy, G. (1982). Kinematics of the foot. *Canadian Journal of Neurological Sciences*, **9**, 119-126.

HICKS, H.J. (1953). The mechanics of the foot. I, The Joints. *Journal of Anatomy*, **87**, 345-357.

LAPIDUS, P.W. (1943). Misconception about the springiness of the longitudinal arch of the foot. *Archives of Surgery*, **46**, 410-444.

SETHI, D.K. (1977). The foot and footwear. *Prosthetics and Orthotics International*, **1**, 173-182.

SOSNA, T., & Sosna, A. (1977). Variability and functional significance of the external ligaments of the ankle for stability of the talocrural joint. *Folia Morphol*, **25**, 371-374.

STOKES, I.A.F., Hutton, W.C., & Scott, J.R.R. (1979). Forces acting in the metatarsals during normal walking. *Journal of Anatomy*, **129**579-590.

YOUM, Y., & Yoon, Y.S. (1979). Analytical development in investigation of the wrist kinematics. *Journal of Biomechanics*, **12**, 613-621.

Mechanical Properties of Some Connective Tissues and Their Components In Vitro

Y. Missirlis
University of Patras, Patras, Greece

The mammalian connective tissues are composite materials. Soft tissues, such as arteries, skin, and so forth, consist of collagen, elastin, mucopolysaccharides (MPS) and lipids as the main passive structural components. Erythrocyte membranes are composed of lipids, glucoproteins and structural proteins (spectrin, actin, ankyrin), not all of which are passive. A very fundamental question to be answered is the following: what is the contribution of each individual component to the overall mechanical characteristics of the whole tissue? A companion question asks: how is the structural integration of any two components expressed in their mechanical properties?

In an attempt to answer such questions we have studied the extensional (tensile, uniaxial) properties of the following tissues: thoracic aorta, aortic valve, skin, fascia lata, Achilles tendon, cornea, and erythrocytes from humans and animals.

Materials and Methods

Collagenous Tissues

Microtensile strips of 2 to 5 mm width and length-to-width ratio of 4:1 were cut from freshly harvested tissues. The strips were obtained from prescribed anatomical positions and in all cases, except the tendons, two perpendicular directions were examined. The animal tissue strips were obtained 1 to 4 hr postmortem while the human tissue was preserved in ice for 6 to 12 hr before it reached us. The tissue was kept moist all the time and either was used in the experiments immediately or was refrigerated (as was the case between experiments) at 3 °C in a Ringer's solution containing penicillin and gentocin.

The tensile stress-strain measurements were performed on a Tensilon testing machine with the strips submerged in Ringer's solution at room temperature. Width and thickness were measured simultaneously and each sample was conditioned before the stress-strain curve in loading was obtained. All the tensile data were taken at a constant crosshead speed which resulted in initial strain rates of 20% to 100%/min. In general 3 strips from the same sample were obtained and tested. The first strip served as the control one. The second strip was subjected to the successive removal of lipids, MPS, and elastin, whereas in the third one collagen instead of elastin was removed. Between successive removals the strips were tested in the Tensilon.

The lipids were removed by extraction with acetone in a Soxhlet apparatus at 37°C for 24 hr (Wood, 1954). MPS, that is, the ground substance, was removed enzymatically with hyaluronidase or hyaluronidase and β-gruronidase (Hoffman, 1973). Collagen was removed by steam autoclaving under water at 120°C until the extract became biuret negative (Newman & Logan, 1950). Finally, selective removal of elastin was accomplished by enzymolysis using a 50-50 mixture of elastase and trypsin inhibitor at pH = 8.6, 25°C for 5 hr (Missirlis, 1977).

Erythrocytes

Human erythrocytes were obtained by finger prick and collected into phosphate buffer saline at a very high dilution. Individual erythrocytes were then partially aspirated into 1 μm internal diameter glass micropipets. The length of the aspirated tongue, observed and measured under a microscope-video system combination, is a function of the aspirating pressure. The length-pressure curve, under certain conditions, gives a measure of the elastic shear modulus of the erythrocyte membrane (Evans, 1973).

A portion of the erythrocytes were made ghost cells, i.e., hemoglobin-free erythrocytes, and another one formed ghost cells which incorporated inside the cells the enzyme trypsin. Trypsin is a general proteolytic enzyme which will attack spectrin and possibly other structural proteins.

Results

Collagenous Tissues

Typical stress (force/initial cross-sectional area)-elongation curves are given in Figures 1 and 2 for the aorta and skin respectively. The general picture which emerged, considering all the data, is that removal of lipids and MPS did not show any measurable difference in the stress-elongation curves. However, removal of collagen had a dramatic effect in all tissues. Aortic valve strips and tendons disintegrated completely while the other tissues studied showed a linear stress-elongation relationship until their break point. These decollagenated tissues exhibited a yellowish color and a rubber-like behavior. Alternatively the deelastination process resulted in the same curve shape as the original strip with one noticeable difference: all the curves were moved to the right.

By considering both the deelastinated and decollagenated stress-elongation curves and correlating them to the original one, it is suggested that the overall stress-elongation curve can be synthesized from two curves: a straight line, originating at zero, of small slope (corresponding to elastin), and a second straight line, originating at the transition region of the initial curve, of much larger slope (corresponding to collagen). Histologic examination of the tissues reveals that, in the unstressed state, elastin bands are straight, while collagen fibers are undulated; also that a lot of intermingling exists between the two components. Therefore, as the native strip elongates, the force is transmitted via the elastin bands while simultaneously the collagen fibers straighten out. Continuing elongation is diminished when the collagen fibers have straightened

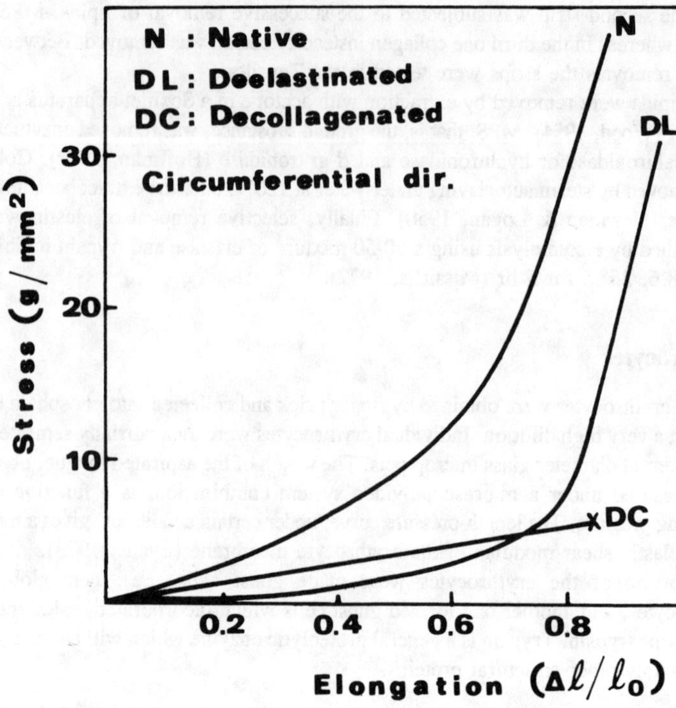

Figure 1—Typical stress-elongation curves of native, deelastinated and decollagenated strips of human ascending thoracic aorta.

completely and the force is transmitted through them. That is why when collagen is removed the strip consists almost entirely of elastin and exhibits the linear stress-elongation character of elastin. When elastin is digested, however, a very small force straightens out the collagen network and then the stiff collagen fibers exhibit diminished further elongation.

The bonds between the two components are not known but obviously play an important role especially in the transition region. What was just outlined is the general pattern of behavior. Individual variations were present due to the relative content of each component of the species and the relative anatomic position (directionality). It is also important to note that the "soft" collagenous tissues were very inhomogeneous, that is, portions of the same strip exhibited different stress-elongation characteristics.

Erythrocyte Membrane

In contrast to the nonlinear behavior of the collagenous tissues the erythrocyte membrane exhibits a linear shear stress-strain relationship over a large extension ratio range. Figure 3 shows this linear relationship for a typical normal erythrocyte and a trypsinized ghost cell. The control ghost cell was indistinguishable from the normal erythrocyte. These results suggest that hemoglobin does not participate in the elastic deformations of the erythrocyte membrane and that partial (?) digestion of spectrin and possibly other proteins diminishes the elastic modulus of the membrane.

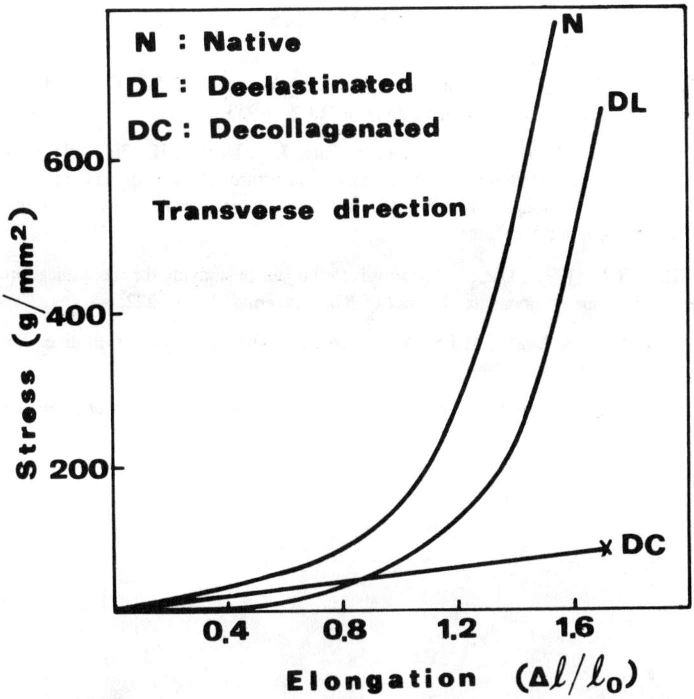

Figure 2—Typical stress-elongation curves of native, deelastinated and decollagenated strips of canine abdominal skin.

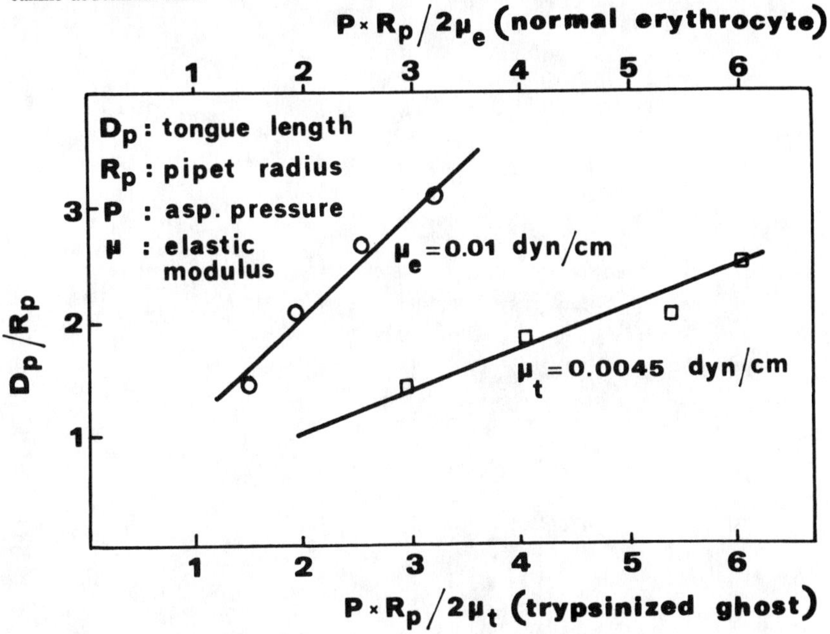

Figure 3—The dimensionless deformation parameter D_p/D_p of normal and trypsinized ghost erythrocytes plotted against the dimensionless membrane tension $P \times R_p/2\mu$.

References

EVANS, E. (1973). New membrane concept applied to analysis of fluid shear and micropipet-deformed red blood cells. *Biophysical Journal,* **13**, 941-953.

HOFFMAN, A.S., Grande, L.A., Gibson, P., Park, J.B., Daly, C.H., Bornstein, P., & Ross, R. (1973) Preliminary studies on mechanochemical-structure relationships in connective tissues using enzymolysis techniques. In R.M. Kenedi (Ed.), *Perspectives in Biomedical Engineering*, pp. 173-176. London: McMillan.

MISSIRLIS, Y.F. (1977). Use of enzymolysis techniques in studying the mechanical properties of connective tissue components. *Journal of Bioengineering,* **1**, 215-222.

NEWMAN, R.E., & Logan, M.E. (1950). The determination of hydroxyproline. *Journal of Biological Chemistry,* **184**, 299-306.

WOOD, G.C. (1954). Some tensile properties of elastic tissue. *Biochimica et Biophysica Acta,* **15**, 311-324.

Prosthetic Joints

H.U. Cameron
University of Toronto, Toronto, Ontario, Canada

It has been said that the artificial hip joint has done more to improve the quality of life for millions of people than any other advance since the polio vaccines. While failures do occur, and while solutions to these failures must be sought, this central fact should be borne in mind. A pygmy sitting on the shoulders of a giant can see further than the giant. The passing of the giants such as John Charnley leaves us pygmies a moral responsibility to carry their work a step further.

Virtually every joint in the body can be replaced, some with more success than others. Hips, knees, and fingers can almost be regarded as a solved problem. Others such as elbows and ankles are not particularly successful; indeed, to paraphrase Cornelius Ryan, the ankle has been called "a joint too far." Most designers have attempted to reproduce normal anatomy. When the size of the anchoring bone is too small and therefore loads on the implant/bone interface are too high, this slavish copying of nature must be questioned. Perhaps radical new designs should be considered which have no counterpart in the mammalian world.

The problems can be defined under certain headings: materials, anchorage, design, and infection. These are all interrelated and cannot be considered in isolation.

Materials

The materials used in the human body must be inert. Tissue fluids are intensely corrosive and some strange enzymes exist. Some polyesters, for example, may be broken down by tissue esterases.

The metals available are stainless steel, titanium alloys and the cobalt chrome alloys. For artificial joints the most widely used alloy is multiphase (MP 35N). To date there are no reports in the literature of fatigue failure of a multiphase implant. There has been a scare recently about the high nickel content of multiphase implants. The total volume of nickel in a multiphase implant is ingested monthly in any normal diet.

The plastics used are silastic and ultra high molecular weight polyethylene. Silastic is used only in low load situations, as in the case of fingers, and is often reinforced with Dacron. Age embrittlement is still a problem with silastic, although improvements continue. Ultra high molecular weight polyethylene is the most commonly used female part of the joint. It should not be used as the male part as the high rate of change of loading results in tensile stresses at the surface and thus surface fraying. Recently, carbon fiber reinforced polyethylene has been introduced. While the wear characteristics of this are improved, designers must be aware that the addition of carbon fibers to this material renders it more sensitive to increased contact pressures. The wear debris is not innocuous and can lead to a granulomatous inflammatory reaction.

Ceramics (fully dense aluminum oxide) are available as bearing surfaces. Once they have lapped into each other, the friction is very low, approaching that of a normal joint. The wear debris is wholly innocuous and produces no local inflammatory reaction. Ceramics bearing on plastics have a lower friction than metal on plastics and they are becoming quite popular in clinical use.

Anchorage

All load bearing implants must be fixed in some manner to bone. Bone, being a living material, has a certain preferential stress window. Loads above that window cause bone resorption and loads below also cause resorption in a manner identical to that seen in astronauts. Currently, the main anchoring material used to keep loads within these stress limits is polymethylmethacrylate (P.M.M.A.).

P.M.M.A. is the average orthopedic surgeon's dream. Bone carpentry need not be exact. The cement is injected as a viscous liquid, the implant inserted and held roughly in the correct position until the cement cures. Improved methods of bone preparation with brushing, lavage, and drying and improved cementation techniques of plugging the canal, pressure injection, pre-pressurization, and hammering down as large an implant as possible have dramatically improved implant durability.

Cement has several unfortunate drawbacks. First, it is weak in tension and shear, and second, it is toxic. Reinforcing cement is illogical as the increased viscosity prevents cement intrusions into bone. Toxicity is manifest by the layer of foreign body giant cells which surrounds the cement and which ultimately causes local bone resorption. Any loaded cemented implant is thus ultimately doomed to fail. The continued local inflammatory reaction renders the implant bed liable to invasion by bacteria, hence the occurrence of late infection.

In the last few years it has become obvious that with appropriate design, implants can transmit loads to bone in a satisfactory manner without the need to use cement. In a tibial component knee replacement, ridged plastic pegs can be used for initial fixation and a grooved undersurface, into which bone will penetrate, absorbs torsional and shear loads. This simple type of fixation has been in clinical use for 7 years with minimal problems (see Figure 1).

As long as the bone is loaded mainly in compression, a modulus mismatch is of no significance if stresses are evenly distributed. Ceramic cups can simply be screwed into the socket of a hip joint at an appropriate angle and the results to date would suggest that the fixation is permanent.

Porous metals into which bone will grow have recently seen wide acceptance in joint replacement design. Unfortunately, if bone ingrowth occurs the perfect bonding may result in extreme bone resorption due to stress relief osteoporosis, the stress passing down the implant rather than being transmitted to the bone. A porous coating should only be applied where a stress transfer is desired. Lack of knowledge of this fact has

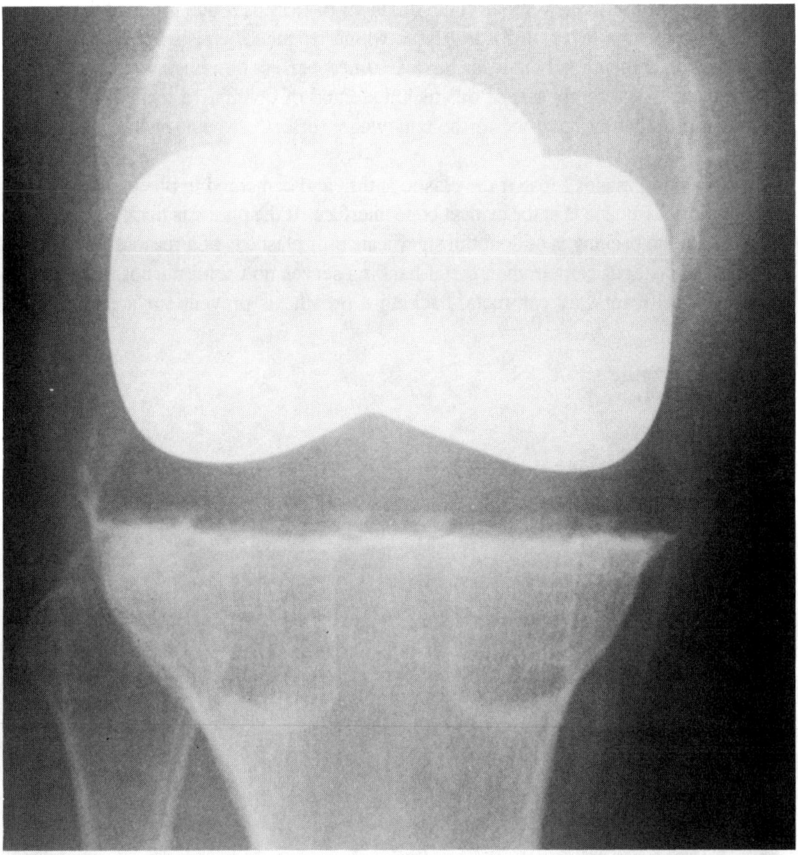

Figure 1—TRICON®. This tibial component is held in place by ridged plastic pegs. Bone condensation around the ridges can clearly be seen. Bone upgrowth into the grooves in the prosthesis resists torsional and shear loads.

led to several frighteningly bad designs being marketed. One prosthesis, after several years of use in Europe, has just undergone a radical design change with the removal of pores.

Design

Designing an artificial joint requires knowledge of materials, loads, bone, and their interactions. In the hip joint, for instance, if a stiff material such as multiphase is used, there is little point in having a collar on the prosthesis. Minimal loads will be transmitted from the collar to the bone. If a more flexible material such as titanium alloy is used, a collar is necessary and that collar must make direct contact with the bone, with no intervening cement.

Most factors such as these are obvious, but some are not so clear-cut. Too much reliance should not be placed on finite element analysis. Currently no such analysis of artificial joints has taken into account the layer of fibrous tissue found at the cement bone interface and a layer of fibrous tissue which frequently exists between the prosthesis and the cement. All analyses have assumed perfect bonding, which is unusual in the extreme. Overemphasis of this technique and of qualitative techniques such as photoelastic modeling have led to the bandwagon effect of metal-backing all plastic implants.

Metal-backing makes sense if the plastic is thin and cemented in place. The backing then helps distribute loads at the cement bone interface. If the plastic is thick and therefore less flexible, the backing is of doubtful significance. If plastic is of a reasonable thickness and inserted without cement then metal-backing serves no useful purpose (see Figure 2). In no significant way can metal-backing a prosthesis prevent surface cold flow.

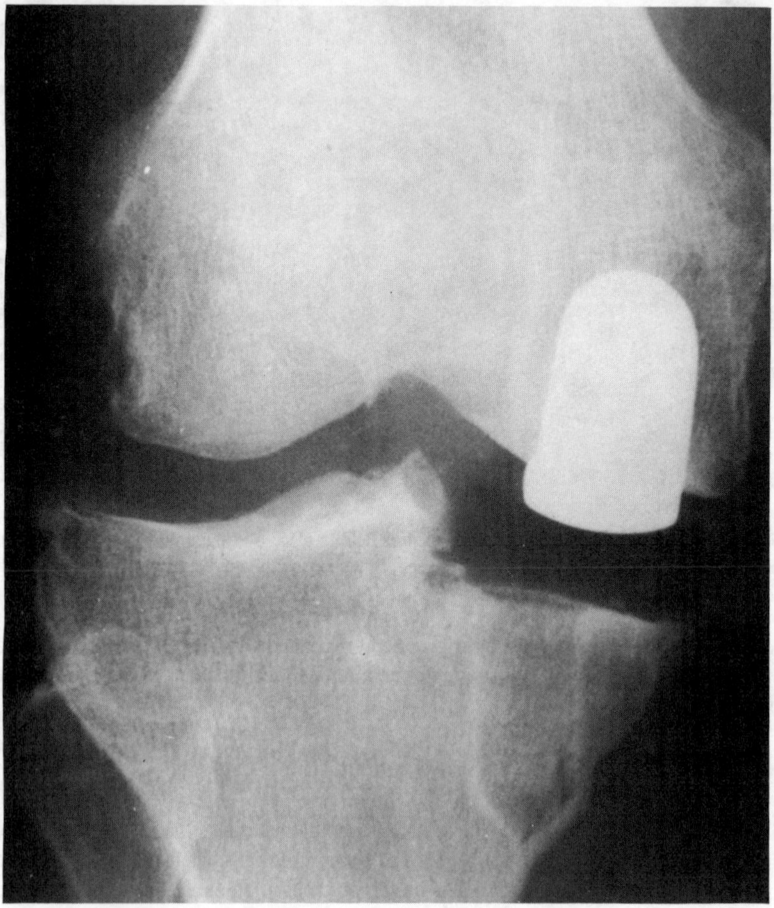

Figure 2—This TRICON® unicompartmental knee is anchored by a ridged peg and grooves on the undersurface. The tibial cut is slightly oblique but the plastic has deformed to exactly mate with the bone. The gently curved upper bearing surface allows total contact with the femoral component in spite of this minor malalignment.

Instrumentation should also fall under the heading of design. In knee replacement, correct angular insertion is vital. Malalignment of as little as 5° may jeopardize the long-term result of the implant. Instrumentation should be sturdy, as simple as possible, and have as few moving parts as possible. A complex, fragile instrument is unlikely to survive more than a few weeks. Orthopedic surgeons like hitting things with hammers. If an instrument cannot stand up to a few good blows with a hammer it is not much of an instrument.

Infection

Infection is still a major concern in surgery. Significant factors, all others being equal, are time, tissue trauma, and bone cement. Operating time and tissue trauma should decrease with increasing experience of the surgeon, and the use of bone cement is slowly decreasing. It is unlikely to disappear, however, as the majority of patients who require new joints are elderly, with poor quality bone. While mixing antibiotics with the cement is extremely useful, a nonreactive, nontoxic, and tough cement is required.

Custom-Tailored Design Specifications of Hip Prosthetic Femoral Components for Implantation

D.N. Ghista, K. Subbaraj, and P. Weiler
McMaster University, Hamilton, Ontario, Canada
Hamilton, Ontario, Canada

This paper provides the basis for custom-design prescription of hip prosthetic femoral components. Biomechanical analysis is carried out for the stresses in the stem, cement, and bone contributing to prosthetic femoral stem loosening, based on radiographic evidences of failure. Optimal analysis of the stem dimensions and cement thickness (normalized in terms of dimensions of the host femur) is then carried out to minimize the failure forces and optimize the performance. The optimal relationships among the design parameters are depicted by nomograms.

Biomechanical Stress Analysis

Forces and Design Parameters

The force system acting on the femoral stems may be predominantly idealized as an axial force and a varus-vulgus bending moment. The design parameters, illustrated in Figure 1, are: the stem length (L), average diameter (d_1, and medullary canal diameter (d_2). The design analysis entails the determination of L_1, d_1, and d_2 in terms of the

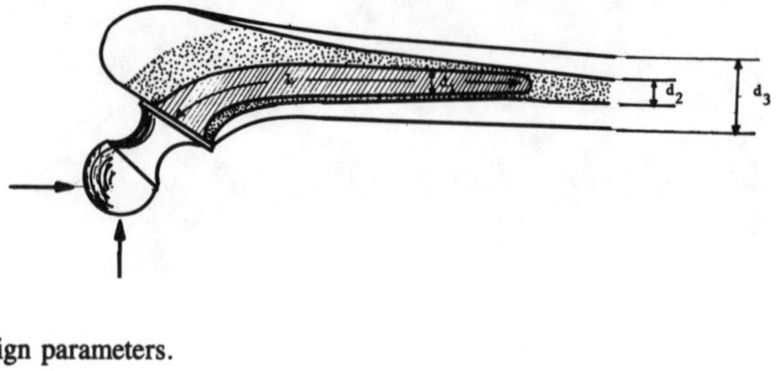

Figure 1—Design parameters.

host femur size (d_3) and the external forces (which are proportional to the weight of the subject).

Failure Mechanism Due to the Axial Force on the Stem (see Figure 2)

The axial force causes a) compressive stress in the cement under the stem tip which, upon exceeding its allowable limit, causes radiographically observed fracture gaps in the cement at the stem tip (see Figure 3) and b) shear stress at the stem-cement interface, whose critical values are responsible for radiolucent zone around the stem-cement interface (see Figure 4). In order to minimize the value of the axial force at which either the compressive stress in the cement at the stem tip or the shear stress at the stem-cement interface reaches its failure limit, we stipulate that the cement compressive and shear stresses reach their failure limits simultaneously to yield the following design parametric relationship among the design parameters of the stem and cement thickness, normalized with respect to the host femur's external dimensions:

$$\frac{\dfrac{2P}{\pi d_1(d_2 - d_1)}}{\dfrac{P}{\pi d_1 L}} = \frac{\text{Failure stress of cement in compression}}{\text{Failure stress of cement in shear}} = \frac{20}{1}, \tag{1}$$

or

$$\frac{f}{e - c} = 20 \tag{2}$$

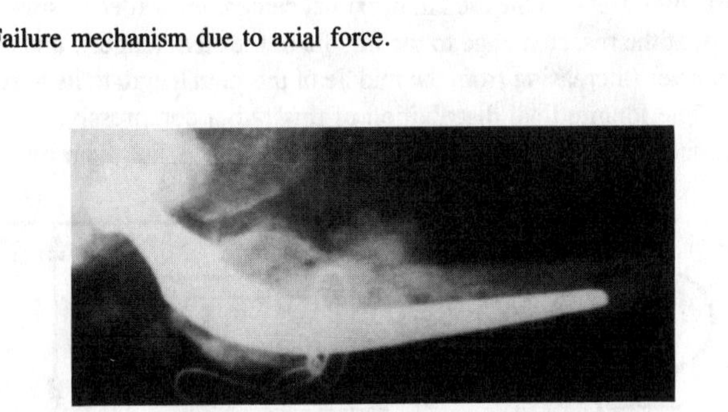

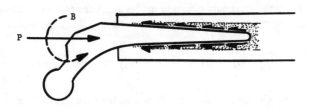

Figure 2—Failure mechanism due to axial force.

Figure 3—Cracks in cement under stem tip (reproduced with permission).

Figure 4—Radiolucent zone along interface (reproduced with permission).

Failure Mechanism Due to Bending Moment on the Stem

Due to the abutment of the stem collar on the resected surface of the cement-filled host-bone medullary canal, the bending moment (B) first induces longitudinally oriented tensile bending stresses (σ_{cl1}) in cement section of the bilayered cement-bone shell (Figure 5):

$$B = \frac{2\sigma_{bl}}{d_3}\left(\frac{\pi d_3^4}{64} - \frac{\pi d_2^4}{64}\right) + \frac{2\pi_{cl1}}{d_2}\left(\frac{\pi d_2^4}{64} - \frac{\pi d_1^4}{64}\right) \tag{3}$$

where σ_{bl}, σ_{cl1} are the stresses in the bone and cement sections, respectively. Upon substituting in the above equation, there are the following relationships:

$$\frac{\epsilon_c}{\epsilon_b} = \frac{d_2}{d_3}; \frac{\sigma_{bl}/E_b}{\sigma_{cl1}/E_c} = \frac{d_3}{d_2}, \text{ i.e., } \sigma_{bl} = \sigma_{cl1}\frac{E_r}{e} \tag{4}$$

where ϵ_b, ϵ_c are the maximum strains in the bone and cement cross-sectional layers; $E_r = E_b/E_c$, are the modulii of bone and cement; $e = d_2/d_3$, we obtain

$$B = \frac{\pi\sigma_{cl1}r_3^3}{4e}[(l - e^4)E_r + (e^4 - c^4)]; C = \frac{d_1}{d_3}, \text{ radius } r_3 = \frac{d_3}{2} \tag{5}$$

Secondly, when the bending moment (B) acts on the stem (see Figure 6), it induces radial compressive stress on the medial proximal cement layer (decreasing from a maximum value p_1 at the resected edge to the middle of the stem length), and on the lateral distal cement layer (increasing from the middle of the stem length to its maximum value p_1 at its tip). The longitudinal distribution of this radial compressive stress is shown in Figure 6, and the magnitude of the maximum stress p_1 is given by

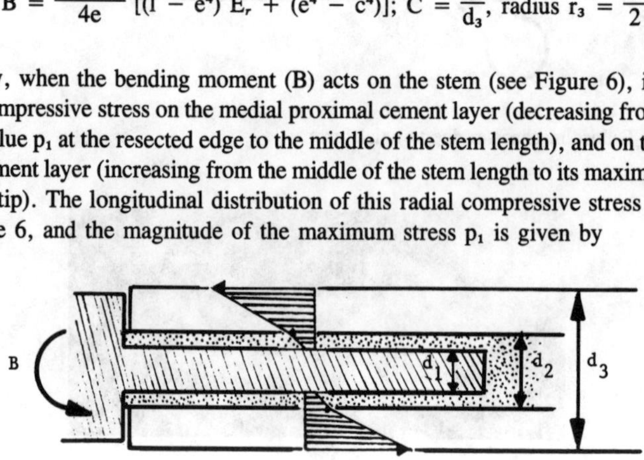

Figure 5—Longitudinal bending stresses in cement due to bending moment (B).

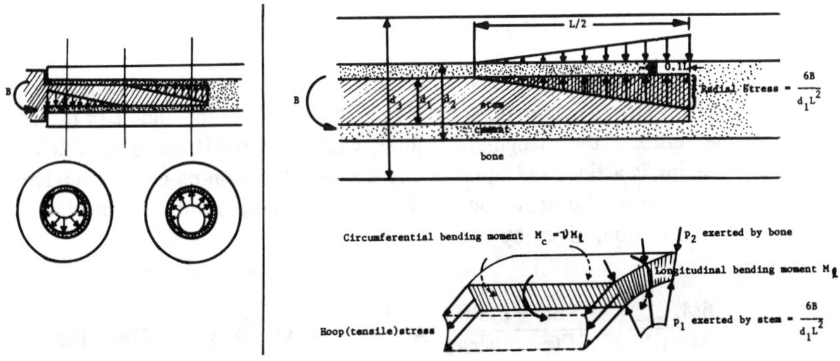

Figure 6—Stresses induced on cement shell element, due to radial compressive stress exerted by stem, due to the bending moment (B).

$$p_1 = 6B/d_1L^2 \tag{6}$$

This stress (p_1) inside the cylindrical bilayered cement-bone shell induces the following stresses and moments on a cylindrical cement element of unit longitudinal and circumferential dimensions:

1) An indeterminate compressive stress (p_2, depicted in Figure 6) acts at the cement-bone interface; its expression is obtained, by equating the radial displacements of the outer edge of the cement shell layer and the inner edge of the bond shell layer, as follows

$$p_2 = p_1p;\ p = \frac{2a^2}{[E_r\{\frac{b^2+1}{b^2-1} + v_b\} - v_c]\ (1-a^2) + (1+a^2)} \tag{7}$$

wherein $E_r = E_b/E_c = 8.33$; poisson's ratio v_b, $v_c = 0.29$ and 0.19, respectively; design parameter $a = d_1/d_2$; design parameter $b = d_3/d_2 = 11e$.

2) A hoop stress (see Figure 6) acts on the longitudinal section. Its expression is obtained by solving the stress equilibrium equations in cylindrical coordinates (with boundary conditions of the radial stresses at the inner and outer circumferential boundaries being equal to p_1 and p_2, respectively). The expression for the tensile hoop stress at the outer edge of the cement cylindrical shell is given by

$$\sigma_{c\theta 1} = \frac{p_1}{1 - a^2}\,[2a^2 - p(1 + a^2)];\ a = \frac{d_1}{d_2} \tag{8}$$

3) A longitudinal bending moment ($M\ell$ per unit circumferential dimension) acts on the circumferential section due to the longitudinal variation of the radial loading acting on the cement shell. Its expression is obtained by approximating a longitudinal strip of unit width of the cylindrical cement shell layer as a beam (of thickness $r_2 - r_1$) on an elastic foundation (provided by the rest of the cement shell to this longitudinal cement strip) under the action of a triangular loading (as shown in Figure 6), varying from a value zero at the middle of the stem to a value q ($= p_1 - p_2$) at the edge of the stem, given by

$$q = p_1 - p_2 = \frac{6B}{d_1 L^2}(1 - p) \tag{9}$$

where p is given by Equation 7. The value of $M\ell$, in the longitudinal beam strip (as depicted in Figure 6), is maximum at a distance of $0.8\ \ell$ from the middle of the stem (or from the vertex of the triangular loading), where ℓ is the length over which the triangular loading is acting, and equals L/2. Based on the beam-on-elastic foundation theory, we have the tensile stress due to $M\ell_{,max}$ at the outer edge of the cement shell's circumferential section, given by

$$\sigma_{cl2} = \frac{6M_{\ell,max}}{(r_2-r_1)^2} = \frac{9B(1-p)c^{1/2}}{2f^3(e-c)^{1/2}r_c r_3^3}[A(0.4\lambda\ell) - A(0.1\lambda\ell) - (\lambda\ell)B(0.1\lambda\ell)]; \tag{10}$$

$f = L/r_3$, $c = r_1/r_3$, $\mu_c = [3(1-v_c^2)]^{3/4}$, $A(x) = e^{-x}(\cos x + \sin x)$, $B(x) = e^{-x}\sin x$

4) A Circumferential Bending Moment $v_c M_\ell$ (per unit length) acts (as shown in Figure 6) on the longitudinal section cylindrical of the cement element. It will cause additional tensile stress $\sigma_{c\theta 2}$ at the upper edge of the cement shell layer, given by

$$\sigma_{c\theta 2} = 6v_c M_\ell/(r_2-r_1)^2 \tag{11}$$

Thus we have tensile longitudinal stresses (σ_{cl1} and σ_{cl2}) at the outer edge of the cement shell layer and tensile circumferences stresses ($\sigma_{c\theta 1}$ and $\sigma_{c\theta 2}$), causing distal-lateral migration of the lower part of the stem and proximal-medial migration of the upper part of the stem (see Figure 7).

In order to determine the optimal values of the stem length and diameter (with respect to the outer cement boundary), we will require the cement to simultaneously fail in circumferential and longitudinal tensions, resulting from Equations 5, 10, 8, and 11 in

$$\frac{4e}{\pi[(1-e^4)E_r + (e^4-c^4)]} - \frac{3[2a^2-p(1+a^2)]}{cf^2(1-a^2)} - \frac{9(1-v_c)(1-p)c^{1/2}\Phi}{2f^3(e-c)^{1/2}\mu_c} = 0 \tag{12}$$

where $\Phi = [A(0.4\lambda L) - A(0.1\lambda L) - (\lambda L)\,B(0.1\lambda L)]$.

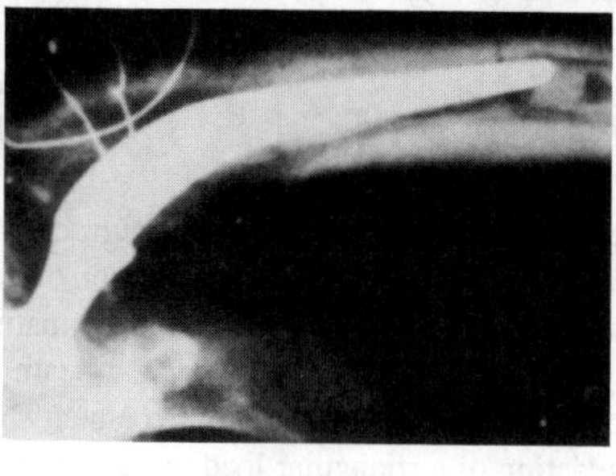

Figure 7—Cracks in cement causing distal-lateral and proximal-medial migration (reproduced with permission).

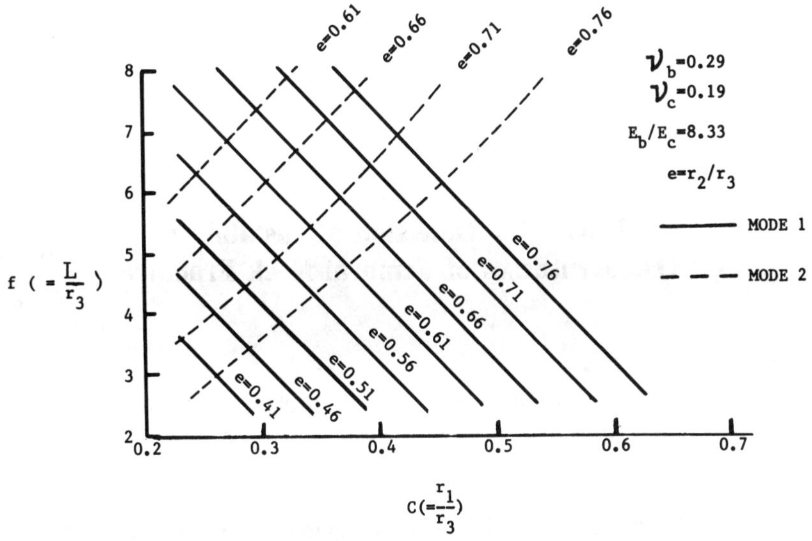

Figure 8—Design parameter space: f (2l/d3) vs. c (d_1/d_3).

Optimization Analysis

For our optimal prosthesis-cement-bone system, no one component should fail before the others or each component must be as fully stressed to its critical value when prosthesis system failure occurs due to stem loosening. In order to determine the optimal design parameters d_1, d_2, L (as shown in Figure 1), we represent the design parameter Equations 2 and 12 on an $f(=2L/d_3)$ vs. $c(d_1/d_3)$ design parameter space (see Figure 8). Therefrom, an optimal set of design parameters satisfying both the design criteria are obtained as:

$$e(=d_2/d_3) = 0.71, \quad f(=2L/d_3) = 7, \quad c(=d_1/d_3) = 0.4$$

Therefrom, we obtain, for representative outer dimension of the femur (d_3) = 3 cm, the stem length (L) = 10.5 cm, the average stem diameter (d_1) = 1.2 cm, the to-be-reamed bone medullary canal (d_2) = 2.1 cm or a 4 mm thick cement layer. Thus these design parameter nomograms enable us to select the optimal stem dimensions and the medullary canal size, to prevent stem loosening and prosthesis failure.

A Device for Detection of Instability on Osteosyntheses of Femoral Neck Fractures

H. Olofsson and L. Rehnberg
University Hospital, Uppsala, Sweden

At the University Hospital in Uppsala, Sweden, we do around 200 operations on femoral neck fractures each year. About 20% of these fail within six months. On these patients a hip replacement arthroplasty is afterwards performed. It should strongly reduce the suffering of these patients if such failures could be avoided, or if the hip replacement arthroplasty—if needed—could be done during the same operation. A device has therefore been built to estimate the stability of the repaired fracture during the operation.

At our clinic we use the method of von Bahr (von Bahr, Syk, & Walheim, 1974) for osteosyntheses of the femoral neck fractures. Two or more screws are inserted in the femoral neck. As the screws have no heads, they do not give rise to any compression. The compression of the fracture is accomplished by an impact hammer after the screws are inserted.

To be able to detect insufficient stability of the osteosynthesis as early as on the operating table, we have designed a device for measuring the stability in vivo during the operation. Our goal is to be able to determine, with the help of this device, if a hip replacement arthroplasty is necessary or not, and if so, to perform it during the same operation.

Method

The device that we have designed for this purpose consists of a rod with a strain gauge attached to it. The rod is covered with a stainless steel tube of length 155 mm and outer diameter 4.5 mm. In one end of the tube there is a sharp tip. In the other end there is a removable handle of length 100 mm and diamter 12 mm attached via a bayonet joint (see Figures 1, 2). To fasten the sensor device laterally in the femur, there is a separate cone, which has a longitudinal cut. Its outer diameter varies between 5 and 6 mm and its length is 25 mm. The strain gauge in the sensor device is connected by a cable and an intermediate amplifier to one channel of an x-t recorder.

Under the shoe, that supports the patient in the operating table, a load cell is attached, which reads the axial load that is applied to the leg. The load cell is connected to an amplifier and to the second channel of the x-t recorder (see Figure 3).

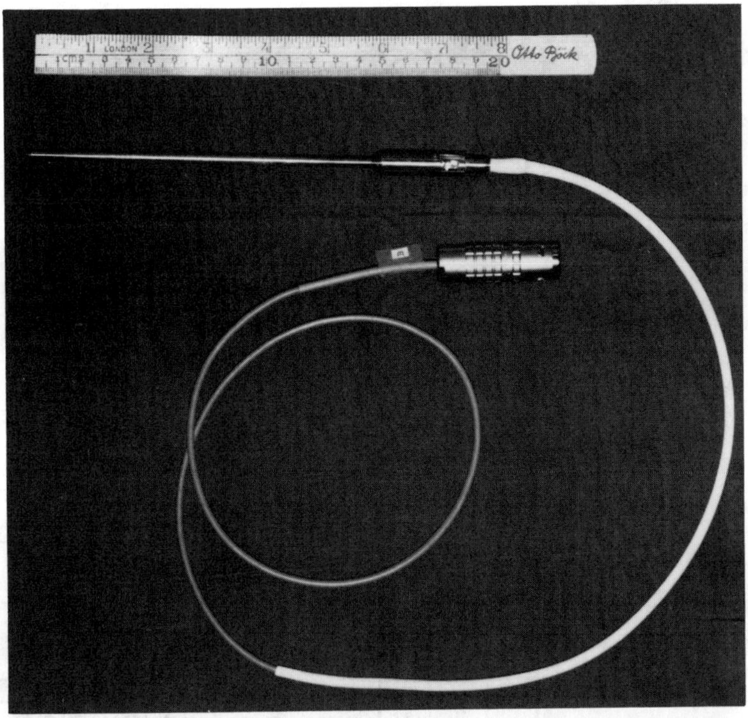

Figure 1—The sensor device.

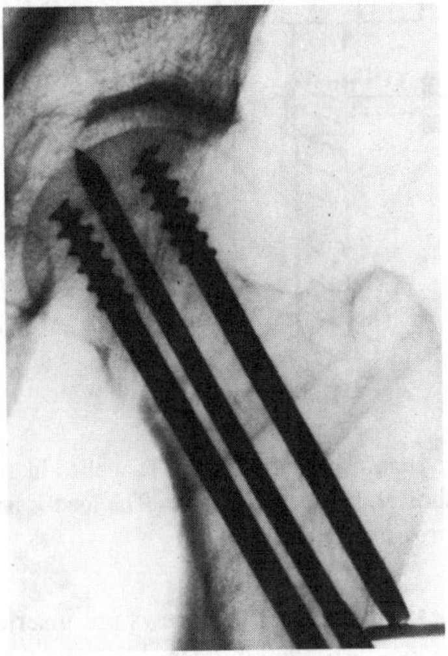

Figure 2—The sensor device inserted in a 56-year-old man with a femoral neck fracture. The method of von Bahr is used for the osteosynthesis.

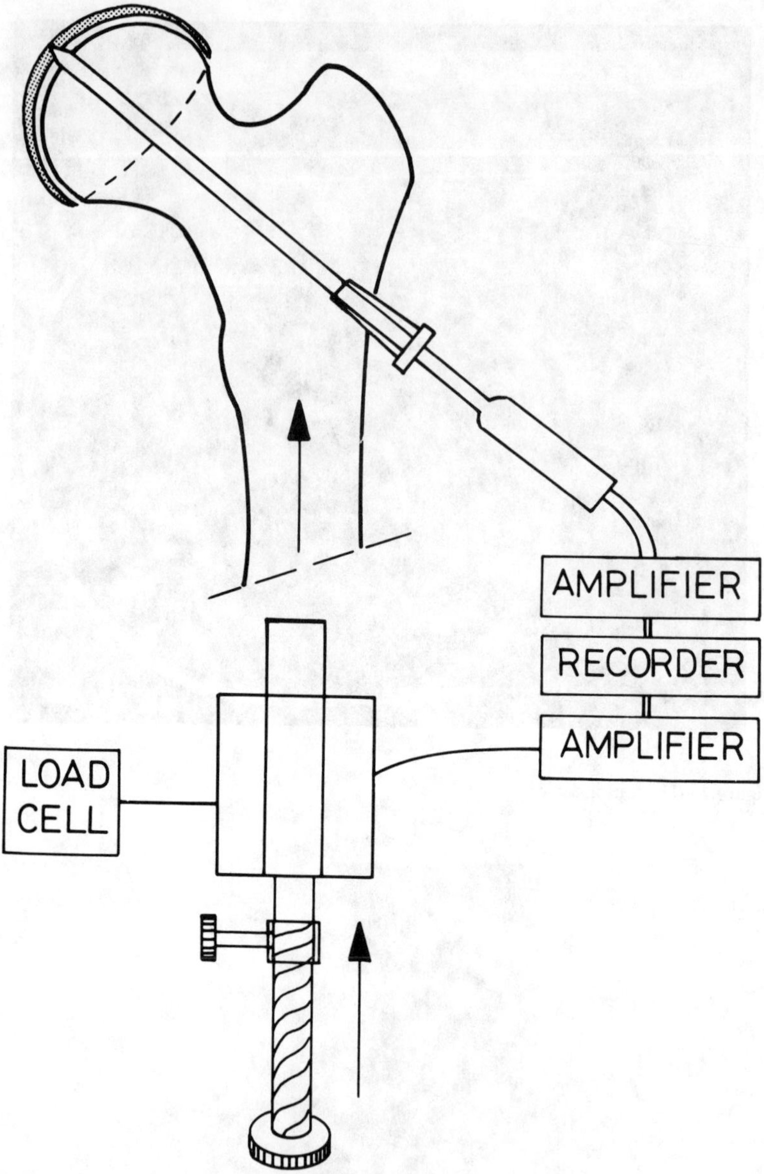

Figure 3—The measuring system. The sensor device is applied in the femoral neck and the load cell is attached under the shoe at the operating table. The load is applied under the shoe in the direction of the shoe and proximally.

During the osteosynthesis, after the two screws are inserted, a third hole is drilled to within 10 mm of the bone-cartilage boundary. We use the same drill as we use for the screws (diameter 5.5 mm). The cone is impacted in place to support the sensor device laterally. The device is then firmly attached in both ends and left unsupported in the middle. Consequently the rod bends when the fractural segments move, and all such movements are recorded by the x-t recorder. Note that it is important to allow sufficient space inside the femur around the sensor device.

The load at the osteosynthesis is obtained by loading the leg axially at the foot proximally. We have found a load of 100 N convenient.

Discussion

Our aim is to obtain a normalized value indicating the degree of instability of the osteosynthesis. The source of errors are pulse, breathing, and measuring errors in the sensor device and in the load cell. These errors are small at a load of 100 N.

If the hole drilled for the sensor device is of too small a diameter, the middle of the sensor will touch the bone while bending and hence give too high a value for the deformation. When using the von Bahr method, we have found that a diameter of the hole of 5.5 mm and a diameter of the sensor device of 4.5 mm is satisfactory.

We insert an extra screw in the hole after the sensor. This, however, can give too high a value for the "instability factor" of the final osteosynthesis.

The sensor device we have described can also be used with other methods of osteosynthesis. However, we have found manufacturing difficulties in making the device any thinner.

Reference

VON BAHR, V., Syk, B., & Walheim, G. (1974). Osteosyntheses of femoral neck fracture using screws. *Acta Chirurgica Scandinavica,* **140**, 277-288.

Biomechanical Guidelines for Effective Application of External Fracture Fixation

D.N. Ghista and K. Subbaraj
McMaster University, Hamilton, Ontario, Canada

An external fracture fixator must apply adequate magnitude of compressive force at the fracture site so as to accelerate healing. It should restore the overall stiffness of the fractured bone to its normal value. Further, no part of the bone should be understressed, and more and more stress should be transferred to the bone as healing progresses so as to prevent osteoporosis.

The approach to fracture fixation has been to provide as rigid fixation as possible. However, in so doing we could be unstressing the fracture site and delaying healing, that is, restoration of the elastic modulus of the fractured segment. Hence the optimal fixation modality may not just necessarily entail design of more and more rigid fixators but rather a provision in the designs to stepwise reduce their rigidity as healing progresses, so that more and more stress is transferred stepwise to the healing fracture site.

We are demonstrating this concept with a simple external fixator for the tibia (see Figure 1), consisting of bilateral bars connected to proximal and distal jigs at the ends and two central turnbuckles. Threaded pins pass from the jigs of the medial bar, through

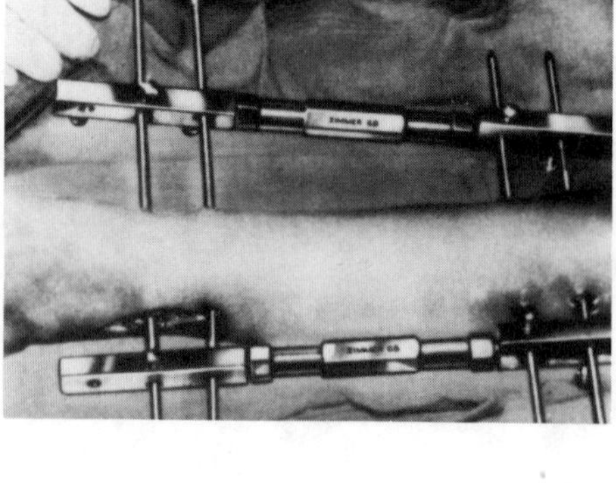

Figure 1—External fixator in use.

the bones, and engage the corresponding threaded holes of the jigs of the laterally placed bar. The number of pins can vary from six to four to two, depending upon the weight of the patient and the stage of bone healing. The turnbuckles facilitate adjustment of the fixator size. After the fixator is in place, by means of the turnbuckles compressive forces are applied to the fractured face to ensure rigid fixation and minimal relative displacement at the fracture site.

In order to implement the above mentioned fixation criteria with this external fixator, we have carried out biomechanical analyses of stresses and deformations in the fractured bone stabilized by the fixator, at different stages of bone healing represented by the ratio (E*) of elastic modulii of the fractured and unfractured bone segment, for two, four, and six pins. Based on the results of stress and deformation of the fractured segment, we are advocating an optimal modality fracture fixation by altering the number of pins as healing proceeds.

Methodology

When a finite element model of the tibia is subject to representative axial force (P) and bending moment (M) resulting from ambulation, tensile bending stresses are induced in it.

A crack in this bone model is simulated by assuming that a small segment of it has a reduced (or in some cases near-zero) elastic modulus ($E^* < 1$). When the ambulatory

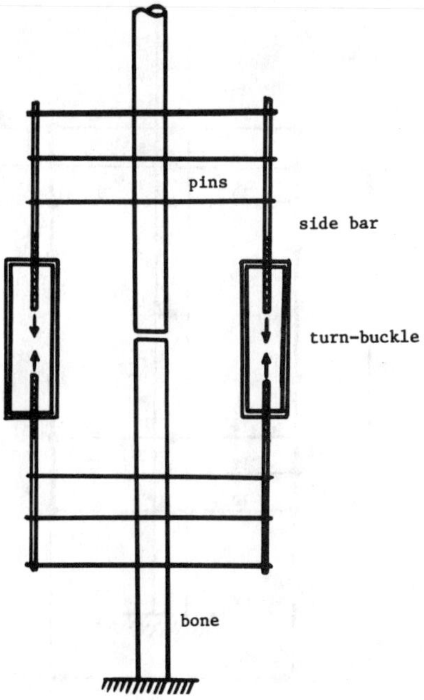

Figure 2—Model of fractured bone and external fixator.

forces are applied to this fractured tibia model, the crack opens (i.e., considerable relative displacements between the ends of the cracked segment occurs) due to the induced bending stresses.

An external fixator is now applied to this cracked tibia (see Figure 2). The stresses and deformation of the cracked segment of the fixator and tibia are determined by finite-element discretization and analysis of this hybrid structure made up of the cracked bone and the fixator (see Figure 3).

The Finite Element Modeling of the External Fixator Fractured Bone

The hybrid structure consisting of the fractured bone with external fixator has been idealized as a rigid-jointed plane frame, for finite element analysis. The bone and the components of the fixator (viz., the pins and the medial and lateral bars) have been assumed to be circular in cross section. A mean value of modulus of elasticity for bone, E_b of 2.1×10^6 psi, has been adopted in the analysis. The material data for different members are given in Table 1. The bone-pin and pin-lateral bar connections are assumed to be rigid.

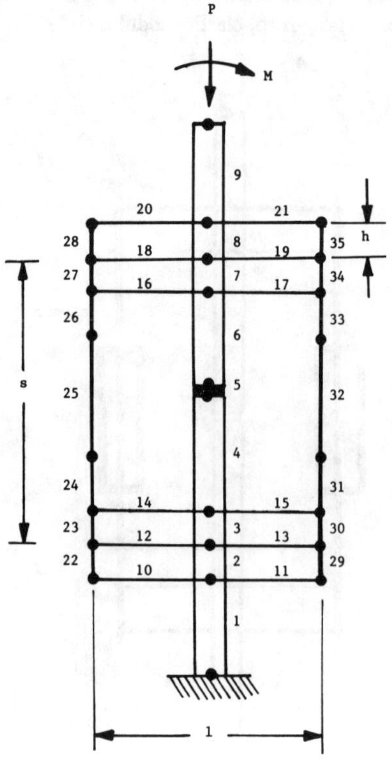

Figure 3—Finite element model of the fractured bone and external fixator.

Table 1

Material Data for Fixator/Bone Components

Member	Diam (in)	L (in)	E (ksi)	
Bone	1.0	14	2,100	Loading: P = 0.1 kip
Pin (s. steel)	0.16	5.0	30,000	M = 0.650 kip-in
Side bar	0.4	7.25	30,000	Pin separation = 0.875 in
Fractured segment	1.0	0.2	0.1-1.0 rel. to intact value	Separation between top and bottom sets of pins = 7.25 in

The Plane Frame Model. The structural model components of the fractured bone and the fixator frame are divided into several plane frame elements, as shown in Figure 3. The plane frame element can both stretch and bend (in its plane) and hence contains the degree of freedom of both the two-dimensional bar and beam elements. At each of its two nodes, there are two translational degrees of freedom and one rotational degree of freedom, giving six degrees of freedom per member. The stiffness matrix of an arbitrarily oriented plane frame element in the local coordinate reference frame is given in Figure 4.

The turnbuckle of the fixator (see Figure 2) is also modeled as a rigid-jointed plane frame member. The tibial bone is considered to be fixed at the bottom (see Figure 2); no translational and rotational degrees of freedom exist at the fixed support. These conditions serve as boundary conditions for the fixator/bone model. The cracked segment is modeled as an element with reduced elastic modulus ($E^* < 1$) relative to its normal value, and bone healing is represented by increase in E^* to a value of 1.0.

$$m = \cos\beta$$
$$n = \sin\beta$$

$$S = AE/L \qquad F = Sm^2 + Dn^2$$
$$A = 4EI/L \qquad G = Smn - Dmn$$
$$B = 2EI/L \qquad H = -Cn$$
$$C = 6EI/L^2 \qquad P = Sn^2 + Dm^2$$
$$D = 12EI/L^3 \qquad Q = Cm$$

Displacement d.o.f. vector
$$\langle u_1, v_1, \theta_1, u_2, v_2, \theta_2 \rangle^T$$

Force vector
$$\langle U_1, V_1, M_1, U_2, V_2, M_2 \rangle^T$$

$$\begin{bmatrix} F & G & H & -F & -G & H \\ G & P & Q & -G & -P & Q \\ H & Q & A & -H & -Q & B \\ -F & -G & -H & F & G & -H \\ -G & -P & -Q & G & P & -Q \\ H & Q & B & -H & -Q & A \end{bmatrix}$$

Nodal Displacement d.o.f. & Forces Stiffness Matrix

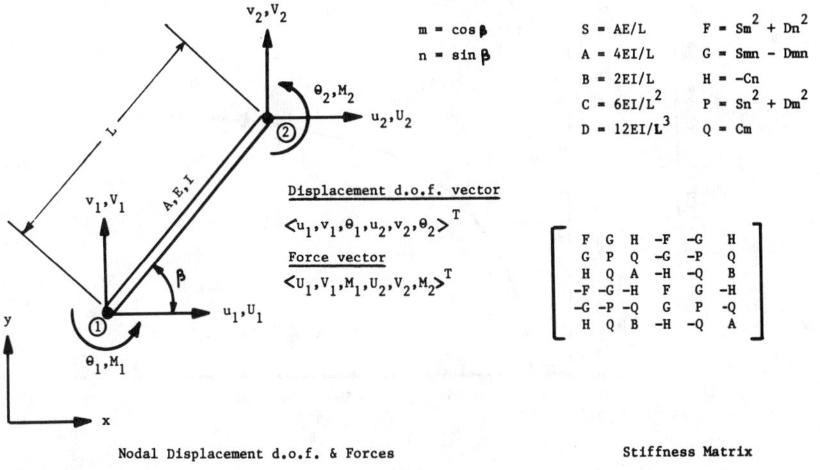

Figure 4—Plane frame element.

The loading on the hybrid bone-fixator frame structure consists of an axial compressive load of P and a bending moment of M in the plane of the frame and is assumed to be acting at the top of the bone (see Figure 3).

The analysis has been carried out for the fixator frame with two, four, and six pins, in order to study the effectiveness of the number of pins on the stress in the cracked element (which is representative of the overall rigidity of the cracked fixator/bone structure).

Results and Discussion

When E* ≅ 0.1, the cracked segment undergoes substantial deformation (or relative horizontal displacement between the nodes of the cracked finite-element segment) due to the induced tensile bending stresses induced by the ambulatory forces. This is not conducive to healing. Hence, to alleviate these tensile stresses and promote healing

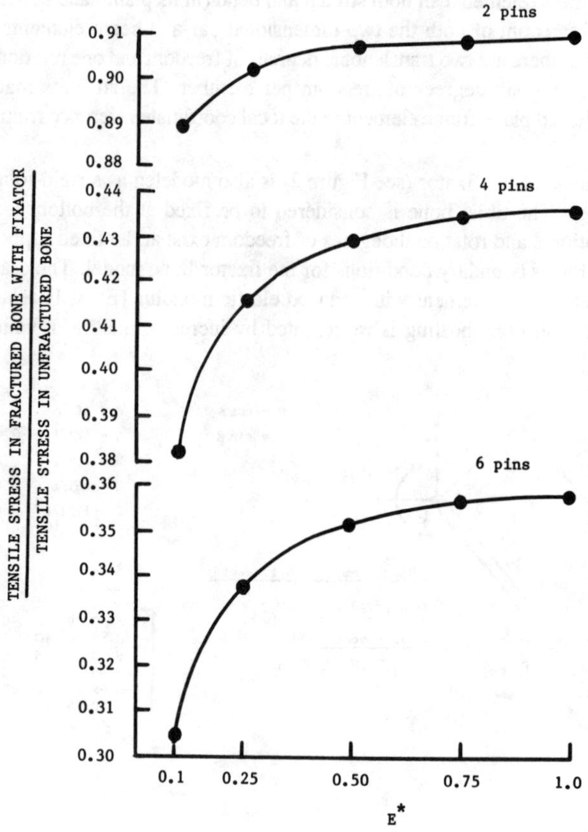

Figure 5—The tensile stress in the cracked bone segment vs. modulus E* of the fractured segment relative to its unfractured value, for fixator with 2, 4, and 6 pins.

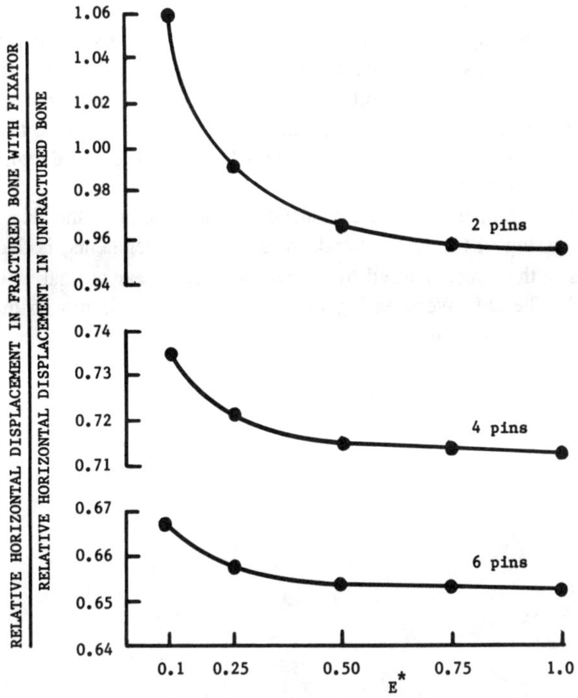

Figure 6—The relative lateral displacements in the cracked bone segment vs. modulus E* of the fractured segment.

in the early stages following fracture (when E* $\cong$ 0.1), the turnbuckles are rotated to apply bilateral compressive forces. This induces compressive stresses in the cracked bone finite element, and ensures the prevalence of only compressive stress across the bone cross-section. The ratio (σ_c^*) of the induced compressive stress in the cracked segment to the stress in the unfractured bone (caused by standing and ambulation) in turn stimulates generation of the E* value in the uncracked segment.

When, in the healing phase the fixator/cracked tibia finite-element model, with this incrementally increased value of E* in the cracked (finite-element) segment, is subjected to the ambulatory forces, additional tensile bending stresses are induced in the cracked segment. However, the cracked segment is now able to sustain the tensile stresses, provided the σ_c^*, the ratio of the tensile stress in the cracked segment to the stress in the unfractured bone, keeps up with the increase in E*.

Figure 5 shows that if two pins are used, the build-up of the modulus E* in the cracked segment in the initial post-fracture phase is not adequate to sustain the tensile stress σ_t^*. Also, based on Figure 6, the relative horizontal or lateral displacement between the nodes of the cracked segment is substantial with the use of two pins and not conducive to bone healing. On the other hand, with six pins, the bone would be understressed (and possibly osteoporotic) at the time of removal of the fixator.

Hence the optimal modality of fracture fixation is to start off with six pins, and eliminate relative displacement between the ends of the cracked finite-element segment caused by the small magnitude of tensile bending stress induced by the standing-ambulatory forces, by rotating the turnbuckles to apply bilateral compressive forces. At about E* = 0.3, two of the six pins can be removed. By this time, E* is adequate to enable the healing fractured finite-element bone segment to enable it to sustain some tensile stress. When E* = 0.6, two further pins can be removed, so that by the time E* = 1.0 and prior to the removal of the external fixator, the cracked segment is already sustaining near-normal stresses.

The implication of this stepwise reduction in the rigidity of the fixator, with concomitant stressing of the cracked and uncracked bone segments, is that distribution of modulus in the bone, induced by the stress in it, is near normal when the fixator is removed. Thereby overstressing of and consequential injury to the destiffened osteoporotic bone is avoided.

An Evaluation of Variables Which Alter the Mechanical Behavior of Fixator Frames

F. Behrens and W.D. Johnson
St. Paul-Ramsey Medical Center and University of Minnesota,
St. Paul, Minnesota, USA

In the past decade external skeletal fixation has become the preferred method for immobilization of severe open tibial fractures and infected bony lesions. Presently one- and two-plane configurations are most popular (Hierholzer et al., 1978; Vidal et al., 1976). However, it has recently become obvious that the transfixion pins which are used with all bilateral frames can cause anterior compartment syndromes, injuries to the anterior tibial artery, and permanent stiffness of the ankle and subtalar joints (Emerson & Grabias, 1983; Green, 1981). One- and two-plane unilateral fixator frames occupy a cross-sectional sector of less than 90° and, therefore, can be applied to the anteromedial aspect of the leg without jeopardizing neurovascular structures or musculotendinous units. While the clinical advantages of these frames are attractive, their mechanical properties are presently unknown. It was the purpose of this study to examine experimentally how such structural variables as the number of pin planes (one-plane versus two-plane frames), the spread between the pins in each bony fragment, the bone-rod distance, and selective pin loosening would influence the behavior of different fixator frames. It was hoped that this investigation would provide some indications about:

(1) The relative rigidity of bilateral and unilateral configurations
(2) The optimal means to construct maximally rigid unilateral frames
(3) The most convenient mechanisms to gradually decrease frame rigidity in order to direct larger forces and moments across the fracture site during the later part of the bony healing process.

All frames were tested under conditions of axial compression as well as sagittal (SAG) and frontal (FRO) bending, recognizing that under most clinical circumstances anteroposterior bending moments are about three to four times larger than those applied in the transverse plane (Bresler & Frankel, 1950; Zarrugh, 1981).

Material and Methods

This study was undertaken with an AO tubular fixator using two 5-mm pins in each main bony fragment, tubular rods of 11-mm outside diameter, and independent articulations which allow for selective pin loosening. All frames were built as equivalent configurations with pin-pin (P-P) spreads of 4.4 and 9.0 cm, and a maximal pin distance of 24.4 cm. Corresponding to the clinical situation, the rods applied to pins inserted in the frontal plane were kept at a bone-rod (B-R) distance of 8.0 cm, while sagittally applied frames had B-R options of 8.0 and 2.5 cm. The two half frames making up the two-plane unilateral frames were applied in a straight sagittal and frontal direction and connected by two standard pins (5 mm). Two modifications of one-plane unilateral frames were erected: (a) using two anterior rods (2 rods) and a B-R distance of 2.5 and 8.0 cm, and (b) using one anterior rod but loosening the pin clamps closest to the fracture site (central pins loose).

All frames were erected on a tubular model bone of laminated linen with an outside diameter of 2.54 cm, a wall thickness of 0.64 cm, and a modulus of elasticity of 31 GPa. All models had a fracture gap of 1 cm and were tested with an MTS materials testing machine. The applied loads and moments and the displacement of the model bone fragments were recorded on an x-y plotter and the stiffness of the fixator frames was determined from the slopes of the recorded curves. Compressive stiffness was expressed as N/mm and bending stiffness in N•m per degree of angulation.

Results

The results are summarized in Figures 1 and 2.

FRAME TYPE	DIMENSIONS		BENDING (NM/°)		COMP
Bilateral	P-P	B-R	SAG	FRO	(N/M)
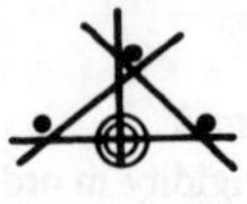 2-PLANE	44 90	80 80	10.4 17.1	21.3 20.4	154 182
 1-PLANE	44 90	80 80	3.9 5.9	19.5 18.3	140 148

Figure 1—Stiffness of bilateral frames in bending and compression. COMP: compression, P-P: pin-pin distance, B-R: bone-rod distance, SAG: sagittal plane, FRO: frontal plane.

FRAME TYPE	DIMENSIONS		BENDING(NM/O)		COMP
Unilateral	P-P	B-R	SAG	FRO	(N/M)
 2-PLANE	44	80	11.0	15.3	63
	90	80	11.5	16.0	77
	90	25 (sag) 80 (fro)	15.0	16.9	350
 1-PLANE (2 rods)	90	80 (fro) / 25 (fro)	20.7	10.6	364
 1-PLANE	44	80	14.2	2.0	28
	44	25	12.4	5.3	161
	90	80	16.0	3.8	31
	90	25	16.0	8.7	252
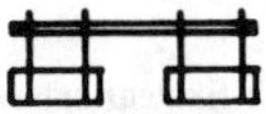 1-PLANE CENTRAL PINS LOOSE	90	80	5.8	1.1	7

Figure 2—Stiffness of unilateral frames in bending and compression.

Conclusions

1. With proper spatial arrangements of pins and rods, unilateral frames can be erected which show stiffness characteristics comparable to bilateral frames.
2. Ratios of sagittal to frontal bending stiffness which correspond to physiological demands (about 3 to 5/1) were only achieved with one-plane unilateral frames.
3. Adding a second half frame (two-plane unilateral frames) had little influence on sagittal bending stiffness, but increased the frontal bending stiffness of unilateral frames by a factor of 2 to 7, and resistance to compression by a factor of 1.5 to 3.
4. The most effective configuration was an anterior one-plane unilateral frame with 9.0 cm pin-pin distance and two connecting rods at 2.5 and 8.0 cm distance.
5. The variables most effective in increasing the resistance to the three tested loading modes were: a) sagittal bending—adding a second anterior rod, (b) frontal bending—adding a second half frame in the transverse plane, and c) compression—adding a second rod and decreasing the bone-rod distance.
6. A gradual decrease of frame rigidity in all three loading modes could be achieved by the sequential transformation of the following frames:

$$\begin{matrix} A_1 \\ A_2 \end{matrix} \rightarrow B \rightarrow C \rightarrow D$$

This transformation decreased resistance to sagittal bending by 74%, to frontal bending by 90%, and to compression by 98%.

Acknowledgment

Supported by Grant #8327 from the Medical Education and Research Foundation of St. Paul-Ramsey Medical Center.

References

BRESLER, B., & Frankel, J.P. (1950). The forces and moments in the leg during level walking. *Transactions of the American Society of Mechanical Engineers,* **72**, 27-36.

EMERSON, R.H., & Grabias, S.L. (1983). A retrospective analysis of severe diaphyseal tibial fractures treated with external fixation. *Orthopedics,* **6**, 43-49.

GREEN, S. (1981). *Complications of external skeletal fixation* (p. 33). Springfield, IL: C. Thomas.

HIERHOLZER, G., et al. (1978). External fixator: Classification and indications. *Archives of Orthopaedic and Traumatologic Surgery,* **92**, 175.

VIDAL, J., et al. (1976). Traitement des fractures ouvertes de jambe par le fixateur externe en double cadre. *Rev. Chir. Orthop.*, **62**, 433.

ZARRUGH, M.Y. (1981). Kinematic prediction of intersegment loads and power at the joints of the leg in walking. *Journal of Biomechanics,* **14**, 713-725.

Prosthetic Ligaments: A Review

G. Drouin
École Polytechnique et Institute de Réadaptation de Montréal, Canada

The knee is probably the most vulnerable articulation in the human body. Its mechanical stability is a delicate equilibrium between muscles and ligament actions and it is subjected to high loads. It is no surprise that the occurrence of ligament injuries is high, especially in activities where the knee is loaded in torsion and lateral bending.

Ligament, in contrast to muscle and bone, cannot be repaired very easily. In clinical treatments of acute ligament tears (Johnson, 1983) different options are offered to the surgeon. Some elect to do no surgery. The surgical procedures that are clinically applied to ligaments are (a) a repair of the traumatized structure only; (b) ligament reconstruction using tendon or fascia lata.

Recently (Kennedy, 1983), an artificial material has been added to ligament grafts. A woven tissue of polypropylene is mechanically bonded to the patellar tendon and this composite structure is introduced in place of the ligament. This technique has been evaluated on goats and is presently undergoing clinical trials. Another concept is to implant a scaffold of carbon fibers embedded in a biodegradable polymer. This material permits the development of new fibrous tissue as the polymer degrades. It is presently undergoing an extensive evaluation both in animals and in humans and the short term results seem promising (Alexander et al., 1981).

Finally, the concept of using a completely artificial prosthesis is also being investigated. The synthetic device aims at a permanent replacement of ligament function. A variety of prosthesis concepts are in research and development stage (Drouin et al., 1980).

This paper will elaborate on the prosthesis concept in general, discussing the design parameters, the methods used to characterize the ligament, the implantation parameters and future areas of research. Although most of the comments are made for the anterior cruciate ligament (ACL), some may be generalized to other ligamentous structures.

Design Considerations

In order to design a suitable prosthesis many other parameters must be considered besides the force at rupture. The ligament is loaded by the displacement of its end attachments. Therefore, it acts as a spring. If the ligament function is to be reestablished the mechanical behavior of the prosthesis should approximate the force-deformation curve of the natural ligament. At implantation, anatomical, geometrical, and mechanical factors should all be considered.

Force at Rupture

Following the poor performance of the polyethylene prosthesis introduced by Richards, many research groups have investigated the load at rupture of the human ligament. Since many variables affect the result it is not easy to obtain a precise value for the load at rupture. This value should be weighed against the maximal fixation force to bone in order to avoid bone avulsion when the device is subjected to excessive loading.

Force-Deformation Curve

The mechanical behavior of canine and human ligaments (Dorlot et al., 1980) have been characterized in the functional range. The force-deformation curve is non-linear and is relatively rigid at about 15% deformation. Curves are obtained using a tensile testing machine. The difficulty is to relate mechanical test results to the in vivo loading situation. Further comments will be made on this subject.

Implantation Sites

In most designs, the prosthesis is fixed to the bones via transtibial and transfemoral channels. Generally, the natural insertion sites determine the origins of these channels. However, the axis of the channels with respect to the prosthesis axis is of prime importance on the loading of the prosthesis. The flexion of the knee creates torsion and bending within the prosthesis. Depending on the characteristics of the implanted device, this additional loading may exceed its inherent strength.

Initial Force at Implantation

The force-deformation curve of the ligament is non-linear. Therefore, the rigidity will vary according to the deformation state. Even if the prosthesis reproduces the same force-deformation characteristics, the force in the prosthesis depends on the initial force at which it was implanted in the knee.

Characterization of Ligament Properties

Macroscopic Characterization

The macroscopic characterization is the evaluation of the mechanical properties of the ligament. Generally, the ligament is tested in bone-ligament-bone preparations on a tensile testing machine. The purpose is to determine the force at rupture and the force-deformation curve. The load at rupture is sensitive to the strain rate, the mounting technique and the mode of preservation of the specimen. The main difficulty is to align the ligament in the machine in such a manner that the fibers are evenly strained. It has been found (Grood et al., 1977) that the force at rupture is a function of the age of the donor. The force-deformation curve of the ACL exhibits less sensitivity to experimental factors. It was described by Grood as containing three zones: the functional, the microrupture, and the rupture zones. The first zone corresponds to a deformation of less than 10%; the second is a zone in which some fibers are broken but the ligament can recover; the third is a zone in which there is irreversible rupture.

Microscopic Characterization

To understand and explain the behavior of the anterior cruciate ligament in a plausible manner, the anatomy of the ligament should be studied and the data integrated in a mathematical model. The ligament is constituted of collagen, elastin, and ground substances. It is therefore a composite material in which the spatial arrangement of the constituents is responsible for its overall mechanical properties. An ongoing study (Yahia et al., 1983) based on a combination of biochemical and microscopic techniques has established the spatial relationship of elastin with respect to collagen. At another level, the orientation of collagen fibers and their aggregation into a bundle is being defined. This information can be used in a three-stage model. The first stage is the modeling of a collagen fiber with elastin, the second stage is the combination of collagen fibers to form a fibril, and the final stage is the modeling of the fiber bundles. This approach will yield pertinent information on the behavior of the ligament based on the interaction of its constituent parts.

Implantation Considerations

Some of the design considerations are also of importance during implantation. Distinct aspects that will be treated briefly are the torsion and bending in a prosthesis and the choice of the implantation sites.

Torsion and Bending in a Prosthesis

The anatomy of the anterior cruciate ligament reveals that this structure is particularly well adapted to the loading it is subjected to. Unfortunately, prosthesis concepts are not as sophisticated, and a closer attention should be given to torsion and bending of artificial devices.

As the knee flexes, torsion and bending are exerted on a prosthesis in amounts that depend on the design of the prosthesis and the orientation of the femoral and tibial attachment sites. This additional loading could lead to early failure of the device. An anatomical study (Drouin et al., 1983) has been combined with a kinematic knee model in order to evaluate the tension, torsion, and bending of the ligament during flexion as a function of attachment site orientation. Seven and six orientations have been investigated on the femur and the tibia, respectively. This study showed that the tibial attachment has less influence on the loading than the femoral attachment sites. For a knee flexion angle of 100°, torsion varied from a minimum of 35° to a maximum of 85° by changing the orientation of the femoral attachment site. The bending imposed on the prosthesis was also analyzed and similar influences were found.

The choice of the implantation site must be made with due consideration of the combined loading. For instance, since the ligament is under tension near full extension it is important that the torsion and bending be reduced at this knee position. The best compromise for the orientation of the femoral attachment channel is an origin at the insertion of the natural ACL and the exit point located on lateral-posterior aspect of the femur.

Initial Force at Implantation

As was discussed, the force-deformation curve is non-linear; therefore the force exerted by the prosthesis on the bone is the sum of its initial force at implantation and the added force caused by the displacement of its extremities. If at implantation the initial force is too high the prosthesis will be too stiff and will be subjected to undesired additional loading; if the prosthesis is not preloaded enough at surgery it will not perform adequately and the knee will remain unstable.

The main problem is to determine the loading on the ligament versus the physiological flexion of the knee. Many studies have been published on the subject (Girgis, 1975; Grood et al., 1977). These data are obtained by measuring the distances between insertion sites during knee motion. However, because of the construction of the ligaments, geometric data are not sufficient and they should be coupled to a force measurement. A protocol is presently under development in our group concerning this problem.

Future Research

The mechanical behavior of the ligament is obtained from tests in a tensile testing machine. It is very difficult to relate this type of loading to the physiological loading on the ligament, especially considering that the ligament is a composite structure that has evolved to fulfill its function. An effort should be made to devise other characterization methods that are better suited to the study of ligamentous structures.

The role of each constituent of the ligament should be established in relation to its spatial arrangement so that in the future the mechanical properties of ligaments could be deduced by microanatomical study.

Finally, there is a lack of diagnostic tools, devices, or protocols to characterize or establish the traumatized state of a ligament. Their development should be based on a better knowledge of the relation between the motion of the knee and the loading on each ligament.

References

ALEXANDER, H., Parsons, J.P., Strauchler, I.D., Corcoran, S.F., Gona, O., Mayott, C.W., & Weiss, A.B. (1981). Canine patellar tendon replacement with a polylactic acid polymer-filamentous carbon tissue scaffold. *Orthopaedic Review,* **10**, 41.

DORLOT, J.M., Ait Ba Sidi, M., Tremblay, G.R., & Drouin, G. (1980). Load elongation behavior of canine anterior cruciate ligament. *Journal of Biomechanical Engineering,* **102**, 190-194.

DROUIN, G., Doré, R., Thiry, P.S., & Jean-Francois, C. (1980). Modelling of a composite prosthesis for quasi-cylindrical ligaments. *Journal of Biomechanical Engineering,* **102**, 194-198.

DROUIN, G., Gely, P., Tiberghien, M., & Tremblay, G.R. (1983). Prosthèse ligamentaire du croisé antérieur: importance de la fixation. *17th Annual Meeting of the Canadian Orthopaedic Research Society*. Juin. Québec.

GIRGIS, F.G., Marshall, J.L., & Al Monajem, A.R.S. (1975). Cruciate ligaments of the knee joint. Anatomical, functional and experimental analysis. *Clinical Orthopaedics and Related Research,* **106**, 216-231.

GROOD, E., Noyes, F., & Butler, D. (1977). Age related changes in the mechanical properties of the knee ligaments. In R. Skalar and A.B. Schultz (Eds.), *Biomechanics Symposium ASME.*

JOHNSON, R.J. (1983). The anterior cruciate ligament problem. *Clinical Orthopedics and Related Research,* **172**, 14-18.

KENNEDY, J.C. (1983). Application of the prosthetics to anterior cruciate ligament reconstruction and repair. *Clinical Orthopedics and Related Research,* **172**, 125-128.

YAHIA, H. (1983). Localization of elastin in canine anterior cruciate ligament by enzyme gold complexes. *9th Annual Meeting of the Society for Biomaterials.* Birmingham.

Evaluation of Foot Deformity in Friedreich's Ataxia Using a Three-Dimensional Geometric Model

J.P. Sirois, P. Allard, P.S. Thiry, and G. Drouin
École Polytechnique Pediatric Research Center, Ste-Justine Hospital
and McGill University, Montreal, Canada

The quantitative description of the changes in the foot's bony structure with disease, treatment, and age has always been of limited statistical value (Gamble & Yale, 1975). This is due to the foot's complex architecture (26 irregularly shaped bones articulating about greatly divergent axes of rotation) as well as the methods of investigation. Classical radiographic techniques have shortcomings in that they are very sensitive to the angle of observation and to relative foot position (Hlavac, 1967). The two most common radiographic views (sagittal and transverse) tend to be observed independently from one another (Gamble & Yale, 1975). Moreover, since x-rays are often produced in non-weight bearing conditions, their clinical significance is limited.

In order to better describe the bony structure of the foot, a three-dimensional geometric model based on a new standardized x-ray technique is proposed. Its application to a Friedreich ataxia patient is presented.

Modeling of the Foot

Modeling of the bony structure of the foot involves four distinct stages: (a) production of x-ray images; (b) anatomical identification and digitization of bony structures on two x-rays; (c) resection of these structures in three-dimensional space and finally (d) description of the geometry of the bones.

Standardized Radiographic Technique

The location of the x-ray source, the condition of weight or non-weight bearing and the position of the foot with respect to the lower leg are all variables influencing x-ray images of the foot. In order to control these variables an apparatus for the standardization of foot x-rays has been designed (Sibille, Tremblay, Thiry, & Allard, 1981). This apparatus, illustrated in Figure 1, immobilizes the patient's feet while x-ray images are produced. The feet are plantar flexed 15° in order to reduce bone image superposi-

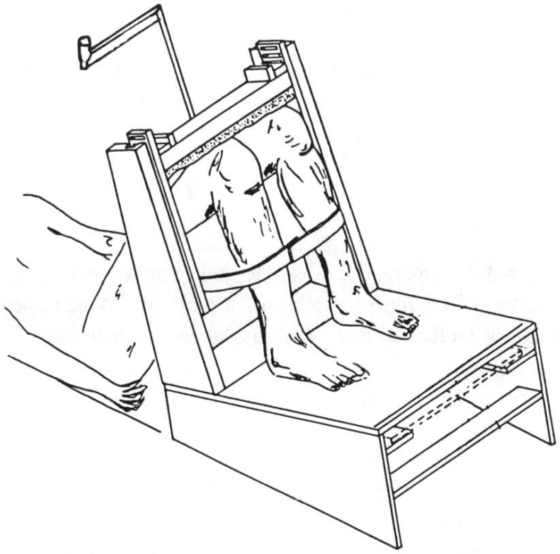

Figure 1—X-ray standardizing apparatus.

tion in the hind foot. Weight bearing is simulated by the application of a calibrated force to the flexed knees, thus the apparatus is equally well suited for ambulant and non-ambulant patients. With this apparatus, it is possible to produce standardized images of the antero-posterior, sagittal, and oblique planes. However, being primarily interested in three-dimensional reconstruction, a 20° divergent stereoimagery technique is used in this model. More specifically, the first radiograph is the antero-posterior view while the second is an oblique view produced by laterally shifting the x-ray source. In these x-ray views all the bones of the foot are distinguishable.

Anatomical Landmark Recognition

Spatial resection of a bone requires recognition of distinct anatomical landmarks on two differently oriented x-ray images. These requirements coupled with the scarcity of easily recognizable features on all bones render the use of special recognition techniques mandatory. Two techniques are used: (a) the midpoint of a contour: when an anatomical structure such as a smooth articular margin can be delineated, the midpoint of this contour is considered as a distinct anatomical point; (b) area centroid: by subdividing images of irregularly shaped bones into finite areas, the centroids of these areas are considered distinct anatomical points. The choice of low divergence (20°) stereoimages minimizes errors inherent in these approximations.

Resection of Anatomical Landmarks in a Three-Dimensional Space

Following their identification in both x-rays, anatomical landmarks as well as twelve calibration objects positioned in the space above the patient's feet are digitized using a Hipad tablet. The internal orientation of each projection as well as the three-dimensional position of anatomical landmarks are obtained by a Direct Linear Transformation Technique (DLT) (Marzan & Karara, 1976). Experiments on a human foot skeleton indicate that the precision obtained on the spatial position of anatomical structures is within 4 mm.

Three-Dimensional Model

The model proposed documents the absolute and relative position and orientation of the bones of the foot. The representation of a particular bone depends on the complexity of its shape as well as of its typical displacement about its articulations. Figure 2 illustrates this model.

Long slender bones, such as the metatarsals and phalanges, articulating primarily about fixed axes of rotation, are modeled by a line segment originating on their head and terminating on their base. Though this two-point schematization of long bones does not completely determine their position, the neglected axial rotation is of much lesser importance than the longitudinal rotation.

The irregularly shaped bones of the hindfoot and midtarsus, whose axes of rotation are not fixed in space, are modeled by three line segments originating on three anatomical landmarks of known position. One can thus completely describe the three-dimensional orientation of these bones. The orientation of these bones is taken as that of the normal to the triangular surface formed by the three line segments originating on the three anatomical landmarks.

Finally, the ankle joint has been included in the model because its configuration has been shown to have far reaching effects on the foot's structure (Hlavac, 1981). It is modeled by a line segment passing through the inferior extremities of the malleoli.

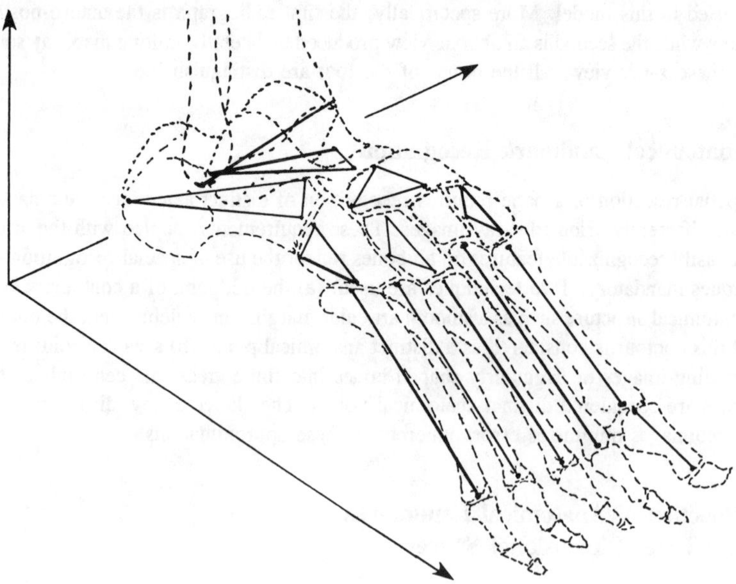

Figure 2—Geometric model of the foot.

From such a "stick diagram" model of the foot, many new indices can be derived. For example, articular subluxation can be estimated from the distance between anatomical landmarks on each of the mating articular surfaces. The relative orientation of two bones is obtained from the spatial angle between them.

Results

The evaluation of evolving foot deformities in Friedreich's ataxia using this three-dimensional model has recently been undertaken. The structures of particular interest in this neuromuscular disease are the metatarsals, the talus and the calcaneus. Three

Table 1

Orientation of the Metatarsals, the Talus and the Calcaneus in a Cavus Foot Patient

Spatial Angle	Interpretation	Cavus Foot	Foot Skeleton	Difference
a) 1st-5th metatarsals	Forefoot torsion	51°	32°	19°
b) 1st metatarsal-calcaneus	Arch elevation and forefoot inversion	95°	121°	−26°
c) Talus-calcaneus	Calcaneal rotation	54°	40°	14°

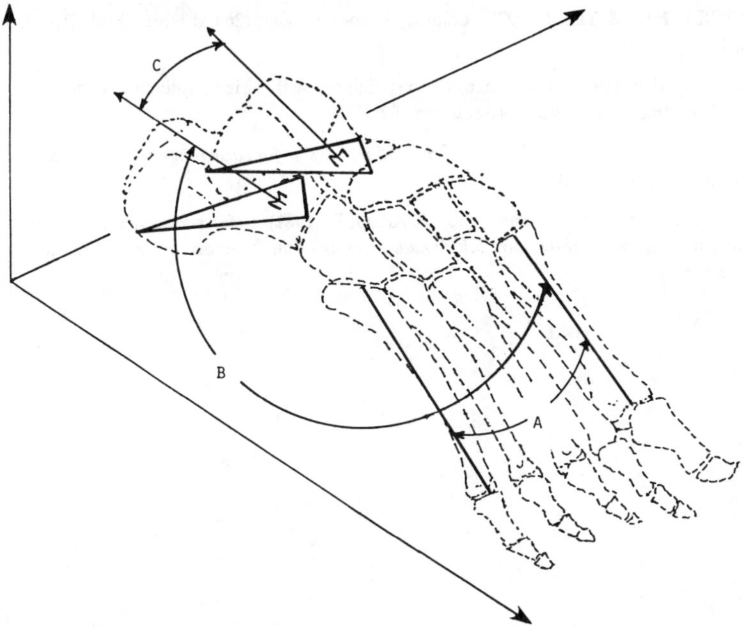

Figure 3—Three-dimensional indices.

related three-dimensional indices are introduced in Table 1 together with their interpretation. Since clinical norms for these new indices have yet to be determined, the preliminary results from the analysis of a patient with cavus feet are compared to those obtained from a foot skeleton. The differences observed indicate by their importance the advanced state of cavus deformity and agree with a previous study (Allard, Sirois, Thiry, Geoffroy, & Duhaime, 1982). Figure 3 illustrates the disposition of these three indices.

Conclusion

The model proposed is relatively simple, yet it satisfies the following clinical requirements: (a) the x-ray exposure required for the standardized technique is equivalent to that of conventional techniques; (b) the execution of the technique is straightforward; (c) most importantly, it brings forth tangible and accurate diagnostic parameters.

Three-dimensional modeling of the foot offers a great number of new indices whose clinical significance is being established through the study of various foot deformities. These studies make the need for a normal population data bank evermore pressing.

References

ALLARD, P., Sirois, J.P., Thiry, P.S., Geoffroy, G., & Duhaime, M. (1982). Roentgenographic study of cavus deformity in Friedreich's Ataxia patients. *Canadian Journal of Neurological Sciences*, **9**, 113.

GAMBLE, F.O., & Yale, I. (1975). *Clinical foot roentgenology* (2nd ed.). New York: R.E. Krieger Co.

HLAVAC, H.F. (1981). Differences in x-ray findings with varied positioning of the foot. *Journal of the American Podiatry Association*, **57**, 465.

MARZAN, G.T., & Karara, H.M. (1976). *Rational design for close range photogrammetry*. U.S. Dept. of Commerce, PB-252-447.

SIBILLE, J., Tremblay, C., Thiry, P.S., & Allard, P. (1981). Reference apparatus for normalized bi-planar radiographs of the foot. *5th Annual Meeting of the American Society of Biomechanics*. Cleveland.

III.
BIOMECHANICS OF SPINE AND ITS REHABILITATION

The Human Spine: Story of Its Biomechanical Functions

M.M. Panjabi
Yale Medical School, New Haven, Connecticut, USA

The human spine is a mechanical structure consisting of 24 vertebrae that connect the head to the trunk and trunk to the pelvis. The size of the vertebrae increases from the first cervical to the last lumbar vertebra. The spine has three curvatures: the cervical and lumbar are convex towards the anterior, while the thoracic is concave. The mechanical design of the human spinal column is such as to provide three basic functions: to allow motion between body parts, to carry loads, and to protect the delicate spinal cord. However, there are situations where the protection is not enough and there is injury of the spine. The purpose of this paper is to describe the three basic functions and how they are carried out; and to present a biomechanical model for the evaluation of the injured spine.

During flexion/extension, lateral bending and axial rotation, the 24 vertebrae allow significant amounts of movement between the head and pelvis: a total of 250° of sagittal plane rotation, 150° of bending from side to side, and 180° of rotation. Although it is interesting to know the total movements of the human spine, it is the relative movements at each vertebral level that are important, both biomechanically as well as clinically. This is done with the help of a functional spinal unit which consists of two adjacent vertebrae connected together by nine spinal ligaments and the disc.

A composite graph (see Figure 1) shows the ranges of motion of the vertebrae in the three planes of movement (White & Panjabi, 1978). In flexion/extension, motion in the cervical spine is uniform with a peak at C5-C6. The motion increases from T1-T2 to L5-S1. In lateral bending, motion is nearly constant within the entire spine except at C1-C2, where it is nearly zero. At Occiput-C1, the axial rotation is very small while at C1-C2, it is rather large. There is a sharp decrease in this motion in the lower thoracic and entire lumbar spine. It may be of interest to note that the highest number of thoracolumbar injuries are concentrated in this region.

The coupling patterns in the spine are such that axial rotation is coupled to lateral bending. This coupling pattern is most strong in the cervical spine, rather weak in the thoracic spine, and again stronger in the lumbar region. The direction of the coupled motion in the lumbar region is opposite to that in the cervical spine. These coupling patterns may be helpful in the diagnosis of internal derangement that might have taken place in the spine due to injury or disease.

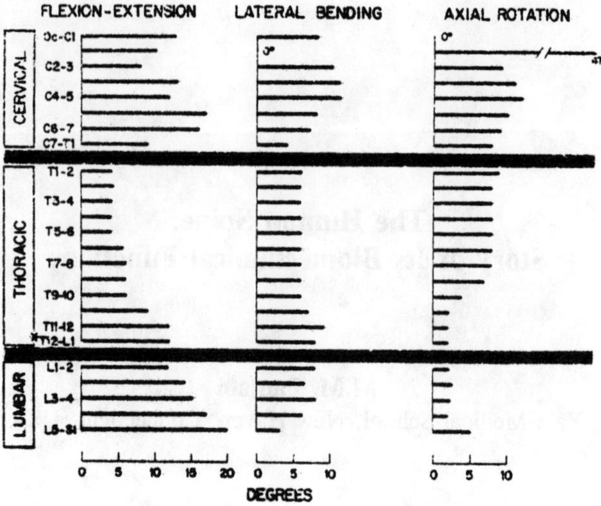

Figure 1—Representative values for rotation at the different vertebral levels of the spine for the three traditional planes of motion (White & Panjabi, 1978).

Axial compression strength of the vertebrae increases from the first cervical vertebra to the last lumbar vertebra. The strongest vertebra can carry a load of about 8 kN at slow speed and up to 12 kN at high speed of trauma.

Load on the spine in standing is estimated to be about 800 to 900 N (Nachemson, 1976). Considering this to be equal to 100%, we see that there is a dramatic increase if we flex and carry loads in our hands (see Figure 2). Also note that the loads during sitting are higher than those during standing. The lowest loads on the spine are obtained by lying down, especially on the back.

Protection of the spinal cord is essential for survival. The intensity of the traumatic energy coming from outside the body is decreased and diffused by the various soft and hard tissues that lie between the skin and the spinal cord. These structures are injured in trauma, thus dissipating the energy of the trauma. The traumatic energy is further dissipated by producing injuries of the components of the spinal column.

The protection does not always work. We know there are spinal injuries when the protection mechanism is overwhelmed by the traumatic forces. Once the patient comes to the emergency room with a broken neck or back, it is important to have information that will be helpful in making the decision regarding the treatment. Should the patient have a simple soft collar and go home, or should a high risk surgical procedure be adopted? The decision is difficult and important.

A biomechanical model has been developed to provide objective information both for the normal as well as the injured spine and thus help in the evaluation of clinical stability of the spine after injury (Panjabi, White, & Johnson, 1975). The cervical spine was chosen as the first region to be studied. Experimental design consisted of functional spinal units and applying flexion and extension loads of 1/4 body weight force applied horizontally. The resulting sagittal plane motion was measured. Ligaments were then transected either from anterior to posterior or from posterior to anterior. With this experimental design we hoped to obtain thresholds of motion, before and after

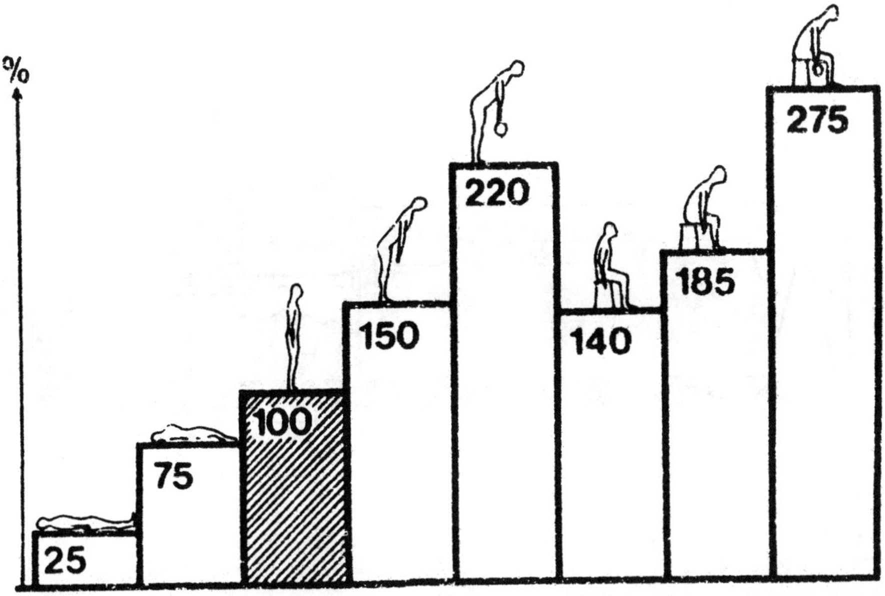

Figure 2—A diagrammatic comparison of in vivo loads in the third lumbar disc during various activities. Note that the load on the spine while sitting is greater than that in standing (Nachemson, 1976).

injury, that would clearly separate the injured spine, requiring surgical stabilization, from the normal or inherently stable spine requiring no surgical treatment. The experimental set up is seen in Figure 3.

A total of 17 functional spinal units were tested. For the intact unit, the range of translation was 0.28 to 2.67 mm. After cutting of ligaments, just prior-to-failure, this range increased to 2.01 to 4.89 mm. Note that there is an overlap between the intact and the prior-to-failure specimens in their mechanical behavior. The same was true for the vertebral angulation. We have also found the same trends in similar studies conducted in the thoracic and lumbar regions.

These observations and other clinical considerations have led us to develop the so-called checklist of clinical instability of the spine (White, Southwick, & Panjabi, 1976). This list consists of several items, some of them based upon our biomechanical studies while others are from the clinical experience, including the neurological status as well as the potential loads to which the spine may be subjected. While examining the patient, each item is given a point value. All the points are added. If the sum of the points is greater than or equal to five, the spine may be considered unstable, requiring surgical stabilization. Significant clinical experience will be required in the future to validate and modify the checklist proposed. We realize the limitations of our studies. However, we believe this is the first attempt at providing objective answers to a problem of great clinical significance.

In summary, the human spine is a chain-like mechanical structure of significant complexity, delicacy, and importance to the survival of the human body. Biomechanical studies of the human spine are not only interesting for their own sake, to provide the

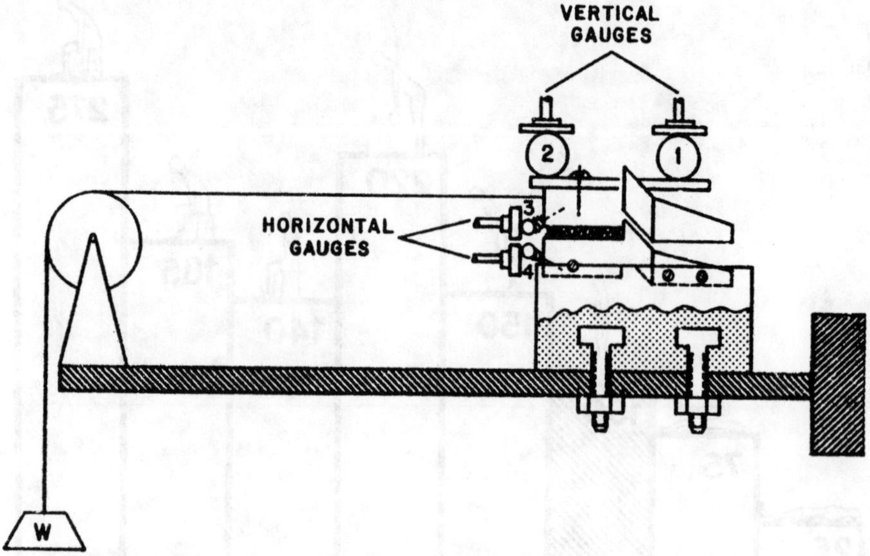

Figure 3—The experimental setup for the biomechanical model to determine the cervical spine threshold of spinal instability (Panjabi, White, & Johnson, 1975).

basic understanding of the mechanics of the spine, but also are important for the clinical understanding and treatment of spinal injuries and diseases. The future holds more exciting research. The spine is a three dimensional structure and this aspect must be explored to the fullest in the future. Modeling of spinal injuries must become more realistic. In vitro models, although anatomically accurate to the human spine, cannot simulate the healing process within the body. Suitably chosen animal models could address this aspect of the problem. Finally, mathematical models must be constructed that are sophisticated, not only mathematically, but also in the depiction of the reality of the human spine. These models will help us to bring the benefits of research to the patient.

Acknowledgments

Supported by N.I.H. Grants NS10174, AM30361, and AM00299, and a grant from the Veterans Administration.

References

NACHEMSON, A.L. (1976). Lumbar spine. An orthopaedic challenge. *Spine*, **1**, 59-71.

PANJABI, M.M., White, A.A., & Johnson, R.M. (1975). Cervical spine mechanics as a function of transection of components. *Journal of Biomechanics*, **8**, 327-336.

WHITE, A.A., & Panjabi, M.M. (1978). *The clinical biomechanics of the spine.* Philadelphia: J.B. Lippincott Co.

WHITE, A.A., Southwick, W.O., & Panjabi, M.M. (1976). Clinical instability in the lower cervical spine: A review of past and current concepts. *Spine,* **1**(1), 86-99.

Estimation of the In Vivo Load in the Lower Spine Based on a Semidirect Approach

F. Schläpfer and B. Nigg
University of Calgary, Calgary, Canada

F. Magerl
Kantonsspital, St. Gallen, Switzerland

S.M. Perren
Laboratory for Experimental Surgery, Davos, Switzerland

Researchers are interested in the magnitude of the load in the spinal column. Theoretically there are three possible approaches: direct, semidirect, and indirect. The direct approach involves direct measurement of the spinal load through a transducer which replaces a load bearing element of the spinal column. The authors are not aware of any laboratories using this approach.

The semidirect approach is based on in vivo measurements of variables related to the spinal load. Although a few methods exist to determine such variables, the relationship between them and the spinal load is generally unknown. In order to determine the spinal load from the measured variables, this relationship has to be derived either theoretically or experimentally by in vitro investigations. Only in the case of the intradiscal pressure measurements has a partial relationship been experimentally derived (Nachemson, 1966). However, this method is limited to conditions in which the axial spinal force is dominant.

The indirect approach requires modeling of the path along which an externally acting load is transmitted to the spine. Various authors have proposed models to describe quasi-static load situations. However, in dynamic loading the influence of the neuromusculoskeletal control system has to be included in the model. The authors are not aware of any such complex model.

In this paper, a semidirect method is described using an instrumented external fixator (Magerl, 1982; Schläpfer, Magerl, Jacobs, Perren, & Weber, 1980) to estimate the spinal loads in the sagittal plane.

Material and Method

An instrumented external fixator was used to measure in vivo variables related to the spinal load. The device is in clinical use to stabilize the spine (see Figure 1) (Magerl, 1982). Screws placed through the pedicles into the bodies of the vertebrae adjacent to the unstable segment are externally connected by a frame. The connecting clamps are designed as load cells which measure the moments acting in the planes defined by the individual clamps and screws.

The estimation of the in vivo spinal load is discussed for one patient. The subject had a spondylitis at the spinal level L1/L2. The affected area included the intervertebral disc L1/L2 and a part of the bodies of L1 and L2. In a first operation the affected tissue was removed. The external fixation device bridged the spine from T12 to L3. In addition, a dorsal spondylodesis was made with cancellous bone grafts between the spinous processes of L1 and L2. Three weeks later a second operation was performed to place two ceramic blocks together with cancellous bone into the defective segment.

Before and after the implantation of the ceramic blocks measurements were recorded as the patient slowly raised the arms in the sagittal plane holding weights of 0, 10, 20, and 40 N (held with both hands) to a horizontal position. The difference of the moments in the clamps between the initial and final position of the arms was recorded.

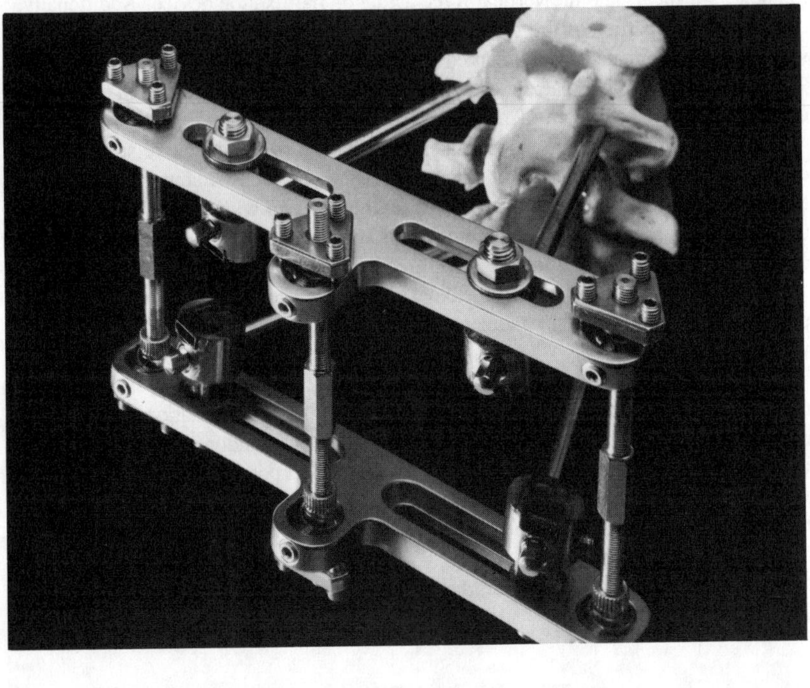

Figure 1—External fixator used to stabilize spinal fractures. The clamps connecting the screws with the frame are instrumented with strain gauges.

Measurements (numbered 1, 2, 3, 4, 5) were performed 1 week before and 4, 6, 11, and 13 weeks after implantation of the ceramic blocks. Measurement 5 was on the day of removal of the fixation device.

A two-dimensional model described the relationship between the moments measured in vivo in the clamps and the three spinal load components: axial force, a-p moment, and a-p shear. The part of the model describing the load-deformation behavior of the fixator was experimentally determined, while the load-deformation behavior of the stabilized spinal segments had to be theoretically deduced. The individual motion segments were considered as behaving in a linear fashion during flexion/extension and shear. The stiffness properties for the intact segments were 200 Nm/rad for flexion/extension (Markolf, 1972) and 550,000 N/m for the a-p shear (Liu, Ray, & Hirsch, 1975). The load-deformation characteristic of the intact segments relative to compression was represented by a nonlinear equation extrapolated from the data of Markolf and Morris (1974). Due to the nonlinearity, the mechanical behavior of the motion segments depends on the axial preload. Based on the data of Nachemson (1966) an axial preload of 900 N was chosen. The material properties of the defective segment were introduced into the model as unknowns.

The mechanical behavior of the system spine-fixator is derived by superimposing the deformations of the various spinal levels while assuming a rigid connection between the fixation device and the spine. The resulting model consists of three nonlinear equations and four unknowns: three spinal load components (axial force, a-p bending moment, a-p shear) and the vertical force acting on the fixation device. Additionally, the material properties of the defective segment are not known. Since the model is statically indeterminate, it was assumed that there was no cocontraction of the muscles. Thus, the axial force was calculated considering the spine as a lever system (lever arm of weight 0.6 m, lever arm of muscles 0.04 m).

The calibration for the unknown material properties of the defective segment was calculated using the data gathered from arm raising with and without 10 N. Calibration is based on the following assumptions: (a) the behavior with respect to flexion/extension and shear is linear; (b) before implantation of the ceramic blocks the force/deformation characteristic in the longitudinal direction is the same as for the intact motion segments; (c) after implantation of the blocks, the segment is rigid with respect to compression; and (d) the spinal a-p moment is zero after the second operation for arm lifting with and without 10 N.

Material properties prior to the second operation were calculated using various moments and material constants applied to the model. Thus, information was obtained on the deformation in the stablized spinal part and on the position of the new fulcrum in the defective segment. A comparison of the estimated position of the fulcrum with x-rays and check of the mechanical significance of the calculated deformation was used to approximate the material data of the defective segment as well as the moments produced by the back muscles.

After the calibration of the model, the spinal load was calculated for arm lifting with 20 and 40 N.

Results and Discussion

Based on the calibration of the model, the stiffness of the defective motion segment without the ceramic blocks was 2 Nm/rad in flexion/extension and 100,000 N/m in a-p shear. After the implantation of the ceramic blocks, the flexion/extension stiffness increased to 100 Nm/rad and shear stiffness to 550,000 N/m.

In Table 1 the calculated maximum values of the axial spinal force, the a-p moment, and the a-p shear are shown.

Measurement 1 (prior to implantation of ceramic blocks) demonstrated a large calculated extension moment which increased as weight increased. It seems that in the case of a large defect like the one studied, the back muscles play an important role in maintaining the anatomical position of the spine. Due to the moment produced by the back muscles, the spinal axial force prior to the implantation of the ceramic blocks was larger. The shear (posterior) force increased with increase in the weight from 0 to 10 N and decreased slightly for weights ranging from 10 to 20 N. It appears that with increasing weight the stabilized spinal segments started to flex in spite of the increasing extension moment.

Measurements 2 to 5 show that implantation of the ceramic blocks caused a drastic decrease of the axial force and the moment. With increasing weight an increased flexion moment was seen, presumably due to increased tendency of the subject to flex. The calculated shear force was posteriorly directed and increased from measurement 2 to 4 while measurement 5 showed a drastic decrease. The various measurements show the initial increase of the shear force with increasing weight but slowed down at larger weights. In the case of measurements 2 and 3 a decrease in shear force was observed,

Table 1

Calculated Maximum Spinal Force (F_{ax}), a-p Moment (M_{ap}) (Positive: Extension) and a-p Shear (S_{ap}) (Positive: Posterior)

Measurement	Weight (N)	F_{ax} (N)	M_{ap} (Nm)	S_{ap} (N)
1 (before)	0	420	4	14
	10	605	5	25
	20	810	6.5	24
2 (after)	0	320	[0]	25-32
	10	480	[0]	28-36
	20	640	−0.5	24-32
	40	960	−0.5	18-30
3 (after)	0	320	[0]	23-35
	10	480	[0]	33-44
	20	640	−0.4	29-41
	40	960	−0.5	29-34
4 (after)	0	320	[0]	31-35
	10	480	[0]	33-43
	20	640	−0.2	30-47
	40	960	−0.3	24-47
5 (after)	0	320	[0]	18-24
	10	480	[0]	22-32
	20	640	−0.6	15-34
	40	960	−0.5	33-37

Values in the brackets are assumed values.

presumably due to increasing cocontraction of the trunk muscles. This up-and-down behavior of the shear force gradually disappeared from measurements 2 to 4. Measurement 5 represents a special case. It seems that during the measurement, cocontraction of the muscles occurred. This assumption is strengthened by the observation that the subject was nervous due to the upcoming removal of the fixator.

Conclusion

A semidirect approach is presented, which is based on in vivo measurements performed with an instrumented external fixator. The relationship between the measured variables and the spinal load is described by a two-dimensional model. In its present state the usefulness of the model is limited due to a lack of information about the vertical force acting on the fixator. Thus, muscle cocontraction cannot be considered. An improved measuring technique will allow consideration of cocontraction.

Only a small part of the spine (2 to 4 segments) has to be modeled to describe the relationship between the measured variables and the spinal load. Thus, it may be assumed that the model is independent of the physiological (muscle activities, intraabdominal prssure, etc.) and mechanical (spring-damping) behavior of the body, that is, the model may be derived from mechanical in vitro tests performed on isolated human spinal specimens. The variables measured with the fixator reflect the influence of the physiological and mechanical behavior of the body on spinal load.

The approach described is based on measurements done on patients. Hence, the mechanical properties of the modeled spinal segments change as healing occurs. It is difficult to estimate these alterations by means of in vitro tests. An in vivo calibration of the model is desirable. Based on special body exercises (e.g., arm raising with small weights, used in this study), at least static calibration may be possible.

The relevance of the information on the spinal load gathered with the above-described semidirect approach is as follows. In certain situations (shortly after surgery or in situations such as measurement 1) the observed load condition corresponds to the one seen in an unstable spine. The information gathered in such cases is of importance, for instance, in the design of fixation devices or the planning of a rehabilitation program. Information about spinal load in a healthy person (important for ergonomics, etc.) may be expected toward the end of the healing process when the situation starts to "normalize."

Acknowledgment

Supported by the Swiss Research National Foundation Grant No. 83.929.1.81.

References

LIU, K.Y., Ray, G., & Hirsch, C. (1975). The resistance of the lumbar spine to direct shear. *Orthopedic Clinics of North America*, **6**, 33.

MAGERL, F. (1982). External skeletal fixation of the lower thoracic and the lumbar spine. In H.K. Uhthoff (Ed.), *Current concepts of external fixation of fractures* (pp. 353-366). Heidelberg: Springer-Verlag.

MARKOLF, K.L. (1972). Deformation of the thoraco-lumbar intervertebral joints in response to external loads. *Journal of Bone and Joint Surgery,* **54A**(3), 511-533.

MARKOLF, K.L., & Morris, J.M. (1974). The structural components of the intervertebral disc. *Journal of Bone and Joint Surgery,* **56A**(4), 675-687.

NACHEMSON, A. (1966). The load on lumbar discs in different positions of the body. *Clinical Orthopedics,* **45**, 107-122.

SCHLÄPFER, F., Magerl, F., Jacobs, R., Perren, S.M., & Weber, B.G. (1980). In vivo measurements of loads on an external fixation device for human lumbar spine fractures. Proceedings of the Conference on Engineering Aspects of the Spine. *I. Mech. E.,* **C131**(80), 59-64.

An Interactive Algorithm to Identify Segmental Stiffness Properties— A 3-D Spine Model

M. Daniele
Illinois Institute of Technology, Chicago, Illinois, USA

R. Vanderby
Northwestern University, Chicago, Illinois, USA

A.G. Patwardhan
Rehabilitative Engineering and Development Center,
VA Hospital, Hines, Illinois, USA

To evaluate and predict surgical correction of spinal deformities, it is a comon procedure to model the spine as a mechanical system. This typically involves the determination of segmental stiffness properties by direct experimentation performed on cadaver specimens. The problem inherent in this approach is that stiffness properties so obtained generally do not reflect the response of the in vivo spine. Trunk musculature, the ribs and intercostal muscles, and even intraabdominal pressure all add additional stiffness to the vertebral column, and are absent during laboratory testing. Moreover, cadavers with pathological spines are extremely rare, resulting in a marked paucity of experimentally obtained stiffness properties.

To go beyond basic understanding and approach clinical usefulness, a spine model must be capable of describing the in vivo response of pathological spines in individual clinical cases. The model described herein uses an existing set of in vivo load-displacement data to identify in vivo segmental stiffness properties for the case in question. These data are available from two orthogonal radiographs showing initial and final geometry of the spine due to a known loading. Thus simple preoperative traction could yield the data to identify the required stiffness properties, and the model could be used as a clinical tool to aid in designing the corrective instrumentation required. The identification of stiffness properties is performed by formulating an optimal design problem, indicating which of the model's constitutive parameters to change to minimize the difference between the model's calculated displacements and the actual displacements as shown by the radiographs.

Methods

A static nonlinear finite element code was developed in which each spinal motion segment is modeled as a single 3-D beam element with end nodes located at the centroids of adjacent vertebral bodies. This creates a piecewise continuous model in which the ambiguous in vivo constitutive factors are "lumped" into the beam elements, and a correct equivalent stiffness is identified. Standard linear elastic elements were used, and the usual finite element equation results:

$$KU = F \tag{1}$$

Geometric nonlinearity was incorporated by a solution algorithm of applying an incremental load, solving Equation 1 for incremental displacements, and reformulating the stiffness matrix before the next increment. Figure 1 shows the finite element representation of the spinal configuration and a typical corrective device. Note that the global X-axis defines the first and last nodes. Thus it is required to identify values of I_x, I_y, and I_z, the principal moments of inertia for each beam element so that the model's displacements match those of the radiograph. A nodal error function is defined at each node as the difference between the model's displacement (U_i) and that of the radiograph (d_i):

$$\omega_i = (U_{xi} - d_{xi})^2 + (U_{yi} - d_{yi})^2 + (U_{zi} - d_{zi})^2 \tag{2}$$

Following the State-space procedure of Haug and Arora (1979), we seek a change in the design variables (b) which minimizes an objective function ω_i. The design variables are the moments of inertia:

$$b^T = [I_{z1}, I_{y1}, I_{x1}, \ldots . I_{xn1}] \tag{3}$$

where n1 is the number of elements. This procedure is interactive, and the first step is to calculate the matrix of design derivatives (L):

$$L = [L_1, L_2, \ldots \ldots L_{nf}]$$

$$L_i = \partial\omega_i/\partial b$$

Here nf is the number of nodal error functions. We use Equation 1 as the state equation:

$$h(b,U) = KU - F = 0 \tag{4}$$

A linear approximation to the state equation is made, and the following equations result:

$$\partial\omega_i/\partial b = -\lambda_i^T \, \partial h/\partial b \tag{5}$$

Where the adjoint (co-state) vector λ is calculated as:

$$K\lambda_i = (\partial\omega_i/\partial U)^T \tag{6}$$

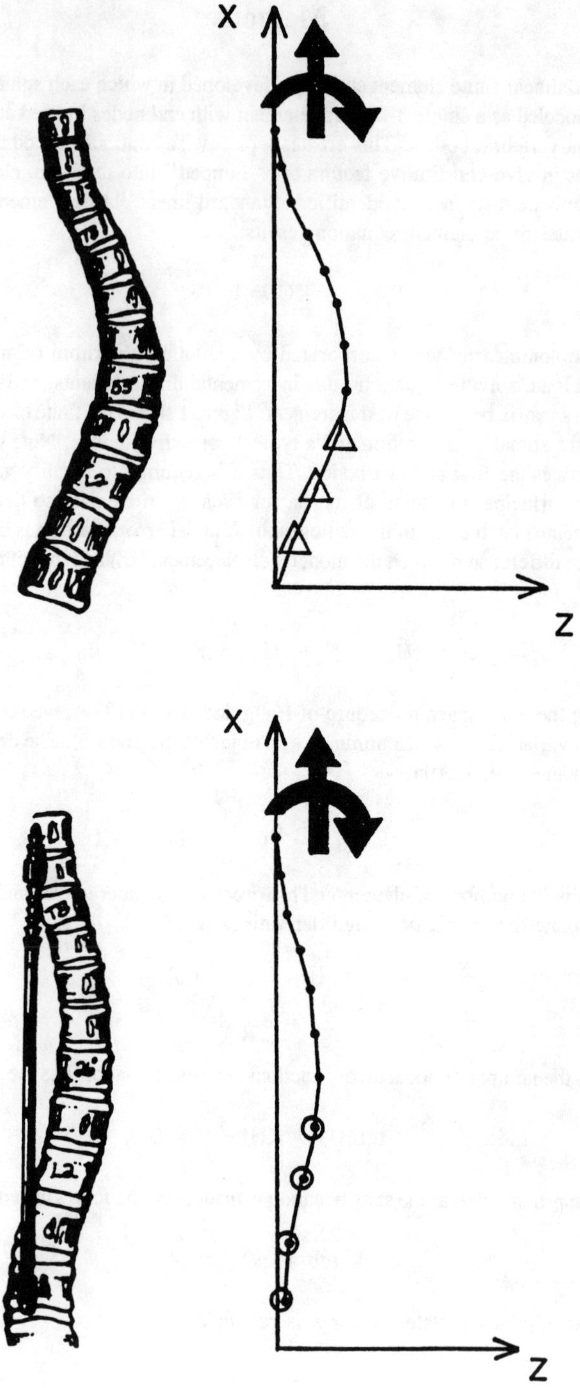

Figure 1—Modeling the instrumented spine.

By considering the stiffness matrix as the sum of the element matrices:

$$\partial h/\partial b_i = [A_i^T(\partial k_i/\partial b_i)A_i]\ U \tag{7}$$

where A is the sparse Boolean transformation matrix, and k is the transformed element stiffness matrix. The relative magnitudes of the derivatives indicate which design variables should be altered to improve the nodal error functions. While any one of these derivatives may be of biomechanical interest, their number can be unwieldy. That is, several hundred derivatives may be generated for a large problem. Thus a gradient projection algorithm was included to calculate a change in b based on the value of the derivatives. Nodal error functions which are near their optimum value can be imposed as constraints, to preclude design changes which would decrease the objective function, yet increase other nodal error functions. Now use only those columns of L corresponding to m constraint nodes:

$$L = [L_1, L_2, \ldots L_m]$$

$$\delta b = -1/2\gamma\ [L_p + Lv] \tag{8}$$

where L_p is the gradient of the objective function, and v is a vector of length m given by:

$$L_T L_p = -L^T L v \tag{9}$$

Note that Equations 6 and 9 involve solving a system of linear algebraic equations. The finite element code must include a subroutine to accomplish the same task (i.e., reducing the equilibrium equations), and hence this optimization procedure is nicely suited to a finite element type of formulation.

Results

Two cases of scoliosis (3-D curvature) and one case of kyphosis (exaggerated 2-D curvature) were modeled. These a posteriori investigations served to validate the model and evaluate past surgery. In all cases, stiffness properties were identified which yielded displacements in good agreement with the clinical data. In the kyphosis case, the model was then used to conduct a parametric study of the effect of surgical reduction of key segmental stiffnesses, indicating an optimum location and amount of reduction of stiffness. Different stiffness values differ significantly from those established by testing of cadaver specimens (Daniele, 1983) again indicating the need for individual in vivo modeling.

An average session of two hours at an interactive terminal is required to establish the stiffness properties of a 10 element model. The process is enhanced by an interactive graphics device to provide visual comparison of the displacements. This, combined with the design derivatives, is usually enough to make good initial progress. Eventually, the gradient projection algorithm is essential. Figure 2 shows the results of modeling one of the scoliosis cases. These preliminary results are encouraging, and work is continuing to develop a clinical tool of real value.

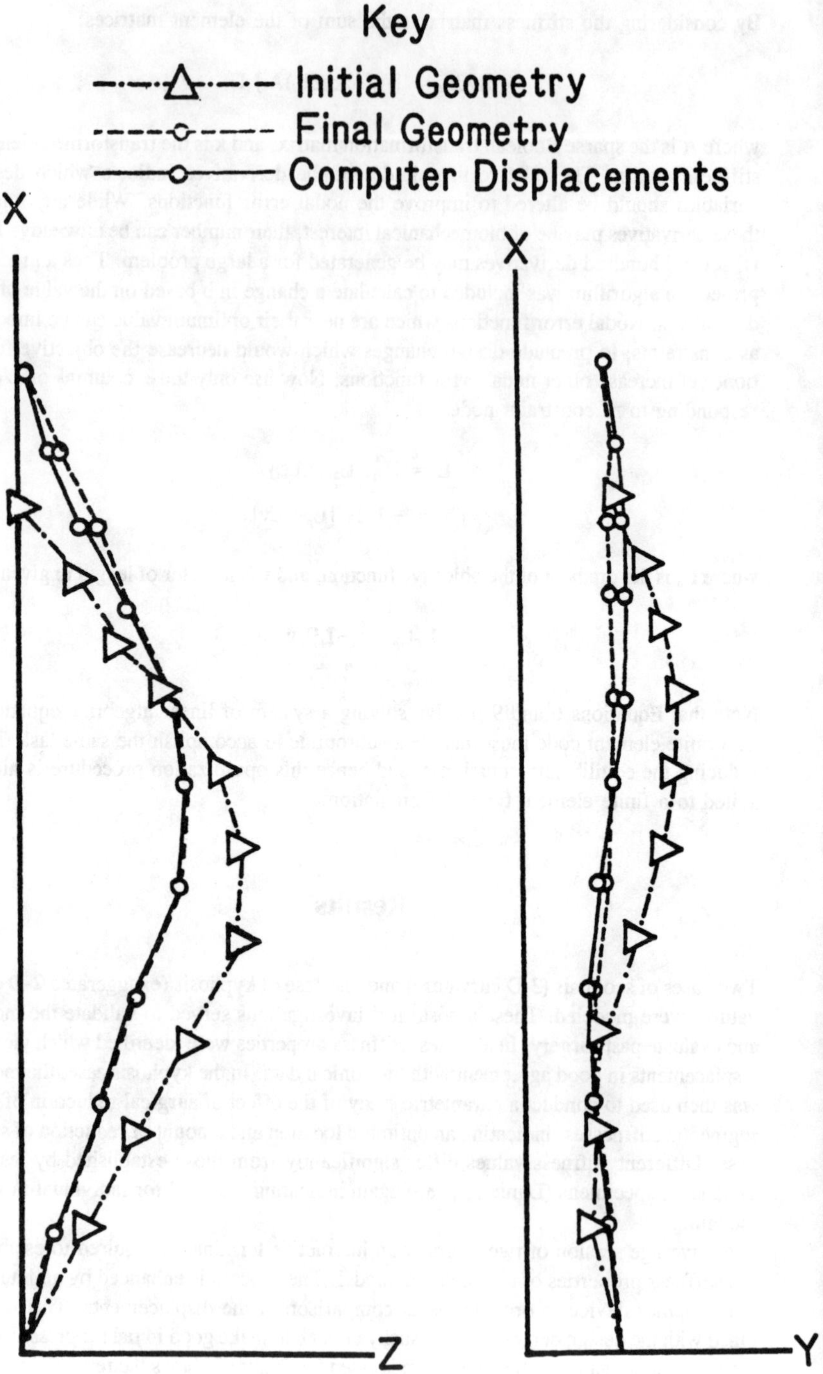

Figure 2—A scoliosis model.

References

DANIELE, M. (1983). *A computer simulation of spinal correction.* Unpublished Master's dissertation, Illinois Institute of Technology. Chicago, Illinois.

HAUG, E.J., & Arora, J.S. (1979). *Applied Optimal Design.* Toronto: Wiley.

Biomechanics of the Harrington Instrumentation for Injuries in Thoracolumbar Region

V.K. Goel
University of Iowa, Iowa City, Iowa, USA

M.M. Panjabi, R. Takeuchi, M.J. Murphy, W.O. Southwick, and R.D. Pelker
Yale Medical School, New Haven, Connecticut, USA

The unstable types of fractures in the thoracolumbar region require operative intervention to stabilize the spine. Various surgical procedures (or techniques) achieve spine stabilization employing either the anterior approach (anterior fracture fixators, plates) or the posterior approach (Harrington instrumentation, plates, Knodt's rod, etc.). Retrospective studies discussing specific treatment procedures or techniques reveal a varying degree of success rate. Although a number of biomechanical in vitro studies have provided some objective evidence regarding the comparative mechanical characteristics of these procedures (Dunn, McBride, & Daniels, 1979; Nagel, Koogle, Piziali, & Perkash, 1981; Purcell, Markolf, & Dawson, 1981; Stauffer & Neil, 1975), the literature lacks data in relation to the effectiveness of different techniques during physiological range of motion, using a truly three-dimensional motion measuring system. This paper describes the methodology and the effect of various techniques on the motion behavior of multilevel spinal segments with particular reference at the injury site. A new procedure involving the use of compression and distraction combination of Harrington rods simultaneously is also evaluated.

Methods

Five ligamentous thoracolumbar spine segments (T8-L4, T10-L5, T9-L4, and two T10-L5) were removed from fresh cadavers and immediately sealed in double plastic bags and stored at −20°C. At a later date the specimens were thawed, cleaned and prepared for testing. A Plastic Padding® block and a loading frame were rigidly secured to the inferior and superior aspects of the specimen respectively for its proper mounting and load applications in a testing rig (see Figure 1). To track the vertebral body position in space, five plates, each carrying three noncolinear markers (Marker Plates,

MP) were secured rigidly to five vertebrae of interest: vertebrae superior and inferior to the vertebral level involved in the injury (no. 3-2), vertebral levels receiving Harrington rods and hooks (no. 4-1) and the vertebra adjacent to the vertebral level receiving the superior hooks (no. 5-4) (see Figure 1). A-P and lateral x-rays were taken to identify the location of markers with respect to the geometric centers (and other points) of vertebral bodies.

The three-dimensional motion behavior of the spine specimens was investigated using the principles of stereo-photogrammetry. The cameras were loaded with 35 mm black and white films and the calibration cage carrying 30 marker points with precisely known 3-D coordinates was mounted in place. The cage was photographed and removed from the setup without disturbing the cameras. The prepared spine specimen (intact) was mounted instead. One of the four clinically relevant load types, flexion/extension, lateral bending or axial torsional moment was applied to the top vertebra through the loading frame. A maximum of 4 N•m moment in four equal steps was applied for each load type with cameras recording the resulting movements of the markers after each step.

The first injury involved cutting of posterior elements (ligaments and facets) and the posterior half of the disc (Half Injury Model, hi). The vertebral levels involved in the injury for various specimens were: T11/T12, T12/L1 (for three specimens) and L1/L2. The 3-D motion behavior for the injured specimen subjected to the same loads was recorded. Thereafter the specimen was stabilized sequentially by a pair of compression Harrington rods (CR), a pair of distraction Harrington rods (DR), and a combination of 2 distraction rods and 1 compression rod (DCR). Harrington rod hooks

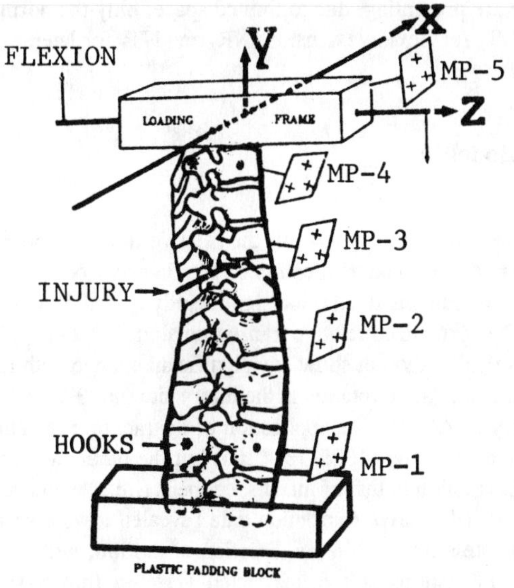

Figure 1—The specimen with a loading frame and a base is shown. The hooks for rods are fixed at levels marked '*.' Marker plates are shown as MP-1 to MP-5. The translation of points '.' were computed. Half injury (hi) and half injury with osteotomy (hio) are shown as — and --- respectively.

were anchored to laminae of vertebrae two levels superior and inferior to the injury. The changes in the motion pattern of the stabilized specimen in response to external loads were recorded after each of these three procedures. Then the specimen was subjected to an additional injury: osteotomy of the vertebra inferior to the involved disc (hio, see Figure 1). The anterior longitudinal ligament was left intact. The testing protocol of the hi model was repeated. Finally, the anterior longitudinal ligament was also cut off with a scalpel (Total Injury, ti) before further testing.

From the photographs the data were analyzed using a Direct Linear Transformation (DLT) program and rigid body mechanics, to yield three rotational (Euler) angles about the X, Y, and Z axes and translations of one point on each vertebra (see Figure 1), along the three axes. These six displacement parameters were calculated with respect to the zero load position. The relative motion: R_X, R_Y, R_Z, and T_X, T_Y, T_Z between no. 4-1, no. 3-2 and no. 5-4 markers for each specimen and each load step were obtained. The percentage change in the relative motion parameters (NR or NT) for the injured and stabilized specimens with respect to *intact specimens* were calculated. Thus the relative motion data for injured and stabilized specimens were nondimensionalized with respect to the respective controls (the *intact specimen data*). This enabled us to compare the results of one specimen with the other. For the final load step student "t-tests" were undertaken to evaluate the significance of the changes in these parameters.

Results

Although the human spine exhibits a three-dimensional motion in response to the load applications, we are presenting, due to limited space, only the normalized major motions: NR_X and NT_Z for flexion/extension, NR_Z and NT_X for lateral bending, and NR_Y for axial rotation.

Half Injury Model (hi)

Flexion. In flexion, after half injury (hi) the relative motion increased by 100% and was significant ($p < .05$). The application of Harrington Rods (any of the three procedures) resulted in reducing the motion significantly ($p < .05$) to −75% with respect to the normal. The normalized relative translation along the Z-axis, NT_Z, between the two vertebrae in flexion did not show any significant change with injury or stabilization. The normalized relative rotation at the injury site (no. 3-2) increased significantly with half injury ($p < .05$). The application of distraction rods alone (DR) resulted in a significant reduction to −100% ($p < .05$) but the other two techniques (CR and DCR) restored the motion to that of intact specimens (i.e., the change was not significant). The normalized relative translation data revealed a wide scatter, although the mean values indicate a reversal in the direction of motion, with respect to the intact specimen. Similar rotations at the unaffected vertebra (no. 5-4) did not increase significantly after half injury but decreased with the application of Harrington instrumentation with the maximum reduction for DR technique and least for DCR technique. The relative translation, NT_Z, increased with half injury (75%) and reduced by −50% with stabilization. The least reduction was observed with the DCR technique.

Right Lateral Bending. The effect of injury in terms of normalized relative rotation, NR_Z, in right lateral bending was insignificant for the no. 4-1 level. The use of compression rods also did not produce any significant alterations with regard to the normal. The other two techniques reduced the motion by −50% ($p < .05$). The change in normalized relative translation, NT_X was found to be insignificant except for DR and DCR techniques (reduced by −30% approximately). The changes in behavior at the injury site (no. 3-2) with half injury and CR technique were insignificant, while for DR and DCR techniques, the decreases in magnitude were significant ($p < .05$). In the latter case the motions reduced by −75%. The effect on the normalized translation, NT_X, was not significant. The production of half injury and thereafter the stabilization did not produce significant alteration in the motion behavior at the unaffected vertebral level.

Right Axial Rotation. The production of half injury increased the relative rotation between no. 4-1 significantly ($p < .05$) while the effect of Harrington rods in stabilizing the injured specimen was to restore the motion to that of intact specimens, that is, the change in behavior was not significant. The motion at the injury site (no. 3-2) was not affected by the injury itself or the rods to any significant level. At the unaffected vertebra (no. 5-4) the use of the DR technique produced a significant increase in motion (+100%) while other procedures did not produce any significant change.

Half Injury Osteotomy Model (hio)

It was not possible to test the specimens after this injury and prior to any stabilization as the specimens were unstable.

Flexion. The normalized relative rotation, NR_X, between the vertebrae receiving hooks (no. 4-1) reduced significantly with the stabilization, irrespective of the type of technique used. The least reduction was provided by the DR technique (−50%) followed by DCR and then CR techniques. The effect of stabilization on the normalized relative translation, NT_Z, was not significant. At the injury site (no. 3-2) the use of DCR technique produced a significant reduction of approximately −100% ($p < .05$). The other two techniques restored the motion to the intact specimen. The normalized relative translation, NT_Z, between these vertebrae revealed that the CR and DCR techniques resulted in significant reduction in the motion (about −100%) while DR technique did not produce any significant alterations in comparison to the normal. The effect of stabilization at the unaffected vertebra (no. 5-4) was to reduce the normalized relative rotation ($p < .05$) significantly. Similar trends were observed for the normalized relative translations, NT_Z, but the level of reduction was −30% only.

Right Lateral Bending. The use of DR and DCR techniques in stabilizing the spine resulted in a significant ($p < .05$) reduction in the normalized relative rotation, at the no. 4-1 level to −50%. The change in rotation with CR technique was not significant. Similar results were obtained for the normalized relative translation, NT_X. At the injury site, the effects of stabilization with regard to rotation were similar to the ones for no. 4-1. In terms of normalized relative translation, NT_x, only the CR and DCR techniques resulted in a significant reduction (−100%). No significant effects were observed at the unaffected vertebra (no. 5-4).

Right Axial Rotation. Due to a wide scatter in the data, it was not possible to draw any conclusion, although statistically no significant alteration with regard to the intact specimen behavior could be observed.

Discussion

This study differs significantly from previous biomechanical studies. For example, Purcell et al. (1981) concentrated on the evaluation of (a) the number of vertebral levels on either side of the injury site at which the hooks for Harrington Distraction Rods (DR) may be attached to achieve better stability; and (b) the determination of flexion loads (or range of motion) which may lead to the dislodgement of the hooks. The injury model investigated by them was similar to our "hio" model. Similarly, Nagel et al. (1981) determined the amount of distraction at the injury site (L1-L2) but the force magnitude applied at the cadaver's head to produce the desired motion was not quantified. Their major emphasis was to evaluate the effectiveness of external braces, DR technique, and spinous wiring. Other studies concentrated on the biomechanical comparisons of various techniques on the basis of the failure strengths of spinal specimens stabilized using different techniques. We believe that the present study is a very comprehensive one dealing with the physiological three-dimensional motion behavior of the normal, injured, and stabilized spine specimens subjected to various load types.

The overall effect of inducing the injury was to increase the relative motions between the vertebrae including the functional spinal unit (no. 5-4) unaffected by the injury. Although for the latter case, the increase is not that significant. For this functional spinal unit the increase reflects the relatively increased bending moment due to the loading frame at this level of an injured specimen in comparison to the intact specimen. After the injury the spinal segment superior to the injured level deflects relatively more. It results in a higher bending moment (due to loading frame) in all levels. Once the specimen is stabilized there is no more increased motion and thus the change in motion behavior is due to the redistribution of forces resulting from Harrington instrumentation of the specimen itself.

The three procedures tested were evaluated against the hypothesis that ideally a procedure should restore the spinal motion behavior as close to the normal as possible, especially at the injury site. If any two procedures reduced the motion significantly ($p < .05$) then the procedure which resulted in a lesser mean value for the parameter was considered better, although the other technique has also reduced that parameter in the specimen significantly. The study clearly shows that as long as the injury is up to half injury (hi), all the procedures are likely to result in satisfactory reduction. But based on the above hypothesis the best stabilization in lateral bending is provided by CR technique but in flexion and axial rotation the use of DCR technique was found to yield better fixation. For an excessively injured spine (half injury plus osteotomy), none of the procedures was able to resist axial rotations effectively; in lateral bending and flexion all the techniques proved equally effective. Thus the results tend to suggest that the use of combination rods in reducing the spine injury may be superior to the other two techniques.

References

DUNN, H.K., McBride, G.G., & Daniels, A.U. (1981). Evaluation of the mechanical stability of fractured spines with various fixation systems. *27th Annual ORS* (pp. 24-26, 146). Las Vegas, NE.

NAGEL, D.A., Koogle, T.A., Piziali, R.L., & Perkash, I. (1981). Stability of the upper lumbar spine following progressive disruptions and the application of individual internal and external fixation devices. *Journal of Bone and Joint Surgery*, **63A**, 62-70.

PURCELL, G.A., Markolf, K.L., & Dawson, E.G. (1981). Twelfth thoracic-first lumbar vertebral mechanical stability of fractures after Harrington-rod instrumentation. *Journal of Bone and Joint Surgery*, **63A**, 71-78.

STAUFFER, E.S., & Neil, J.L. (1975). Biomechanical analysis of structural stability of internal fixation in fracture of the thoracolumbar spine. *Clinical Orthopedics and Related Research*, **112**, 159-164.

Mechanical Tests of Spinal Fixation Devices

**V.J. Raso, M.J. Moreau,
D. Budney, W.A. Chandler, and D.K. Kiel**
Glenrose Hospital, Edmonton, Canada

Thoracolumbar spine fractures are often aggressively treated with internal instrumentation. Although the aim of operative intervention is well documented, the actual instrumentation used varies considerably: Harrington distraction, Harrington compression, Weiss springs, and titanium mesh are but a few of the choices available. Recently Luque's segmental spinal instrumentation (SSI) or SSI combined with Harrington rods have been advocated for the treatment of thoracolumbar fractures.

There appears to be a proliferation of designs with no means available to compare the efficacy of each design. Late angular deformity is a common problem because inadequate fixation allows movement around the fracture site, delaying healing. These problems have led investigators (Jacobs, Nordwall, & Nachemson, 1982; Laborde, Bahniuk, Bohlman, & Samson, 1980; Nagel, Koogel, Piziali, & Perkash, 1981; Stauffer & Neil, 1975) to devise testing schemes for the evaluation of different designs. The experiments involved loading an instrumented spinal model to failure. Although this approach provides information regarding the overall mechanical strength and stiffness of a particular fixation device, it provides no information pertaining to motion at the fracture site. Certainly it is important that any device contemplated for fracture fixation provides rigidity sufficient to prevent fatigue or acute failure of the instrument under load; it is also important that the instrument minimizes movement of the fracture site if the conditions for healing are to be optimized (Perrin, 1979).

The purpose of this study was to prepare a model of the instrumented thoracolumbar fracture where movements at the fracture site could be monitored. Each device was evaluated in terms of its ability to prevent motion at the fracture site under load. The aim of this stage of the study was to determine the efficacy of three of the more popular devices: (a) Harrington distraction rods, (b) Harrington compression rods, (c) segmental spinal instrumentation in protecting a spine with a compression fracture at L1 under low levels of load.

Prototype Fracture

Mature sow spines, removed intact from freshly slaughtered animals and immediately frozen, were chosen as the experimental model. The T10-L4 segment of sow spine was excised and thawed. A 1.5 cm groove was created in the anterior body of L1 with a hacksaw making L1 the body most prone to failure. The spine was then enveloped by two cylindrical aluminum sleeves, and subjected to an increasing compressive load until a marked drop in the load deflection curve was observed. A total of seven spines were prepared and fractured in this manner after which each was ready for the application of a fixation device.

Fracture Instrumentation

Six spines were instrumented, tested, instrumented again with a different fixation device and retested. In this manner, each type of instrumentation was tested four times, twice when it was applied first, twice when it was applied a second time. Each specimen acted as its own control. A seventh spine was instrumented twice with the same instrumentation to determine repeatability of the technique.

Once instrumented, new aluminum sleeves similar to those used during the creation of the fracture were attached. In addition to the two sleeves, three aluminum rings were attached to separate vertebral bodies. The upper two rings were mounted in the instrumented region and had normal disc and a normal vertebra between them (normal region). The lower two rings encompassed a normal disc and a fractured vertebra (fracture region). The rings provided an attachment point for the four screws used as markers to follow the motion of the spine during testing. On all rings the markers were positioned at 90° to each other, each equidistant from the center of the ring.

A photographic technique was utilized to record the motion of the instrumented spine as it was loaded. The method consisted of using two cameras to photograph the spine from views 90° apart. This allowed monitoring of spinal motion in any direction. Two tripod-mounted Nikon 35 mm cameras equipped with 200 mm lenses were used. The motor drive of each camera was connected to an intervalometer to facilitate firing of both cameras simultaneously from one location. These photographs were synchronized with load application.

Testing Procedure

With normal activity the major load acting on the spine is eccentric compression, that is, a combination of flexion moments plus axial compression (Jacobs et al., 1982). The spinal column itself is sensitive to torsion (Roaf, 1960). For these reasons the following three types of loads were applied to the spinal segments: (a) axial compression, (b) torsion, and (c) compression bending (flexion moments).

Compression loading required no special equipment. A solid cap was placed on the end of the aluminum sleeves and the spine held in an Instron using the adaptor brackets. The fractures were stressed to approximately 50% of fracture load. Photographs were taken intermittently at different load increments. After completion of the compression test, the solid end-cap was removed and a torsion bracket attached to the aluminum sleeves. The spine was fixed at the bottom and pinned at the top preventing all but rotational movement. Pure torsion was applied to the instrumented spine through a pulley arrangement. The torque was increased in 2.6 MN increments to a maximum of 10.4 mm.

Compression bending was the final test performed on each device. A knife edge attachment to the Instron was placed in a notch in the end cap arm providing a 2.5 cm lever arm.

Data Analysis

All information was retained in the form of 35 mm glass mounted slides. The translation of the photographic information to true physical positions was accomplished with the aid of a computer. Approximately one thousand photographic images were obtained and analyzed. For each set of photographs pertaining to a specific instrumented spine and load configuration, two datum points and nine marker points were digitized. Four position variables were determined: (a) the vertical position Z, which is the vertical distance from the datum plane to the centroid of a particular ring, (b) the rotational position, the angle of rotation about a vertical axis, (c) the frontal slope, and (d) the lateral slope.

Displacements of any of the four position variables relative to the zero load position were determined by calculating the difference between a given load position and a zero load position. This procedure was applied to both views of each ring under all load conditions for all tests.

Changes in vertical, rotational, and angular displacements were determined for both the normal and the fractured regions. The difference in the displacements of the two regions provided a measure of the fracture site displacement alone. This method of determining displacements was applied to both left and right view.

Results

The results were presented graphically for each spine in the form of load-deflection curves for both the normal and the fractured region. Each instrumentation was analyzed and compared.

The performance of the devices in compression indicate that segmental wiring allows less movement at the fracture site than Harrington distraction. The Harrington compression and segmental wiring performed similarly. Results in torsion were similar for all devices. Results in compression bending showed very little difference between Harrington distraction and segmental wiring, although segmental wiring may protect the fracture site better at higher loads. Harrington distraction allowed less movement than Harrington compression. The segmental wiring protected the fracture site better than Harrington compression.

Error Analysis

The following were analyzed as sources of potential errors.

Data Collection. Spine stiffness and size varied from specimen to specimen. To avoid this problem, each spine was instrumented with two different devices permitting the two devices to be compared independently. The major shortcomings of this method were that aging effects and first test manipulations may have changed the stiffness properties of the spine causing the second test results to be in error. These were overcome by: first, applying the two instrumentations in reverse order to a different spine and second, by carrying out a repeatability test. This consisted of instrumenting a specimen with a particular device followed by a standard set of tests. The device was then removed and the spine was refrozen for a period of time. The specimen was then reinstrumented with the same device and retested.

The results give a measure of overall repeatability including aging, setup inconsistencies, effects of first test on spine properties, and errors in digitizing and analysis. The torsion test indicated that both normal and fracture regions showed more movement during the second tests, implying greater spine flexibility. The compression test indicated excellent repeatability in th fracture region and improved performance in the normal region during the second test.

Errors involving camera field foreshortening were minimized by using long (200 mm) focal length lenses.

Digitizing. Errors involving curser positioning were tested by repeatedly projecting and digitizing a single slide. Each of 11 points: 2 data points and 9 marker points were digitized 10 times. The horizontal and vertical position was reproducible to within ± 1 mm.

Data Analysis. The transformation of the two dimensional coordinates to the three dimensional ring positions required two assumptions. The ring markers had to be at the same radius from the centroid and they had to be orthogonal. Measurements of these values showed that the radius was correct to within ± 0.25 mm and that the deviation was from 90° resulted in a vertical or horizontal error of the marker points of ± 0.25 mm.

Combining all sources of coordinate position error discussed above suggests an uncertainty of approximately ± 1.285 mm (1.0 for digitizing, 0.035 for foreshortening, and 0.25 mm for ring geometry). This figure is improved upon during the analysis by combining the data from the left and right views.

Conclusions

1. A repeatable scheme for production of L1 compression fractures was developed.
2. A suitable model for recording spinal movement after instrumentation was devised.
3. A method for analysis of the photographic data was devised.
4. Preliminary results indicate:
   (a) Harrington compression and segmental wiring perform equally well in compression, better than Harrington distraction.
   (b) No difference in device performance in torsion testing.
   (c) Harrington distraction and segmental wiring protect equally well in compression bending, better than Harrington compression.
5. Errors were found to be due to digitizing, parallax and ring geometry.

Repeated testing is necessary in order to offer statistically reliable results.

Acknowledgments

We would like to acknowledge the MSI Foundation for their support in funding this research.

References

JACOBS, R.R., Nordwall, A., & Nachemson, A. (1982). Reduction, stability and strength provided by internal fixation system for dorso-lumbar spine injuries. *Clinical Orthopedics and Related Research,* **171**, 300-308.

LABORDE, J.M., Bahniuk, E., Bohlman, H.H., & Samson, B. (1980). Comparison of fixation of spinal fractures. *Clinical Orthopedics and Related Research,* **152**, 303-310.

NAGEL, D.A., Koogel, T.A., Piziali, R.L., & Perkash, I. (1981). Stability of the upper lumbar spine following progressive disruption in application of individual internal and external fixation devices. *Journal of Bone and Joint Surgery,* **63A**(1), 62-70.

PERRIN, S.M. (1979). Physical and biological aspects of fracture healing with special reference to internal fixation. *Clinical Orthopedics and Related Research,* **138**, 175-195.

PURCELL, G.A., Markolf, K.L., & Dawson, E.G. (1981). Twelfth thoracic first lumbar vertebral mechanical stability of fractures after Harrington rod instrumentation. *Journal of Bone and Joint Surgery,* **63A**(1), 71-78.

ROAF, R. (1960). A study of the mechanics of spinal injuries. *Journal of Bone and Joint Surgery,* **42B**(4), 810-823.

STAUFFER, E.S., & Neil, J.L. (1975). Biomechanical analysis of structural stability of internal fixation in fractures of the thoraco-lumbar spine. *Clinical Orthopedics and Related Research,* **112**, 159-164.

Spinal Deformities Evolution Assessment in Friedreich's Ataxia by Means of Intrinsic Parameters

J. Dansereau, P.S. Thiry, P. Allard, and G. Drouin
Ecole Polytechnique, Ste-Justine Hospital, and McGill University
Montreal, Canada

J.V. Raso
Glenrose Hospital, Edmonton, Canada

Serious spinal deformity is associated with Friedreich's ataxia. The evolution of the deformity is usually measured by the Cobb angle which gives a first approximation of the curvature of the scoliosis. However, this measure does not reflect the tridimensional aspect of the deformity. A recent study (Allard et al., 1982) has investigated the spinal deformity in Friedreich's ataxia with a tridimensional reconstruction program using cubic spline functions. It computes spatial parameters of a geometric curve based on the location of the centroid of each thoraco-lumbar vertebra obtained from standardized biplanar radiographs (McNeice, Koreska, & Raso, 1975). One of these parameters is the projected surface area (PSA). This parameter is defined as the measurement of the virtual surface observed when the spine is projected on a plane perpendicular to the axis formed by superimposing its extremities.

The investigation of the evolution of the spine deformity in Friedreich's ataxia revealed that the PSA index increased considerably more than the Cobb angle. It was thus assumed that a significant torsion might be superimposed on the scoliosis of the spine. This hypothesis was confirmed by a simulation of preset torsion of the thoracic spine with respect to the lumbar spine introduced in the geometric curve model (Dansereau, Allard, Raso, & Thiry, 1982). But the simulation developed in that study gives, at the most, a rough indication as to what could be the repartition of the torsion along the thoracic spine. Hierholzer and Lüxmann (1982) have proposed the use of invariant shape parameters to describe the scoliotic spine. This paper presents a more thorough description of the shape of the spine curve by means of its intrinsic curvature $\varkappa$ and torsion τ. Such an approach was also suggested by Patwardhan and Vanderby (1981).

Definition of the Intrinsic Curve Parameters

The basic equations to determine the intrinsic $\varkappa(s)$ and $\tau(s)$ of a spatial curve, function of s, its arc length parameter, are deduced from the well known Frenet formulas:

$$\frac{d\vec{T}(s)}{ds} = \varkappa(s)\,\vec{N}(s), \quad \frac{d\vec{B}(s)}{ds} = -\tau(s)\,\vec{N}(s) \tag{1}$$

which are expressed in terms of the Frenet frame field, composed of $\vec{T}(s)$, unit tangent vector, $\vec{N}(s)$, unit principal normal vector and $\vec{B}(s)$, unit binormal vector (O'Neill, 1966). The same Frenet formulae can also be applied to obtain the intrinsic curvatures $\varkappa_{AP}$ and $\varkappa_{LAT}$ of the antero-posterior and lateral projections, respectively, from the spatial curve.

Methods

To obtain the numerical values of $\varkappa(s)$ and $\tau(s)$, it is essential to know the first, second, and third derivatives of the spine curve at each point. These derivatives can be obtained by interpolation methods using spline functions. But cubic splines are not satisfactory because they don't have continuous third derivatives. So it was necessary to develop an original quintic spline interpolation for the spine curve with imposed slope at one end (T1) and simply supported at the other (S1). The details of the implementation of this numerical method are beyond the scope of this paper and are to be published elsewhere.

Results

The determination of these intrinsic parameters was applied to study the evolution of the deformity of a Friedreich ataxia patient over a 12-month period (three visits). Figure 1 shows the evaluation of the three curvature intrinsic parameters $\varkappa_{AP}$, $\varkappa_{LAT}$ and $\varkappa$ at the first and at the third visit. These graphs give an idea of the evolution of the scoliosis, the kyphosis and the lordosis. Figure 2 presents the intrinsic torsion results also at the first and at the third visit. The peaks of torsion occurring at L1 and T3 don't have much diagnostic significance (they reflect a localized phenomenon associated with a region of negligible curvature).

An attempt was made to extract, from these intrinsic parameters obtained at every point on the spine curve, the most diagnostically relevant information. It was felt that in the case of Friedreich's ataxia, the following specific parameters were of clinical significance in the scoliotic thoracic region: the maximum curvature of the thoracic spine ($\varkappa_{max}$), the maximum curvature of its antero-posterior plane projection ($\varkappa_{AP}$max), the maximum curvature of its lateral plane projection ($\varkappa_{LAT}$max) and the maximum torsion (τmax). The results are shown in Table 1 where the evolution of the PSA index was also recorded. The most interesting observations of these results are that:

(1) $\varkappa_{AP}$max has an evolution which can be put in correspondence with the Cobb angle,
(2) $\varkappa_{max}$ represents not only the extent of scoliosis but also of kyphosis,
(3) τ_{max} shows a dramatic increase at the second and third visit,
(4) the PSA index is a very sensitive parameter which responds not only to a change

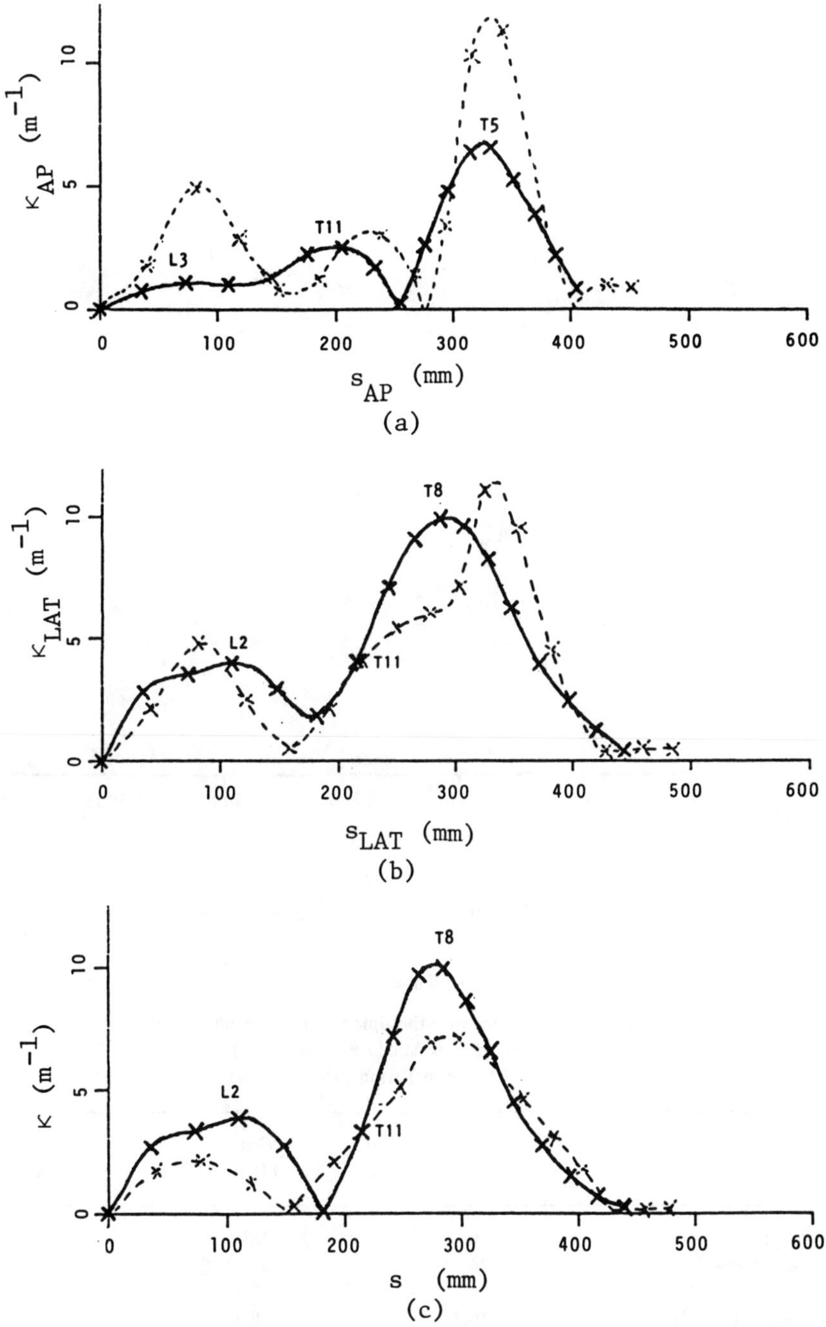

Figure 1—Comparison between the first (—) and third (---) visit: (a) antero-posterior curvature versus length, (b) lateral curvature versus length, (c) tridimensional curvature versus length.

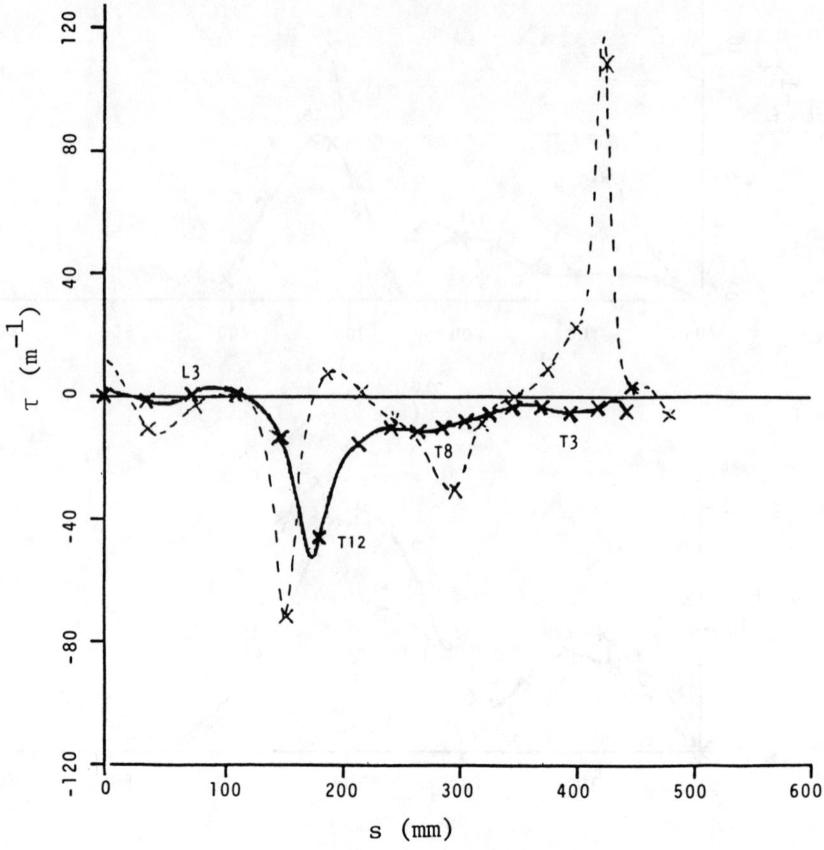

Figure 2—Torsion versus length. Comparison between the first (—) and third (---) visit.

Table 1

Clinical Data Obtained on the Spine Curve Reconstruction for a Friedreich Ataxia Patient (G.C.) Over a 12-Month Period (Three Visits)

	Visit I	Visit II	Visit III
Age	16	16½	17
Cobb angle (°)	24	33	38
$\varkappa_{AP}$max (m^{-1})	6.7	7.1	11.7
$\varkappa_{LAT}$max (m^{-1})	10.1	8.1	7.1
$\varkappa_{max}$ (m^{-1})	10.0	7.55	11.5
τ_{max} (m^{-1})	−11.5	−14.1	−30.9
P.S.A.	1,135	1,000	2,100

in the thoracic torsion (between second, third and fourth visits) as was previously shown (Dansereau et al., 1982) but also to a change in the $\varkappa_{LAT}$max or kyphosis angle (between first and second visit).

Conclusion

This paper has shown that the intrinsic spine curve parameters (curvature and torsion) are a powerful diagnostic tool to assess the evolution of scoliosis in Friedreich's ataxia. It is felt that this diagnostic tool should be applied to a larger population of Friedreich ataxia patients as well as to other cases of paralytic or idiopathic scoliosis. It could also be concentrated to the analysis of the lumbar section of the spinal column.

Acknowledgments

This research was funded by the Canadian Friedreich Ataxia Association and by the Institut de Recherche en Santé et en Sécurité du Travail du Québec.

References

ALLARD, P., Dansereau, J., Thiry, P.S., Raso, J.V., Duhaime, M., & Geoffroy, G. (1982). Scoliosis in Friedreich's Ataxia, *Canadian Journal of Neurological Sciences*, **9**, 105-111.

DANSEREAU, J., Allard, P., Raso, J.V., & Thiry, P.S. (1982). The influence of the projected surface area index on spinal rotation in Friedreich's Ataxia. *Proceedings of the Tenth Annual Northeast Bioengineering Conference* (pp. 196-199). Hanover, New Hampshire.

HIERHOLZER, E., & Lüxmann, G. (1982). Three-dimensional shape analysis of the scoliotic spine using invariant shape parameters. *Journal of Biomechanics*, **15**, 583-598.

McNEICE, G., Koreska, J., & Raso, J. (1975). Spatial description of the spine in scoliosis. *Advances in Bioengineering* (pp. 76-86). ASME, Winter Annual Meeting, Houston.

O'NEILL, B. (1966). *Elementary differential geometry.* New York: Academic Press.

PATWARDHAN, A., & Vanderby, Jr., R. (1981). A three-dimensional large displacement, continuum model of a human spine. *Advances in Bioengineering* (pp. 33-35). ASME, Winter Annual Meeting, Washington, D.C.

Coupled Motion of the Spine: A Study of Abnormal Behavior

J. Koreska, J.R. Seebacher, and C.F. Moseley
Hospital for Sick Children, Toronto, Ontario, Canada

Biomechanics of the spine has become an important factor in the clinical management of scoliosis. The response of the normal spine to a variety of loading conditions has been examined in vitro (King & Vulcan, 1971; Lin, Lui, & Adams, 1978; Markolf, 1972) as well as in vivo (Andersson, 1974; Reuben, Brown, Nash, & Brower, 1979). A dominant, and clinically relevant characteristic of normal behavior is the mechanical coupling which occurs when the spine bends laterally (White, 1971). Structural features of the spinal column cause individual vertebrae to rotate about the longitudinal axis of the spine. This is referred to as coupled motion.

We used this knowledge as the basis for a clinical study of coupled motion in non-ambulatory children with Duchenne Muscular Dystrophy (DMD). These boys become wheelchair dependent at about age ten. Within a few years they develop significant spinal deformities, often at an unpredictable rate. Total collapse of the spine may occur in less than one year. Clinical management of such spines is extremely difficult and often ineffective. For the past decade, our approach has been to stabilize the spine prophylactically using a custom fitted seat in the wheelchair to maintain proper alignment of the spine. We also attempt to predict the onset of collapse, on the basis of regular clinical follow-up, including three-dimensional analysis of the spinal shape using Antero-posterior (A-P) and Lateral roentgenograms of the spine (Koreska & Smith, 1982).

Methods and Subjects

A method for measuring vertebral rotation as part of our routine analysis of the spine was developed and implemented recently (Monji & Koreska, in press). Four anatomical points (see Figure 1) are identified on an A-P x-ray film and used to determine three lengths as shown in Figure 2. Axial rotation of the vertebrae relative to the plane of the film is then calculated using a simple ratio relationship:

Axial rotation = (A-B)/Bw/C, where C is a constant.

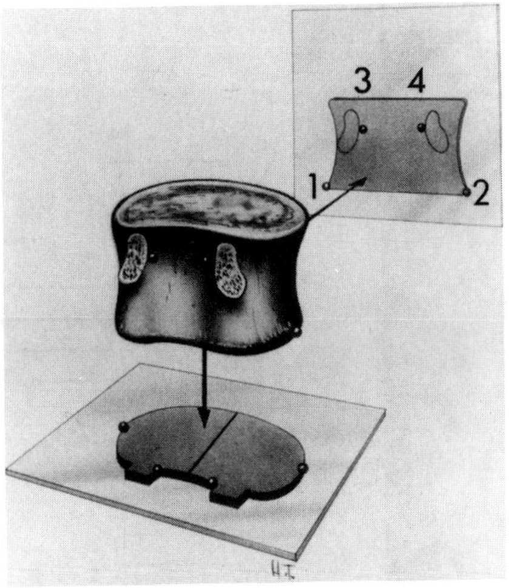

Figure 1—Drawing of vertebral body showing the four anatomical points used in determining axial rotation of the vertebrae.

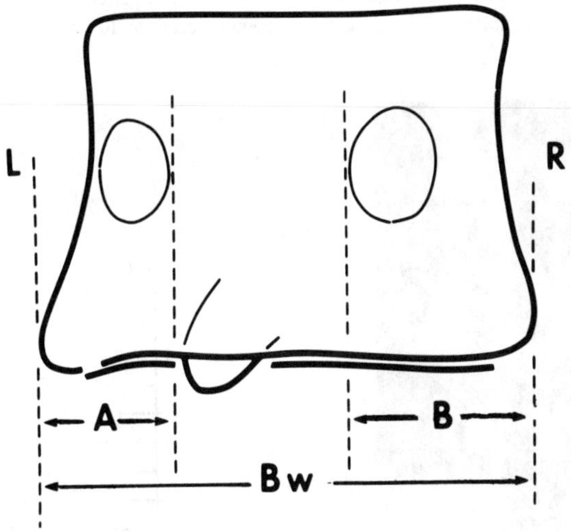

Figure 2—Diagrammatic representation of the A-P x-ray film image of a vertebral body showing the lengths calculated from the four anatomical points identified in Figure 1. Axial rotation is determined using the relationship given in the text.

The method was verified clinically by analyzing A-P spine x-rays from patients undergoing active treatment for spinal deformities. Examples are given for a patient with idiopathic scoliosis (see Figure 3) and one with paralytic scoliosis (see Figure 4). The spine x-ray image is shown on the left. Axial rotation at each vertebral level

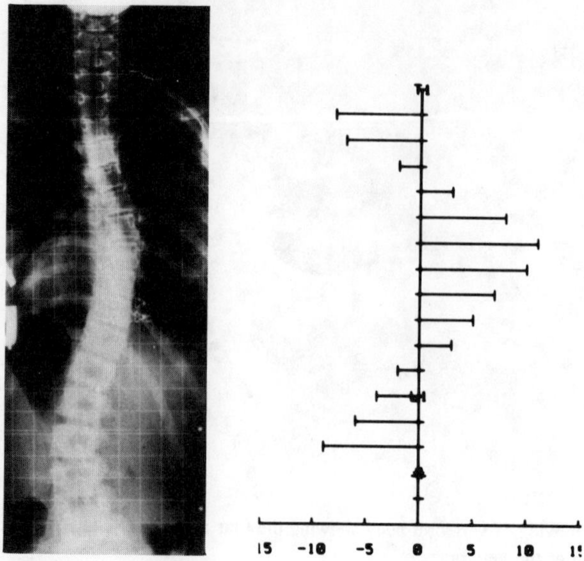

Figure 3—Example of rotation measurement on x-ray film of patient with idiopathic scoliosis. Absolute rotation is plotted using L4 as the reference vertebrae (0° rotation).

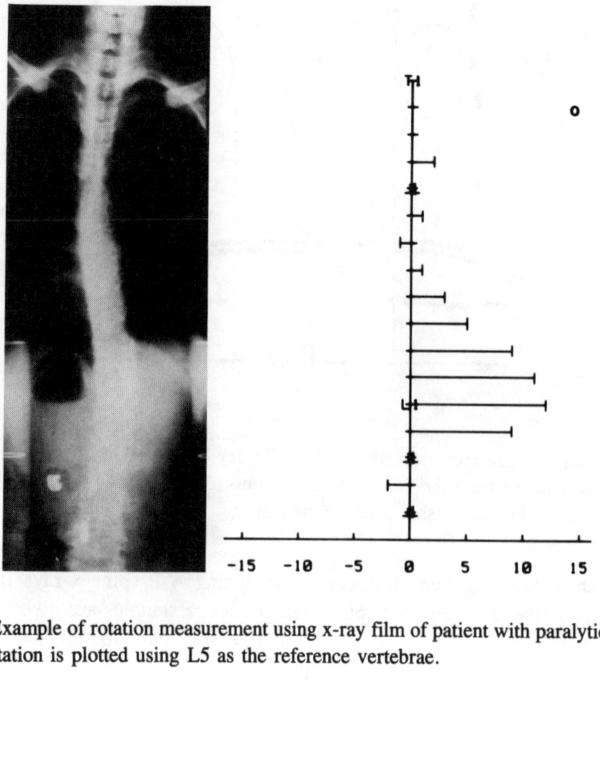

Figure 4—Example of rotation measurement using x-ray film of patient with paralytic scoliosis. Absolute rotation is plotted using L5 as the reference vertebrae.

is plotted as a horizontal line shown on the right side of each figure. Rotation was defined as positive if clockwise when viewing the vertebrae from above.

For the present study, we examined sixteen normal subjects and thirty patients with DMD. Each individual contributed three A-P views of the spine, one in the normal upright sitting position and one each with the spine bending to the left and to the right. A positioning device was used to ensure standardized alignment of pelvis and shoulders. Lateral bending views were obtained by inserting a 30° wedge in the seat of the positioning device.

Vertebral rotation was assessed as the change in segmental rotation which occurred when the spine moved from its neutral position to the end of the bend.

Results

In the normal subjects, we observed coupling which was anti-directional in the lumbar spine, that is, bending to the left caused the spine to rotate to the right and co-directional in the thoracic spine (see Figure 5). This, however, was not the case in the paralytic spines. Here, there was a general lack of consistency in the coupling mechanism with large variations in segmental coupling behavior (see Figure 6). Even in the neutral position of the spine, we observed significant segmental rotations not seen in the normal spines studied.

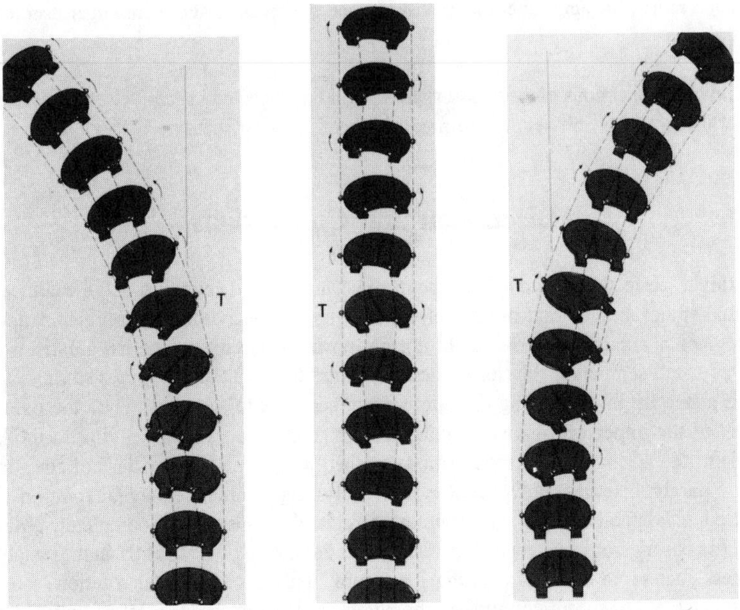

Figure 5—Artist's representation of coupled motion in a normal spine bending laterally. Segmental coupling is anti-directional in the lower (lumbar) region and co-directional in the upper (thoracic) section. Note that segmental rotation is present in the neutral position as well.

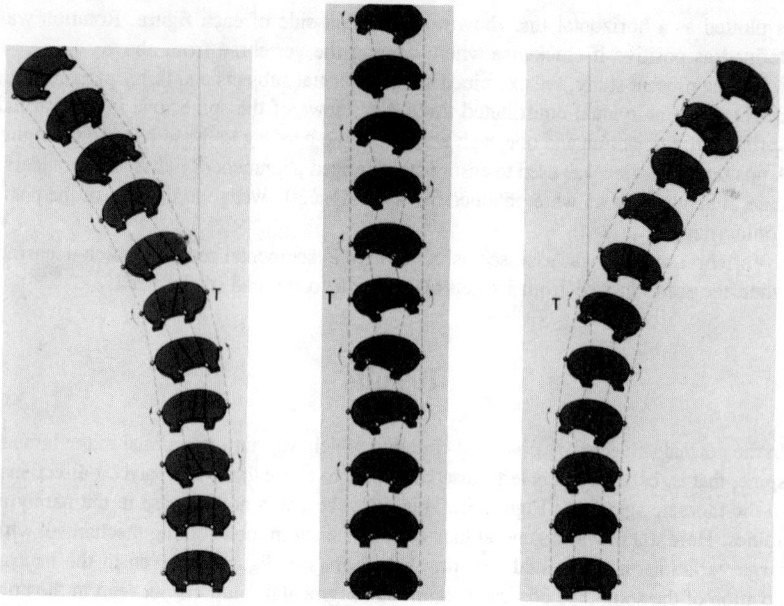

Figure 6—Illustration of irregular coupling of axial rotation with lateral bending in an unstable paralytic spine. Coupling pattern is inconsistent and segmental rotations are larger than in the normal spine.

Clinical implications of these observations will be presented in detail elsewhere. Some general comments, however, are made in the following section.

Discussion and Conclusions

A straight, perfectly aligned spine appears to be an idealized condition. All the clinically normal spines studied had minor defects in coupling behavior and irregular patterns of segmental rotation in the straight upright position. However, the irregularities involved small segmental rotations when compared to the paralytic spines studied, and were generally in alternating directions from segment to segment. Thus the overall effect of the imperfections was a balanced spinal column. When bending to the left or right, the normal spine irregularities vanished into a consistent pattern of coupling. In the paralytic spines, however, the noticeably larger initial segmental rotations increased with lateral bending, and coupling was either inconsistent or completely absent.

These observations suggest that the facet joints are not involved directly in the coupling process, but serve as blocks setting maximum limits for segmental rotation.

The lack of consistent coupling in the paralytic spines studied, most of which were mildly deformed, suggests that the coupling mechanism is an important aspect of normal as well as abnormal spine behavior. In the DMD patients studied with long-term follow-up, a lack of consistent coupling preceded rapid progression of the deformity leading to structural collapse of the spine.

Acknowledgment

We gratefully acknowledge financial support from The Muscular Dystrophy Association of Canada.

References

ANDERSSON, G.B.J. (1974). On myoelectric back muscle activity and lumbar disc pressure in sitting postures. Unpublished doctoral dissertation. University of Goteborg, Sweden.

KING, A.I., & Vulcan, A.P. (1971). Elastic deformation characteristics of the spine. *Journal of Biomechanics,* **4**, 413-429.

KORESKA, J., & Smith, J.M. (1982). Portable desktop computer-aided digitiser system for the analysis of spinal deformities. *Medical and Biological Engineering and Computers,* **20**, 715-726.

LIN, H.S., Liu, Y.K., & Adams, K.H. (1978). Mechanical response of the lumbar intervertebral joint under physiological (complex) loading. *Journal of Bone and Joint Surgery,* **60A**, 41-55.

MARKOLF, K.L. (1972). Deformation of the thoracolumbar intervertebral joints in response to external loads. *Journal of Bone and Joint Surgery,* **54A**, 511-533.

MONJI, J., & Koreska, J. (in press). Analysis of spine rotation—A new accurate method for clinical use. *Spine.*

REUBEN, J.D., Brown, R.H., Nash, C.L., & Brower, E.M. (1979). In vivo effects of axial loading on healthy, adolescent spines. *Clinical Orthopedics and Related Research,* **139**, 17-27.

WHITE, A.A. III. (1971). Kinematics of the normal spine as related to scoliosis. *Journal of Biomechanics,* **4**, 405-411.

The Use of Electrical Stimulation of Muscle to Treat Scoliosis in Children (ESI)

M.A. Herbert and W.P. Bobechko
Hospital for Sick Children, Toronto, Ontario, Canada

Scoliosis (lateral curvature of the spine) is a common disease occurring in 15% of the school age population (Brooks, Azen, Gerberg, Brooks, & Chan, 1975). It is of unknown etiology, and in 0.3% of the children, it is highly progressive (3 to 5°/month) and requires aggressive treatment.

The treatment of scoliosis depends on the age of the patient and the degree of curvature (0° is normal). Curve progression parallels the rate of growth in children, but slows or stops at skeletal maturity. In adults, curves of more than 40° progress relentlessly, and it is necessary to fuse the spine (with either Harrington or Dwyer instrumentation) to stabilize it and prevent its continuing collapse. Fusions are carried out in adults with progressive curves, or in children where the curves measure more than 40°.

In children whose curves measure less than 40°, treatment is usually by bracing, or more recently, by electrical stimulation. The majority of bracing programs use the Milwaukee brace (Mellencamp, Blount, & Anderson, 1977). The brace surrounds the body with a tight-fitting pelvic bucket and three vertical steel rods that join in a neck ring. Pads attached to the bars exert three point bending forces on the spine. The patient does exercises in conjunction with brace wearing. Bracing is carried out for 22 hr/day. Since the majority of patients who require brace treatment are adolescent females, this is a very difficult time for the patients, and the need to wear the brace continuously can lead to significant psychological problems; brace programs have a large number of noncompliant patients. Because of these problems, many centers have concentrated on the design of new braces that are easier to hide under clothes.

Many years ago, we began investigating the use of electrical stimulation to control scoliosis in children. The goal of the work was to establish a program of treatment based on the concept of electrical stimulation of the paraspinal (back) muscles at night while the child slept. There would be no requirement for bracing or auxiliary exercise programs. Initial animal studies established that a program of night-time stimulation would affect the direction of growth of a growing spinal column. Further studies were directed to understanding the effect of long-term intermittent stimulation on skeletal muscles.

Animal Experimentation

Stimulation was carried out using two different implantable systems: (a) a cardiac pulse generator (70 pulses/min; 24 hr/day) connected to stainless steel electrodes looped through the paraspinal muscles on one side of the spine, or (b) a pulse train generator (1 s trains at 30 pulses/s, with either 4 or 9 s off between trains, 15 hr/day) connected to platinum corkscrew electrodes implanted on one side of the spine. Fifty growing pigs were stimulated for 8 weeks, with x-rays of their spines taken at regular intervals. On x-ray, 76% of the animals developed scoliotic curves (average curve: 12°) during this time. Five animals were studies for an 8-week period following the stimulation, during which the stimulation induced curves were maintained.

Histological and histochemical studies of muscle taken at implantation, after 8 weeks of stimulation, and after a further 8 weeks of no stimulation, showed an alteration in the ratio of fiber types in the normally mixed paraspinal muscle. At 8 weeks, there was an increase in the concentration of the fast red fiber type, and a corresponding decrease in the fast white fiber numbers (with no change in the slow red concentration). The measurements at 16 weeks showed a partial reversal of the fiber changes, with the ratios tending towards those of the control animals. There was an increase in the thickness of the intramuscular septa, and some degeneration and regeneration of muscle fibers was evident in the area of the electrodes.

The animal work showed that unilateral stimulation of the paraspinal muscles could alter the direction of growth of the spine, with the spine curving towards the stimulation. This meant that if the stimulation were applied about the apex of the curve on the convex side, it would produce a force countering the increasing scoliosis.

Equipment

In adapting the experimental work to human use, a radio-frequency coupled system was developed in conjunction with Medtronic, Inc. (Minneapolis). The system consists of an implantable receiver and leads, and an external transmitter and antenna. The actual stimulation pulses are transmitted from the transmitter (via the antenna) noninvasively through the skin to the receiver. This system allows the stimulation parameters and battery to be changed at any time.

The receiver is a passive (batteryless) device designed to receive only the signals from the transmitter, modify them, and then conduct them through the leads into the appropriate muscles to produce stimulation. The receiver components are embedded in an epoxy disc (4 cm diameter, 1 cm thick) coated with silicone rubber for tissue compatibility. The receiver is attached to three leads of platinum-iridium "tinsel" wire terminating in platinum corkscrew electrodes. The electrodes are placed into the appropriate paraspinal muscles at the time of surgery. The three electrodes are placed above, at, and below the apex on the convex side in a negative-positive-negative polarity configuration. This arrangement reduces the current spread to other areas of the muscle, helping to localize the contraction to the region of the apex of the curve. The receiver is placed in a subcutaneous pouch on the convex side of the curve.

The external transmitter develops the stimulating signals, and has adjustments for all the stimulation parameters (although generally only amplitude is adjusted). The antenna is a flat silicone rubber covered disc, which is taped on the skin over the subcutaneously placed receiver with a disposable, double sided adhesive patch. The actual stimulation signal is carried over a 6 ft extension cord from the receiver to the antenna, providing the patient with enough lead to move around in bed at night.

Stimulation Parameters

The stimulator is adjusted to provide a muscle contraction lasting 1.5 s with a rest period of 9 s between contractions. The actual stimulation is a 1.5 s train of AC coupled pulses 220 μs wide, repeating every 33 ms (30 pulses/s). Since the paraspinal muscle does not relax during the interpulse time the contraction follows the train timing, i.e., on for 1.5 s and off for 9 s.

The pulse bursts are then transmitted on a 460 kHz carrier wave through the skin to the receiver. In dual channel devices (used to treat double curves), the second channel uses a carrier of 185 kHz. For both channels, the pulse trains are cycled at the same time, but the amplitudes are independently adjustable. The receiver integrates the carrier signal to provide the stimulation energy. The output amplitude is adjustable to a maximum of 15 V. The system is powered by a 9 V battery in the transmitter.

Surgical Insertion of the Receiver

The receiver is inserted under general anesthetic using a midline skin incision. The skin is reflected back on the convex side of the curve exposing the paraspinal muscles on the convex side. A subcutaneous pouch is made inferior to the twelfth rib to accommodate the receiver.

To determine the electrode placement, disposable needle electrodes are used to test various stimulating positions located about the apex of the curve. The needle tip is placed deep, to within 5 mm of the depth of the rib or transverse process. Trial stimulation is carried out under x-ray and visual control to locate the electrode positions producing maximal muscle contraction and correction of the curve. Once the optimal electrode positions are found, the needles are replaced by the corkscrew electrodes, using a specially designed system of insertion instruments. The skin is then closed. The patient is discharged from the hospital and treatment is started 10 days later.

Clinical Results

Patients are suitable for treatment by stimulation if they have curves between 25° and 40° and are not skeletally mature. Both single and double curves can be treated. After implantation and the first couple of visits (to set stimulation levels), the children are followed as outpatients at 9-month intervals. At the time of the return visit, they turn off the stimulator in the morning and then are seen in the afternoon clinic. A standing 3-ft AP radiograph is taken (no stimulation) and the curve measured and compared to the last prestimulation x-ray measurements. If the changes are less than 5° (either increased or decreased), the patient is classified as P.A. (progression arrested), while improvements of 5° or more are IMP (improved) and increases in the curve of more than 5° are PROG (progressed). Patients as young as 3½ years have been implanted.

The results in Table 1 are for 134 patients with 184 curves under treatment (50 double curves). All the curves were greater than 25°, except where the patient had a double curve with one greater and one less than 25°. Seventy-five percent of the curves were greater than 30°. Nearly 50% of the patients had at least 2 years of followup, with the average being 2 years and 4 months. From the table, it is seen that the success rate (the sum of the P.A. and Imp. columns) ranges from 100% down to 80% for curves less than 40°. As the curve increases, the success starts to fall and is down to 50% for curves over 50°. Figure 1 shows the results of treatment with one patient.

Table 1

Clinical Results of ESI

Curve (°)	No.	Prog. (%)	P.A. (%)	Imp. (%)
10-19	3	0	67	33
20-29	39	15	54	31
30-39	99	19	51	29
40-49	39	44	28	28
50-59	4	50	50	0

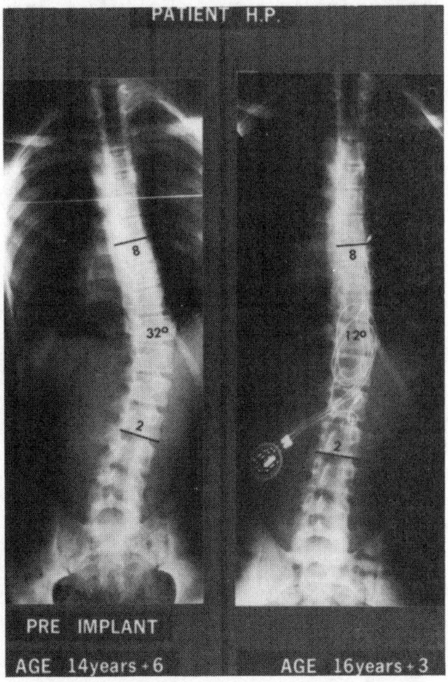

Figure 1—Results of treatment on a 14½-year-old male after nearly 2 years of ESI treatment.

Twenty-eight curves have been followed for an average of 18 months after skeletal maturity. At that time 86% of them were improved or arrested, when compared to the start of their treatment. No changes in the patient's naturally occurring kyphosis or lordosis was measured (lateral x-rays are taken every 18 months). While lead lifetime was a problem at the start of the program, recent improvements in the lead design has produced a lead whose reliability is currently 100%. No other serious complications have arisen.

Summary

The use of electrical stimulation (ESI) to treat scoliosis in growing children is a good alternative to the common bracing procedures. It is highly successful in arresting curve progression and in a large number of patients there is significant improvement in the curve (an average of 8°). Most importantly, patient compliance has been 100%, indicating acceptance by the patients.

References

BROOKS, H.L., Azen, E., Gerberg, E., Brooks, R., & Chan, L. (1975). Scoliosis: A prospective epidemiological study. *Journal of Bone and Joint Surgery,* **57A**, 968-972.

MELLENCAMP, D.D., Blount, W.P., & Anderson, A.J. (1977). Milwaukee brace treatment of idiopathic scoliosis-late results. *Clinical Orthopedics,* **126**, 47-57.

Classification of the Anatomical Spinal Curves of Female Students in Standing Position

D. Wielki, X. Sturbois, and Cz. Wielki
Catholic University of Louvain, Louvain-la-Neuve, Belgium

Although methods used to measure the curve of the rachis are evolving, their precision and the interpretation of the results are not satisfactory. This measuring method was established by using an "electronic spherosomatograph," an apparatus conceived in the JECO Laboratory (1974) and presented at the Medical Congress of Sport in Marseille (Wielki, 1978).

Methods

To analyze the results of these measurements, we utilized the "radius method with intersection points" which allows proper expression of the cyphotic curves and lordotic curves by calculating the size of the radius of the circle that is the closest to the curve registered. This method of analysis and the method of measuring the anatomical curves of the spine with the apparatus were perfected in the JECO Laboratory and were presented previously (Wielki, 1981).

Using this method an attempt was made to establish the principal types of curves of the spinal column in order to enable easier comparisons with research results.

The subjects for this research study were 170 healthy female students beginning their studies in the physical education program at the Catholic University of Louvain, Louvain-la-Neuve. All students had passed a complete medical exam prior to their participation in this study. The recordings occurred a) in October, 1978—50 subjects; b) in December, 1979—43 subjects; c) in November, 1980—37 subjects; d) in December, 1981—40 subjects. There were no pathological cases among these students; therefore they have been considered as being in good health.

The recording occurs while the subject is in a free standing position, with the heels at the same height, being stabilized at three levels (feet-hips-forehead), the head at the "Frankfurt position" during a period of "controlled breathing."

Besides the measurements of height and weight of every subject, we have also measured the anatomical curves of the spine from the C_7 (upper point 7th cervical vertebra) to the L_5 + 4 cm (lower point 5th lumbar vertebra + 4 cm) allowing us to take into consideration the position of the sacrum which is related to the spine and its curves (Wielki, 1979).

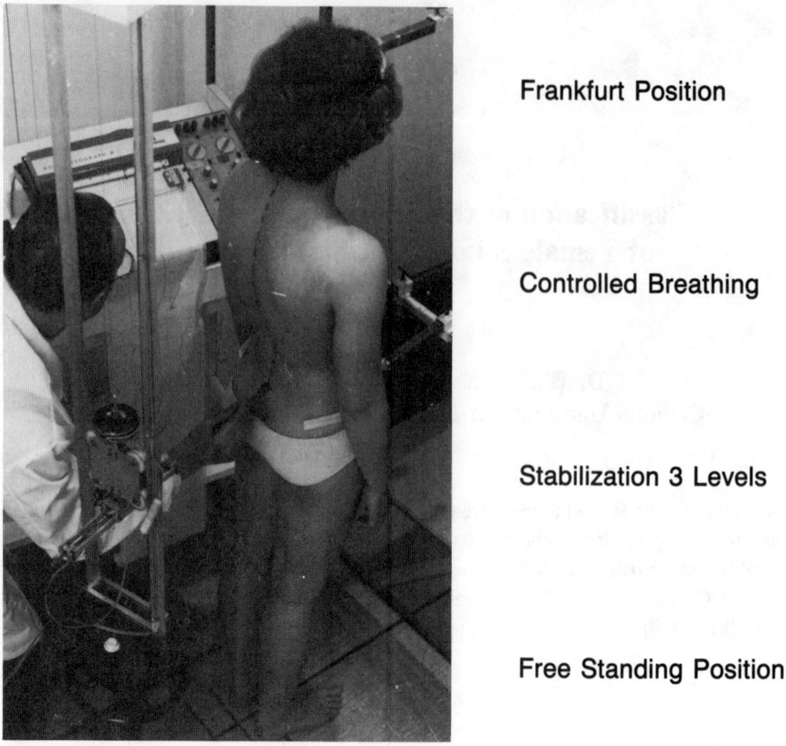

Figure 1—Recording of the physiological curves of the spine.

Results and Discussion

The results presented in Figure 2 are the most important variables of the spinal column and are averages of the variables of the group.

If we reproduce the average subject of the total group, we note that the lumbar and dorsal curves are regularly curvilinear and that they present curved lines of 483 mm and 191 mm (r_1 and r_2) respectively. The proportion of these curves rises to 2 to 1, with the height of the dorsal curve at two-thirds of the length of the spinal column and the height of the lumbar curve at one-third of the total length and is expressed by IDL = 228, with the top of the dorsal curve D lower than the middle of the curve ID = 181 and the top of the lumbar curve E nearly at the middle of the curve IL = 121.

The high point A in relation to the lower point B is at an inclination of 30, 9 mm. This result is termed the index of inclination (II = [HB:HA] × 100).

For the dorsal-lumbar index, we have constructed a curve of accumulated frequencies. This curve demonstrates a Gaussian distribution. On the basis of this curve the total population has been divided into 3 subgroups using the points of inflexion as the basis of the graphic determination. These three subgroups correspond to different profiles of the spinal column and divide the subjects into highly disproportionate groupings.

A→C = 32.4 mm
h₁ = 29 mm
r₁ = 483 mm

C→B = 156 mm
h₂ = 17.9 mm
r₂ = 191 mm

i₁ = 31 mm
i₂ = 479 mm

$IDL = \left(\frac{AC}{CB} \times 100\right) = 228$

$II = \left(\frac{HB}{HA} \times 100\right) = 6.4$

$ID = \left(\frac{AF}{FC} \times 100\right) = 185$

$IL = \left(\frac{CG}{GB} \times 100\right) = 121$

A Upper point
C Intersection point
A→C Cord of kyphosis
h₁ Height of kyphosis
r₁ Radius of kyphosis
D Top of kyphosis

B Lower point L₅ + 4 cm
E Top of lordosis
C→B Cord of lordosis
h₂ Height of lordosis
r₂ Radius of lordosis

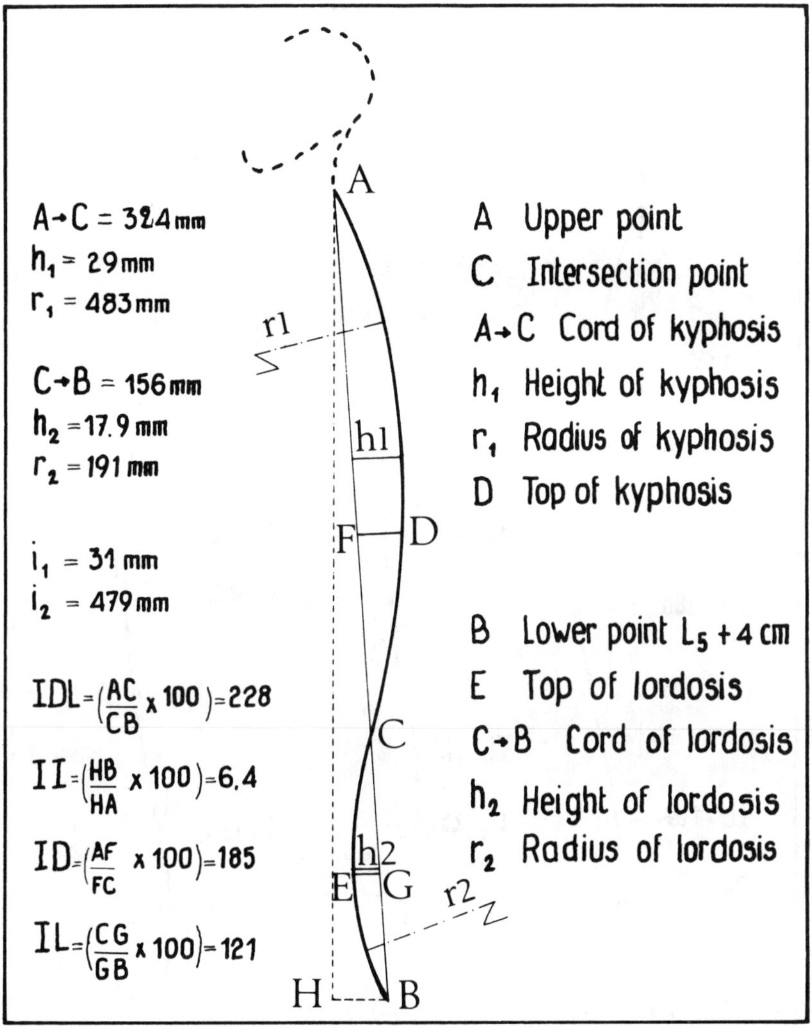

Figure 2—Means of the 170 female students of the physiological curves of the spine.

We have given for these three profiles of curves of the spinal column the denomination Type A, Type B, and Type C.

Type A (Normal). One hundred thirty-nine subjects belong to this group which represented 82% of the total group. This group is composed of subjects with a dorsolumbar index between 130 and 330, of which the average is 215 of nearly the whole group IDL = 228.

The value of the lordosis and kyphosis radii are also very close to the general mean, that is, $r_1 = 472$, whereas, $r_1 = 483$ mm for entire group of subjects, and $r_2 = 187$ whereas $r_2 = 191$, also close to the mean values of the entire group.

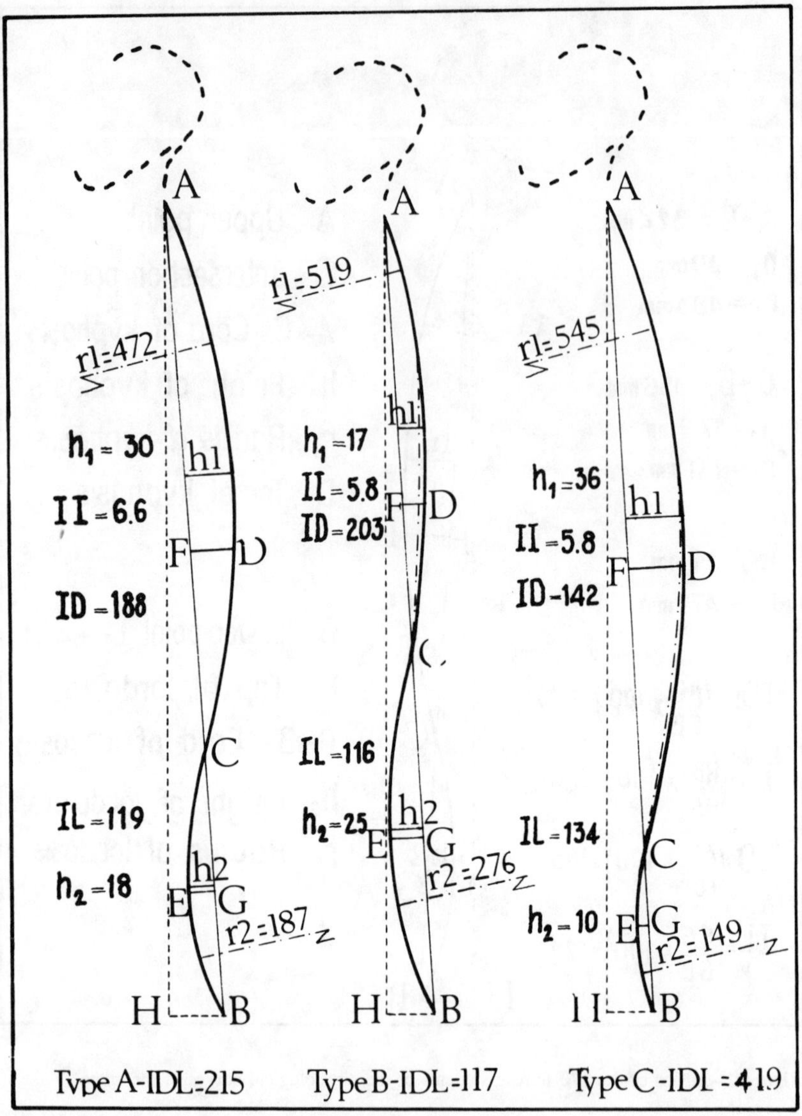

Figure 3—Profiles of the 3 types of the spinal column.

Type B (Lordotic). This group represents 8% of the subjects and is characterized by a mean relation of 8 to 7 between the dorsal and lumbar curves, with the dorso-lumbar index between 94 and 132, which gives us IDL = 117. Note that the curve of the spine is not regular, but we can express it by radius r_1 = 519, which is greater than the whole group (r_1 = 483), with r_2 = 276 greater than the whole group (r_2 = 191), and with h_1 = smaller 17 mm (30) and h_2 = greater 25 mm (18 mm = whole group). This revealed an accentuation of the lumbar lordosis.

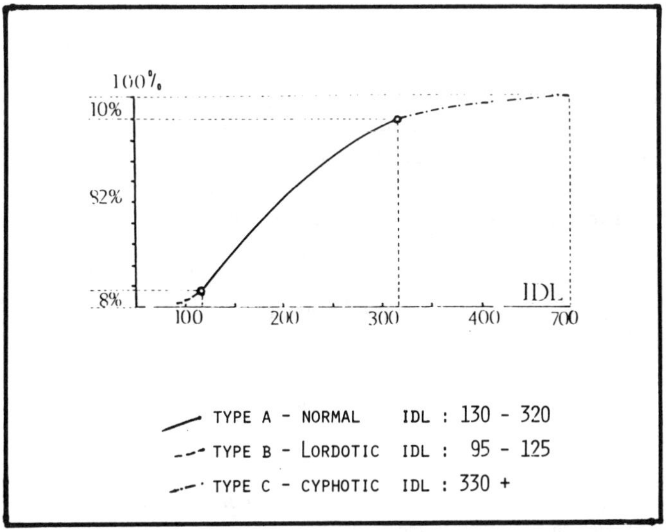

Figure 4—Cumulative curve lumbo-dorsal index (N = 170♀).

Type C (Cyphotic). With mean relation of 4/5 to 1/5 between the dorsal cord and the lumbar cord, this group was composed of 17 subjects which represented 10% of the sample. The dorso-lumbar index of this group is between 330 and 710, which is greater than the mean value of whole group (228). IDL is 419, which determined a relation of 4 to 1 between the dorsal cord and the lumbar cord. The value of the kyphosis radius was greater than the mean, that is, $r_1 = 545$, whereas $r_1 = 483$ for the entire group. This revealed a majoration of the dorsal kyphosis. The value of the lordosis radius ($r_2 = 149$) is less than $r_2 = 191$ for the entire group. The dorsal index is less than the mean, that is, ID = 141, whereas for the entire sample ID = 185; on the other hand, the lumbar index is greater in the Type C group (IL = 134), whereas for the entire group IL = 121. This indicates the rise of the lordosis apex, and the descent of the kyphosis apex in relation to their respective cords. The characteristics of this group were: an increase in the kyphosis cord, that is, 381 mm, whereas for the entire group $A \rightarrow C = 325$, and shortening of the lordosis cord, that is, $C \rightarrow B = 100$, whereas for the group it is 156 mm. At the same time, we notice an increase of the kyphosis height (36.5), the whole group height (29 mm) and shortening of the lordosis height ($h_2 = 9.91$) and for the sample $h_2 = 17, 9$ mm.

Conclusion

The elaboration of this typology seems very useful for the evaluation of the curves of rachis during growth, for the evaluation of pathological states, for reeducation purposes, and even for preventive examinations of elite sportsmen. Furthermore, we have investigated a typology for a group of 271 male students. It is remarkable that the group-division, also done according to the mean IDL, was respectively 200 (Type A), 100 (Type B) and 500 (Type C).

The discrimination analysis confirmed the foundation of this typology and validates the method of division based on the radius method with intersection points of the curve of accumulated frequencies. This typology is also interesting from the clinical point of view as a means of detection, and could contribute as well to the analysis of the development and deformation of the rachis in certain work and sitting conditions or sports practice. Finally, this standardized method can be adopted in other laboratories, it can give reference values for other studies, and we will be able to make better comparisons of results obtained in this area of research.

References

WIELKI, Cz. (1979). Vers une méthode électronique de mesure des courbures de la colonne vertébrale, Lyon Méditerranée, Médical Médicine du Sud-Est, Paris, Nv. 14, T.XV (pp. 1223-1227).

WIELKI, Cz. (1981). Method for measuring the curve of the spine by "electronic spherosomatograph." In H. Matsui & K. Kobayashi (Eds.), *Biomechanics VIII-B* (pp. 1190-1197). Champaign: Human Kinetics Publishers.

Pressure Distribution in Orthotic Devices for Treatment of the Spine

M. Yücel
Hospital of Brakel, Federal Republic of Germany

F. Liebscher
University of Siegen, Federal Republic of Germany

K. Nicol
University of Münster, Federal Republic of Germany

In order to examine the spatial distribution of pressure applied to the human body a measuring device for pressure distribution was used as described by Nicol and Körner (1983). Some investigations will be presented in order to display the efficiency and wide range of application.

Overview of Investigations of Pressure Distribution

Riding Motion and Saddle Pressure in Medical Therapy of Riding. For getting an objective basis for the up-to-now controversial discussions on the effects of medically prescribed riding a special device was needed: A pressure distribution mat with 512 channels was applied layer by layer on to a biaxially bent saddle. The multiplex electronics needed were carried in the saddle bag, the information could easily be registered as the horse was exercised, and cables could be gathered in the center of the exercise circle. One hundred ten measurements were taken with 1.5 s duration and 20 pressure distributions/s each. Furthermore, the motor behavior was filmed. The main idea of the investigation is to find out about common characteristics of the four paces of a horse and their variation. The authors hope to promote the recognition of hippotherapy with their results.

Objective Adjustment of Prosthesis Sockets. Prosthesis sockets with an enlarged inner radius were covered with pressure distribution elements to enable objective measuring of the fitting qualities of the socket. The number of sensors varied between 200 and

500 and were subject to extensive biaxial bendings, especially at the edges. This problem could be solved by arranging the mat segments in stripes. The patient carried the multiplex electronics which were connected to the microcomputer by cable. The investigation was carried out on six patients with different prosthesis sockets. The main results of the 100 transient measurements were (a) confirmation of the expected pressure distribution at the edges of the prosthesis, (b) insights into the control of prosthesis by the tip of the amputated limb. The project should be continued by measurements in standard prostheses where the thickness of the mat and the measuring expenses should be reduced.

Operation Technique on the Patella. As there are different views on the effects of operations on the patella in cases of chronic problems, concerning pressure distributions and their relation to different knee angles, an investigation was made in which a model of the knee, four times enlarged, was equipped with 32 × 16 capacitive sensors. As usual, the capacitor plates were controlled by 32 plus 16 hardware channels and a microcomputer.

Ski Shoe and Fracture. The form of the tongue of a ski shoe is considered an important factor in tibia fractures after forward falls. Some representative shoe tongues were furnished with a tight-fitting pressure distribution mat of 1 mm thickness, in order to get data on simulated falls. The data are registered by a battery operated microcomputer with integrated multiplex electronics carried by the skier. The information was transformed into a video signal, telemetrically transmitted and recorded by a video recorder. The results in this project should help identify parameters responsible for injuries caused by shoe tongues.

Injuries by Safety Belts. The pressure distribution between safety belt and dummy was statically recorded to find an optimum elasticity in horizontal and vertical directions of a safety belt. For this purpose a pressure distribution mat was developed which was fitted to the almost uniaxially bent surface of the dummy. One of the major results showed that the pressure on the collarbone seems to be 10 times as large as the average pressure under the belt.

Comfort of the Saddle of a Bicycle. With a saddle-shaped mat the pressure distribution on newly developed saddles with variable shapes was examined. Besides transient and short static measures, pressure rates were accumulated for pressure periods of 10 min. Although no norm has been worked out yet, some obvious misadjustments could be identified.

Adhesive Characteristics of Working Shoes. The adhesion of working shoes is influenced by the size of the sole and the pressure distribution on it. In order to register high pressures at the edge of the heel, the range of measurement in the system mentioned above was enlarged to 2 MPa and the size of the multiplexer and interface reduced to 8 ICs.

Comparison of Truss Pads

Orthotic devices are constructed to exert forces on selected areas of the human body, for instance to a section of the spine. When large surface orthotic devices are applied, the force imprint at every point should be known in order to (a) predict the efficiency of the device with respect to its unloading, supporting, and correcting function and (b) to avoid overloading the tissue. When conventional truss pads (i.e., the force-exerting part of the orthotic device) for the spine are used, the problem arises that the muscles of the back are weakened as the task of supporting is transferred from the muscles to the orthotic device. In order to overcome the problem a new flexible truss pad furnished with knobs was developed by W. Krause and R. Windhart, called "lumbotrain support." It was designed to feature an actively supportive function. After adjustment to the required pressure, the knobs apply a point massage of the musculature of the lumbosacral area, particularly during motion. Aside from the supporting function not only a massaging but also a stimulating effect is applied to the paraspinal musculature.

In order to objectively establish the effective principle of lumbotrain support, especially with reference to its supporting function of the lumbosacral transition and simultaneous stimulation of the paraspinals, the pressure distribution in the pad area was measured in a male and a female patient (Yücel, Breitenfelder, Liebscher, & Nicol, 1984). We used a slim sensor (51 × 4 cm) consisting of 128 single pressure sensors 16 × 10 mm. In this way the biaxial curvature of the truss pad could be treated as uniaxial. This mat was placed between the pad and the body. The pressure distribution was determined by two successive measurements, while no pressure and the normal correcting pressure, respectively, were applied to the pads. A pilot study showed a great influence of the breathing condition and the posture on the results. Therefore the measurements had to be normalized to these aspects. As the measuring program provides the possibility of continuous measurement over half an hour, long term measurements are planned. We hope to gain information about the correcting function and tissue load while performing various kinds of physical activities.

Figures 1 and 2 show how the new truss pad was applied and provide an example of a measurement of pressure distribution. The pressure is indicated by extended HEX

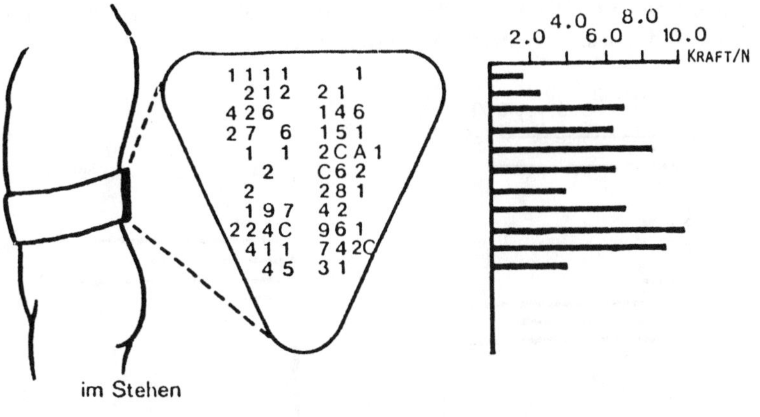

Figure 1—New truss pad, standing patient. Application of the truss pad, distribution of pressure, and force (kraft) acting on horizontal lines.

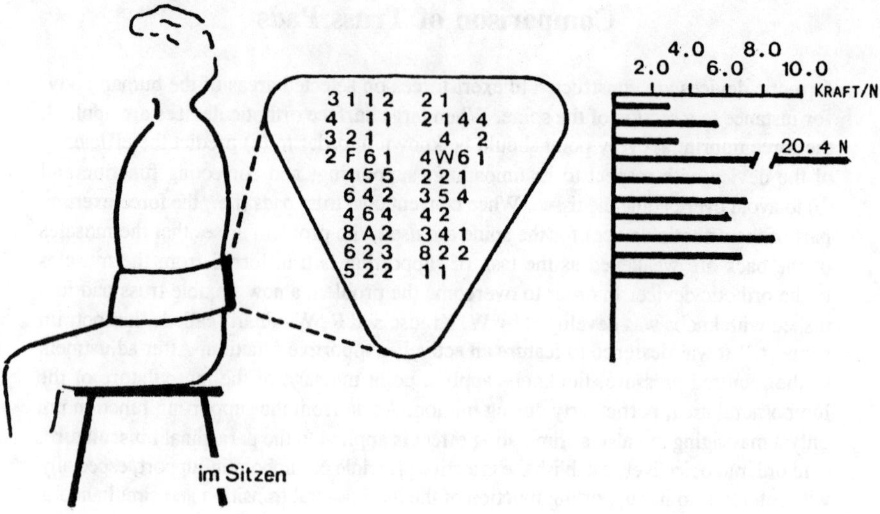

Figure 2—Same as Figure 1, sitting patient.

code (1 . . . 9, A . . . W). Figure 1, which was taken while standing, shows a nearly C-shaped presure distribution with maximum rates at top left, middle right, and bottom left, indicating an asymmetry in the patient's back. When the patient is seated (see Figure 2), the pressure is increased, but the characteristics do not change significantly. On the right hand side of Figures 1 and 2 the force acting on all sensors mounted in horizontal lines are summed up, thus showing the vertical distribution of pressure. Both in sitting and standing, pressure can be observed at all vertical positions, though the maximum rates are in the middle. In contrast the conventional truss pad (see Figure 3) shows (a) much lower homogeneity of pressure and (b) great differences between the sitting and standing position. While standing, nearly all pressure is concentrated in the upper and lower rim, whereas in sitting position all pressure is concentrated in the central part of the truss pad. Peak pressure rates of more than 100 kPa were measured, thus indicating overloading of the tissue.

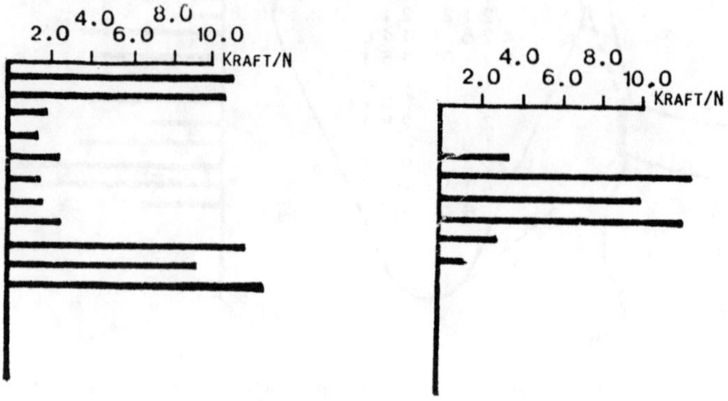

Figure 3—Conventional truss pad, standing patient (left) and sitting patient (right).

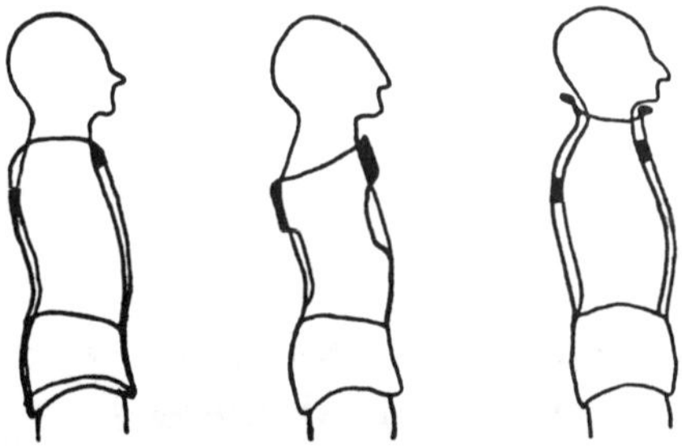

Figure 4—Orthoses Brakel, Münster, Milwaukee.

Comparison of Constructions

It can be shown that orthoses for kyphosis have to apply an angular momentum to the pelvic areas that have different directions for standing and sitting conditions. Orthotic devices constructed in Brakel, Münster and Milwaukee (see Figure 4) were compared. Measurement of pressure distribution was carried out on five subjects, each subject using two of the three constructions.

In order to avoid the influence of posture and breathing, each measurement was repeated four times. The measurement showed that all orthosis under investigation fulfilled the demand to correct kyphosis while sitting. The Brakel orthosis applies an additional force to the lumbar area while sitting, a force that is not seen in the standing position. In the Münster orthosis, the force on the lower part of the back increased. This was compensated by an increased force to the chest area. In the Milwaukee orthosis the patient is forced to stretch actively and to minimize kyphosis in this way. Consequently the forces acting on the chest, back, belly, and the pelvic area are much lower compared to the other constructions, but there is pressure against the chin.

References

NICOL, K., & Körner, U. (1984). Pressure distribution on a chair for disabled subjects. In D.A. Winter, R. Norman, R. Wells, K. Hayes, & A. Patla (Eds.), *Biomechanics IX-A*, pp. 274-280. Champaign, IL: Human Kinetics Publishers.

YÜCEL, M., Breitenfelder, J., Liebscher, F., & Nicol, K. (1984). Die Messung der Druckverteilung bei der Lumbotrainbandage. *Zeitschrift für Orthopädie*, **122**, 287-289.

Pressure Distribution on a Chair for Disabled Subjects

K. Nicol
University of Münster, Münster, Federal Republic of Germany

U. Koerner
Orthopädische Anstalten Volmarstein,
Wetter, Federal Republic of Germany

Measuring Technique

Since 1975 a capacitance type measuring system for normal force has been developed, the progress of which has been reported at recent Congresses of Biomechanics (Aisslinger, Nicol, & Preiss, 1981; Hennig & Nicol, 1978; Nicol & Hennig, 1978; Nicol, Preiss, & Albert, 1981). In order to minimize measuring error such as inhomogeneity of sensitivity, linearity, hysteresis, and relaxation, about 100 different foam rubber dielectric materials were investigated. Best results were obtained with a material with low compressibility which was shown in Nicol, Preiss and Albert. This material was used in a two channel measuring system for vertical force, where a homogeneous layer of dielectric was glued between two steel plates. In order to gain a suitable signal in this device, a highly sensitive bridge type capacity voltage transducer had to be developed, which offers 10^{-7} sensitivity and 200 Hz bandwidth. The main problem was to achieve equal sensitivity on every spot of the surface of the force platform.

The homogeneous layer was replaced by discrete small pieces of dielectric, therefore increasing the pressure on the material. In this way the signal was increased and the material was used in its optimum pressure range. Moreover, the problem of uniform sensitivity was eliminated, as suitable pieces of dielectric can be easily selected.

Using this platform type, the high resolution transducer could be replaced by a low resolution type which offers easy interfacing to computers. The measuring capacity is loaded via a resistor from a constant voltage source. A timer connected to the capacitor sends an impulse when a certain voltage level is reached. The impulse is set to the interrupt line of the microcomputer, thus stopping a software counter of 2 μs resolution. As 4 timers are included in one single IC, even a four channel interface (resolution 5,000; measuring rate 100 Hz) is extremely simple.

In addition to this two channel system, multichannel systems for pressure distribution with up to 32,000 sensors have been developed. One of them is a platform with 2,000 sensors for measurement of pressure distribution under the foot of a walking, running, or jumping person. Thirty-two and 64 capacitor plates, respectively, are joined to rows of three-layer stripes in order to simplify wiring and to provide shielding.

Sensors for pressure distribution can also be constructed in a flexible way. Though the sensor was glued to form a plane, it can be bent to fit cylindrically shaped bodies of nearly any curvature. When this sensor is used on spherically or saddle shaped bodies, severe problems arose, as some areas of the sensor have to be stretched by 10% or even more, whereas other areas have to be compressed by the same amount. These problems have been overcome in 3 different ways. First the sensor was glued to fit the shape of the body on which it will be used. This construction is used on bodies of complicated surface structure. Secondly, the sensor is slim so that the curvature is nearly uniaxial, even if several sensors are used side by side on biaxial curved surfaces. Thirdly, the sensors are constructed in thin layers using metallic weaving.

A multiplexer interface has been developed for use with Apple microcomputers and the circuitry can be reduced to two small printed boards plugged into slots of the Apple. Investigations in various fields have been carried out, some of which are reported in Yücel, Liebscher, and Nicol (1984).

Methods of Investigation

One special application shall now be described in more detail. Many physically handicapped people have to face severe problems while sitting. As their body often is not normally shaped, very high peak pressure can be observed in normal chairs. Moreover,

Figure 1—Arrangement of the pressure chair and computer system.

they often are not able to change their sitting position frequently in order to allow the loaded tissue to relax. The following investigation was conducted to get an idea as to whether chairs can be designed to fit the individual problems and if it is possible to discover groups of abnormal pressure distributions for groups of handicapped. The chair used for investigation had large supporting areas for the arms, and the seat was divided into a front and a rear part. Altogether, 6 flexible sensor mats for pressure distribution covered the contact area. The total number of sensors was 474. The mats were connected to an Apple computer via a universal type multiplexer, as shown in Figure 1.

Results

First studies were carried out which showed great differences between individual patients. The results obtained with ten normal subjects were all similar to Figure 2, left side. Three patients suffering from hemiplegia showed extremely asymmetrical pressure distribution with high peak pressure under the ischial tuberosity, such as shown in Figure

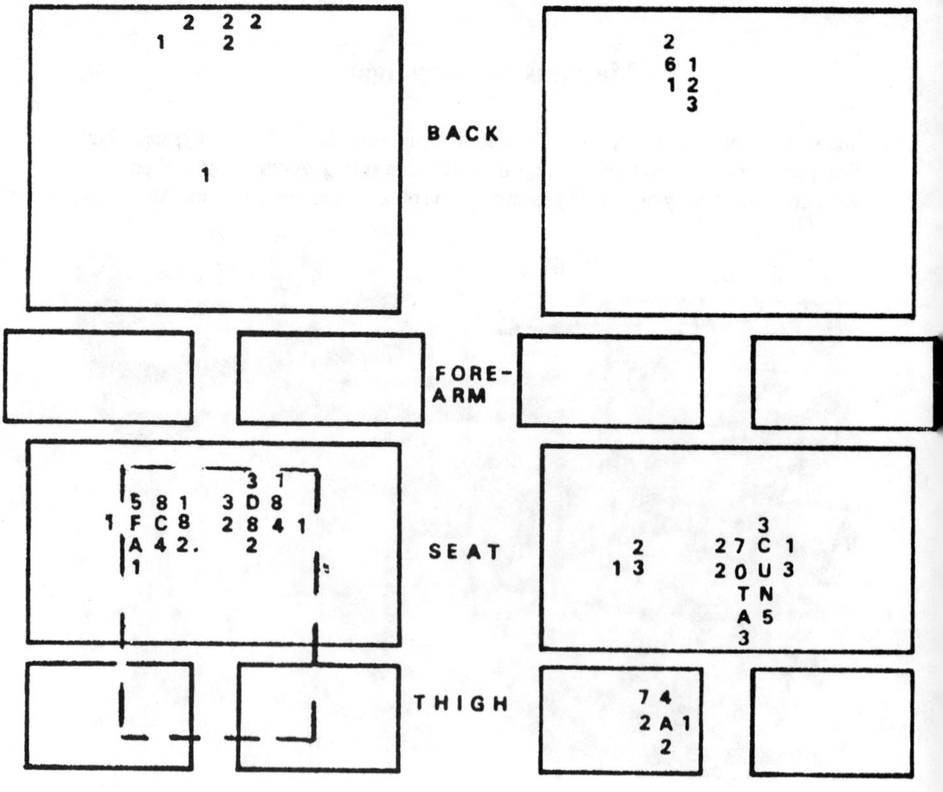

Figure 2—Pressure distribution on the measuring chair, indicated by an extended HEX-code. Normal at left, hemiplegia at right.

2 right side. For display purposes an extended HEX-code was used: A = 10 . . . F = 15 . . . K = 20 . . . U = 30. Six patients with scoliosis were tested, two of which showed very small but heavily loaded ischial areas which were asymmetrically loaded in every case. In most cases the peak pressure at the seat and back were not on the same side of the body. In contrast, for nine patients suffering from dismelia and other handicaps of the upper and lower extremities, peak pressures on seat and back pressure can be observed on the same side; with three patients the peak back pressure was observed in the middle. For 22 patients suffering from dysmelia, brain injuries, and other handicaps characteristic pressure distribution could now be identified.

Recently we started an investigation of the sitting behavior of 400 physically handicapped patients. Some examples shall be given as preliminary results. In Tables 1 to 6 the pressure exerted to the center part of the seat is plotted, as indicated by the dashed lines in Figure 2. The scale is normalized to give 99 units for the maximum pressure.

Tables 1 and 2 compare the pressure in two patients suffering from dysmelia and arthrogryposis, respectively. Though only the arms are affected in both cases, the

Table 1

Dysmelia

.	1	2	.	.	.	.	.
.	2	1	.	.	.	.	.
.	.	5	1	.	.	.	.
.	4	4	4	.	.	1	0
.	5	7	.	.	.	2	.
.	7	3	.	.	.	.	.
.	5	1	.	.	.	.	.
.	2	3	.	.	.	.	.
.	.	.	.	.	.	.	.
.	4	.	.	.	.	.	.
.	5	2	.	.	.	.	.

Table 2

Arthrogryposis

.	.	.	.	.	.	.	.
.	.	.	.	.	.	.	.
.	.	1	.	1	.	7	3
.	.	28	3	.	.	15	4
.	.	45	7	.	.	18	2
.	.	.	.	.	.	.	.
.	.	.	.	.	.	.	.
.	.	.	.	.	.	.	.
.	.	.	.	.	.	.	.
.	.	.	.	.	.	.	.
.	.	.	.	.	.	.	.

pressure distribution is extremely different. Whereas in the dysmelia the left side of the body, in particular the thigh, carries the whole body weight, in the arthrogryposis case no abnormality can be encountered. Tables 3 and 4 show two further examples for asymmetric pressure distributions. In the one patient (cerebral palsy, Table 3) the right thigh carried nearly as much load as the ischial area. The other patient (paraplegia, Table 4) loads the right ischial area only. As a consequence, the peak pressure was extremely high; up to 300 kPa have been measured. Table 5 (cerebral palsy) and Table 6 (one sided amputation of both extremities) give two other examples of high peak pressures. Although the cerebral palsy loads both thighs the ischial pressure of the loaded side is very high. The amputee had the maximum pressure encountered up till now, which was double the value observed for normals.

Table 3

Cerebral Palsy

.	.	.	.	.	.	.	.
.	.	.	.	.	.	.	.
.	.	.	.	.	.	.	.
.	.	.	.	5	2	.	.
22	37	.	19	49	3	.	.
6	6	.	2	4	.	.	.
.	.	.	.	.	.	.	.
.	.	.	.	.	.	.	.
.	.	.	.	.	.	.	.
.	.	.	.	.	.	.	.
1	.	.	.	.	26	18	7

Table 4

Paraplegia

3	3	3	10	4	.	.	.
.	.	.	.	.	.	.	.
.	.	.	.	.	8	88	.
.	.	.	3	19	81	43	.
.	.	.	.	.	.	.	.
.	.	.	.	.	.	.	.
.	.	.	.	.	.	.	.
.	.	.	.	.	.	.	.
.	.	.	.	.	.	.	.
.	.	.	.	.	.	.	.
.	.	.	.	.	.	.	.

Table 5

Cerebral Palsy

.	.	.	.	.	.	.	.
.	.	10	.	.	.	.	.
.	14	85	25	.	.	.	.
.	8	38	14	.	.	.	.
2	7	6	.	.	.	.	.
.	.	.	.	.	.	.	.
.	.	.	.	.	.	.	.
.	.	.	.	.	.	.	.
.	.	.	.	.	.	.	.
3	17	2	.	.	7	3	.
1	13	.	.	.	7	3	.

Table 6

Amputee

.	.	.	.	.	.	2	.
.	.	.	.	.	6	5	.
.	.	.	.	.	.	.	.
.	.	.	.	.	.	.	.
.	.	.	.	.	.	.	.
.	.	.	.	.	31	1	.
.	.	.	.	3	29	2	.
24	99	.	.	.	3	.	.
.	11	.	.	.	.	.	.
.	.	.	.	.	.	.	.
.	.	.	.	.	.	.	.

References

AISSLINGER, U., Nicol, K., & Preiss, R. (1981). Device for high resolution force distribution measurement. In A. Morecki, K. Fidelus, K. Kedzior, & A. Wit (Eds.), *Biomechanics VII-A* (pp. 548-552). Baltimore: University Park Press.

HENNIG, E.M., & Nicol, K. (1978). Registration methods for time-dependent pressure distribution measurements with mats working as capacitors. In E. Asmussen & K. Jorgensen (Eds.), *Biomechanics VI-A* (pp. 361-367). Baltimore: University Park Press.

NICOL, K., & Hennig, E.M. (1978). Measurement of pressure distribution by means of a flexible, large surface mat. In E. Asmussen & K. Jorgensen (Eds.), *Biomechanics VI-A* (pp. 374-380). Baltimore: University Park Press.

NICOL, K., Preiss, R., & Albert, H. (1981). Capacitance type force measuring system—methods and applications. In A. Morecki, K. Fidelus, K. Kedzior, & A. Wit (Eds.), *Biomechanics VII-A* (pp. 553-557). Baltimore: University Park Press.

YÜCEL, M., Liebscher, F., & Nicol, K. (1985). Pressure distribution in orthotic devices for treatment of the spine. In D.A. Winter, R. Norman, R. Wells, K. Hayes, & A. Patla (Eds.), *Biomechanics IX-A*, pp. 269-274. Champaign, IL: Human Kinetic Publishers.

The Utilization of Moire Topography in Physical Therapy

M.M. El-Sayyad
Cairo University, Cairo, Egypt

Both the initial screening of patients with scoliosis and their follow-up during treatment present a considerable challenge to the physical therapist. The lack of accurate and scientifically acceptable assessment tools has made it extremely difficult to document exactly the various parameters of improvement.

Since the Moire method is accurate and yields graphical data immediately, it seems an ideal tool in defining the surface contour of the back and relating this topographical information to back deformities. Moire topography is not a new technique, having been described as early as the late 19th century. Its applicability to the body surface shape was first introduced by Takasaki in the early 1970s. Recently a Moire study was published by Adair, van Wijk, and Armstrong (1978). They found that the Moire method registered x-ray diagnosed scoliosis with greater accuracy than clinical observation.

Many researchers have developed an interest in measuring the contour changes in the backs of patients with scoliosis (El-Sayyad & Syed, 1981) in which actual measurements of the body surface can be made and expressed mathematically. The need for actual measurements is evident when one examines back deformities and considers possible therapeutic measures to improve it and keep a follow-up of the improvement. This study was therefore designed to produce and describe a technique for determining prognosis in back deformities.

Method

The Moire equipment (see Figure 1), constructed as simply and as mobile as possible, consists of a specially constructed screen 75 $\times$ 50 cm. The screen is made of grids which are separated by a distance equal to their thickness. Two cameras (an SX-170 sonal focusing Polaroid and Canon AV-1) located 170 cm from the screen were used to obtain Polaroid pictures and slides. A 1000 W tungsten light source was used 50 cm above the cameras. The positioning of patients is the most important point in taking a good picture for Moire analysis. A Moire topogram taken with the patient misaligned cannot be used for quantitative measurement. Tests of the projection type Moire method were repeatedly made, and consequently it was considered that the most suitable Moire photographing system applied to Moire analysis was to photograph the Moire fringes of the erect positioned examinee's back in parallel with the plane of the grating. The patient was asked to stand looking straight ahead, arms at the side in a relaxed position. This was done as close as possible to the grid without actually touching it. If required, the patient's position was adjusted until the fringe patterns demonstrated by the buttocks were symmetrical. No attempt was made to position the upper trunk except that the patient was asked to relax.

Twelve children, between the ages of four and seven years, with back deformity (scoliosis) were selected for the study. Each child was examined clinically by physical examination, and photographed monthly by the Moire method and by x-ray for three months during intensive physical therapy.

To obtain the angle of spinal curvature, measurements were performed at the points of maximum and minimum asymmetry of Moire fringes and used in the mathematical relation to calculate the angle (see Figure 2). The level of maximum asymmetry for each of these asymmetric areas was judged and the fringe difference between two equidistant points opposite the midline determined. Each separate area of asymmetry was measured. θ is the angle of spinal curvature. A reference line AB is drawn by joining the midpoint of the neck to the midpoint of the waist. From this line, the distances to the first visible Moire fringe on both sides is measured at different points. The position of the spine is at the midpoint of these fringes. From the position of the spine at a given point, the distance to the line AB is obtained as d. At the point of maximum asymmetry from line AB, the distance is noted as d_1. At the point A above the point C, where the Moire fringes show minimum asymmetry, the distance is d_2. At point B below the point C, where the Moire fringes again show minimum asymmetry, the distance is d_3. The angle of spinal curvature is then given by:

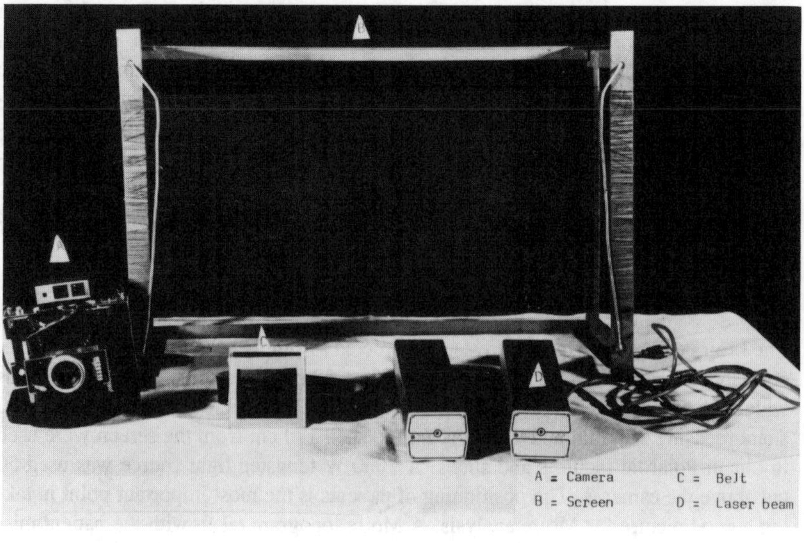

Figure 1—Equipment used in the study.

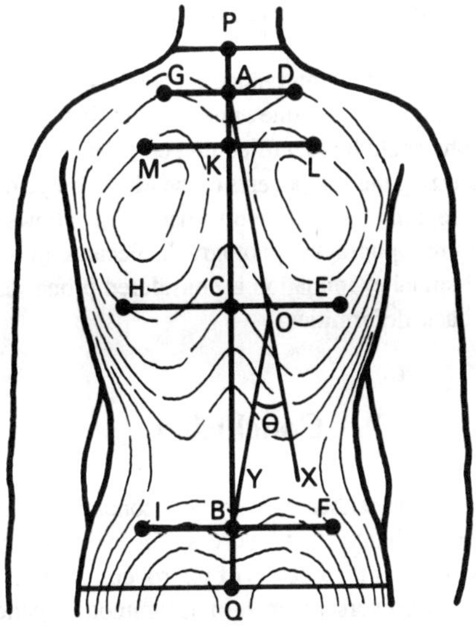

Figure 2—Measurement of the angle of spinal curvature by Moire topographs.

$$\theta = \tan^{-1}(d_1 - d_2/AC) + \tan^{-1}(d_1 - d_3/BC).$$

X-rays were measured by the method of Cobb and all visible curves measured. For the purposes of this study all Moire topographs were measured with no reference to the patients' x-rays.

Clinical Results and Discussion

The Moire topography offers several advantages over visual inspection at the time of examination. First, it allows future analysis of the surface shape without the actual presence of the patient, thereby allowing the physiotherapists to review a permanent record. Second, it allows for a precise quantification of the amount of asymmetry on the surface of the back.

The angle results gave a correlation coefficient of 0.85. However, this parameter must be treated with care, as systematic changes may alter the shape of the back, and therefore the values of the angles, without altering the correlation. Systematic errors have been shown to distort the results, and there is evidence that these are mainly caused by changes of posture. This indicated that a greater degree of postural control may be necessary. Also on the evidence of a few recordings showing changes over the shoulder blades, it is thought that bringing the arms forward so that the rib cage (rather than the shoulder blades) is recorded might improve results.

The advantage of this method in comparison with clinical ones is the possibility to document the condition of the back. Furthermore, the Moire method is easy to apply and can be managed by physical therapists. Additional research is required to determine the best method of quantification of surface topography and the correlation of the back surface topography to the clinical standard for measuring deformity by the Cobb angle.

In the process of following mild and moderate scoliosis curvatures with both x-ray and Moire fringe topography, it became apparent that distinct pattern shapes as seen on the Moire topographs could be identified that correlated with the anatomic regions of the scoliosis. By analyzing the components of the Moire fringe topogram, it is possible to identify clearly the region of spinal involvement. This allows for a more objective examination of the Moire topographic photograph. Because of its objectivity and efficiency, this system of spinal examination is considered proper and suitable for determining prognosis in back deformities.

Conclusion

Roentgenological screening is neither ethically nor economically justifiable. With Moire topography a new, safe, reliable, fast, and easy-to-interpret method has been introduced to document the various parameters of improvement. The instrument should be simple to operate, have a universal screen to accommodate patients of all sizes, and be lightweight and completely portable.

References

ADAIR, I.V., van Wijk, M.C., & Armstrong, G.W.D. (1978). Moire topography in scoliosis screening. *Clinical Orthopedics*, **129**, 165-171.

EL-SAYYAD, M.M., & Syed, A. (1981). Cobb's angle measurement by Moire topographs. *The 34th Annual Conference on Engineering in Medicine and Biology, Houston, Texas.*

IV. ELECTRO-MYOGRAPHIC KINESIOLOGY AND NEURAL CONTROL

Interrelationships Among Muscle Fiber Types, Electromyogram, and Blood Pressure During Fatiguing Isometric Contraction

T. Moritani
Texas A & M University, College Station, Texas, USA

F.D. Gaffney, T. Carmichael, and J. Hargis
University of Texas Health Science Center, Dallas, Texas, USA

In animal studies, skeletal muscles with a predominance of fast-twitch (FT) fibers have been shown to possess shorter contraction times, higher twitch and tetanic tensions, and greater susceptibility to fatigue than muscles with predominantly slow-twitch (ST) fibers (Burke & Edgerton, 1975). Similar results have also been demonstrated in human gastrocnemius by Garnett, O'Donovan, Stephens, and Taylor (1978) using controlled intramuscular microstimulation, glycogen depletion, and muscle biopsy techniques. In agreement with these findings, the percent FT fiber distribution of the vastus lateralis was shown to correlate with muscle fatigability as measured by the decline in maximal force during 50 repeated isokinetic knee extensions ($r = 0.86$) or by the increase in EMG/torque ratio ($r = 0.84$) (Nilsson, Tesch, & Thorstensson, 1977; Thorstensson & Karlsson, 1976). More recently, Komi and Tesch (1979) and Moritani, Nagata, and Muro (1982) have demonstrated that muscle with a high proportion of FT fibers shows a significantly greater decline in the mean power frequency (MPF) of the surface EMG during fatigue than ST fiber dominant muscle.

These data strongly suggest the possibility of developing multiple regression equations by which human skeletal muscle fiber types can be predicted. To develop such equations, the present investigation was undertaken to establish interrelationships among muscle fiber types, myoelectric signal characteristics (EMG frequency power spectra), and systemic blood pressure during fatiguing isometric plantar flexion at 50% MVC.

Methods

Muscle Fiber Types Determination

Fourteen men were tested in this study. Muscle biopsy samples were taken from the lateral head of the gastrocnemius muscle using the percutaneous needle biopsy technique (Bergstrom, 1962). The fibers were stained histochemically for myofibrillar ATPase after preincubation at pH 10.3, 4.6, and 4.3, respectively (Brooke & Kaiser, 1970). Since some FT fibers are fatigue-resistant and some are highly fatigable, muscle fibers were identified as Type I, Type IIA, and Type IIB (Brooke & Kaiser).

Experimental Protocol and EMG Analysis

The maximal voluntary contraction (MVC) of the plantar flexor muscle group was tested according to the methods described elsewhere (Moritani et al., 1982). The exerted isometric force was transmitted through a load cell force transducer and recorded on a conventional X-Y recorder. Then the subject was instructed to maintain an isometric contraction at 50% of MVC, since the greatest accumulation of lactate (+ pyruvate) occurs when this level of force is sustained to fatigue (Ahlborg et al., 1972).

Myoelectric signals were picked up by two miniature size electrodes (4 mm pick up area, 6 mm interelectrode distance). These active electrodes were placed over the lateral head of the gastrocnemius 10 mm medial to the biopsy site. The reference electrode was attached over the medial malleolus. The myoelectric signal was amplified and recorded on an analog FM data recorder and then low-pass filtered, digitized at a sampling rate of 2048 Hz by the use of LSI 11/23 minicomputer. The digitized data were processed with a tapered window function and 512-point fast Fourier transform to obtain a power spectrum periodgram. Ten consecutive periodgrams were averaged to calculate mean power frequency (MPF) during a nonfatiguing 5-s plantar flexion at 50% of MVC. Muscle fatigability was evaluated during the 50% MVC fatiguing contraction by determining the rate of increase in the myoelectric signal amplitude (RMS) and the rate of decline in mean power frequency (FINDEX) as a function of sustaining time. The systemic blood pressure (systolic blood pressure: SBP) was measured during the first minute of the fatiguing contraction.

Results

Results indicated that MPF obtained during the 5-s nonfatiguing contraction showed a significant negative correlation ($r = -.791$, $P < .01$) with percent Type I and a positive correlation ($r = .794$, $P < .01$) with percent Type IIB fibers (see Figure 1). It was also found that the rate of changes in myoelectric signal amplitude and mean power frequency during muscle fatigue as largely accounted for by the differences in muscle fiber compositions. Percent Type I showed significant correlations with FINDEX ($R = .813$, $P < .01$) and with RMS ($r = -.872$, $P < .01$) (see Figure 2). Thus, individuals with predominantly Type I fibers would have a smaller increase in EMG amplitude and a smaller decline in mean power frequency during fatigue. Interestingly, systolic blood pressure measured at the first minute of the fatiguing contraction showed moderate, but significant correlations with percent Type I ($r = -.649$, $P < .05$) and with percent Type IIB ($r = .572$, $P < .05$). It is suggested that there could be a possible difference in the intramuscular transmural pressure upon the blood vessels in different muscle fiber types.

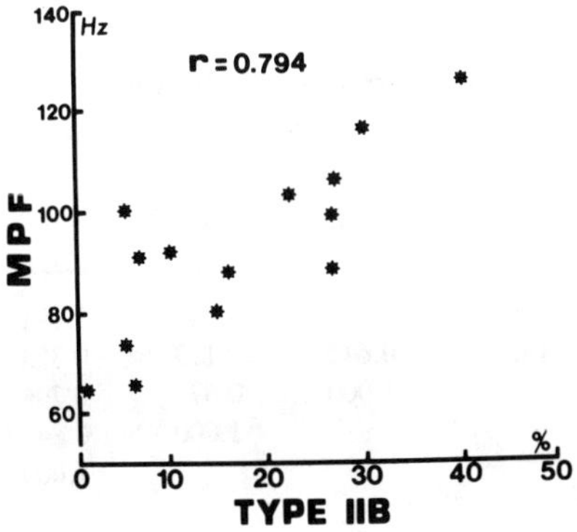

Figure 1—Relationship between percent Type IIB fibers and MPF obtained during nonfatigue 5-s contraction.

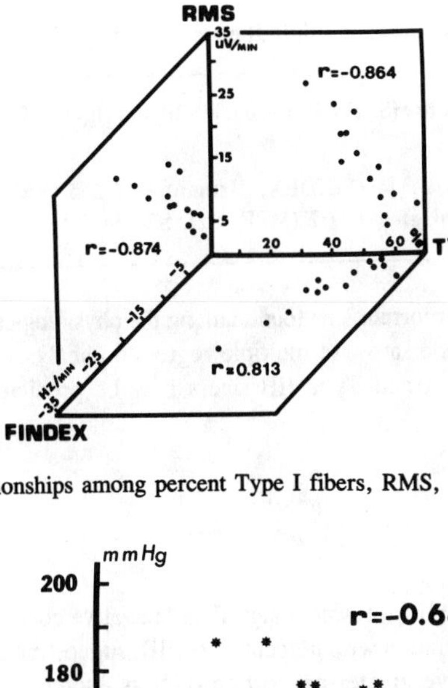

Figure 2—Interrelationships among percent Type I fibers, RMS, and FINDEX.

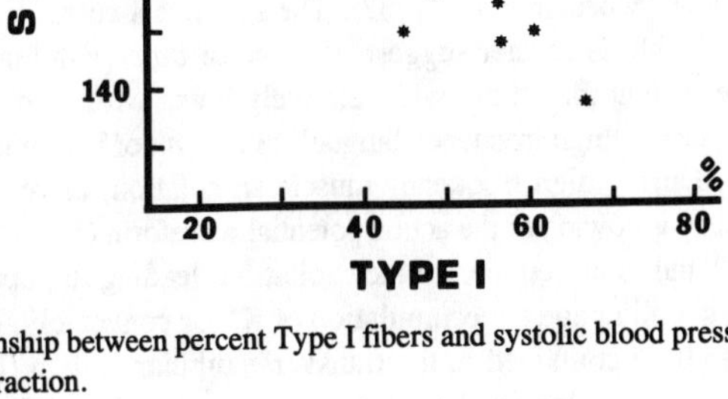

Figure 3—Relationship between percent Type I fibers and systolic blood pressure measured during fatiguing contraction.

Table 1

Correlation Coefficients

	Type I	Type IIA	Type IIB	SBP	MPF	FINDEX	RMS
Type I	1.000	0.128	−0.840	−0.649	−0.791	0.813	−0.864
Type IIA		1.000	−0.642	−0.173	−0.353	0.054	−0.395
Type IIB			1.000	0.572	0.794	−0.899	0.872
SBP				1.000	0.842	−0.722	0.624
MPF					1.000	−0.852	0.836
FINDEX						1.000	−0.874

$r = .532, P < .05$
$r = .661, P < .01$

Stepwise Multiple Regression Analysis

Type I (%) = 77.095 − 0.831 (RMS, uV/min) − 0.08 (SBP, mmHg) + 0.216 (FINDEX, Hz/min), SEE = 4.89

Type IIB (%) = 16.792 − 1.374 (FINDEX, Hz/min) + 0.555 (RMS, uV/min) − 0.154 (SBP, mmHg) + 0.072 (MPF, Hz), SEE = 5.11

Table 1 summarizes intercorrelations found among the physiological parameters tested in this study. Results of the stepwise multiple regression analysis indicated that percent distribution of Type I and Type IIB fibers may be predicted with reasonable accuracy.

Discussion

Our results indicated that MPF showed a significant negative correlation with percent Type I and a positive correlation with percent Type IIB, suggesting that muscle groups with higher MPF may have greater proportion of "fast-twitch" fibers.

These results support our earlier findings that the biceps brachii had a significantly higher MPF (mean difference of 35.7 Hz, $P < .01$) than the soleus during maximal isometric contractions (Moritani et al., 1982). The significant correlation between MPF and FINDEX (see Table 1) further suggests that motor units with higher MPF would fatigue to a greater extent than those with relatively lower MPF. Edwards (1981) has shown some evidence of "high frequency fatigue" as a result of neuromuscular transmission impairments. During high frequency muscle stimulation, there is a rapid loss of force accompanied by a slowing of the action potential waveform (Jones, Bigland-Ritchie, & Edwards, 1979) and impaired membrane excitation, leading to impaired excitation-contraction (E-C) coupling due to accumulation of K^+ (or conversely, depletion of Na^+ in the extracellular fluid contained in the transverse tubular system (Bigland-Ritchie, Jones, & Edwards, 1979). Our results seem to suggest that Type IIB fibers with high MPF would be more susceptible to follow the high frequency fatigue patterns.

Our data on the myoelectric signal changes during fatigue are in good agreement with the previous investigators (Komi & Tesch, 1979; Mills, 1982; Moritani et al., 1981, 1982) and further suggest that percent distribution of Type IIB fibers may play a significant role in the myoelectric signal changes seen during muscle fatigue. Since slow-twitch fibers have a greater potential for lactate utilization as fuel (Armstrong et al., 1974) and have a smaller amount of lactate accumulated (Tesch, 1980), the effects of reduced pH upon E-C coupling (Ca^{++} binding to troponin and the affinity of the sarcoplasmic reticulum for Ca^{++}) would be more markedly manifested in the fast-twitch, particularly Type IIB, fibers. Lago and Jones (1977) have shown that synchronization of MU activity and MU firing frequency can significantly alter the lower frequencies of the EMG power spectrum. These considerations and the possibility of reduced MU firing frequency (Grimby, Hannerz, & Hedman, 1981) during fatigue strongly suggest that myoelectric signals may become synchronized at low frequencies as fatigue progresses. Thus, muscles with predominantly fast-twitch fibers seem most likely to show large, low-frequency EMG oscillations together with high frequency decay which in turn increase the lower EMG frequency spectral energy and myoelectric signal amplitude.

Acknowledgments

This study was supported in part by NHLBI Young Investigators Awards #HL25710 and by grants from the Texas Engineering Experiment Stations (Project #9058E).

References

AHLBORG, B., Bergstrom, J., Guarnieri, L.G., Harris, R.C., Hultman, E., & Nordesjo, L.O. (1972). Muscle metabolism during isometric exercise performed at constant force. *Journal of Applied Physiology,* **33**, 224-228.

ARMSTRONG, R.B., Saubert, C.W., Sembrowich, W.L., Shephard, R.E., & Gollnick, P.D. (1974). Glycogen depletion in rat skeletal muscle fibers at different intensities and durations of exercise. *Pflügers Archives,* **352**, 243-256.

BERGSTROM, J. (1962). Muscle electrolytes in man. *Scandinavian Journal of Clinical and Laboratory Investigation,* Suppl. 68.

BIGLAND-RITCHIE, B., Jones, D.A., & Wood, J.J. (1979). Excitation frequency and muscle fatigue: Electrical responses during human voluntary and stimulated contractions. *Experimental Neurology,* **64**, 414-427.

BROOKE, M.H., & Kaiser, K.K. (1970). Muscle fiber types: How many and what kind? *Archives of Neurology,* **23**, 369-379.

BURKE, R.E., & Edgerton, V.R. (1975). Motor unit properties and selective involvement in movement. *Exercise and Sports Sciences Review,* **3**, 31-81.

EDWARDS, R.H.T. (1981). Human muscle function and fatigue. In R. Porter & J. Whelan (Eds.), *Ciba Foundation symposium on human muscle fatigue: Physiological mechanisms* (pp. 1-18). London: Pittman Medical.

GARNETT, R.A.F., O'Donovan, M.J., Stephens, J.A., & Taylor, A. (1978). Motor unit organization of human medial gastrocnemius. *Journal of Physiology,* **287**, 33-43.

GRIMBY, L., Hannerz, J., & Hedman, B. (1981). The fatigue and voluntary discharge properties of single motor units in man. *Journal of Physiology,* **316**, 545-554.

JONES, D.A., Bigland-Ritchie, B., & Edwards, R.H.T. (1979). Excitation frequency and muscle fatigue: Mechanical responses during voluntary and stimulated contractions. *Experimental Neurology,* **64**, 401-413.

KOMI, P.V., & Tesch, P. (1979). EMG frequency spectrum, muscle structure, and fatigue during dynamic contractions in man. *European Journal of Applied Physiology,* **42**, 41-50.

LAGO, P., & Jones, N.B. (1977). Effect of motor unit firing time statistics on EMG spectra. *Medical and Biological Engineering and Computers,* **15**, 648-655.

MILLS, K.R. (1982). Power spectral analysis of electromyogram and compound muscle action potential during muscle fatigue and recovery. *Journal of Physiology,* **326**, 401-409.

MORITANI, T., Nagata, A., & Muro, M. (1981). Electromyographic manifestations of neuromuscular fatigue of different muscle groups during exercise and arterial occlusion. *Japanese Journal of Physical Fitness and Sport Medicine,* **30**, 183-192.

MORITANI, T., Nagata, A., & Muro, M. (1982). Electromyographic manifestations of muscular fatigue. *Medicine Science in Sports and Exercise,* **14**, 198-202.

NILSSON, J., Tesch, P., & Thorstensson, A. (1977). Fatigue and EMG of repeated fast voluntary contractions in man. *Acta Physiologica Scandinavica,* **101**, 194-198.

TESCH, P. (1980). Muscle fatigue in man with special reference to lactate accumulation during short term intense exercise. *Acta Physiologica Scandinavica* (Suppl.), **48**, 1-40.

THORSTENSSON, A., & Karlsson, J. (1976). Fatigability and fiber composition of human skeletal muscle. *Acta Physiologica Scandinavica,* **98**, 318-322.

Tension-EMG Relationship During Maximum Voluntary Contraction

Y. Nakamura, H. Ohmichi, and M. Miyashita
University of Tokyo, Tokyo, Japan

Many investigations have indicated that there exists a positive relationship between myoelectrical activity and muscular force. Therefore, if the ratio of muscular force to the magnitude of the electromyogram (EMG) is known, it is possible to estimate the muscular force from the EMG. However, there are several factors which influence the relationship between the muscular force and the amount of the EMG (Jonsson, 1978). The present study was designed to investigate the relationship between muscular force and myoelectrical activities from the viewpoints of muscle length during isometric contraction and of shortening velocity during isokinetic contraction.

Methods

Six healthy males of our laboratory participated in this study. Mean value (range) of age, height, and weight for the subjects was 24.6 (22 to 27) years, 173 (167 to 180) cm, and 70 (60 to 83) kg, respectively. All subjects were fully informed of the purpose and the procedures associated with this study, and consented to participate.

The left knee extensors were tested. The Cybex II dynamometer (Lumex, Inc., New York) was used for measuring isometric and isokinetic knee extension torque. The subject was seated on a bench and belted to the bench at his waist. The input axis of the dynamometer was adjusted to the center of the subject's knee joint and his lower leg was strapped to the input lever at his ankle. Since the exerted torques were influenced by the weight of his lower leg and the machine lever, the moment caused by gravity was measured at each angle of the knee joint before the experiments.

The subject was asked to perform maximum isometric knee extensions three times at each of eleven different angles of the knee joint from 70° to 170° (180° = full extension). The subject was also asked to repeat a maximum isokinetic knee extension continuously 30 times with 10 s intervals at each of four different angular velocities: 60°, 120°, 180°, and 240°/s. The range of motion was from about 70° to 180°, that is, from an almost completely flexed position to a fully extended position in this device.

EMGs were recorded from the vastus medialis, rectus femoris, and vastus lateralis muscles by surface bipolar silver disk electrodes (10 mm in diameter). A pair of electrodes was place 50 mm apart over the center of the muscle belly. The EMG signals were amplified, filtered (5 to 300 Hz), and stored on an FM tape recorder (NFR-3915, Sony Co., Tokyo) with the torque signal.

After the experiments, the recorded signals were played back for processing. To obtain an averaged magnitude of EMG during isometric trials, the EMG signals were full wave rectified and averaged for a given period during which the exerted torque seemed to be constant. The mean value of three trials at each angle of the knee joint was used. As for isokinetic trials, the torque signal and the rectified EMG were averaged during the extension from 85° to 175° of the knee joint angle, and are expressed below as the averaged torque and averaged EMG, respectively. The mean value of 30 repetitions was used for the results.

Results

During isometric contraction, the maximum torque was exerted at approximately 110° knee joint angle with little inter-individual difference. The values ranged from 186 to 259 N•m among the subjects. The torque decreased toward more the flexed or extended position, 50% at 170° and 70% at 70° of the knee joint angle. A definite tendency could not be found in EMGs related to muscle length and angle of the knee joint, even

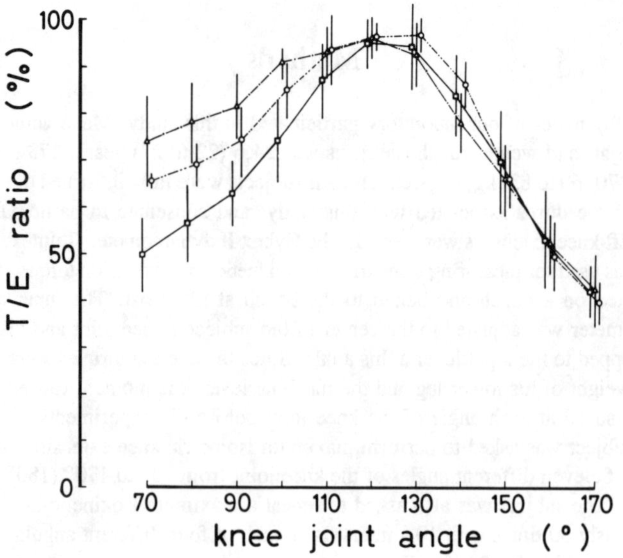

Figure 1—T/E ratios of vastus medialis (□), rectus femoris (△), and vastus lateralis (○) muscles, related to knee joint angle during maximum isometric contraction.

though a maximum effort was exerted in all cases. However, the ratio of the knee extension torque to average EMG (T/E ratio) was the highest at 110 to 130° of the knee joint angle for all subjects and all muscles. The ranges of the maximum T/E ratios among the subjects were 449 to 819 N•m/mV for the vastus medialis, 346 to 1066 N•m/mV for the rectus femoris, and 546 to 947 N•m/mV for the vastus lateralis. The T/E ratios decreased as the knee was in either more flexed or extended position than 110° to 130° (see Figure 1). The ratio was reduced to 50 to 74% of the maximum value in the most flexed position (70°), and to 40 to 42% in the most extended position (170°).

The torques and average EMGs during isometric contraction (angular velocity = 0°/s) at nine angles from 90° to 170° of the knee joint were averaged in order to determine the reference value to averaged torque and averaged EMG during isokinetic contraction. Averaged torques at 0°/s ranged from 141 to 184 N•m for six subjects, and the ranges of the averaged EMGs at 0°/s were 0.257 to 0.655 mV, 0.202 to 0.499 mV, and 0.227 to 0.339 mV for the vastus medialis, rectus femoris, and vastus lateralis muscles, respectively. As the angular velocity increased from 0°/s to 240°/s, the mean value of averaged torque decreased and the mean value of averaged EMG slightly increased (see Figure 2). The T/E ratio decreased with the angular velocity (see Figure 3). The T/E ratios at 240°/s were less than a half of those at 0°/s, ranging from 224 to 630 N•m/mV for the vastus medialis, 282 to 828 N•m for the rectus femoris, and 416 to 783 N•m/mV for the vastus lateralis.

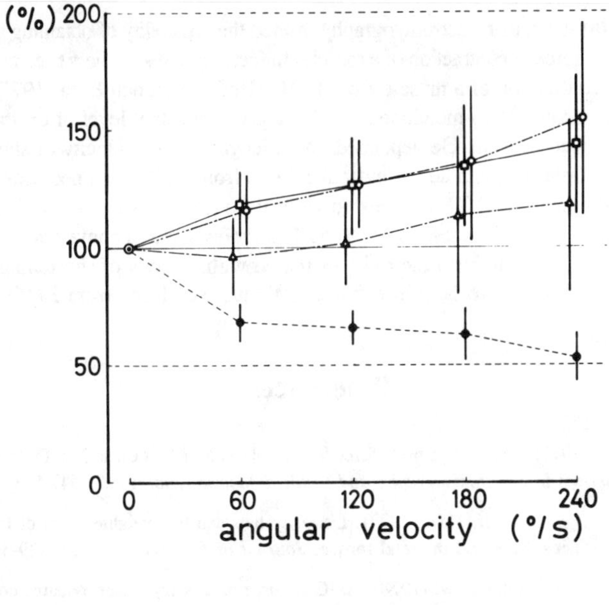

Figure 2—Averaged torque (•) and averaged EMGs (vastus medialis: □, rectus femoris: Δ, and vastus lateralis: O), related to angular velocity of maximum isokinetic contraction (0°/s = isometric contraction).

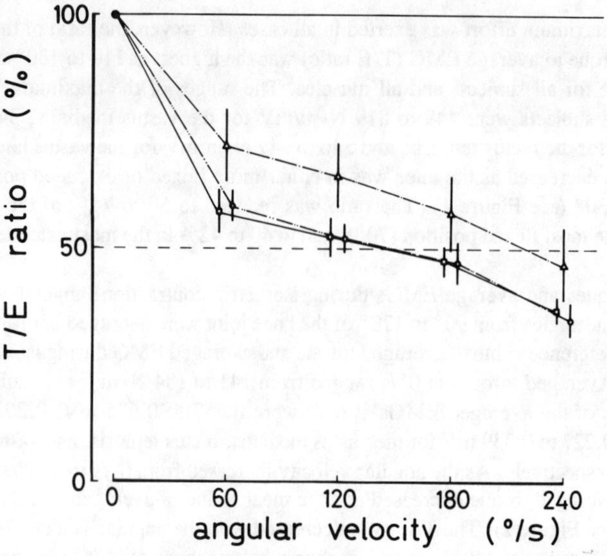

Figure 3—T/E ratios of vastus medialis (□), rectus femoris (Δ), and vastus lateralis (O) muscles, related to angular velocity of isokinetic contraction (0°/s = isometric contraction).

Discussion

Jonsson (1978) stated that electromyography offered the possibility of obtaining a relative measure of the force of contraction of a muscle. In fact, some researchers tried to estimate the force of contraction of a muscle from EMG (Hof & Van den Berg, 1977, 1978). However, Bouisset (1973) mentioned that for a given constant level of excitation the tension developed by a muscle depended on its length and its velocity of shortening. Therefore, in order to estimate the muscular force from EMG, it is necessary to take account of the tension-EMG relationship from the viewpoints of muscle length and of shortening velocity. The results—that the T/E ratio showed a convex curve in relation to muscle length, and that the ratio decreased with velocity of shortening—might give usable information to help in estimating a muscular force from EMG.

References

BOUISSET, S. (1973). EMG and muscle force in normal motor activities. In J.E. Desmedt (Ed.), *New Developments in Electromyography and Clinical Neurophysiology*, **1**, 547-583.

HOF, A.L., & Van den Berg, Jw. (1977). Linearity between the weighted sum of the EMGs of the human triceps surae and the total torque. *Journal of Biomechanics*, **10**, 529-540.

HOF, A.L., & Van den Berg, Jw. (1978). EMG to force processing under dynamic conditions. In E. Asmussen & K. Jorgensen (Eds.), *Biomechanics VI-A* (pp. 221-228). Baltimore: University Park Press.

JONSSON, B. (1978). Kinesiology: With special reference to electromyographic kinesiology. *Electroencephalography and Clinical Neurophysiology* (Suppl.), **34**, 417-428.

Signal Characteristics of EMG: Effects of Ballistic Forearm Flexion Practice

J.P. Boucher and M.S. Flieger
University of Massachusetts, Amherst, Massachusetts, USA

Practice effects upon neuromuscular control mechanisms underlying maximum speed human movements are traditionally investigated through the analysis of temporal components of agonist and antagonist muscle EMG activity (Lagasse, 1979; Person, 1958). Such temporal analyses, however, yield little information regarding changes occurring in the activation pattern of motor unit pools, whereas EMG signal characteristic analyses, such as power spectral analysis, have been shown to be useful techniques to assess such changes. Power spectral analysis has proven helpful in investigating localized fatigue (Viitasalo & Komi, 1977), neuromuscular disorders (Muro, Nagata, Murakanii, & Moritani, 1982), and surface electrode placement (Boucher & James, 1982), but has scarcely ever been applied to the investigation of neuromuscular control mechanisms responsible for maximum speed ballistic human movements.

The present study was undertaken to assess long-term (day-to-day) alterations of surface EMG signal characteristics due to practice of ballistic maximum speed forearm flexion task.

Methods

Subjects and Procedures

Ten college age students (5 male, 5 female) participated in this study. Each student performed ten maximum speed forearm flexion trials under each of three different inertial loads (0, 4 and 8 times the forearm moment of inertia) on 8 separate practice days. A specially designed apparatus (see Figure 1) was utilized to isolate and standardize the experimental movement, which consisted of a maximum speed flexion of the right forearm from a 15° resting position to a 90° target along the sagittal plane. The forearm of each subject was positioned parallel and attached with a wrist cuff to a wooden bar allowed to rotate freely around an axis which coincides with the elbow joint center of rotation. On every trial, a subject was asked to execute the movement following verbal commands by the experimenter, and volitionally stop the movement as close as possible to the 90° target.

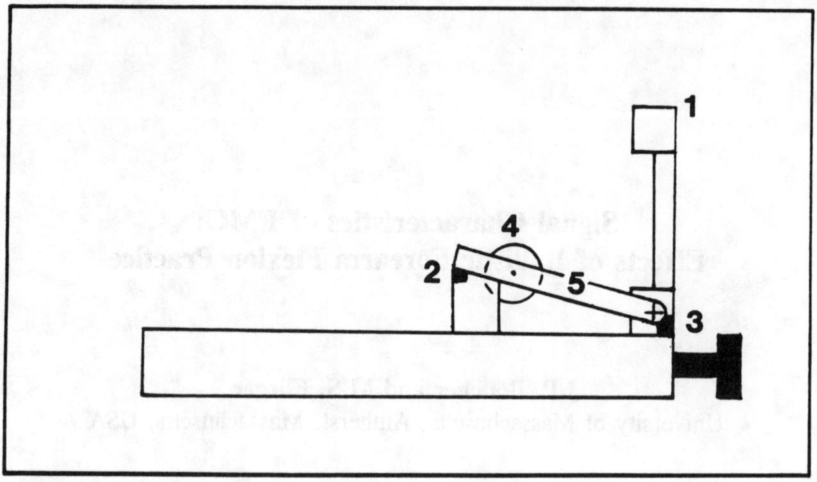

Figure 1—Movement apparatus. (1) 90° target, (2) movement initiation and (3) completion microswitches, (4) load, (5) wooden bar.

EMG Signal Processing

Beckman bipolar surface electrodes (Ag-AgCl) were used to simultaneously pick up EMG signals from the long head of the biceps brachii (agonist) and the lateral head of the triceps brachii (antagonist) during each flexion trial on 3 of the 8 practice days; that is, days 1, 4, and 8. Active electrodes were placed 4.25 cm apart in a position parallel to muscle fiber direction. These electrodes remained in place over each muscle's motor point only after skin-electrode impedance was reduced to 5 kΩ or less. The ground electrode was attached to the skin overlying the right clavicle. EMG signals were differentially amplified and visually inspected for extraneous noise and/or artifact during recording with a two channel Medic electromyograph (Medic Flexline-S, model SNV2H4, spectrum range 2 Hz to 20 kHz). The EMG signals that passed visual inspection, along with movement initiation and completion event markers, were then stored on a Sony analog recorder to await analog-to-digital conversion (2 kHz conversion rate) and quantification.

EMG Signal Quantification

The EMG parameters studied were measured on four components of the triphasic pattern distinctive of ballistic flexions of the forearm (Angel, 1981). Figure 2 illustrates the four components investigated: onset of the first biceps brachii EMG burst (C1), end of the first biceps brachii EMG burst (C2), second biceps brachii EMG burst (C3), and triceps brachii EMG burst (C4). On every recorded flexion trial, two EMG parameters were quantified: the fast Fourier power spectral density function (Hamming Window) mean power frequency (MPF), calculated according to the method of Kwatny, Thomas, and Kwatny (1970), and peak power frequency (PPF), as defined by Muro et al. (1982).

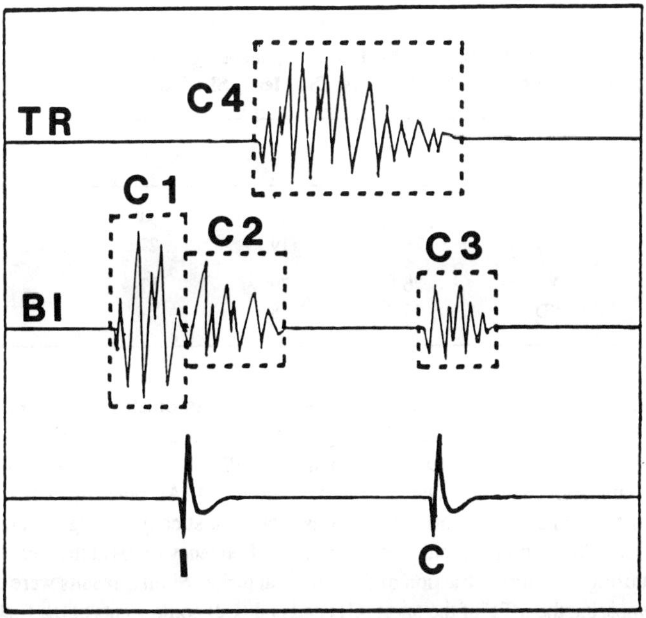

Figure 2—Schematic representation of the EMG recorded for a typical ballistic forearm flexion trial. BI and TR: biceps and triceps brachii raw EMG; I and C: initiation and completion (target) event markers (see text for description of C1 to C4).

The effects of practice across days and the influence of sex and inertial loading upon the MPF and PPF were assessed using a factorial analysis of variance model with repeated measures. Data collected on days 4 and 8 were further compared using an intraclass reliability analysis of variance model in order to test for the reproducibility of the parameters monitored.

Results and Discussion

Reproducibility of the data collected on C3 and C4 was found to be too low to warrant further analysis. However, reliability values for the MPF and PPF for agonist EMG burst components C1 and C2 ranged from fair to good (intraclass R = .73 to .88). These results corroborate the findings of Viitasalo and Komi (1975) and Muro et al. (1982) on the reliability of these EMG parameters measured during isometric contractions. All PPF results paralleled the results for the MPF; therefore, this section will focus on the MPF data only.

Table 1 presents the MPF means and standard deviations for the day's main effect for the two components of the first agonist EMG burst (C1 and C2). The 8 Hz (9%) MPF drop occurring from the beginning to the end of the first agonist EMG burst (from C1 to C2) represents a statistically significant difference ($p < .01$). In C2, the increase in MPF due to practice (13% increase from day 1 to day 8) also is a significant dif-

Table 1

MPF (Hz) for the Two Components (C1 and C2) of the First Agonist EMG Burst (X: Mean; SD: Standard Deviation)

Days		1	4	8	Grand Mean
C1	$\bar{X}$	86	85	94	89
	SD	28	19	27	25
C2	$\bar{X}$	76	79	87	81
	SD	17	18	24	20

ference ($p < .05$), whereas in C1 the 9% increase due to practice failed to reach the significant level. A Duncan multiple range test revealed that the MPF on day 8 for C2 was significantly greater ($p < .01$) than the MPF on days 1 and 4, whereas the MPF remained virtually unchanged from day 1 to day 4. All other main effects (sex, load, and trial) and interaction comparisons were statistically nonsignificant.

Desmedt (1981) reported rapid single motor unit frequency drops at the end of agonist activity during ballistic contraction of the tibialis anterior. Similar results were obtained when comparing the MPF of the onset (C1) and end (C2) components of the first agonist EMG burst. Such decreases in frequency, often accompanied by increases in EMG potential amplitude (Kots, 1977), reveal that motor unit synchronization may occur at higher levels of ballistic or fast isotonic muscle contractions. Temporal EMG analyses have demonstrated that the duration of the onset component (C1) of the agonist muscle remains unchanged with practice, while the total duration of the first agonist burst (C1 + C2) decreases significantly (Boucher & Lagasse, 1980). The consistency of the C1 MPF along with the increase in the C2 MPF that occurs with practice support the findings that the onset or static component (C1) of the agonist activity is not affected by practice, and the end or dynamic component (C2) of the first agonist EMG burst represents a labile component of the triphasic EMG pattern. Furthermore, the results also suggest that the dynamic component (C2) is somewhat resistant to change since MPF significantly increased on the eighth practice day only. Kots (1977) observed an increase in agonist muscle EMG amplitude with practice. Hence, the increase observed in MPF that occurs with practice suggests that more fast-twitch motor units are recruited with practice. However, firing rate modulation of active motor units remains a plausible neuromuscular control mechanism that may be involved in practice.

Finally, the low reliability of signal characteristics measured in the antagonist EMG burst (C4) and the lack of consistency in the occurrence of the second agonist EMG burst (C3) remain a puzzling question which warrants further investigation.

Acknowledgment

The authors would like to express their sincere appreciation to Dr. W. Kroll. This study was supported by a grant from the U.S. Army Medical Research and Development Command (contract DAMD-17-80-C-0101).

References

ANGEL, R.W. (1981). Electromyographic patterns during ballistic movements in normals and hemiplegic patients. In J.E. Desmedt (Ed.), *Progress in Clinical Neurophysiology*, **9**, 347-357. Basel: Karger.

BOUCHER, J.P., & Lagasse, P.P. (1980). Effects of functional electrical stimulation (FES) on neuromuscular coordination mechanisms. *Journal of Biomechanics*, **14**, 498-499.

BOUCHER, J.P., & James, R.J. (1982). Effects of ground electrode position on electromyographic potentials. In *Proceedings of the 5th Congress of the International Society for Electrophysiological Kinesiology*. Ljublana, Yugoslavia.

DESMEDT, J.E. (1981). The size principle of motorneuron recruitment in ballistic or ramp voluntary contractions in man. In J.E. Desmedt (Ed.), *Progress in Clinical Neurophysiology*, **9**, 97-136. Basel: Karger.

KOTS, Y.M. (1977). *The organization of voluntary movement, neurophysiological mechanisms.* London: Plenum Press.

KWATNY, E., Thomas, D.H., & Kwatny, H.G. (1970). An application of signal processing techniques to the study of myoelectric signals. *IEEE Transactions on Bio-Medical Engineering*, **BME-17**(4), 303-312.

LAGASSE, P.P. (1979). Prediction of maximum speed of human movement by two selected muscular coordination mechanisms and by maximum static strength. *Perceptual and Motor Skills*, **49**, 151-161.

MURO, M., Nagata, A., Murakanii, K., & Moritani, T. (1982). Surface EMG power spectral analysis of neuromuscular disorders during isometric and isotonic contractions. *American Journal of Physical Medicine*, **61**(5), 244-254.

PERSON, R.S. (1958). An electromyographic investigation on coordination of the activity of antagonist muscles in man during the development of a motor habit. *Pavlovian Journal of Higher Nervous Activity*, **8**, 13-23.

VIITASALO, J.H., & Komi, P.V. (1975). Signal characteristics of EMG with special reference to the reproducibility of measurements. *Acta Physiologica Scandinavica*, **93**, 531-539.

VIITASALO, J.H., & Komi, P.V. (1977). Signal characteristics of EMG during fatigue. *European Journal of Applied Physiology*, **37**, 111-127.

Height and Slope of IEMG as Predictors of Torque Parameters

R.N. Robertson, M.J. Carlton, L.G. Carlton, and K.M. Newell
University of Illinois, Urbana, Illinois, USA

The relationship between the electrical activation of intact human muscle and the resulting torque output for isotonic contractions has not been clearly defined. A systematic relationship between the electrical activity of a muscle and mechanical output could only be established when movement velocity was held constant (Bigland & Lippold, 1954), when the mechanical properties of the muscle were considered (Hof & VandenBerg, 1981), or when the EMG was modeled as discrete input pulses producing a series of muscle tension twitches (Kilmer, Kroll, & Congdon, 1982). The major difficulty in relating electrical activity to mechanical output for in situ isotonic contractions is the result of moment-to-moment changes in the mechanical and electrical state of the muscle. These changes could result in disproportionate increases or decreases in the output of the system when compared to changes in the electrical activity of the muscle.

The technique employed in the present study was designed to minimize the effect of factors which produce nonproportionalities over the course of the contraction, and between conditions of varying mechanical output. This was accomplished by dividing the rectified, integrated EMG trace into discrete intervals and comparing these to specific measures of mechanical output. The rationale for this approach stems from the notion that the history of activation is one important criterion for determining tension production, in contrast to the total activity over the entire contraction.

Specifically, this study examined the relationship between IEMG and various measures of mechanical output over different movement velocities and task constraints within a single trial. Further, the question of whether discrete measures of EMG provide a useful way of predicting mechanical output was addressed.

Method

Subjects and Apparatus

Fourteen subjects produced movements utilizing a horizontal displacement bar (see Newell, Carlton, & Carlton, 1982). Electrical activity from the pectoralis and posterior deltoid were monitored using silver Narco Bio-System surface electrodes. Low and high frequency cutoffs of 10 and 500 Hz, respectively, were used. Acceleration, displacement, and EMG data were collected with a sampling rate of 1000 Hz.

Procedures

Subjects participated in a single testing session during which they completed supported horizontal flexion movement at a single movement time (400 ms) and four distances (4°, 8°, 16°, 32°) for two movement termination conditions (follow-through or stop).

Experimental Design and Data Analysis

Fourteen subjects, divided into two groups, participated in one testing session completing 80 trials for each of the four distances. The statistical analysis was conducted over the last 40 trials in each condition. Stepwise multiple regression analyses were performed for each condition of both experiments. The analyses determined the best 5 EMG predictors for each of the four mechanical output variables and calculated the R^2 terms. Figure 1 depicts the 9 independent variables entered into the model as predictors and the four dependent variables chosen as the parameters to be predicted.

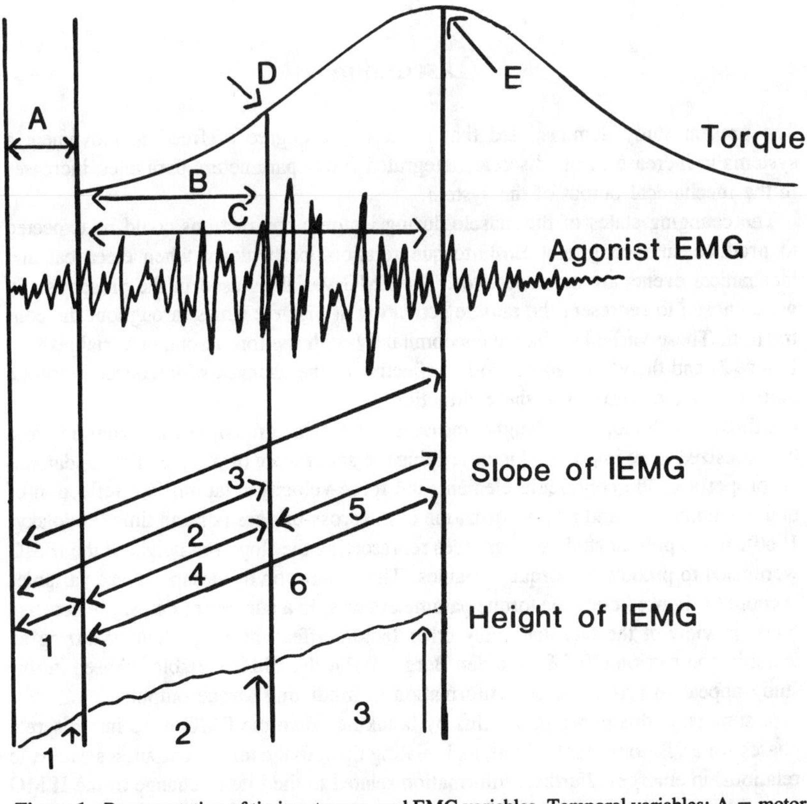

Figure 1—Representation of timing, torque, and EMG variables. Temporal variables: A = motor time; B = time to peak rate of change of torque; C = time to peak torque. Force variables: D = peak rate of change of torque; E = peak torque. The IEMG slopes were calculated over the 6 designated time periods and the height of the IEMG was calculated over 3 time periods.

Results

Figure 2 represents the relationship between the EMG slope and height variables and each of the mechanical output variables. These figures illustrate that the EMG parameters of height and slope increase systematically with increases in mechanical output. It should be noted that positive torque duration decreases with higher velocities of movement due to the inverse relationship between peak torque and torque duration (Newell et al., 1982). Correlations determined between the EMG variables and peak torque, peak rate of change of torque, and impulse were greater than .95 ($p < .01$) for all comparisons. For torque duration, the correlations were greater than −.95 ($p < .01$) for the follow-through condition and greater than −.70 ($p < .01$) for the stop condition.

Figure 3 depicts the proportion of variance (R^2) the EMG parameters account for in predicting the mechanical output variables. Generally, for the follow-through condition the ability to predict mechanical output was greatest when the average velocity of movement was larger. In the stop condition, this trend was not observed, and in fact, little differences between conditions were noted. The largest R^2 values were found for predicting the duration of the positive torque phase. The output variable least predictable by the EMG parameters was the torque impulse.

Discussion

The present study demonstrated that for a single degree of freedom movement a systematic increase in the discrete, integrated EMG parameters paralleled increases in the mechanical output of the system.

The changing states of the muscle during isotonic contractions could be expected to produce an incongruent EMG-torque relation, particularly when electrical and mechanical events are summed over. The EMG variables used in the present study were chosen to represent the muscle activation at discrete times throughout the contraction. These variables change concomitantly with the torque output variables (see Figure 2) and therefore, appear to be reflective of the changes which occur in torque output over the duration of the contraction.

Utilizing both slope and height, measures of the integrated electromyogram were hypothesized to be importannt in describing the active state of muscle. The fundamental property of the contractile element, the force-velocity relation, is a reflection of both the number of and rate of formation of the cross-bridges per unit time (Podolsky, 1960). In the present study the variables representing the slope and height of the IEMG were used to predict the torque variables. The proportion of variance these variables account for in predicting the torque parameters was, in a number of cases, greater than 50%. In view of the fact that many other factors affect the torque output during an isotonic contraction (Hof & Van den Berg, 1981), the EMG variables chosen in this study appear to provide useful information in predicting torque output.

In summary, this paper shows that by breaking down the EMG trace into discrete phases for an isotonic contraction, and relating them to the torque output, a systematic relationship emerges. Further, information related to the rate of change of the IEMG trace (slope) in conjunction with the total amount of electrical activity (height) account for a significant proportion of the mechanical output variance.

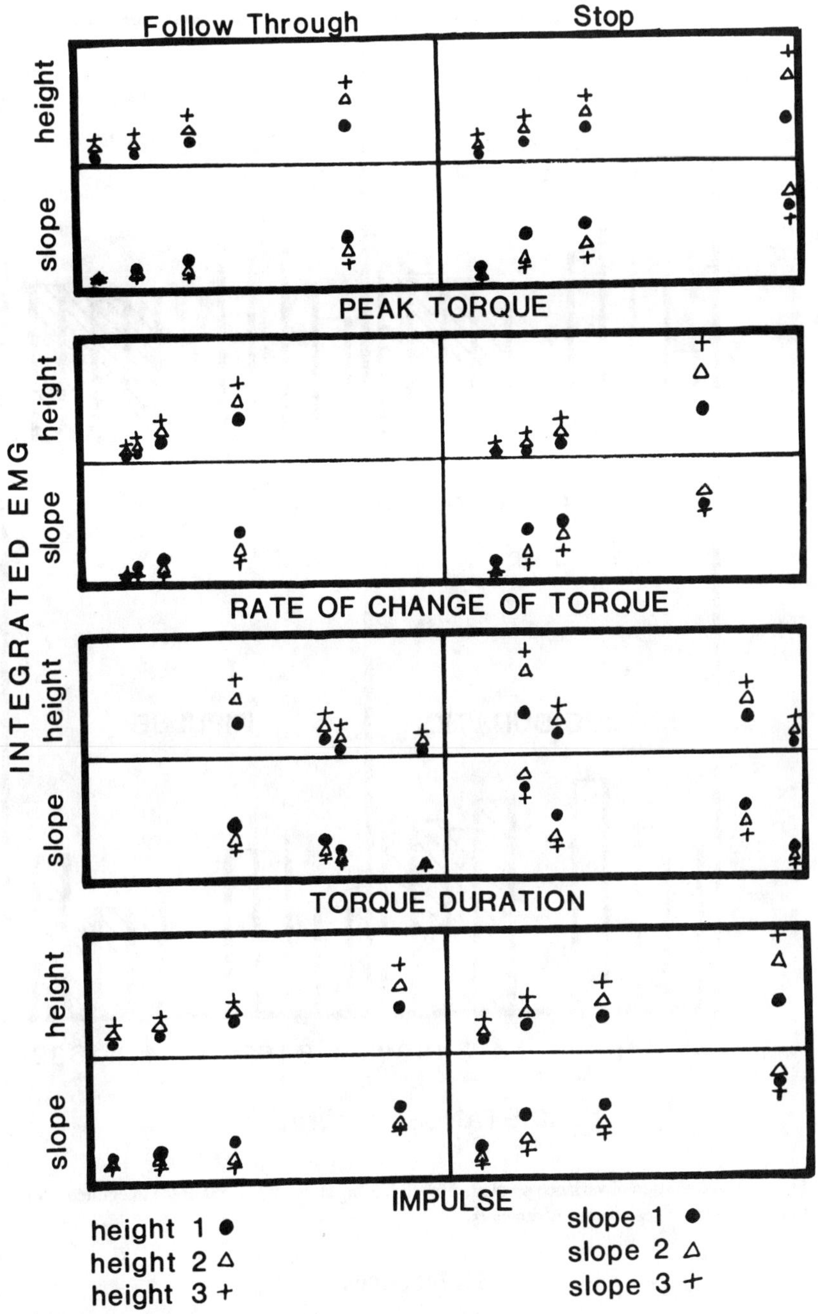

Figure 2—Height and slope of the IEMG (averaged across subjects) plotted against four torque variables. Height and slope variables are defined in Figure 1.

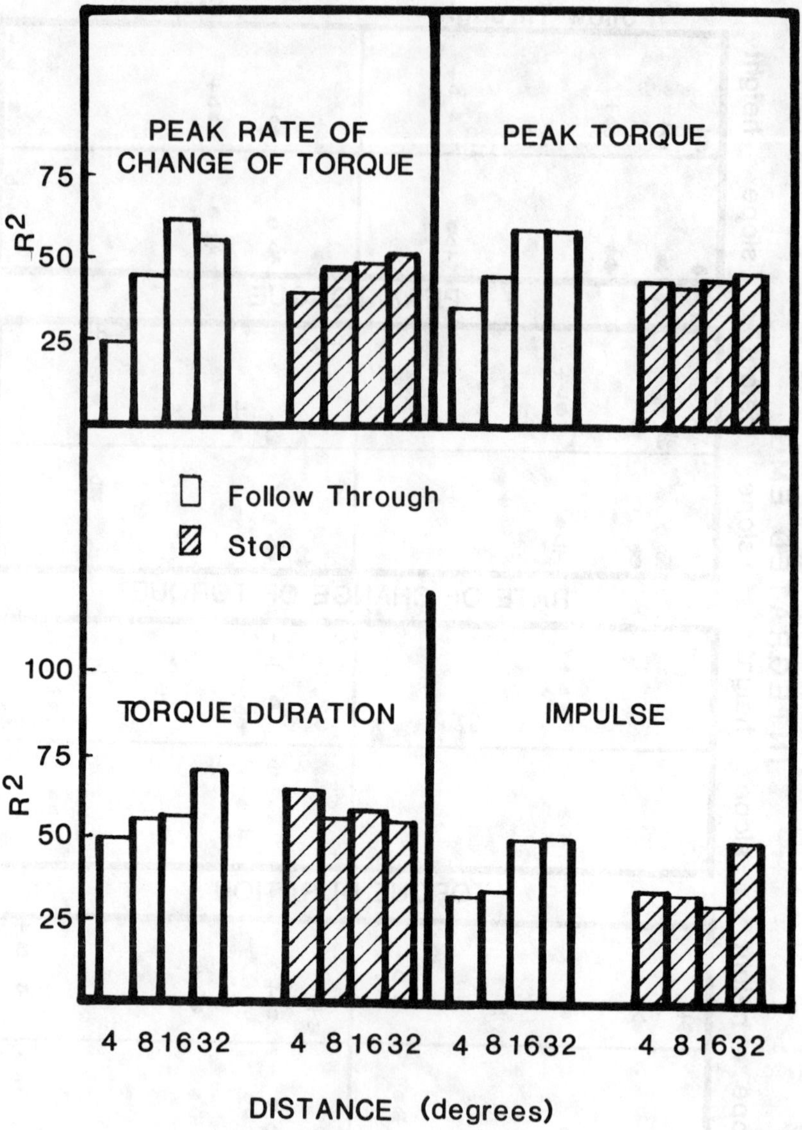

Figure 3—Proportion of variance accounted for (R^2) using slope and height of the integrated EMG to predict mechanical output parameters.

References

BIGLAND, B., & Lippold, O.C.J. (1954). The relation between force, velocity and integrated electrical activity in human muscles. *Journal of Physiology,* **123**, 214-224.

HOF, A.L., & Van den Berg, Jw. (1977). Linearity between the weighted sum of the EMGs of the human triceps surae and the total torque. *Journal of Biomechanics,* **10**, 529-539.

HOF, A.L., & Van den Berg, Jw. (1981). EMG to force processing III: Estimation of model parameters for the human triceps surae muscle and assessment of the accuracy by means of a torque plate. *Journal of Biomechanics,* **14**, 771-785.

KILMER, W., Kroll, W., & Congdon, V. (1982). An EMG-level muscle model for a fast arm movement to target. *Biological Cybernetics,* **44**, 17-26.

NEWELL, K.M., Carlton, L.G., & Carlton, M.J. (1982). The relationship of impulse to timing error. *Journal of Motor Behavior,* **14**, 24-45.

PODOLSKY, R.J. (1960). The kinetics of muscular contraction: The approach to the steady state. *Nature,* **188**, 666-668.

Investigation and Modeling of the Relationship Between Integrated Surface EMG and Muscle Tension

A. Dąbrowska and K. Kędzior
Technical University of Warsaw, Warsaw, Poland

For the last 30 years the relationship between surface rectified-integrated myoelectric signals (IEMG) and tension developed by a skeletal muscle has been a problem in biomechanics. Usually an algebraic relationship for static conditions (steady state of isometric contraction) is used to describe this relationship. This is caused by a need to operate on simple mathematical models explaining the cooperation of a set of muscles serving a joint (Dąbrowska & Kędzior, 1981; Morecki, Ekiel, & Fidelus, 1971).

However, even for isometric contraction the real process has a dynamic character, due to a buildup or drop of tension, very similar to transients in control systems. For this reason the relationship between IEMG and tension can be expressed more precisely by a differential equation or corresponding transfer function. It means that the muscle can be represented by a cybernetic model shown in Figure 1.

Based on previous results (Bieżanowska & Kędzior, 1981; Kędzior, 1973) it is assumed in this work that skeletal muscle in situ can be treated as an inertial object of high order of an automatic control system with two inputs (excitation represented by IEMG, and the length of muscle represented by joint angle α) and one output (tension represented by torque M exerted by muscle with respect to the joint axis).

The purpose of this study was to determine the time constant of the above mentioned inertial object and to investigate if and to what degree it depends on muscle length, that is, joint angle α. A method known in automatic control practice consisting of the determination of the gain vs. phase characteristics will be used. This method has been successfully used for this purpose by other authors (Soechting & Roberts, 1975).

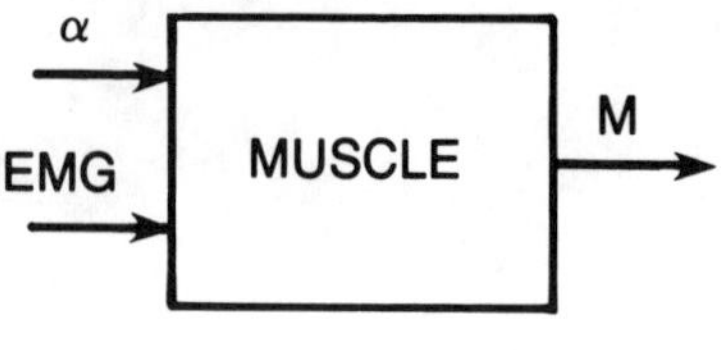

Figure 1—Model of the skeletal muscle in situ. For symbols, see the text.

Methods

The investigation was carried out on four male subjects and conducted on a special test rig (Kedzior, 1978; Morecki, Kedzior, Biezanowska, Dabrowska, & Paśniczek, 1983). The muscles of the elbow joint were investigated during isometric flexion and extension of a horizontally placed forearm for five different values of the joint angle $\alpha = 30°, 50°, 70°, 90°$, and $110°$. The experiment consisted of the generation by a subject of a sinusoidally variable torque M of mean value 0.5 M_{max} (where M_{max} = the maximum static torque exerted by the subject in flexion or extension), of amplitude $\Delta M \approx 15\%\ M_{max}$, and frequency $\omega = 1, 2, 3, 4$, and 5 Hz.

The subject observed a pattern signal from a generator displayed on a screen of a two-channel oscilloscope and tried to shape the trace of M into a similar wave. Simultaneously the IEMGs for the most active flexors and extensors were recorded (see Figure 2). Each measurement was repeated four times.

It was assumed that the most active muscle represents all the muscles in the group of flexors or extensors. In the course of the experiment the only recorded trials were those corresponding to exclusive involvement of either flexors or extensors.

Data Processing

It was assumed that under isometric conditions (α = const) the relationship between the IEMG and M, given in the form of the spectrum transmittance is of the type

$$G(j\omega) = \frac{M(j\omega)}{IEMG(j\omega)} = \frac{k}{(Tj\omega+1)^n} \qquad (1)$$

where $G(j\omega)$ = spectrum transmittance, ω = frequency, k = amplification coefficient, T = time constant, n = order of the inertial object, and $j = \sqrt{-1}$.

Numerical processing of the recorded runs, similar to those shown in Figure 2, was performed on-line and consisted in determination of the first harmonic for both IEMG and M runs, using the Fast Fourier Transformation method. This in turn permitted us to determine gain $|G(j\omega)|$ and phase $\phi(\omega)$ for the experimental characteristics (see Figure 3).

It should be noted that Equation 1 can be rewritten as:

$$G(j\omega) = |G(j\omega|\bullet e^{j\phi(\omega)} \qquad (2)$$

where e = natural logarithm base.

The experimental characteristics were approximated by the least square method using theoretical Equation 1 with $n = 1, 2, 3, 4$. It was shown that $n = 2$ ensures a sufficiently good approximation for flexor muscles and $n = 4$ for extensor muscles, and that the value T and k depends strongly on the joint angle α.

Data for $T(\alpha)$ for one of the investigated subjects are collected in Table 1. The data for other subjects are of the same order, the relationships $T(\alpha)$ and $k(\alpha)$ being similar.

It should be noted that $T(\alpha)$ pertains more to the muscle itself, whereas $k(\alpha)$ is much more associated with the instrumentation used. For this reason numerical values of $k(\alpha)$ are omitted.

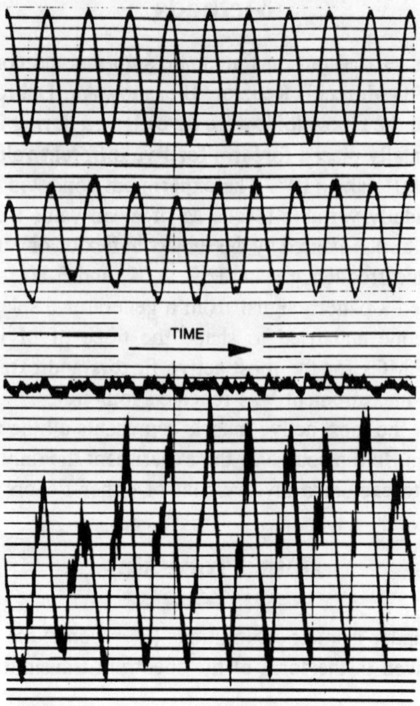

Figure 2—Example of recorded measurements. For symbols, see the text.

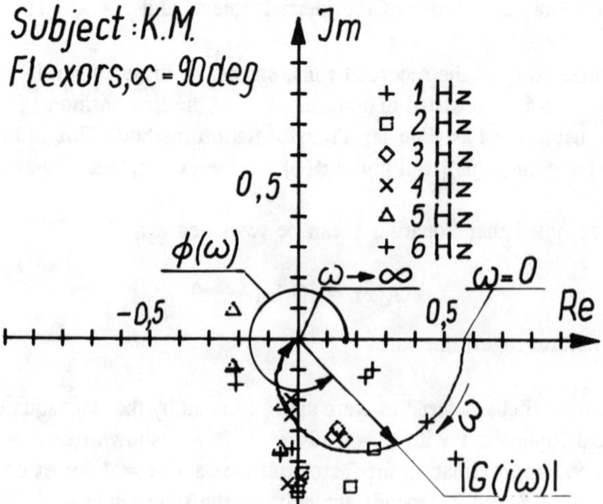

Figure 3—Example of approximation of experimental data with analytical curve G(jω): Re = real number axis, Im = imaginary number axis. For the remaining symbols, see the text.

Table 1

Data for Subject K.M.

α (deg)	30	50	70	90	110
Flexors, T, (s), object order $n = 2$	0.042	0.063	0.060	0.045	0.043
Extensors, T, (s), object order $n = 4$	0.25	0.23	0.25	0.25	0.14

Conclusions

The obtained results prove that T and k depend on the length of the muscle, that is, on joint angle α, which means that the relationship between the input IEMG and output M parameters has the form of a differential equation with variable coefficients. This is consistent with the results obtained previously for isolated muscles (Biezanowska & Kędzior, 1981; Kędzior, 1973).

However, the actual numerical values have no general meaning, since they depend on the instrumentation used, in particular the properties of the device used for receiving and converting the myoelectric signals.

The numerical values of results obtained for a specific individual can be utilized for a description of the cooperation of his muscles in performing a movement, under specified measurement conditions.

The results obtained can be treated as the next step towards developing an adequate mathematical model of mechanical activity of skeletal muscle in situ.

Acknowledgment

Supported in part by Polish Academy of Sciences, Key—problem 06.9.

References

BIEZANOWSKA, E., & Kedzior, K. (1981). Simulation approach to modelling and investigation of static and dynamic properties of skeletal muscles. In A. Morecki, K. Fidelus, K. Kedzior, & A. Wit (Eds.), *Biomechanics VII-A* (pp. 208-214). Baltimore: University Park Press.

DABROWSKA, A., & Kedzior, K. (1981). Cooperation of muscles under dynamic conditions. In A. Morecki, K. Fidelus, K. Kedzior, & A. Wit (Eds.), *Biomechanics VII-A* (pp. 215-222). Baltimore: University Park Press.

KEDZIOR, K. (1973). Investigation of dynamic properties of isolated skeletal muscle. *The Archive of Mechanical Engineering (Warsaw)*, **20**, 219-238.

KEDZIOR, K. (1978). Modelling of biomechanical properties of actuators of muscle drive (in Polish). Warsaw Technical University Publications. *Mechanika*, **56**.

MORECKI, A., Ekiel, J., & Fidelus, K. (1971). *Bionics of motion* (in Polish). Warsaw: Polish Scientific Publishers.

MORECKI, A., Kedzior, K., Biezanowska, E., Dabrowska, A., & Paśniczek, R. (1983). Cooperation of muscles under dynamic conditions with stimulation control. In R.M. Campbell (Ed.), Control aspects of prosthetics and orthotics, Proceedings of the IFAC Symposium, *IFAC Proceedings Series* (pp. 7-15).

SOECHTING, J.F., & Roberts, W.J. (1975). Transfer characteristics between EMG activity and muscle tension under isometric conditions in man. *Journal of Physiology* (Paris), **70**, 779-793.

EMG Activity Level Comparisons in Quadriceps and Hamstrings in Five Dynamic Activities

C.L. Richards
Université Laval, Quebec, Canada

Dynamic activities place a variety of demands upon muscles in order to accomplish motor tasks. The intensity of this demand in different activities can be inferred from electromyographic (EMG) profile comparisons in participating muscles since electromyograms reflect motor unit activity patterns. Although EMG profiles have been studied for various activities or calibrated to EMG levels in isometric contractions (Winter, 1976), comparisons of EMG activity levels in the same muscle during several dynamic activities have not been described. This study describes EMG activity levels relative to a maximal activation level, defined during maximal voluntary isokinetic contractions, in the quadriceps and hamstring muscles during five dynamic activities (gait, stair ascent and descent, rising from and sitting on a chair).

Subjects and Methods

Electromyograms were taken with surface electrodes fixed to the skin over five muscles of one lower extremity in 12 normal women. The mean age was 37.8 years (*S.D.* = 13.0) and the weight 59.8 Kg (*S.D.* = 6.8). As illustrated in Figure 1, the EMG activity was recorded simultaneously to torque and knee angle during maximal voluntary isokinetic knee extension and flexion movements (Cybex II, isokinetic dynamometer). To reduce movement artifacts to an acceptable level, electrode leads were raised and held about 0.5 cm above the skin surface by tape attachments (Knutsson & Richards, 1979), connected to an electrode board, and by a 9.6 m shielded cable to the electrode selector unit of a polygraph (Grass, model 7D). The myosignals were fed to a.c. preamplifiers (type 7P3B) and recorded first as unintegrated (raw) EMG to check for movement artifacts, and if the records were judged acceptable, subsequently the EMG signals were rectified and time averaged ("integrated", IEMG) with a time constant of .2 s to allow for visual estimation of the amplitude. After completion of the isokinetic movements (three repetitions at 30°/s, 90°/s and 180°/s for extension and flexion), the electrodes were left in place, and three electronic footswitches were taped to the ipsilateral shoe sole. The EMG activity was recorded during different activities with

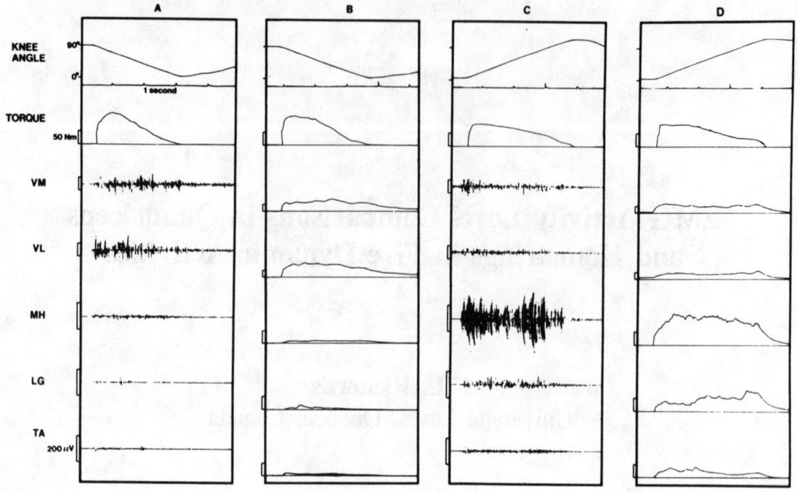

Figure 1—Records of concomitant knee angle, torque and EMG during knee extension (A, B) and knee flexion (C, D) in a normal subject. The traces indicate from above: knee angle, torque and EMG in vastus medialis (VM), vastus lateralis (VL), medial hamstrings (MH), lateral gastrocnemius (LG) and tibialis anterior (TA). Two repetitions of each movements are shown: raw EMG (A, C) and "integrated" EMG with time constant of 0.2 s (B, D). Calibration bars indicate 50 Nm (torque) and 200 uV (EMG).

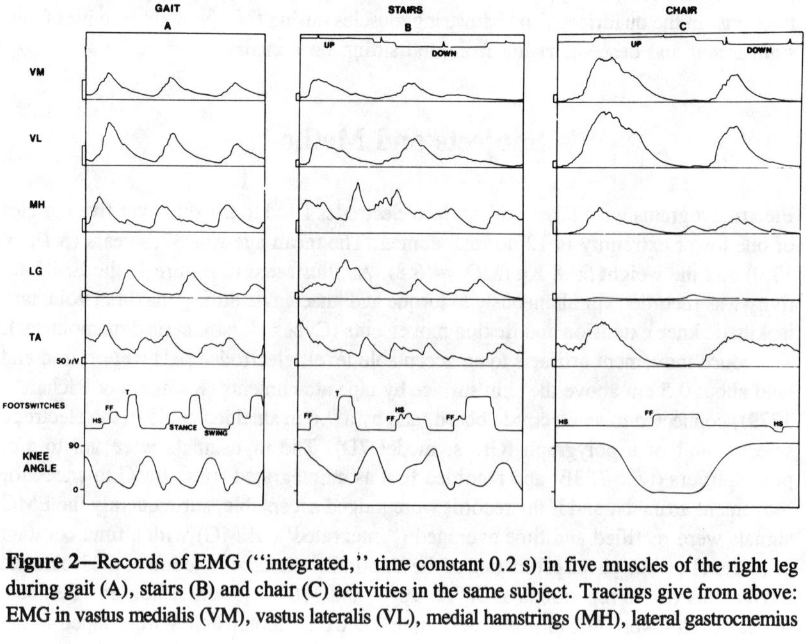

Figure 2—Records of EMG ("integrated," time constant 0.2 s) in five muscles of the right leg during gait (A), stairs (B) and chair (C) activities in the same subject. Tracings give from above: EMG in vastus medialis (VM), vastus lateralis (VL), medial hamstrings (MH), lateral gastrocnemius

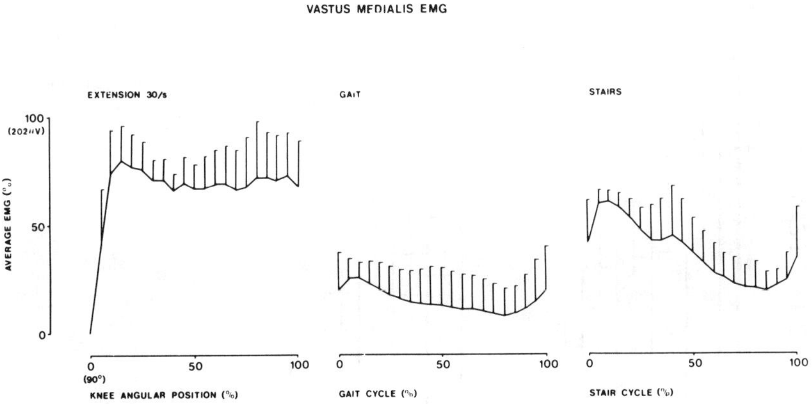

Figure 3—Comparison of mean EMG activity in the vastus medialis (VM) during isokinetic knee extension (A), gait (B) and stair ascent movements (C) in 9 normal subjects. The average peak EMG amplitude in isokinetic movements is given in brackets on y-axis and gait and stair curves are plotted relative to this peak amplitude. Isokinetic and stair curves derived from 1 movement or cycle and gait from 10 cycles in each subject. Vertical bars indicate + 1 S.D.

and without an electrogoniometer attached to the leg. Subjects were requested to walk at their "normal" speed and several gait cycles recorded, they then climbed a flight of three stairs (stair height, 17. 1cm), crossed a joining platform, and descended a similar flight. Finally, eletromyograms were taken during the act of sitting and rising from a standard armless chair (seat height, 43.2 cm).

Figure 2 gives an example of the EMG records in one subject during five activities (gait, stair ascent, stair descent, and chair up and down). Mean EMG profiles were calculated for these activities in the vastus medialis (VM), vastus lateralis (VL) and medial hamstrings (MH) and in addition in each subject the peak EMG amplitude recorded during these five activities was compared to the peak EMG amplitude in the same muscle during the maximal voluntary isokinetic contractions at 30°/s. Comparisons were thus made between representative movement cycles for the stair and chair movements defined by the electrogonimeter, footswitch and marker signals, and the average of 10 gait cycles with the third repetition of the isokinetic movement. The amplitude of the EMG activity during the gait, stair and chair movements was expressed as percent of the peak EMG activity recorded during the isokinetic knee extension (VM and VL) and flexion (MH) movement at 30°/s.

Results

The mean patterns of EMG activity in the VM during isokinetic knee extension, gait, and stair ascent cycles in the same nine subjects are given in Figure 3. Since the amplitude of the EMG in this figure for the gait and stair cycles are plotted relative to the mean peak amplitude recorded in the maximal voluntary movements, it is possible to visual-

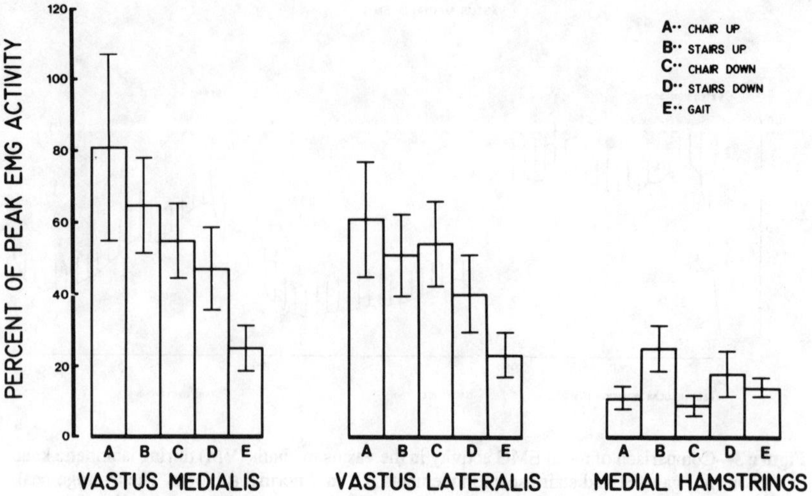

Figure 4—Comparison of percent of peak EMG activity in the vastus medialis (VM), vastus lateralis (VL), and medial hamstrings (MH) during 5 dynamic activities. Amplitude of peak EMG activity is expressed as percent of EMG activity in isokinetic knee extension (VM and VL) and knee flexion (MH) movements at 30°/s. Vertical bars indicate mean $\pm$ 2 S.E.M., $N = 12$.

ly estimate the level of activation in the VM during the three movements. It can also be seen that although the peak amplitude occurs when the knee is relatively flexed in the isokinetic movements, the level of EMG activity remains high throughout the knee movement. The mean peak amplitude in gait occurred in early stance when the knee has a flexion phase (knee angle about 14°) related to weight acceptance (Knutsson & Richards, 1979) and in stair ascent at about 10% of the cycle, corresponding approximately in time to the acceptance of weight on the leading leg when the knee is relatively flexed (Richards, 1980).

Figure 4 gives a summary of the EMG peak activity level comparisons for the three muscles. The order of the activities on the x-axis is kept constant and based on the descending percent utilization levels in VM, thus permitting a visual comparison of the degree of EMG activity in the different muscles during the activities. The VM and VL present similar profiles which differ from that in the MH. Rising from a chair elicits the highest level of activation, and gait the least in the knee extensors. In contrast in the MH the highest percent utilization is in stair climbing and the lowest in sitting down.

Discussion

In this study the EMG activity in the VM, VL, and MH was compared in five dynamic activities. The EMG activity in the different activities, expressed in percent of maximal voluntary activation in isokinetic contractions, gives an estimate of reserves in maximal voluntary capacity as judged from surface EMG activity. Comparisons made using peak EMG amplitude did not, however, control for the type of contraction (Komi, 1973) muscle length (Haffajee, Moritz & Svantesson, 1972) or the velocity of the contraction (Bigland and Lippold, 1954) known to influence muscle activation profiles.

The nature of the motor demand also varied in many respects since movements under strong volitional drive were used as the basis of comparison for highly automatized preprogrammed movements requiring a minimal of volitional drive such as gait (Grillner, 1975). The peak EMG comparisons described in the present study can, nevertheless, be applied to the evaluation of dynamic motor capacity in patients with motor disturbances. These EMG comparisons also give data pertinent to the study of functional anatomy and the relative roles of muscles in different movements.

Acknowledgments

This work was supported in part by grants from the Medical Research Council of Canada and "Fonds de la recherche en santé du Québec." Thanks are due in particular to Kevin Mullarky and Daniel Tardif for technical assistance and to Lyse Laroche for typing the manuscript.

References

BIGLAND, B., & Lippold, O.C.J. (1954). The relation between force, velocity and integrated electrical activity in human muscles. *Journal of Physiology (London)*, **123**, 214-224.

GRILLNER, S. (1975). Locomotion in vertebrates: Central mechanisms and reflex interaction. *Physiological Review*, **55**, 247-304.

HAFFAJEE, D., Moritz, V., & Svantesson, G. (1972). Isometric knee extension strength as a function of joint angle, muscle length and motor unit activity. *Acta Orthopedica Scandinavia*, **43**, 138-147.

KNUTSSON, E., & Richards, C. (1979). Different types of disturbed motor control in gait of hemiparetic patients. *Brain*, **102**, 405-430.

KOMI, P.V. (1973). Relationship between muscle tension, EMG and velocity of contraction under concentric and eccentric work. J.E. Desmedt (Ed.), *New Developments in Electromyography and Clinical Neurophysiology: Vol. 1* (pp. 596-606).

RICHARDS, C. (1980). Dynamic muscle function in human normal, pathological and prosthetic knee joints. Unpublished doctoral dissertation, McGill University, Montreal.

WINTER, D.A. (1976). The locomotion laboratory as a clinical assessment system. *Medical Progress Through Technology*, **4**, 95-106.

Surface Electromyogram (EMG)/Muscle Force: An Experimental Approach

U.P. Wyss and V.A. Pollak
University of Saskatchewan, Saskatoon, Saskatchewan, Canada

Several researchers have reported that the relationship between isometric muscle forces and the integrated EMG is linear. Some investigators have also pointed out, however, that this might not be true, owing to very high muscle contractions. Not many findings have been reported on the relationship between the EMG and dynamically changing muscle forces. In the present study, the EMG processing was based on the peaks from the depolarization and the following repolarization waves of the acting motor units as they are seen in the surface EMG.

Methods

Experimental Setup

It was found necessary to choose a muscle group with a minimal number of muscles involved in a specific movement. Elbow flexion in a neutral forearm position was thus chosen for all experiments. The upper arm was strapped to a support of adjustable height as shown in Figure 1. The forearm and the hand lay on a plexiglas plate so that the influence of the gravitational force was minimal. In this position, the subject was asked to flex the arm with a load attached by means of a nylon string to a wrist cuff. The string included a force transducer for direct force measurements. To calculate the moment acting at the elbow, the motion had to be recorded as well. For this purpose an optical system recorded the flashes of the light emitting diodes (LED) attached to specific anatomical and experimental landmarks (Wyss & Pollak, 1981). The recordings for the kinematics were synchronized with the EMG recordings of the activity of the biceps, brachialis, and triceps muscles. The biceps and the brachialis are the major flexors, and the triceps recordings were required to estimate the influence of the antagonist muscle. The three EMG signals and the force transducer signal were recorded on an FM recorder.

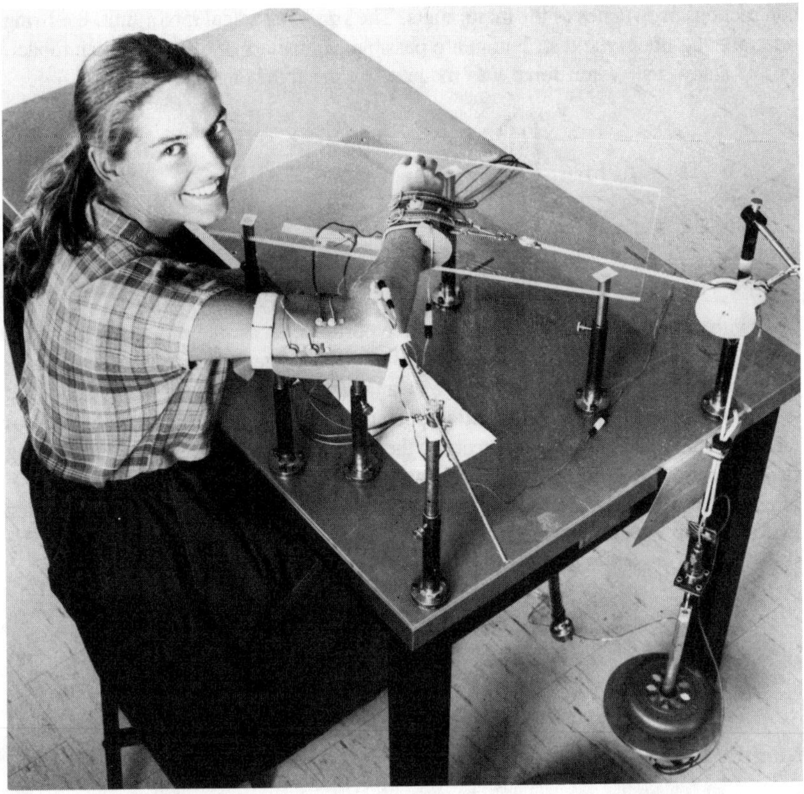

Figure 1—Experimental setup with seated subject.

Muscle Model

The muscle model is basically similar to that described by Hof and Van den Berg (1981). The major difference is the determination of the active state, which is based in this model on the peaks detected in the filtered EMG, and not on the rectified integrated EMG. No mathematical explanation is given, but it is clear that a motor unit firing further away from the surface is seen in the surface EMG with a smaller amplitude than a motor unit firing closer to the surface. As long as the peaks can be detected, they trigger the same twitch function. For medium and strong contractions, the EMG peaks may overlap one another at least partially, thus leading to lower IEMG values. When using the peaks, only action potential synchronization does not allow the recording of all peaks. The model further consists of the muscle force/muscle length relation, a series elastic component, and the force velocity relation.

Active State

The peaks detected in the surface EMG occur as a result of depolarization and repolarization "waves" of the motor units and are known as action potentials. The latter trigger contractions or twitches of the motor units. The size of an actual motor unit, the firing rate, and the fiber type distribution are possible parameters of a single twitch model. In this study, the twitch force was modeled on the following equation:

$$TF = f(FE) \cdot \begin{cases} (\frac{X}{B})^2 \cdot A & O < X < B \\ A & B < X < (B+C) \\ (\frac{D-X}{D})^2 \cdot A & (B+C) < X < (B+C+D) \end{cases} \quad (1)$$

$D = 1.5 \cdot (B+C)$

The twitch function is characterized by three portions, an ascending, a constant maximum period, and a descending portion. B, C, and D stand for a mean duration of the three portions, while A represents a mean maximum of a single twitch, and X the time after the occurrence of the action potential. The function f(FE) is a factor less than or equal to 1. It takes into account the decreasing twitch force with increasing firing events. For different motor unit types (slow, fast), it is necessary to use different values for A, B, and C. Each peak triggers a new twitch function curve, which is summed up with the other curves, ultimately resulting in the processed peak EMG (PPE), as expressed in Equation 2:

$$PPE = \sum_{i=1}^{i=n} TF_i \quad (2)$$

Modified "Hill Relation"

The relation between the force and the speed of a contracting muscle, which was first described by Hill, is in our terms written in Equation 3, with n and b used as the Hill parameters:

$$FA = PPE \cdot \frac{f(PHC) + n \cdot DPHC/b}{1 - DPHC/b} \quad (3)$$

The output force FA can be calculated by multiplying PPE, derived from Equation 2, by the modified "Hill relation." The function of the contractile joint angle [f(PHC)] is required to integrate the force/muscle length relation into the model. The results obtained by Williams and Stutzman (1959) and our own measurements have led to Equation 4 for the force/muscle length relation:

$$f(PHC) = .3 + 0.00636 \cdot (PHC-30°) \qquad 30° < PHC < 140° \quad (4)$$

DPHC is the differential of PHC, which is the sum of the effective joint angle PH and the series elastic component PHSEC. The latter as expressed by an angle, is determined by the force FA, which is developed by the contractile element. Equation 5 has been developed to fit the characteristic curve found by Ritchie and Wilkie (1958). The values of PHSEC max. were, however, determined by the muscle parameters of the biceps and brachialis found by own measurements:

$$PHSEC = PHSEC\ max. \cdot (1 - e^{-(\frac{FA+1}{70+0.15 \cdot FA})}) \tag{5}$$

Electrode and Muscle Parameters

The surface EMGs of the three muscles investigated in this study were recorded by means of two electrode sizes. Large electrodes were used for the biceps and the triceps, as no other muscles are close, and their EMGs could significantly interfere with the EMG activity of the measured muscles. However, it is much more difficult to measure the activity of the brachialis, because it is largely covered by the biceps. To minimize the interference of the biceps EMG, small electrodes were used to record the activity of the brachialis. For the present study it was important to know the approximate relation between the EMG recordings by means of the two electrode sizes, because both the biceps and the brachialis contributed to the elbow moment. Consequently it was necessary to multiply the filtered EMG of the brachialis by a scaling factor. Since only the action potentials of the motor units located close to the electrodes can be recorded, the processed peak EMG (PPE) must be multiplied by another scaling factor. The main contributors to the elbow flexion in a neutral position are the biceps, the brachialis and the brachioradialis. An, Hui, Morrey, Linscheid, and Chao (1981) used the term PCSA (physiological cross-sectional area) that is an indicator of the possible tensile strength that can be developed by the individual muscle during maximum contraction. On the basis of these values they found the strength of the brachialis to be approximately 50% greater than that of the biceps.

Results

The results of one experiment show how the muscle model is applied. It is one of 15 experiments performed on a 42-year-old male, flexing the arm with a 50 N weight attached to the wrist. The purpose of this study has been to compare the calculated moment with the moment measured at the elbow. The measured moment was derived by multiplying the rectangular force component of the force transducer by the distance between the center of the trochlea (joint axis) and the attachment of the nylon string to the wrist cuff. Figure 2 shows the moment contribution of the biceps and the brachialis to the calculated moment in curves 1 to 3. Curve 4 illustrates the measured moment, while curve 5 shows the difference between the calculated and the measured moment. Considering all model assumptions and the fact that the EMGs of only two major flexors were measured, it may be stated that the resulting deviations are rather small. Without the inaccuracies encountered in our force measurements, which explain the irregularities in the calculated moment, the difference between the measured and the calculated moment would be even smaller. Similar results have been found in other experiments with the same subject as well as with other subjects. The idea that at higher muscle contractions the influence of the partial overlapping of individual action potentials could be neglected by considering the action potential peaks only appears to be more advantageous than using the integrated EMG as the basis for a muscle model. More experiments with other muscles and with other movements are needed to obtain more conclusive evidence.

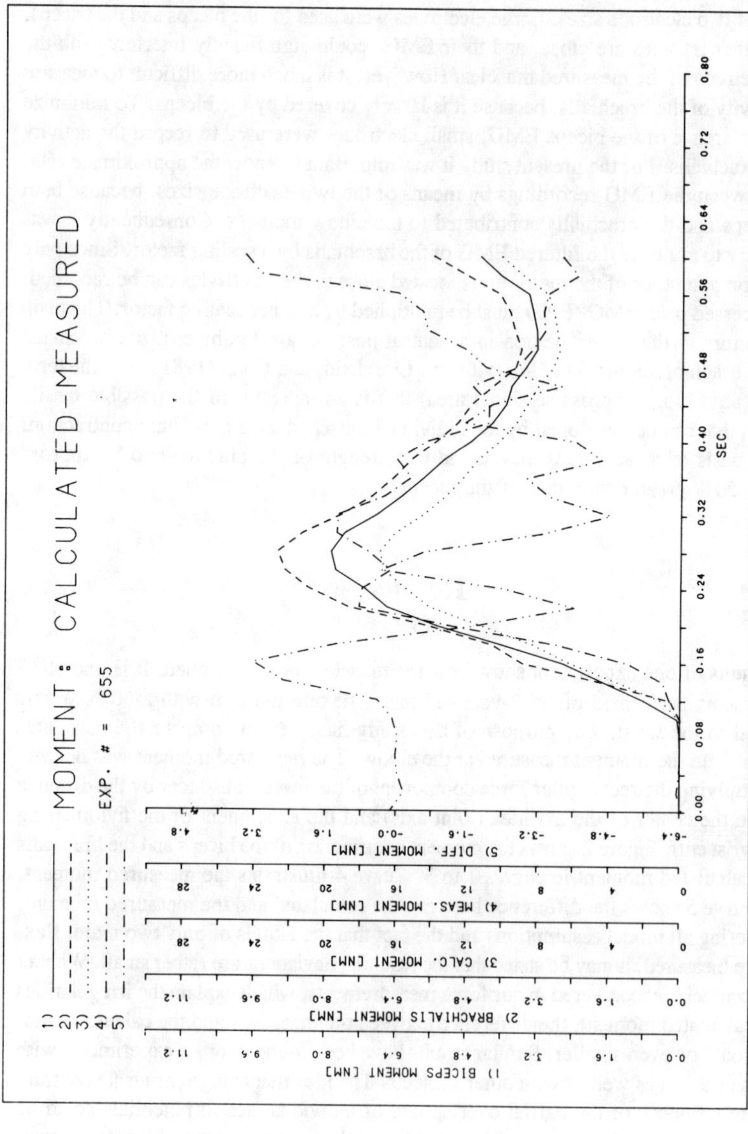

Figure 2—Results of one experiment with the calculated and measured arm flexion moment and the difference between the two moments.

Acknowledgment

Supported in part by a grant of the Natural Sciences and Engineering Research Council of Canada and the Medical Engineering Department of Sulzer Brothers Limited, Winterthur, Switzerland.

References

AN, K.N., Hui, F.C., Morrey, B.F., Linscheid, R.L., & Chao, E.Y. (1981). Muscles across the elbow joint: A biomechanical analysis. *Journal of Biomechanics,* **14**, 659-669.

HOF, A.L., & Van den Berg, Jw. (1981). EMG to force processing I: An electrical analogue of the Hill muscle model. *Journal of Biomechanics,* **14**, 747-758.

RITCHIE, J.M., & Wilkie, D.R. (1958). The dynamics of muscular contraction. *Journal of Physiology,* **143**, 104-113.

WILLIAMS, M., & Stutzman, L. (1959). Strength variation through the range of joint motion. *The Physical Therapy Review,* **39**, 145-152.

WYSS, U.P., & Pollak, V.A. (1981). Kinematic data acquisition system for two- or three-dimensional motion analysis. *Medical and Biological Engineering and Computing,* **19**, 287-290.

Electrical Discharge Patterns of Leg Muscles Reflecting Dynamic Features During Simultaneous Hip and Knee Extension Movements

M. Kumamoto and N. Yamashita
Kyoto University, Kyoto, Japan

H. Maruyama
Seibo Junior College, Kyoto, Japan

N. Kazai
Bukkyo University, Kyoto, Japan

Y. Tokuhara
Teikoku Women's University, Osaka, Japan

F. Hashimoto
Osaka Electro-communications University, Osaka, Japan

Human movements generally consist of complex plural joint movements where joints are connected in series and in parallel. In serial joint movements of pulling with the arm, with simultaneous shoulder extension and elbow flexion, and of extending the leg with simultaneous hip and knee extensions, such as occurs in rowing with both arms, or walking, the mechanism of generation and transmission of forces is not simple (Yamashita, 1975). In previous reports, where the hip and knee joints were extended simultaneously under isometric conditions with maximal effort, the resultant force output exerted at the sole was found not to be the summation of the forces developed by the hip extension and the knee extension along the functional force direction of the movement. Instead it was found to be limited by the weaker one of these forces (limiting joint). The limiting joint shifted from one joint to the other despite no postural change when the functional force direction was altered. The electrical activities of the bifunctional leg antagonists, the Rectus femoris (Rf) and the Biceps femoris (Bf), showed reversed patterns depending on whether the resultant force was limited by either the knee joint or the hip joint force.

The present experiments were designed to elucidate further detailed relationship between the dynamic features and quantitative changes in electrical activities of the leg muscles in the biarticular movement, hip, and knee extension.

Methods

Subjects cmployed in the experiments were 4 healthy male adults. The subjects were requested to lie down in a supine position with the hip and knee joints kept at 90° and to perform simultaneous hip and knee extensions under isometric conditions with maximal effort along various force directions from the extreme condition of the Line H_0 to the other extreme one of the Line K_0 as shown in Figure 1A. The force directions were kept to always pass through the ankle joint in order to discard the influence of the ankle joint force. Resultant force output exerted at the sole was recorded with the strain gauge set along the functional force line. Electrical activities of leg muscles during the simultaneous hip and knee extensions were recorded with conventional methods utilizing surface electrodes 10 mm in diameter. Muscles tested were the Gluteus max-

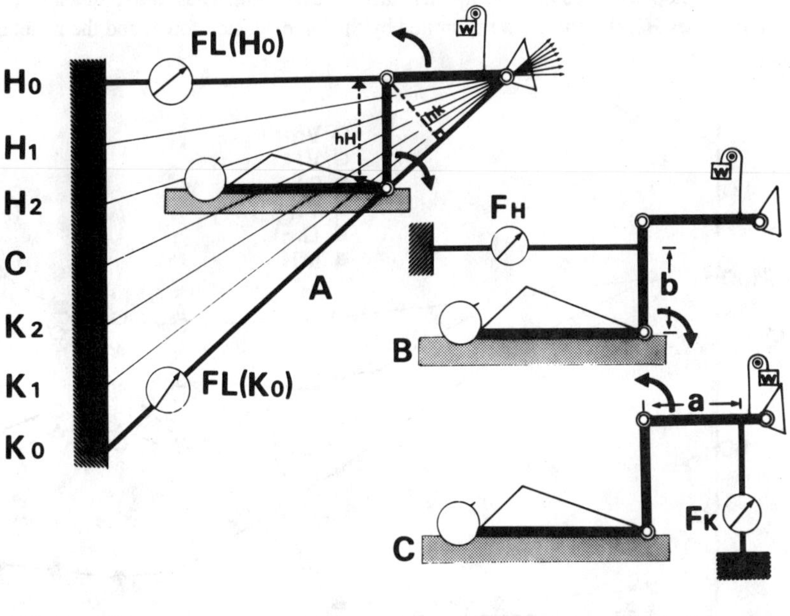

Figure 1—Experimental conditions. Subject was requested to lie down in the supine position with the hip and knee joints kept at 90°. The lower leg weight was compensated by a small amount of weight (W) eliminating the electrical activities from the knee extension muscles. A: Simultaneous hip and knee extensions were performed along Lines H_0 to K_0 utilizing fine steel chain in order to maintain isometric conditions. A load cell (strain gauge) was set in between the chain. B: Maximal moment around the hip joint was individually measured as Fh•b. C: Maximal moment around the knee joint was individually measured as Fk•a.

imus (Gm) and the Vasti medialis (Vm) and lateralis (Vl) of the monoarticular muscles, and the Rectus femoris (Rf), the Tensor fasciae latae (Tfl), and the Medial Hamstrings (MH) of the biarticular muscles. The resultant forces were simultaneously recorded with the electrical activities.

Results

FL(H_0) or FL(K_0) in Figure 1A, the resultant force in the extreme condition where the force direction passed through the knee joint or the hip joint respectively, was very close to the force developed by the hip extension along Line H_0 or to the force developed by the knee extension along Line K_0. The value of the force developed by the hip extension along Line H_0 or by the knee extension along Line K_0 was obtained when the individual maximal moment around the hip (FH•b in Figure 1B) or the knee joint (FK•a in Figure 1C) was divided by hH or hK in Figure 1A, respectively. As to the functional force directions between the extremes (Line H_0 and K_0), the resultant force was close to the smallest value obtained when the individual maximal moment around the hip or the knee joint was divided by the distance between the hip or the knee joint and the functional force line, the moment arm for each joint. Thus, the resultant forces, along Lines H_0, H_1, and H_2 were limited by the hip extension force, and the resultant

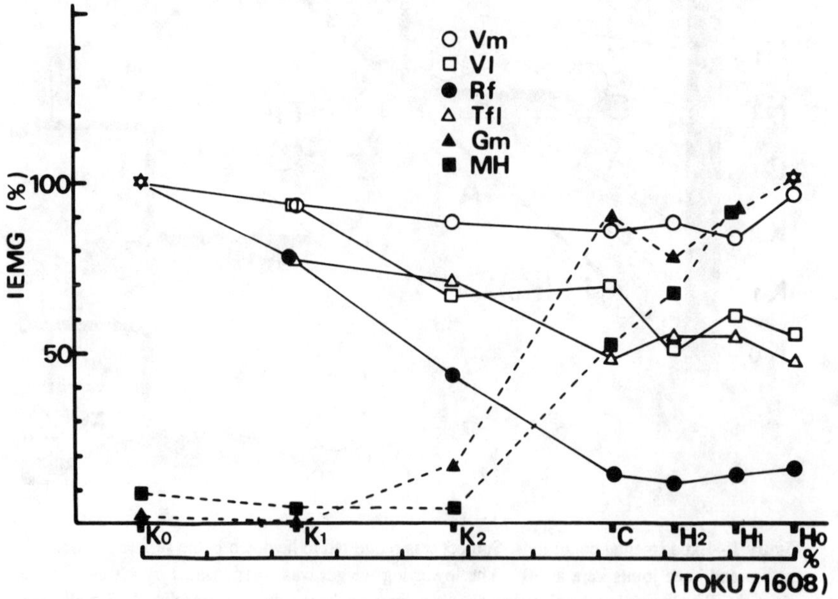

Figure 2—Changes in integrated values of electrical activities with changes in functional force direction. Ordinate: Integrated values of electrical activities (IEMG) (shown in percentage). Abscissa: Proportional positions of the points where the functional force directions crossed the thigh between the hip and knee joints. Points K_1 and K_2 were determined by dividing the distance between Point K_0 (hip joint) and Point C (critical Line C) by 3, and Points H_1 and H_2, by dividing the distance between Point H_0 (knee joint) and Point C by 3.

forces along Lines K_0, K_1, and K_2, by the knee extension force. Direction C was critical where the forces developed by both joints along Line C were equal.

Integrated values of the electrical discharges (IEMG) from the leg muscles tested were plotted in percentages as shown in Figure 2. The IEMGs of the Gm and the MH during the simultaneous hip and knee extensions in the extreme case of Line H_0 were 100%, and the IEMGs of the Vm, the Vl, the Rf, and the Tfl during the simultaneous hip and knee extensions in the other extreme case of Line K_0 were 100%. The IEMG of the Rf of the biarticular muscle decreased when the functional force direction changed from Line K_0 to Line H_0, and the one of the biarticular MH decreased from Line H_0 to Line K_0 (Figure 2). The activities of both muscles reversed around Line C reflecting which joint force was limiting the resultant force output in all subjects. The Tfl showed the same tendency as the Rf in one of the four subjects, but showed not less than 50% value at Line H_0 in the other three cases. The Vm and the Vl of the monoarticular muscles were less inhibited than the Rf, and the Gm showed the same tendency as the Vm and the Vl between Lines H_0 and C but beyond Line C the Gm steeply decreased as the MH (Figure 2).

Discussion

Quantitative analyses of changes in the electrical activities of the leg muscles revealed that the biarticular Rf and MH clearly reversed their activities around dynamically critical Line C. In the same experimental conditions, changes in amplitude of evoked potential of the Rf by the tendon tap were examined, and the results obtained suggested the existence of an antagonistic inhibition between the antagonistic biarticular muscles (Yamashita, Kumamoto, Tokuhara, & Hashimoto, 1983). Further, even in the monoarticular muscles of the Vm, Vl, and Gm, although their activities were less inhibited than in the case of the biarticular muscles, the knee extensors, Vm and Vl, and the hip extensor, Gm, reversed their activities around critical Line C. Thus, even though there was no postural change, the electrical discharge patterns of the leg biarticular and monoarticular muscles could alter their activities, reflecting the dynamic condition of which joint was limiting the resultant force output in the simultaneous hip and knee extensions.

Eccles and Lundberg (1958) reported the pattern of Ia-excitatory and Ia-inhibitory actions to motor nuclei of hip and knee muscles in the cat. However, the results obtained in the present experiments indicate that, if a dynamic condition could be given in their cat experiments, such a pattern may be altered, even though there is no postural change, reflecting which joint is limiting a resultant force output of the simultaneous hip and knee extension movements.

Dynamic calculation in the present experiments is only possible under static conditions with maximal effort. However, the electrical discharge patterns of the leg antagonistic biarticular muscles during a gait cycle (Kumamoto et al., 1981) and jumping (Kameyama, Oka, Hashimoto, & Kumamoto, 1981) suggested that the dynamic features revealed in the experiments could be applicable even under kinetic and submaximal conditions (Yamashita & Kumamoto, 1976).

Thus, as far as a resultant force output of complex joints movement is concerned, the dynamic and neuromuscular functional features of serial joint movement play an important role in improving performance of human complex plural joint movements. Further, results obtained in the present experiments suggest that neuromuscular features of such serial joint movements are not explained from the viewpoint of a single joint movement alone.

References

ECCLES, R.M., & Lundberg, A. (1958). Integrative pattern of Ia synaptic actions on motoneurones of hip and knee muscles. *Journal of Physiology,* **144**, 271-298.

KAMEYAMA, O., Oka, H., Hashimoto, F., & Kumamoto, M. (1981). Electromyographic study of the ankle joint muscles in normal and pathological gaits. In A. Morecki, K. Fidelus, K. Kedzior, & A. Wit (Eds.), *Biomechanics VII-B* (pp. 50-54). Baltimore: University Park Press.

KUMAMOTO, M., Oka, H., Kameyama, O., Okamoto, T., Yoshizawa, M., & Horn, L. (1981). Possible existence of antagonistic inhibition in double-joint leg muscles during normal gait cycle. In A. Morecki, K. Fidelus, K. Kedzior, & A. Wit (Eds.), *Biomechanics VII-B* (pp. 157-162). Baltimore: University Park Press.

YAMASHITA, N. (1975). The mechanism of generation and transmission of forces in leg extension. *Journal of Human Ergology,* **4**, 43-52.

YAMASHITA, N., & Kumamoto, M. (1976). Force generation in leg extension. In P. Komi (Ed.), *Biomechanics V-B* (pp. 41-45). Baltimore: University Park Press.

YAMASHITA, N., Kumamoto, M., Tokuhara, Y., & Hashimoto, F. (1983). Electrical inhibition on bifunctional muscle during double joint movement. In H. Matsui & K. Kobayashi (Eds.), *Biomechanics VIII-A* (pp. 440-443). Champaign: Human Kinetics Publishers.

The Mechanism of the Silent Period During the Contraction of the Masseter of Myo-Dystrophy Patients Using the Stimulus of Tap-Percussion

A. Nagata, M. Muro, and H. Sakuma
Tokyo Metropolitan University, Meguro-Ku, Japan

Ever since Hoffman showed the phenomenon of a silent period (SP) in the EMG during the human phasic reflex, numerous studies both in humans and animals have been carried out. In these studies, the mechanisms of the EMG of the SP have been linked either with the central nervous system of motor control or with the peripheral motor control system below the spinal cord level. The former includes the brain stem reticular formation inhibition (Ikai, 1955) and the central nervous system control programming theory (Iwase et al., 1981). The latter includes the autogenic inhibition (Granit, Kellerth, & Szumsky, 1966), the Renshaw cell's recurrent inhibition (Anastasijevic & Voco, 1980), the antagonist's reciprocal inhibition (Merton, 1951), and a temporal cessation of the muscle spindle afferent discharge (Denny-Brown, 1928).

At present, it seems, therefore, that the physiological mechanism of SP occurrence in the masticatory muscle EMG have not been clearly established. Furthermore, some contradictory results have been reported with respect to the duration of SP appearance among the patients with abnormal occlusion and TMJ syndrome. Several investigators have reported that the latency and the duration of SP appearance increases with these pathological changes, while others have found it to be decreased.

It was the purpose of this study to establish the relationship between the initial biting force levels and duration of EMG SP during isometric voluntary contraction at various force levels by tap percussion to the mentum. Furthermore, we have analyzed the SP occurrence mechanism and EMG frequency power spectrum in order to determine the qualitative differences, if any, in the motor unit discharge frequency patterns before and after the onset of SP as induced by the tap. The computer analysis of the EMG waveforms by means of fast Fourier transform and its mathematical basis have been fully described elsewhere.

Methods

Six healthy adult male subjects and eight patients volunteered for this study. Patients consisted of three Duchenne-type persons, two Facioscapulohumeral persons and three

Myotonic Dystrophy persons. The surface EMG with a bipolar lead system was recorded from the belly of the masseter 2.5 cm apart. The maximal biting force (MBF) was determined as the highest value obtained during trials.

EMG power spectrum was analyzed for a period of 40 ms before and after the onset of the SP. The computer analysis of the EMG waveform was carried out according to the method described by Hamba, in which a Fourier transform based upon Uhlich's algorithm was used.

In order to clearly establish the amplitude of the SP quantitatively, we have defined the onset of the SP as a myoelectric signal amplitude below 20 μV. For this reason, we have termed the newly defined EMG SP the "Quiet Period" (QP).

Results and Discussion

Figure 1 represents a typical set of masseter EMG data of a healthy person showing the time course of changes in the duration of QP and the corresponding biting force during mentum tap (3.2 G). At the right side of this figure relationships among three physiological parameters (percentage of MBF, QP duration, and force reduction following QP) are presented, for heavy (3.2 G) and light (1.8 G) tap percussion. Figure 2 shows two typical sets of the patients' EMG data for Myotonic Dystrophy (MD) and Duchenne (D) types.

As is seen from these figures, the QP appeared shortly after the onset of tap stimulus with a latency period of approximately 12 ms (healthy) or 22 ms (patient). It was apparent that the QP duration became longer with lower initial percentage of MBF and that the relative decline of biting force was correspondingly greater with longer

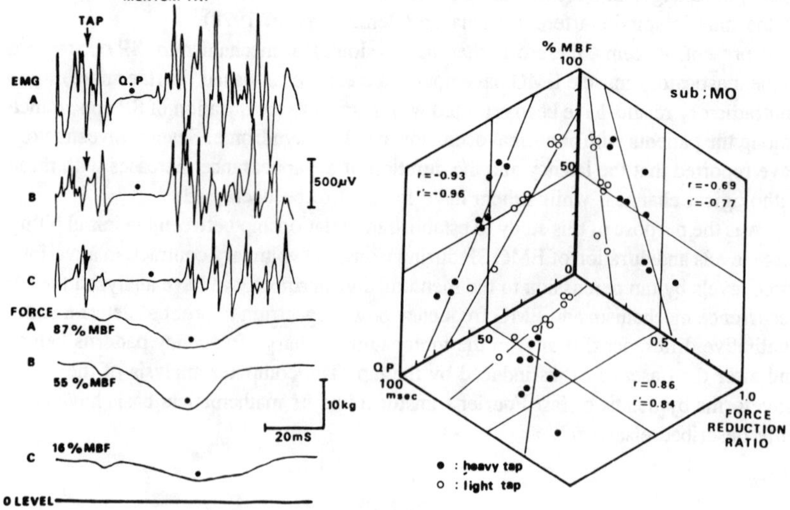

Figure 1—Myoelectric signals during isometric contraction (% MBF) of the healthy subject's masseter. Right side: relationship among % MBF, QP duration, and force reduction ratio.

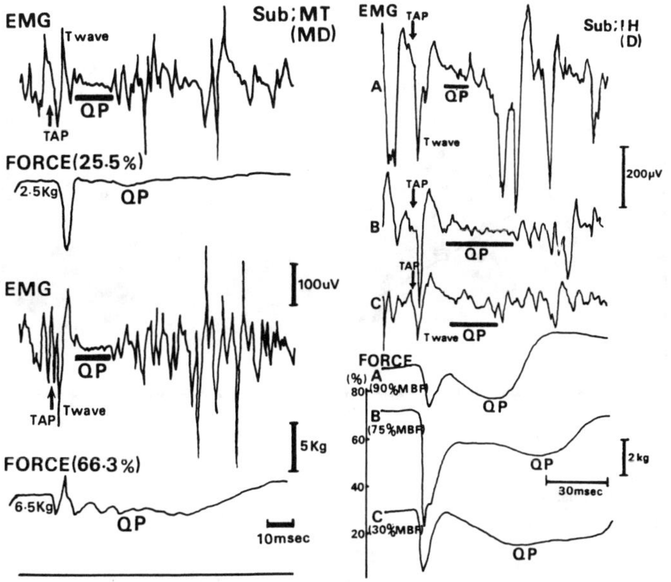

Figure 2—Myoelectric signals during isometric contraction of the patient's masseter. Left side: Myotonic Dystrophy. Right side: Duchenne types.

QP and heavy class of syndrome. However, during 100% MBF, the QP could hardly be observed.

The correlation between QP duration and percentage of MBF was found to be $r = -.703$ ($p < .01$) of healthy subjects and $r = -.215$ ($p < .05$) of patients (see Figure 3). Note that in this particular figure, QP duration for each subject has been standardized by using each subject QP – %MBF regression Y intercept as 100% QP.

Figure 4 shows a typical set of masseter EMG power spectral data obtained during 40 ms periods before and after the QP by the mentum tap during 30, 45, and 60% MBF. When these power spectra were compared, there were no systematic differences in the power spectral density and spectral energy of different frequency components. These results indicate that the predetermined biting force and the degree of muscle contraction were well maintained after the onset of QP without more recruitment of motor units. These spectra suggest that the recurrence of motor unit activity did exist following the onset of QP without qualitative and quantitative changes in the MU firing pattern. This finding strongly argues against the notion that the occurrence of the QP is related to the brain-stem reticular formation inhibition or the central nervous system motor programming.

Interestingly, Szentagothai (1948) has reported that the primary Ia afferent fibers from the masseter spindle have their cells of origin in the mesencephalic V nucleus and form monosynaptic excitatory synapses on the homonymous masseter motoneuron. For this reflex arc, the latency of the T-reflex of the masseter was found to be 7.5 ms. Our results and the aforementioned studies seem to differ slightly by 2 ms. However, this slight discrepancy might have been the result of different experimental procedures employed to induce a T-reflex.

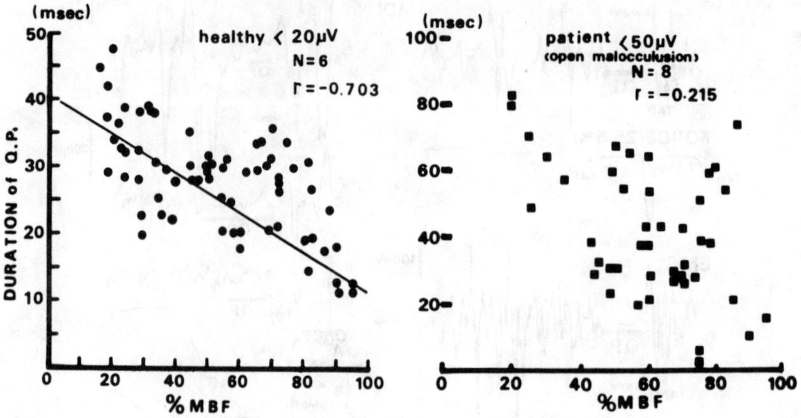

Figure 3—Relationship between QP duration and the max. biting force (% MBF) of healthy subjects and patients.

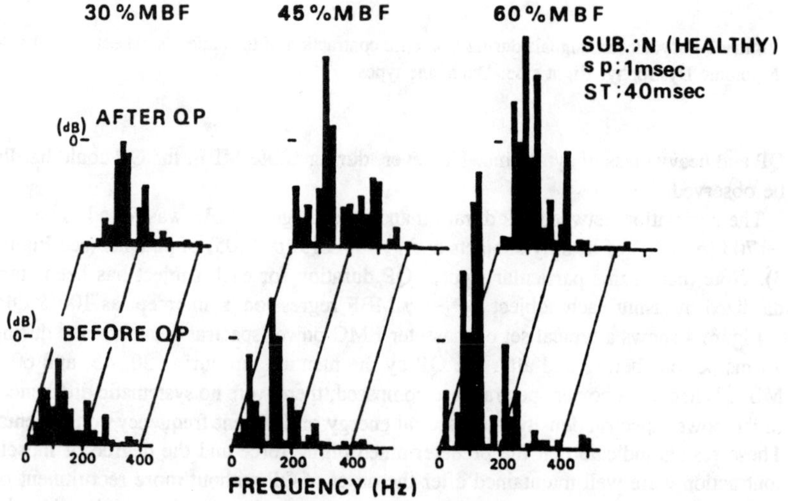

Figure 4—EMG frequency power spectra obtained from the masseter during a 40-ms period before and immediately after the onset of QP for a healthy subject. Vertical bars represent spectral energy at each frequency band analyzed.

As we have shown in Table 1, the patient EMG QP was not more evident during the mentum tap than that of the healthy subjects. And patients with the open malocclusion have shown the threshold of 50 μV of EMG QP occurrence. On the other hand, healthy persons and patients with normal malocclusion have shown the threshold of 20 μV. Therefore, the QP occurrence is not necessarily related to the periodontal mechanoreceptor input and neuromuscular control. Furthermore, the QP duration was actually diminished with increasing percentage MBF levels and with abnormal maloc-

Table 1

Comparative Results of EMG QP, T-Wave, and Biting Force Between Healthy and Patient Subjects

Group	N	Latency (ms)	Duration (ms)	T. Amplitude (μV)	Biting Force Decrease (%)
QP 50 μV patients (open malocclusion)	4	25 (8.2)	39.8 (19)	472.2 (357)	70.4 (45)
QP 20 μV patients (normal malocclusion)	4	18 (3.2)	31.8 (8)	666.8 (313)	40.2 (14)
Healthy	6	12 (1.2)	23.4 (6)	302.5 (129)	28.0 (8)

( ): Standard Deviation

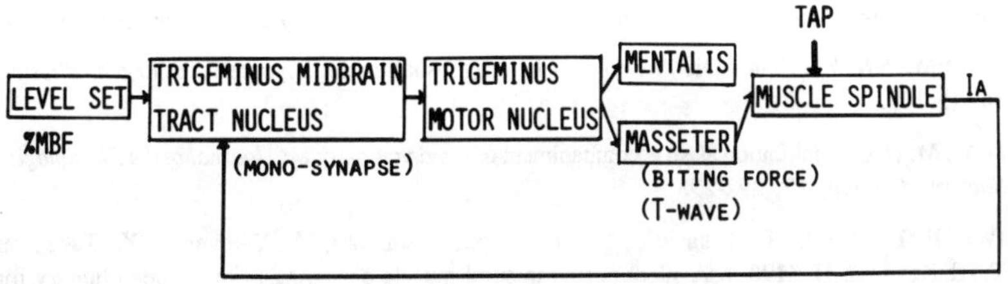

Figure 5—A proposed block diagram of the QP occurrence mechanism.

clusion. This evidence seems to support the notion that the autogenic Golgi inhibition is also not a primary factor for regulating the onset of the QP. Our evidence seems to suggest that the human cranial motor axons seldom possess such recurrent axon branches and the Renshaw cells' recurrent inhibition therefore seems remote in intact human masticatory muscle.

As shown in Figure 5, EMG activities disappeared shortly after a brief activation (T-wave) with monosynaptic reflex arc of Ia discharge. Since an initial EMG amplitude of about 500 μV was suddenly reduced below 20 μV or 50 μV, a strong inhibitory input must have been brought about, thus suggesting the existence of "presynaptic" inhibition (Eccles, 1964). The appearance of EMG immediately after the QP may be interpreted as a result of an abolished presynaptic inhibitory effect and refacilitation of motoneuron triggered by the reactivated muscle spindles.

Conclusion

A QP was induced by means of the mentum tap, and EMG activities disappeared shortly after the T-reflex. Significant negative correlations were found between the initial percentage MBF and duration of the QP. Analysis of EMG frequency power spectra before and immediately after the onset of QP revealed that there were no apparent changes in the power spectral parameters. Patients' QP with open malocclusion did not appear on the threshold of 20 μV EMG level. These results seem to support the notion that the mechanism of QP occurrence during a mechanical tap has close links with defacilitation of the alpha motoneurons caused by the presynaptic inhibition originating in the muscle spindle Ia afferent fibers.

References

ANASTASIJEVIC, R., & Voco, J. (1980). Renshaw cell discharge at the beginning of muscular contraction and its relation to the silent period. *Experimental Neurology,* **69**, 589-598.

DENNY-BROWN, D. (1928). On inhibition as a reflex accompaniment of the tendon jerk and of other forms of active muscular response. *Proceedings of the Royal Society,* **103**, 321-326.

ECCLES, J.C. (1964). Presynaptic inhibition in the spinal cord. *Progress in Brain Research,* **12**, 65-91.

GRANIT, R., Kellerth, J.O., & Szumsky, A.J. (1966). Intercellular autogenic effects of muscular contraction on extensor motoneurons: The silent period. *Journal of Physiology,* **182**, 484-503.

HOFFMANN, P. (1934). Die Physiologischen Eigenshaften der Eigenreflexe. *Ergen. Physiol.,* **36**, 15-108.

IKAI, M. (1955). Inhibition as an accompaniment of rapid voluntary act. *Journal of the Physiological Society of Japan,* **17**, 292-298.

IWASE, T., Uchida, T., Takanashi, Y., Suzuki, N., Hashimoto, M., Yamamoto, Y., Takegami, T., & Koyama, H. (1981). A silent period in sural muscle occurring prior to the voluntary forward inclination of the body. *Neuroscience Letters,* **, 21**, 183-188.

MERTON, P.A. (1951). The silent period in a muscle of the human hand. *Journal of Physiology,* **114**, 183-198.

SZENTAGOTHAI, J. (1948). Anatomical considerations of monosynaptic reflex arcs. *Journal of Neurophysiology,* **11**, 445-454.

Analysis of Electromyographic Silent Period Prior to a Rapid Voluntary Movement in Humans

K. Mita, H. Aoki, R. Tsukahara, and K. Yabe
Institute for Developmental Research, Aichi Prefectural Colony,
Kasugai, Aichi, Japan

The electromyographic signal (EMG) which is an indication of the electrical activities in contracting muscle is of value in the study of the motor control mechanism. Figure 1A shows an EMG pattern of a rapid voluntary movement following a slight sustained contraction. An electromyographic silence is observed just before the initiation of the phasic activity. This phenomenon is referred to as the "premotion silent period" (p.s.p.). The p.s.p. was first noted by Ikai (1955) on the EMG record during an elbow extension. Afterwards, Gatev (1972) and Yabe and Murachi (1975) verified this phenomenon in the muscle activities during extension and flexion of elbow joint and knee extension. It has been suggested that the p.s.p. is due to an inhibition from the upper center of the central nervous system and plays a major role in the motor control. However, the physiological mechanism of the p.s.p. appearance has not been defined clearly.

A negative slow potential which was named the "movement-related cortical potential" (MRP) is observed on the electroencephalographic record from the scalp accompanying a rapid voluntary movement (see Figure 1B). This cortical activity is specifically related to the physiological process associated with the initiation and the motor control

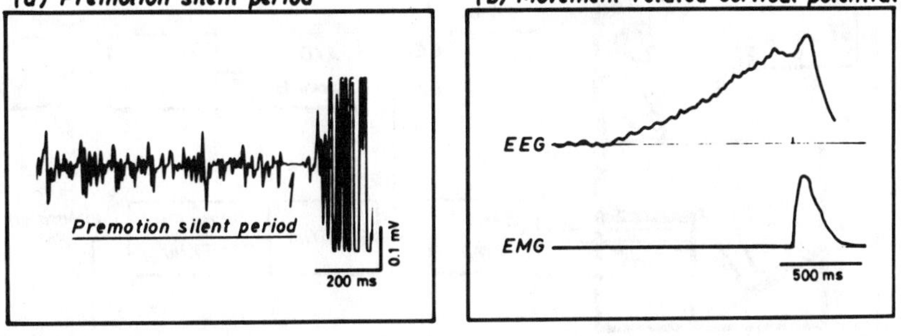

Figure 1—Schematic representation of premotion silent period and movement-related cortical potential.

of the voluntary movement (Bates, 1951; Kornhuber, & Deecke, 1965; Vaughan, Costa, Gilden, & Schimmel, 1965).

The present study was undertaken to investigate the relationship between appearance of the p.s.p. and activities in the upper center, and to estimate the physiological mechanism of the p.s.p.

Experiments

The experiment was performed on five healthy adults aged 22 to 42 years. A rapid plantar flexion of the right foot following a slight and stable contraction was chosen as a voluntary movement. The experimental trials were repeated about 300 trials on each subject. Each subject was asked to contract the right triceps surae muscle slightly and to maintain the stable contraction corresponding to a level of 10% of the maximum strength for 3 to 5 s. Afterwards, he was instructed to perform a voluntary plantar flexion as quickly as possible. The EMG activities were led off from the triceps surae and anterior tibial muscles of the right foot using a time constant of 0.03 s. The surface electrodes (10 mm diameter) were placed on the long axis of the muscles about 30 mm apart. The cortical activities were recorded from the scalp by use of a time constant of 2.0 s. Chlorided silver disc electrodes (7 mm diameter) were arranged on three points as follows: a) the vertex (C_z); b) the midpoint between the left central (C_3) and the vertex (C_z) (termed $C_{3'}$) near the motor area corresponding to the triceps surae muscle of the right foot; and c) contralateral point equivalent to the $C_{3'}$ point (termed C4′) (see Figure 2).

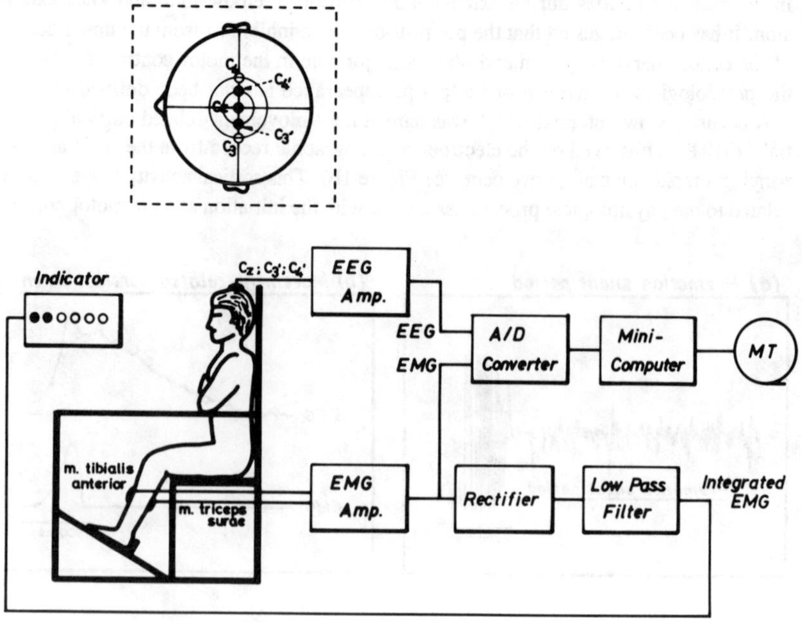

Figure 2—Experimental arrangement.

The raw EMG and EEG records were sampled at a rate of 1000 samples/s for a period of 3.0 s from 2.5 s before the initiation of the phasic EMG and digitized with 12 bit accuracy using A/D converter. Digitized EMG and EEG data were stored on the digital magnetic tape.

Data Processing

Let us consider a mathematical model for describing the EEG response and its statistical parameters prior to the rapid voluntary movement. The EEG response Xo(t) may be modeled as a linear sum of the cortical potential related to the movement Xm(t) and the background EEG which is thought to be the spontaneous activity Xs(t).

$$Xo(t) = Xm(t) + Xs(t) \tag{1}$$

Since the mean value of the spontaneous activity Xs(t) is considered to be zero, the mean EEG response $\eta_{Xo}(t)$ is given by

$$\eta_{Xo}(t) = E[Xo(t)] = \eta_{Xm}(t) \tag{2}$$

That is, the mean of the EEG response Xo(t) is equal to that of the cortical activity evoked by the movement Xm(t). As to the variance of the EEG response $\sigma^2_{Xo}(t)$, if it is assumed that Xs(t) is independent of Xm(t), the variance of Xo(t) can be represented by a linear sum of those of the evoked potential Xm(t) and the spontaneous activity Xs(t).

$$\sigma^2_{Xo}(t) = \sigma^2_{Xm}(t) + \sigma^2_{Xs}(t) \tag{3}$$

Therefore, the above description indicates that the cortical potential owing to the voluntary movement can be statistically defined by the mean value of the EEG response $\eta_{Xo}(t)$ and that the spontaneous EEG activity can be detected by the variance of Xo(t).

In order to clarify the relationship between the p.s.p. and the MRP, it is also necessary to examine the mean and variance of the EMG response. Then, the processings were performed on both the EMG and EEG, simultaneously.

Results

Figure 3 demonstrates variability of the EMG and EEG activities obtained by superimposing 10 records for a period of 3 s associated with the rapid voluntary movement. The top traces show the rectified EMGs derived from the triceps surae muscle and the other traces reveal the EEG responses from the areas C_z, $C_{3'}$ and $C_{4'}$, respectively. As to the EEG responses, complex temporal patterns consisting of the spontaneous activities and the slow negative potentials were observed. Then, the statistical processings were applied to a set of these experimental data for a period of 2 s before the onset of the phasic EMG activity.

Figure 4A shows the mean (ensemble average) of the rectified EMG and EEG. The solid traces indicate the processed results computed from the original data with and

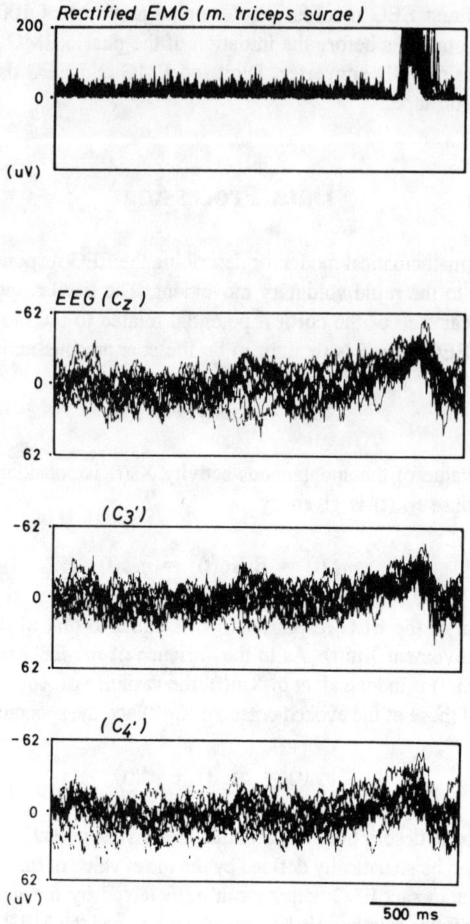

Figure 3—Variability of EMG and EEG activities associated with a rapid voluntary movement.

without the p.s.p., respectively. The mean EMG activity remained stable up to 100 ms before the initiation of the phasic discharge. Afterward, the mean activity began to decrease regardless of appearance of the p.s.p. However, the depression for the case of the EMG with the p.s.p. was more pronounced than that without the p.s.p. As to the mean of the EEG response, negative slow variations began from 1.2 s before the onset of the EMG burst over all three areas and were considered to be consistent with the MRP previously reported. There was a significant difference in the amplitude of the MRP between the cortical activities with and without the p.s.p. The EEG activity with the p.s.p. provided the MRP with lower level in comparison with that without the p.s.p.

Figure 4B shows the standard deviation (SD) of the EMG and EEG. The SD of the EMG activity decreased from 100 ms prior to the phasic discharge and had similar features to the mean value. On the other hand, the SD of the EEG response maintained a stable level over all periods processed. No significant difference in value of the SD was led between the EEG activities with and without the p.s.p.

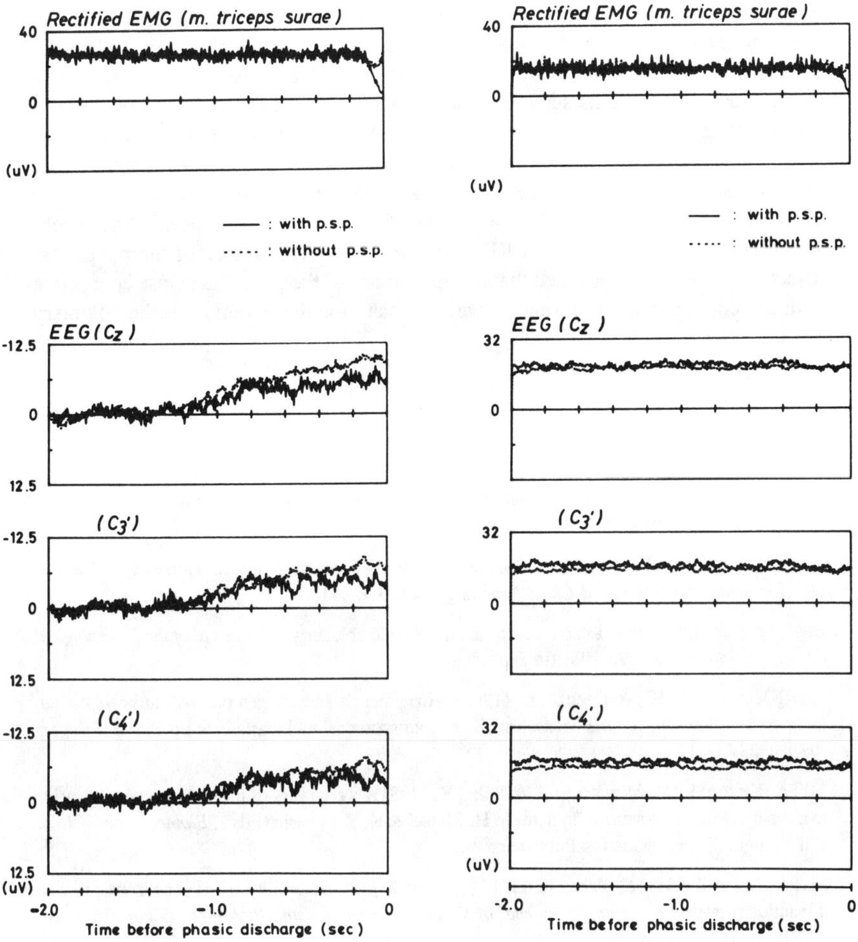

Figure 4—Mean (ensemble average) (a) and standard deviation (b) of rectified EMG and EEG preceding a rapid voluntary movement.

Discussion

The p.s.p. is characterized by the property of occurring just before the onset of a rapid voluntary movement. It has been suggested that the p.s.p. is owing to an inhibitory effect in the central nervous system and is closely related to development and skillfulness of the voluntary movement. However, the mechanism of the p.s.p. appearance has been unsolved until now. In order to analyze the p.s.p., the authors statistically examined the EMG activity in a quick voluntary movement as a nonstationary stochastic process (Mita, Aoki, Mimatsu, & Yabe, 1981). The EMG activity decreased from 100 ms before the initiation of the movement. The rate of depression in the EMG activity with the p.s.p. was more remarkable than that without the p.s.p. Since the electrical activity of the muscle was controlled by the motoneurons, variability in the excitability of the motoneurons prior to a reaction movement was investigated by means of the Hoffmann reflex (H reflex) (Mita, Aoki, & Yabe, 1982). The amplitude of the H reflex began to change with a slight decrease and a progressive increase followed the depression period. These results suggested that the inhibitory influence in the upper center had effect on the muscle activity through the motoneurons in the spinal cord.

It became necessary to confirm the relationship between the appearance of the p.s.p. and the activity in the upper center. Therefore, the present investigation concerning the EEG activity associated with appearance of the p.s.p. was conducted. As a result, a significant difference in the MRP in relation to the appearance of the p.s.p. was detected. It might be concluded that the appearance of the p.s.p. had close connection with the cortical activity during preparatory state for the initiation of the voluntary movement.

References

BATES. J.A.V. (1951). Electrical activity of the cortex accompanying movement. *Journal of Physiology,* **113**, 240-257.

GATEV, V. (1972). Role of inhibition in the development of motor coordination in early childhood. *Developmental Medicine and Child Neurology,* **14**, 336-341.

IKAI, M. (1955). Inhibition as an accompaniment of rapid voluntary act. *Journal of the Physiological Society of Japan,* **17**, 292-298. (in Japanese)

KORNHUBER, H.H., & Deecke, L. (1965). Hirnpotentialänderungen bei Willkurbewegungen und passiven Bewegungen des Menschen: Bereischaftspotential und reafferente Potentiale. *Pflügers Archiv,* **284**, 1-17.

MITA, K., Aoki, H., Mimatsu, K., & Yabe, K. (1981). Variability of the electromyogram prior to a rapid voluntary movement in man. In H. Matsui & K. Kobayashi (Eds.), *Biomechanics VIII-A*. Champaign: Human Kinetics Publishers.

MITA, K., Aoki, H., & Yabe, K. (1982). Variability of excitability of motoneurons prior to a reaction movement. *Japanese Journal of Medical Electronics and Biological Engineering,* **20**, 24-31. (in Japanese)

VAUGHAN, H.G., Jr., Costa, L.D., Gilden, L., & Schimmel, H. (1965). Identification of sensory and motor components of cerebral activity in simple reaction-time tasks. *Proceedings of the 73rd Conference of the American Psychological Association,* **1**, 179-180.

YABE, K., & Murachi, S. (1975). Role of the silent period preceding the rapid voluntary movement. *Journal of the Physiological Society of Japan,* **37**, 91-98. (in Japanese)

Study of the Response to Nociceptive Stimulus Applied During Walking in Humans

A.E. Patla & M. Belanger
University of Waterloo,
Waterloo, Ontario, Canada

During locomotion the central and peripheral influences on the motoneurons must interact appropriately to produce a well adapted movement. Rapid adjustments to unexpected perturbations are provided by fast compensatory mechanisms such as the stretch and the flexor reflexes. To provide appropriate corrective response to any unexpected perturbation, the stereotypical reflex responses are modulated during locomotion. For example, Akazawa, Aldridge, Steeves, and Stein (1982) have shown that the stretch reflex response is modulated during the stepcycle. The study of the stretch reflex response during movement involves isolation of the muscle and hence is difficult to study. Because of the relative simplicity in the experimental protocol, the modulation of the flexor reflex response during movement such as locomotion has been studied extensively and is discussed next.

In mammals such as cats and in lower vertebrates like the dogfish, it has been found that the response to the nociceptive stimulus is dependent upon the phase of the stepcycle during which the stimulus is applied (Duysens & Stein, 1978; Forssberg, 1979; Forssberg, Grillner, & Rossignol, 1975; Grillner, Rossignol, & Wallen, 1977; Wand, Prochazka, & Sontag, 1980). When the stimulus is applied during the swing phase, a short latency activation of the ipsilateral flexor muscles and contralateral extensor muscles is elicited. This produces additional flexion of the limb lifting the paw over the obstacle. In contrast, when the stimulus is applied during the stance phase, it evokes an inhibition followed by an excitation of the extensor muscles and an increased flexor activity in the succeeding swing phase. This has the effect of the paw being thrust backward and then lifted above the obstacle by the brisk swing produced by the subsequent enhanced flexor burst. Because the latency of this enhanced flexor burst varies it cannot be evoked by fixed reflex pathways, but is probably caused by an increase in activity of the central pattern generator responsible for locomotion (Grillner, 1975). These corrective responses to unexpected perturbations are adapted to the ongoing locomotor activity permitting the animal to maintain stability and forward progression.

In humans, the flexor reflex has been studied by a number of researchers (Dmitrijevic, 1973; Faganel, 1973; Hagbarth, 1960; Kugelberg, Eklund, & Grimby, 1960; and Shahani & Young, 1971). The polysynaptic flexor reflex, which serves to withdraw the limb from the noxious stimuli, involves excitation of the synergists and inhibition of the antagonists. Research has shown that the flexor reflex may be influenced by a number of factors: (a) the subject's expectancy and apprehension, (b) habituation, (c) voluntary contraction, and (d) stimulation site, stimulus modality, and intensity.

Although the flexor reflex has been studied during static conditions in humans, how, if at all, this normal defense reaction gets modified during movement has not been investigated in detail. Lisin, Frankstein, and Rechtmann (1973), based on one subject's results, found that flexor reflex became weaker as the subject walked. Susuki, Watanabe, Miyazashi, and Homma (1981), using a rope tied around the torso to provide disturbance, found phase dependent compensatory reactions. The nature of the perturbation which induces alteration in the torso position makes their results difficult to interpret. The focus of this paper is to study the response to nociceptive stimulus applied during the stance and swing phase of the stepcycle in humans.

Experimental Protocol

To study the response to nociceptive stimulus, myoelectric signals were recorded using bipolar silver/silver chloride surface electrodes from the following muscles: tibialis anterior, gastrocnemius (medial head), biceps femoris, and vastus lateralis. Full-wave-rectified and low-pass-filtered (6 Hz) myoelectric signals, along with the stimulus onset signal (a 5V, 100 ms pulse), were digitized on a Hewlett-Packard 9845B computer at 500 Hz per channel and stored on a floppy disk for analysis.

The subjects ($n = 4$) were walking at their natural self-paced speed on a Collins compact treadmill. Electrical stimulus was used in the study, because it is repeatable and quantifiable. The stimulus (2 ms rectangular pulse) was applied unexpectedly via a copper ring electrode on the second toe and a copper ground electrode on the dorsum of the foot placed a few centimeters apart. This site of stimulation is most appropriate for eliciting a flexor reflex (Kugelberg, et al., 1960). Intensity of the stimulus was adjusted to approximately three times the threshold value. The stimulus parameters were similar to the ones used in the animal experiments mentioned above. A heel-strike switch was used to trigger the Grass Stimulator. By adjusting the delay on the stimulator appropriately, the stimulus was applied in the swing and the stance phase of the step cycle. The experimental protocol was as follows: threshold determination, application of the stimulus during standing, and finally in the swing and the stance phases.

Four seconds of data for all five channels were collected. This provides a normal and a disturbed cycle EMG data on the same graph for comparison during the data analysis. An interactive program provided the latency of the response (cursor visually guided by the operator), duration of the phases, and peak values if required.

Results

The modulation of the response to nociceptive stimulus during the step cycle is clearly seen in Figure 1. When the stimulus is applied during the stance phase (see Figure 1a), inhibition followed by excitation of the ankle extensor to normal locomotor activity level (gastrocnemius) is seen. The subsequent ankle flexor activity is enhanced, although this was not as marked as the inhibition of the extensor activity. In the swing phase,

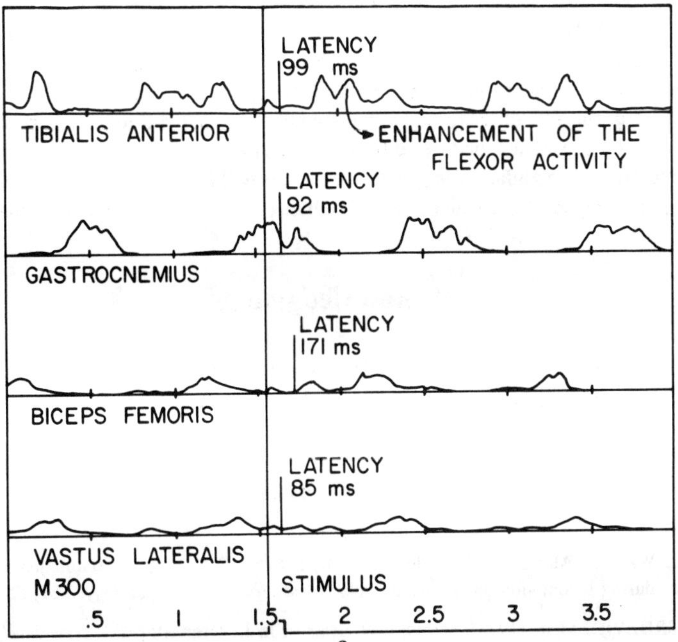

a

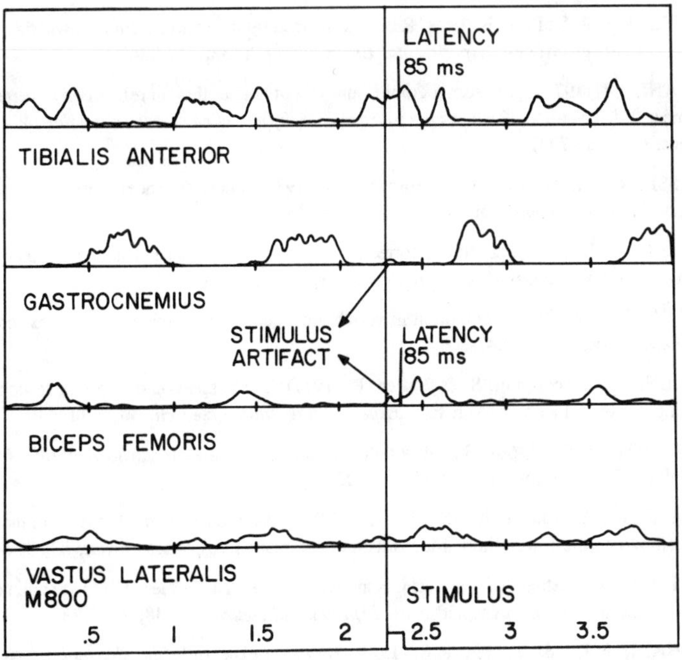

b

Figure 1—The response to nociceptive stimulus during (a) the stance phase and (b) the swing phase of the stepcycle.

the same stimulus produced increased knee flexor activity, indicating the subject was trying to withdraw the limb from the stimulus (Figure 1b). The response in the ankle flexor was masked in the normal walking EMG activity. Similar modulation was seen in three of the four subjects studied. Because of continuous background EMG activity in the fourth subject, the response to the stimulus was difficult to decipher. Repeated trials done on one subject on three different days shows similar results. Further studies will explore this modulation in greater detail and will include pre- and post-stimulus averaging of repeated trials along with simultaneous recording of the limb kinematics.

Acknowledgments

This work was supported by NSERC Grant No. A0070. The authors gratefully acknowledge the assistance of Paul Guy and John Pezzack in programing.

References

AKAZAWA, K., Aldridge, J.W., Steeves, J.D., & Stein, R.B. (1982). Modulation of stretch reflexes during locomotion in the Mesencephalic cat. *Journal of Physiology,* **329,** 553-567.

DIMITRIJEVIC, M.R. (1973). Withdrawal reflexes. In J.. Desmedt (Ed.), *New developments in electromyography and clinical neurophysiology, Vol.* **3** Basel. Karger: (pp. 744-750).

DUYSENS, J. & Stein, R.B. 1978. Reflexes induced by nerve stimulation in walking cats with implanted cuff electrodes. *Experimental Brain Research,* **32,** 213-224.

FAGANEL, J. (1973). Electromyographic analysis of human flexion reflex components. In J.E. Desmedt (Ed.,) *New developments in electromyography and clinical neurophysiology Vol.* **3** Basel. Karger: (pp. 730-733).

FORSSBERG, H., Grillner, S., & Rossignol, S. (1975). Phase dependent reflex reversal during walking in chronic spinal cats. *Brain Research,* **85,** 103-107.

FORSSBERG, H. (1979). Stumbling corrective reaction: A phase-dependent compensatory reaction during locomotion. *Journal of Neurophysiology,* **42,** 936-953.

GRILLNER, S. (1975). Locomotion in vertebrates: central mechanisms and reflex interaction. *Physiology Reviews,* **55,** 247-304.

GRILLNER, S., Rossignol, S. & Wallen, P. (1977). The adaptation of a reflex response to the ongoing phase of locomotion in fish. *Experimental Brain Research,* **30,** 1-11.

HAGBARTH, K.E. (1960). Spinal withdrawal reflexes in the human lower limb, *Journal of Neurology, Neurosurgery, and Psychiatry,* **23,** 222-227.

KUGELBERG, E., Eklund, K., & Grimby, L. (1960). An electromyographic study of the nociceptive reflexes of the lower limb. Mechanism of the plantar responses. *Brain,* **83,** 394-410.

LISIN, V.V., Frankstein, S.I., & Rechtmann, M.B. (1973). The influence of locomotion on flexor reflex of the hind limb in cat and man. *Experimental Neurology,* **38;** 180-183.

SHAHANI, B.T., & Young, R.R. 1971. Human flexor reflexes. *Journal of Neurology, Neurosurgery, and Psychiatry,* **34,** 16-627.

SUZUKI, S., Watanabe, S., Miyazaki, M., & Homma, S. (1983). Emg activity and kinematics of human stumbling corrective reaction during running. In H. Matsui & K. Kobayashi (Eds.), *Biomechanics VIII-A* (pp. 444-454). Champaign, IL: Human Kinetics Publishers.

WAND, P., Prochazka, A., & Sontag, K.H. (1980). Neuromuscular responses to gait perturbations in freely moving cats. *Experimental Brain Research,* **38,** 109-114.

Contribution of Stretch Reflex to Knee Bouncing Movement

Y. Yamazaki
Institute of Technology, Nagoya, Japan

T. Mano, G. Mitarai, and N. Kito
Research Institute of Environmental Medicine, Nagoya, Japan

Where the lengthening of a preactivated muscle precedes its immediate shortening, performance is significantly greater than that produced only by a shortening contraction (Cavagna, Dusman, & Margaria, 1968). This performance enhancement has been suggested to be mainly due to the combined effects of both the reuse of the potential energy stored in the muscular elastic component and the potentiation of the stretch reflex (Bosco, Viitasalo, Komi, & Luhtanen, 1982). However, the contribution of the stretch reflex is still unclear because of the difficulty in examining the movement produced by this reflex in natural human movements. The purpose of the present study was to determine the role of the stretch reflex during knee bouncing movements for the enhancement of performance.

Methods

Five healthy males ranging in age between 18 and 33 years were subjects in the experiments. The subject was seated in a chair in such a way that the hip, knee, and ankle joint formed a right angle, and then he was asked to bounce rhythmically his right leg up and down while still sitting in the chair. The bouncing was done by activating only the triceps surae muscle so as not to contract the quadriceps femoris (QF) muscle. The bouncing movement was attained after several training sessions with the aid of visual feedback from electromyograms (EMGs) in the QF muscle.

The bouncing was first obtained at the subject's preferred frequencies with small, medium, and large movement amplitudes. Secondly, the subject bounced in rhythm with 3 Hz auditory stimuli under three load conditions (with no load, 12 kg, and 23 kg load) where the loads were firmly attached to the thigh. In the case of 3 Hz bouncing, the amplitude was kept steady by means of visual feedback of the hip angle. The auditory stimuli were 1 kHz since wave tone burst of a duration of 25 ms and intensity of 80 dB.

The EMGs of the soleus (SOL), gastrocnemius, tibialis anterior and rectus femoris were recorded with bipolar surface electrodes which were securely fastened on the muscle belly and small size preamplifiers closely attached to the electrodes. Ankle and hip angles, platform force, and platform contact of the ball of the foot were also recorded, respectively, with potentiometers, strain gauges, and an electrical switch.

Results

Figure 1 shows the unfiltered SOL EMG and the ankle angle variation during preferred frequency bouncing with small (A), medium (B) and large (C) amplitudes. These curves were triggered at the moment when the ball of the foot touched the ground. Immediately after the foot touchdown, the variation of the ankle angle showed a forced dorsi-flexion, i.e., stretching of the triceps surae muscle, followed by a plantar flexion during which the foot left the ground (arrows). The SOL EMG showed two activities. One had already begun before the onset of dorsi-flexion and did not synchronize with the foot touchdown and was prominent in Figure 1-C (EMG-1); the other was a large triphasic wave with a short and constant delay after the onset of dorsi-flexion (EMG-2). The MEG-2 in small bouncing was seen during the dorsi-flexion (see Figure 1A), while in medium and large bouncing it appeared during the plantar flexion (see Figure 1B,

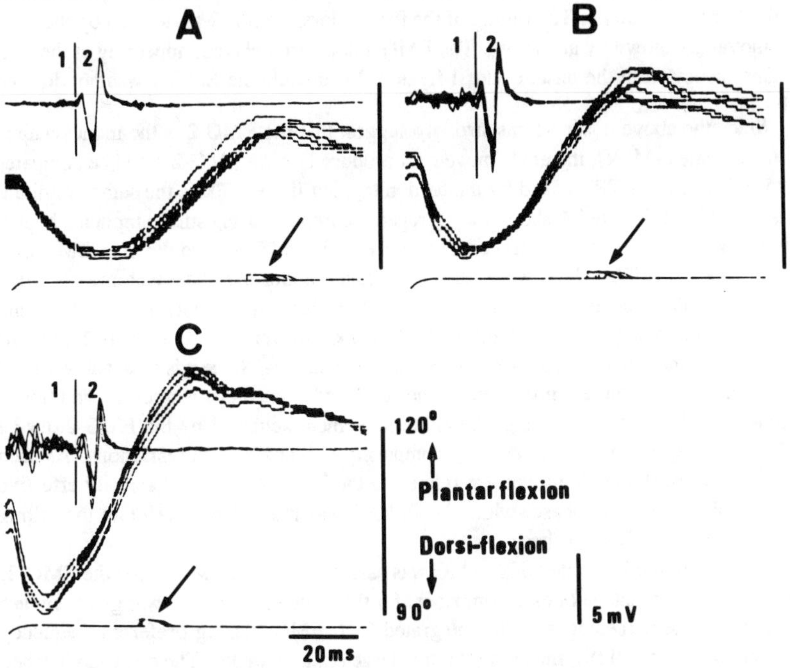

Figure 1—Unfiltered soleus EMG and ankle joint angle during small (A), medium (B), and large (C) amplitude bouncing. Five consecutive recordings triggered by the touchdown of the ball of the foot were superimposed. Arrows show the take-off of the foot. 1: EMG-1 2: EMG-2.

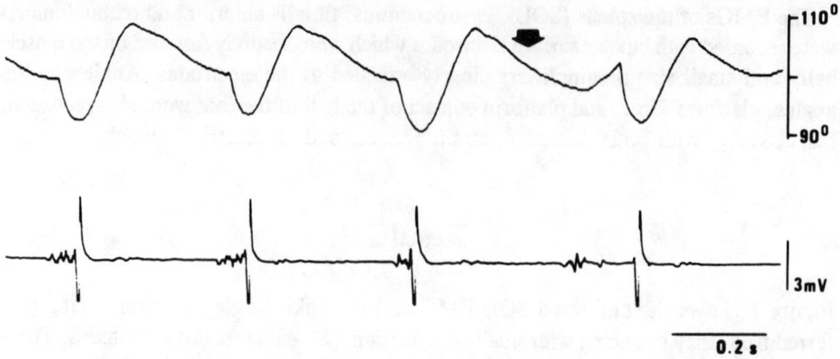

Figure 2—Ankle joint angle and unfiltered soleus EMG showing effects of unexpected removal of the plate, on which a subject was bouncing. An arrow shows the moment when the plate was removed.

C). The latency measured from the onset of dorsi-flexion and the wave form of EMG-2 were almost the same as those of the Achilles tendon reflex (ATR).

In an experiment designed to differentiate the character of EMG-1 and EMG-2, the plate (52 x 40 x 16 cm), on which a subject was bouncing in 3 Hz, was suddenly removed from the as yet unaware subject. Figure 2 shows an example of recording of the SOL EMG and ankle angle. The timing of the forced dorsi-flexion was delayed by the plate removal (as shown by an arrow). The EMG-2 was also delayed, appearing at the constant latency from the onset of dorsi-flexion. Meanwhile the EMG-1 was not delayed by the plate removal.

Since the above observations strongly suggest that the EMG-2 is the monosynaptic stretch reflex (MSR), the ankle movement produced by the EMG-2 could be compared with that of the ATR evoked by the tendon tapping if the ATR is the same amplitude as the EMG-2. Figure 3A shows the averaged recordings during small amplitude bouncing. Figure 3B shows the averaged ankle movement and force and the SOL EMG produced by the ATRs which were carefully elicited under the quasi-isotonic condition for each ATR to be the same amplitude as each triphasic wave in the raw EMG during the small bouncing shown in Figure 3A. The ankle movement and force of Figure 3B were superimposed on those of Figure 3A as dotted curves. Since it is reasonably argued that the same movement produced by the ATR under the quasi-isometric condition is generated during the bouncing, the ankle movement achieved by the EMG-2 during the bouncing could be estimated by comparing two ankle angle variations. As seen in Figure 3B, the ankle movement related to the EMG-2 was mechanically effective in the plantar flexion phase, while the EMG-2 was too late to decelerate the falling leg in the dorsi-flexion phase.

The area of the integrated SOL EMG was calculated for the EMG-1 and the EMG-2, and then the percentage of each component for the value of both areas was given. Figure 4A shows the percentages of the integrated SOL EMGs during preferred frequency bouncing with small (S), medium (M), and large (L) amplitudes. The open and hatched areas show the percentage of the EMG-1 and EMG-2, respectively. For the small bouncing the percentage EMG-2 was greater than the percentage EMG-1 in all subjects. With increasing bouncing amplitude, the percentage EMG-2 decreased, while the percentage EMG-1 increased alternately. The plantar flexion which can be seen before the touchdown amplified in accordance with the increased EMG-1 percentage.

Since the pattern of EMG activities during bouncing depends both on its amplitude and frequency, the effect of load on the SOL EMG activities was examined at the same

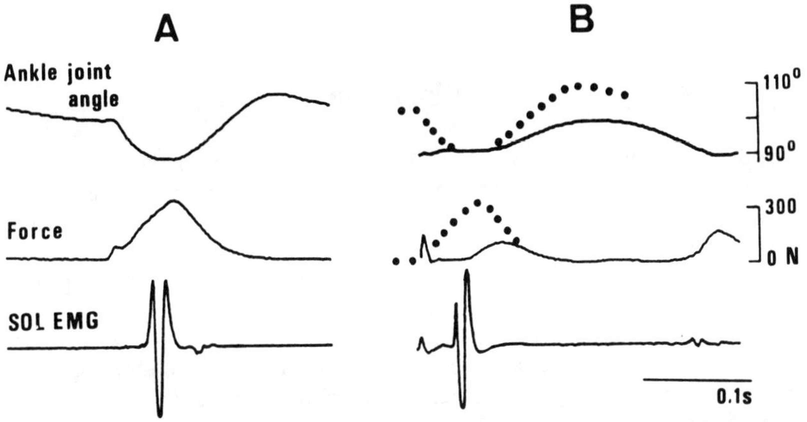

Figure 3—Averaged recordings of ankle joint angle, platform force, and soleus EMG during small amplitude bouncing (A) and those of Achilles tendon reflex under quasi-isotonic condition (B). SOL EMG: averaged raw EMG of the soleus muscle. The ankle joint angle and force of Figure 3A were shown in B as dotted curves.

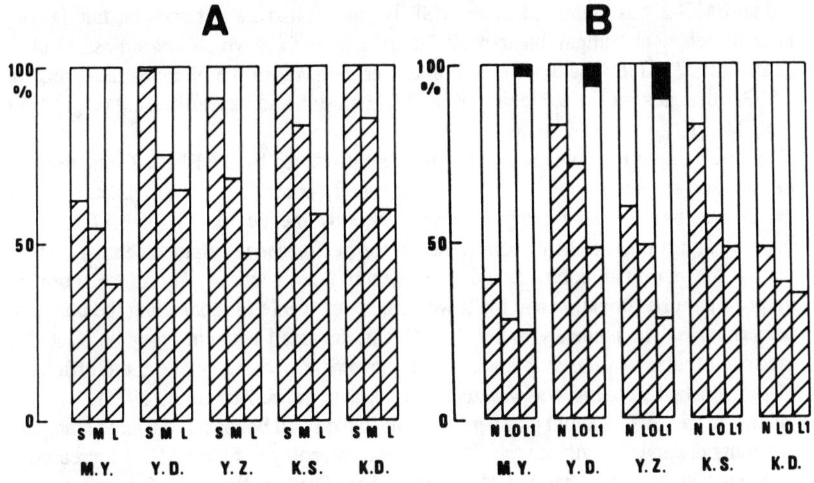

Figure 4—Percentage areas of rectified and averaged soleus EMG during bouncing. Open and hatched blocks show the percentages of EMG-1 and EMG-2, respectively. Shaded blocks show the percentage appeared only in 23 kg loaded bouncing. A: During small (S), medium (M), and large (L) amplitude bouncing with preferred frequencies. B: during 3 Hz bouncing under no load (N), 13 kg loaded (LO), and 23 kg loaded (L1) conditions.

frequency and bouncing amplitude. Figure 4B shows the percentages of the integrated EMGs during 3 Hz bouncing under the amplitude regulation with visual feedback of the hip joint angle. The EMG-2 percentages decreased in accordance with the degree of the subject's loads, while an alternating increase could be seen in the EMG-1. With increase of the EMG-1 percentage, the plantar flexion before touchdown was amplified proportionally to the loads.

Discussion

The EMG activity obtained from the SOL muscle during bouncing showed two components, EMG-1 and EMG-2. The EMG-1 is attributable to the preprogrammed activity and its primary role would be to decelerate the falling leg protecting from an injurious jolt. The facts that the slight plantar flexion was seen before the foot touched down, and that the plantar flexion was amplified with the increased EMG-1 percentage which occurred in larger bouncing or when the subject was subjected to loads, suggest that the role of the EMG-1 is partly to ensure the following stretching by the forthcoming muscle shortening. Also, the EMG-1 component is probably responsible for accelerating the leg upward with reutilization of the stored elastic energy (Cavagna et al., 1968).

The EMG-2 is not produced by the movement artifact which sometimes disturbs the EMG recordings during dynamic movements. Such artifact was greatly suppressed by setting carefully the surface electrodes and using specially designed amplifiers and high-pass filtering before analysis. Further, the property of the EMG-2 strongly indicates that the EMG-2 is true muscle activity.

The EMG-2 was identified as the MSR because it had a short and constant latency after stretching as comparable to the ATR and a wave form which was almost the same as that of the ATR. The experiment of sudden removal of the plate revealed that the EMG-2 depends on the stretching of the SOL muscle and has a different origin from the EMG-1.

In small amplitude bouncing, the movement produced by the EMG-2 was effective in time and magnitude (see Figure 3), and mainly contributed to the plantar flexion of the foot, based upon a comparison with the movement evoked by the ATR. These results suggest that the MSR contributes to the performance enhancement by following up the muscular contraction in the plantar flexion phase. Therefore, both the preprogrammed activation and MSR would be responsible for enhancing the bouncing performance. The tendency to use the MSR in the small bouncing may be related to the result that when the amplitude of the movement became smaller, the efficiency of the constant jumping was greater (Thys, Cavagna, & Margaria, 1975).

Though the EMG-2 could play a greater role in the small bouncing, the large amplitude bouncing depends heavily on the EMG-1. The percentage of the EMG-1 increased in large amplitude, while that of EMG-2 decreased. Further, the EMG-2 was not effective in large amplitude bouncing, where the EMG-2 was too late to influence the bouncing because of its appearance in the late plantar flexion phase. The greater dependency on the EMG-1 during large bouncing is due to the increased need for decelerating the falling leg mass. The above discussion could also be applicable to bouncing under load. Thus in large or loaded bouncing the EMG-1 increases its contribution to the deceleration and acceleration of the leg, and it would also do for the performance enhancement.

References

BOSCO, C., Viitasalo, J.T., Komi, P.V., & Luhtanen, P. (1982). Combined effect of elastic energy and myoelectrical potentiation during stretch-shortening cycle exercise. *Acta Physiologica Scandinavica,* **114**, 557-565.

CAVAGNA, G.A., Dusman, B., & Margaria, R. (1968). Positive work done by a previously stretched muscle. *Journal of Applied Physiology,* **24**(1), 21-32.

THYS, H., Cavagna, G.A., & Margaria, R. (1975). The role played by elasticity in an exercise involving movements of small amplitude. *Pflügers Archives,* **354**, 281-286.

The Acute Effects of Static Stretching on Alpha Motoneuron Excitability

L.K. Thigpen, T. Moritani, R. Thiebaud, and J.L. Hargis
College Station, Texas, USA

Sport enthusiasts presently use various stretching techniques to warm up prior to strenuous participation. These warm-up procedures supposedly ready the muscles and thereby prevent injury and muscle soreness. Research has failed to identify the exact benefits of warm-up stretches, but it has associated muscle soreness, fatigue, and spasms with elevated muscle action potentials (MAP) (deVries, 1968). It would therefore seem that muscle soreness, particularly delayed muscle soreness, could be prevented if stretching reduced MAPs of a fatigued muscle. Reductions in neuromuscular tension have been reported following moderate aerobic exercise by some researchers while employing the H/M ratio (deVris, Simard, Wiswell, Heckathorne, & Carabetta, 1982; deVries, Wiswell, Bulbulian, & Moritani, 1981), which appeared to provide a more reliable measure of alpha motoneuron excitability. It was, therefore, the purpose of this study to investigate the acute effects of static stretching upon alpha motoneuron excitability using the H/M ratio criterion.

Methods and Procedures

Subjects

Eight volunteers were selected after being informed as to the procedures involved in this study. The six males and two females (age 19 to 33) were taking no medication and had no history of neurological disorders. Subjects were also instructed to refrain from the intake of alcohol, aspirin, and caffeine since these drugs have been found to affect the Hoffman reflex (Eke-Okoro, 1982).

Experimental Design

Each subject acted as his or her own control. One randomly assigned leg (experimental) was fatigued 24 hours prior to testing. Fatigue consisted of toe raises (10 sets of 10 repetitions) performed at 70% MVC using a universal shoulder press. Prior to each testing session, testing order was randomly assigned. Pre- and post-testing consisted of measurement of the max H wave followed by max M wave measurement. Treatment consisted of three sets of 20 toe touches statically performed on a 20° inclined board interspersed with equal rest intervals for the experimental leg. A 12-inch step placed beside the inclined board ensured knee flexion and ankle neutrality of the control leg, thereby eliminating any possible stretch of the triceps surae. The control situation consisted of an equal time interval relaxing.

Measurement of H and M Waves

Electrical stimulation of the posterior tibial nerve percutaneously in the popliteal fossa evokes two discrete MAPs (deVries et al., 1981). The first MAP, the M wave, is evoked from direct stimulation (orthodromic) of the alpha efferents. The second MAP is propagated through the large Ia afferents (antidromic) and back through the efferents of the same muscle (see Figure 1). Since the antidromic current bypasses the muscle spindles, the H wave is independent of spindle activity and gama motoneuron excitability (Bishop, Hoffman, Wallis, & Shindell, 1975). This therefore provides a measure of alpha motoneuron excitability (Bishop et. al., 1975; Bonnet, Requin, & Semjen, 1981).

To determine the best placement for the stimulating electrode (4 mm Ag-AgCl, Beckman Instruments) a felt-tipped exploring electrode was used to find the site at which the largest M wave was evoked. The anode (ECG plate electrode) was secured approximately six inches inferior to the glueal fold on the posterior midline of the ipsilateral thigh. Shock intensity was varied from subthreshold to supramaximal in 5-V increments (see Figure 1). The 10- to 15-V range in which the max H waves were evoked was further explored in 1 volt increments to obtain more precise measures. Maximal amplitude of M wave was determined according to the methods of deVries et al. (1981), that is, no change in M wave amplitude despite increase in

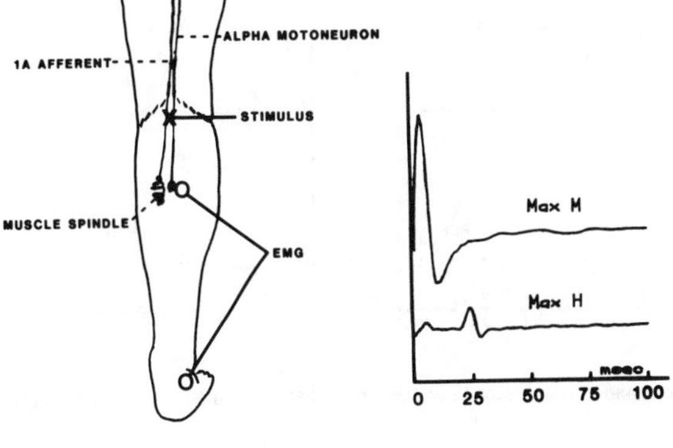

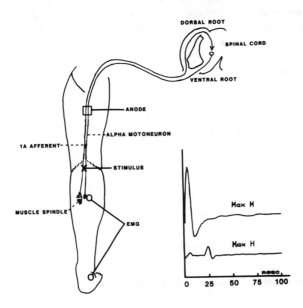

Figure 1—Experimental design for the elicitations and study of M and H waves.

stimulus voltage in three consecutive stimulations. Single rectangular pulses of 0.5 ms duration were delivered from a Grass S-48 stimulator through an isolation unit. Evoked MAPs were amplified (Grass P511) and stored for monitoring on a Tektronix-564 storage oscilloscope and recorded simultaneously on an analog FM data recorder for subsequent computer analysis. Appropriate gain settings of 100 or 1000 ensured optimal resolution of the MAPs from the 10 mm unipolar electrodes (Ag-AgCl, Beckman Instruments). The reference electrode was secured approximately 4 cm medially superior and oblique to the lateral malleolus. Placement of the active electrode was approximately 15 cm distal to the popliteal fossa on the midline of the posterior surface of the calf. Prior to electrode placement each site was cleaned and abraded to reduce source impedence to 3000 ohms and 5000 ohms for EMG and simulator electrodes respectively.

Reliability for determining the max M waves was $r = 0.995$ for the experimental leg and $r = .999$ for the control leg. Any testing session in which the max M wave varied was repeated since these fluctuations would indicate electrode movement and not motoneuron excitability changes. The H/M ratio was computed by dividing the max H amplitude by the max M amplitude for any test session in which the H wave changed while the M wave remained constant. The M wave mean differences (pre – post) of the 16 pairs of analyzed data was 0.25 ± 0.79 mV.

Results

A typical set of data regarding the changes in pre- and posttreatment H waves are shown in Figure 2. We have consistently found that there was very little change, if any, in the maximal amplitude of the H waves for the control leg while the experimental leg post H wave was markedly reduced. The group data indicated that the control leg pre and post H/M ratios were 19.76 ± 5.26 and 19.70 ± 5.62, respectively. The corresponding values for the experimental leg were 19.53 ± 5.67 and 15.19 ± 4.57, respectively. The average relative reduction (percent reduction) in the H/M ratio from pre- to post-test for the control and experimental legs were 0.63 ± 6.64 and 21.49 ± 14.29%, respectively. When the paired *t* tests were performed, no differences were found in the pre- and post-test H/M ratios for the control leg ($t = 0.12$, $P > .05$), nor the pre-test H/M ratios on both legs ($t = 0.18$, $P > .05$). However, static stretch-

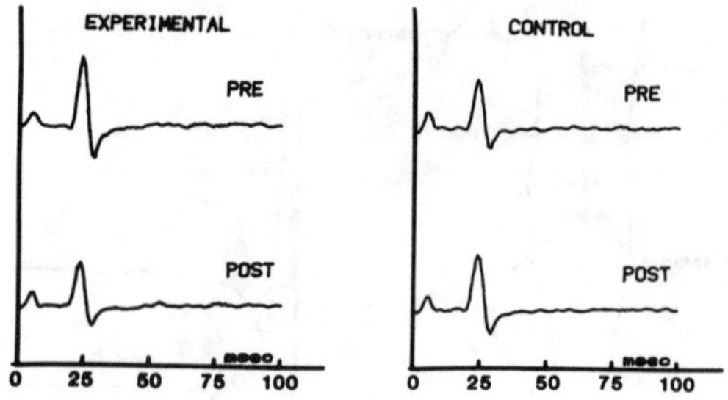

Figure 2—Representative maximum H waves propragated from cutaneous stimulations of the mixed tibial nerve.

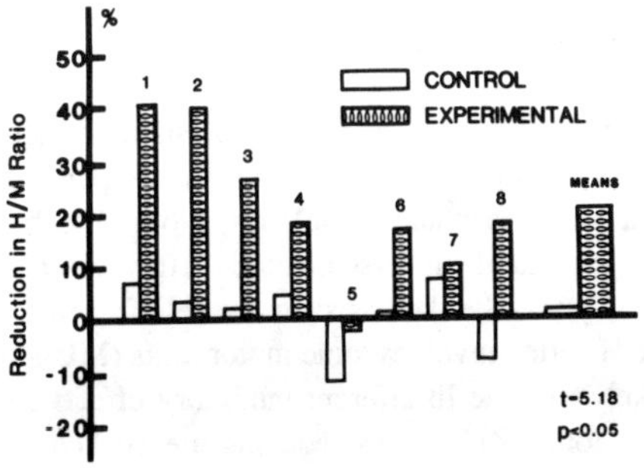

Figure 3—Percent reduction in the H/M ratio following treatment intervention.

ing was found to reduce the H/M ratio in the experimental leg (t = 3.42, P < .05). The post-test H/M ratio on both legs also differed (t = 3.30, P < .05). The greatest difference was found when comparing the percent reduction in the post-test H/M ratio of the control (x = 0.63%) and experimental (x = 21.49%) legs t = 5.18, P < .05).

Discussion

Sport participants are currently using preexercise stretching to "warm up" the muscles and prevent injury. Postexercise stretching is used to "cool down" the muscles and prevent soreness. While these are very common practices among sport enthusiasts, the exact benefits of stretching have not yet been identified. Since muscle soreness and spasm have been associated with an elevated resting MAP (muscle action potential) (deVries, 1968), stretching may in fact provide all these benefits by reducing the electrical activity within the muscle, thereby reducing excess muscular tension. deVries (1968) found that moderate bouts of exercise did produce a tranquilizer effect, in that they reduced the resting MAP (muscle action potential) and H/M ratio (deVries et al., 1981).

There are two major theoretical positions with respect to the causation of delayed muscle soreness: the muscle spasm theory and the structural damage theory. While there is no question that violent trauma can result in the rupture of a muscle and/or connective tissue, most of the exercise (including the present exercise regimen) that is known to cause soreness does not fall in this category. DeVries (1980) has proposed the spasm theory: delayed localized soreness that occurs after unaccustomed exercise is caused by tonic, localized spasm of motor units. A rationale based upon considerable physiological evidence can be constructed to support this hypothesis: (a) intense exercise causes some degree of ischemia in active muscles (Rohter & Hyman, 1962), (b) ischemia causes muscle pain, probably by transfer of "P substance" (Rodbard, Rodbard, & Farbstein, 1972), and (c) a reflex tonic muscle contraction prolongs the ischemia. Through the series of experiments, it has been shown that static stretching markedly reduced resting EMG activity in six of seven subjects who had chronic muscular problems of the shin-split type (deVries, 1961b). Symptomatic pain relief seemed to parallel reduced EMG values (deVries, 1961a, 1961b). One of his subjects who was atypical showed a marked rise in EMG activity and increased level of pain with stretching. It was hypothesized that, in this case, structural damage had indeed occurred (deVries, 1961b).

In the present investigation, it was found that static stretching brought about a statistically significant reduction in the H/M ratio of the experimental leg. These results are entirely consistent with earlier studies (deVries, 1961a, 1961b; deVries et. al., 1981, 1982) and further suggest that the inverse myotatic reflex which originates in the Golgi tendon organs (G.T.O.) may be the basis for the relief of muscle soreness by static stretching. Since the H reflex involves tonic motor units (MU's) (McIlwain & Hayes, 1977), it is most likely that the Ib afferent inhibitory effects from G.T.O. could be mediated through the tonic MU's, thus reducing the evoked H wave amplitude.

Acknowledgments

This study was supported in part by grants from the Texas Engineering Experiment Stations (project 39058E).

References

BISHOP, B., Hoffman, H., Wallis, I, & Shindell, D. (1975). Effects of increased ambient pressure and nitrogen on man's monosynaptic reflexes. *Journal of Applied Physiology*, **38**, 86-90.

BONNET, M., Requin, J., & Semjen, A. (1981). Human reflexology and motor performance. *Exercise and Sport Sciences Review*, **9**, 119-158.

DEVRIES, H.A. (1961a). Electromyographic observations of the effects of static stretching upon muscular distress. *Research Quarterly*, **32**, 468-479.

DEVRIES, H.A. (1961b). Prevention of muscular distress after exercise. *Research Quarterly*, **32**, 177-185.

DEVRIES, H.A. (1968). Immediate and long-term effects of exercise upon resting muscle action potential level. *Journal of Sports Medicine*, **8**, 1-11.

DEVRIES, H.A. (1980). *Physiology of exercise for physical education and athletes* (pp. 474-485). Iowa: W.C. Brown Company Publishers.

DEVRIES, H.A., Wiswell, R.A., Bulbulian, R., & Moritani, T. (1981). Tranquilizer effect of exercise. *American Journal of Physical Medicine*, **60**, 57-66.

DEVRIES, H.A., Simard, C.P., Wiswell, R.A., Heckathorne, E., Carabetta, V. (1982). Fusimotor system involvement in the tranquilizer effect of exercise. *American Journal of Physical Medicine*, **61**, 111-122.

EKE-OKORO, S.T. (1982). The H-reflex studied in the presence of alcohol, aspirin, caffeine, force, and fatigue. *Electromyography and Clinical Neurophysiology*, **22**. 579-589.

McILWAIN, J.S., & Hayes, K.C. (1977). Dynamic properties of human motor units in the Hoffmann-reflex and M response. *American Journal of Physical Medicine*, **56**, 704-710.

RODBARD, S. (1975). Pain associated with muscular activity. *American Heart Journal*, **90**, 84-92.

RODBARD, S. & Farbstein, M. (1972). Improved exercise tolerance during venous congestion. *Journal of Applied Physiology*, **33**, 704-710.

ROHTER, F.D., Hyman, C. (1982). Blood flow in arm and finger during muscle contraction and joint position changes. *Journal of Applied Physiology,* **17,** 819-823.

Do the Firing Statistics of Motor Units Modify the Frequency Content of the EMG Signal During Sustained Contractions?

C.J. De Luca and J.L. Creigh
Boston University, Boston, Massachusetts, USA

It is well known that during sustained muscle contractions, the power density spectrum of the EMG or myoelectric signal detected with surface electrodes displays a frequency shift toward the low-frequency end. The high-frequency components decrease and the low-frequency components increase in amplitude. During the past two decades, various studies have attempted to determine whether the cause of the frequency shift originated from "physical properties" of muscle fibers, such as conduction velocity, or from "control properties," such as firing statistics. This question was investigated by deriving mathematical expressions for the power density spectrum of the myoelectric signal which contained separate functions expressing the individual effect of the firing statistics and the shape of the motor unit action potentials.

Model

The myoelectric signal may be considered to consist of a superposition of individual motor unit action potential trains (MUAPTs). Hence, the power density spectrum of the myoelectric signal may be expressed as:

$$S_m(\omega) = \sum_{i=1}^{p} S_{u_i}(\omega) + \sum_{\substack{i,j=1 \\ i \neq j}}^{q} S_{u_i u_j}(\omega) \tag{1}$$

where $S_{u_i}(\omega)$ = the power density of the MUAPT, $u_i(t)$
$S_{u_i u_j}(\omega)$ = the cross-power spectrum of MUAPTs $u_i(t)$ and $u_j(t)$
p = the total number of MUAPTs in the ME signal
q = the number of MUAPTs with correlated discharges.

For each individual MUAPT, the power density spectrum may be expressed as:

$$S_{u_i}(\omega) = S_{\delta_i}(\omega) \cdot |H_i(j\omega)|^2 \tag{2}$$

where $S_{\delta_i}(\omega)$ = power density spectrum of the interpulse intervals
$H_i(j\omega)$ = the Fourier transform of the motor unit action potential.

It is well known that the time duration of motor unit action potentials increases during sustained contractions (Broman, 1973; De Luca & Forrest, 1973). This modification in the motor unit action potential is directly linked to the decreasing conduction velocities in the muscle fibers. Lindstrom (1970) has shown via a mathematical model that a decrease in the conduction velocity will cause a frequency shift (towards lower frequencies) of the function $H_i(j\omega)$. The question that now arises is: do thc firing statistics of the MUAPTs also cause a modification in the power density spectrum of the myoelectric signal? In other words, are there features of the individual $S_{\delta_i}(\omega)$ that vary with the interpulse intervals (IPI) statistics and that are reflected in the overall spectrum when the $S_{\delta_i}(\omega)$s are combined?

Methods and Results

This point may now be verified empirically. By using the computer assisted decomposition technique developed by LeFever and De Luca (1982) and Mambrito and De Luca (1983), it is possible to obtain highly accurate IPI measurements of MUAPTs many seconds long. The Fourier transform of the IPIs may then be computed directly. Figures 1 and 2 each present the magnitude of such Fourier transforms for three

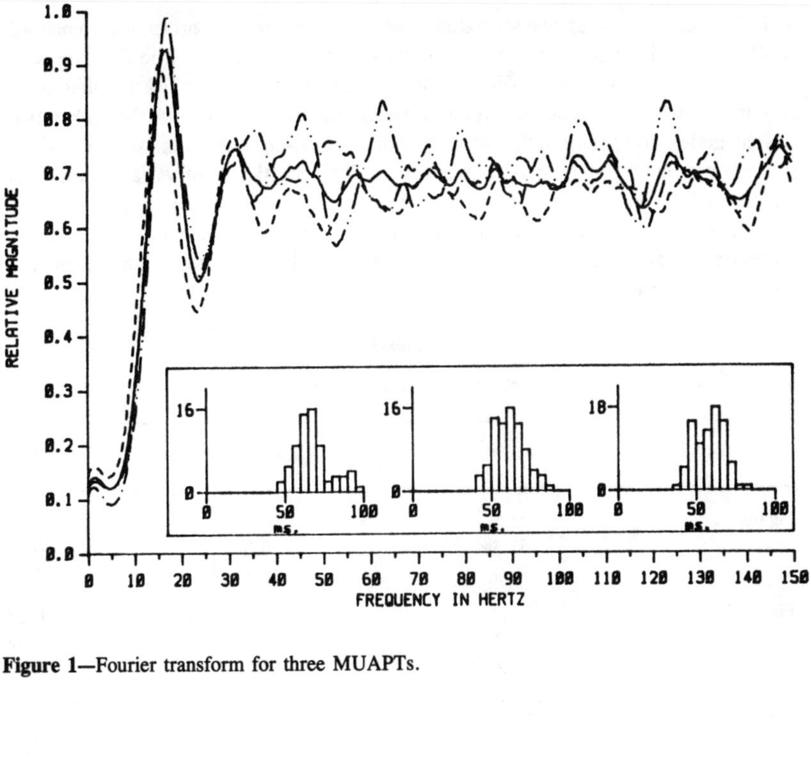

Figure 1—Fourier transform for three MUAPTs.

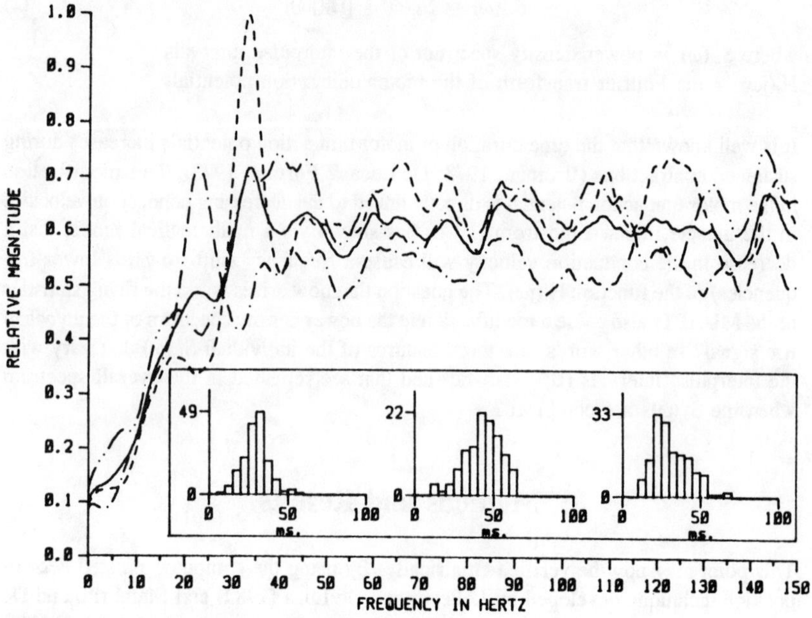

Figure 2—Fourier transform for three MUAPTs.

MUAPTs detected during two separate isometric constant-force contractions maintained at 50% of maximal force in the first dorsal interosseous muscle. The function with the solid line represents the average. The histograms present the IPI distribution of each motor unit, the one on the left corresponding to the function with the broken line and the middle histogram to the function with the dash-dot line. Some statistics of the IPIs are presented in the accompanying table. The coefficient of variation which is the ratio of the standard deviation to the mean value is a measure of the regularity with which the motor unit is discharging. The smaller the coefficient of variation, the sharper and higher will be the peak corresponding to the firing rate in the magnitude of the Fourier transform.

Table 1

Interpulse Interval Statistics

	μ (ms)	s.d. (ms)	c.v.
Figure 1	58.5	9.3	0.20
	61.8	10.0	0.16
	69.3	13.7	0.16
Figure 2	29.7	8.3	0.28
	31.3	11.8	0.38
	43.4	11.2	0.26

Discussion

When the coefficient of variation of the IPIs is small (0.16 to 0.20), the peak associated with the firing rates is clearly distinguishable (see Figure 1), whereas, when the coefficient of variation is higher (0.26, 0.28), then peak is less sharp and has lower amplitude (see Figure 2). The combined effect of the sharpness of the peak (due to the coefficient of variation of the IPIs) and the location of the peak (due to the value of the average firing rate) determines the extent to which a peak is present in the average Fourier transforms. (Compare Figures 1 and 2.)

It should be noted that the two parameters that have been identified as affecting the presence of the firing rate peak are both related to synchronization. That is, the smaller the coefficient of variation and the closer the average firing rate values, the greater the probability of two or more motor units discharging during a specific time interval. It should also be added that physiological events may also occur that may render the MUAPTs dependent and thereby introduce a third parameter to the concept of synchronization.

Also note that in either case the value of the magnitude of the Fourier transform is essentially constant beyond 40 Hz. The fluctuations beyond this point are due to the random nature of the IPIs.

The examples presented in Figures 1 and 2 indicate that the pattern and possibly the energy content of the power density spectrum of the myoelectric signal below 40 Hz may be altered by the statistics of the discharge properties of the motor units. Note that the average spectra shown do not take into account the shaping effects of the $H_i(j\omega)$ on each $S_{\delta_i}(\omega)$, but assuming that the motor unit action potential shapes are independent of the MUAPTs, the average shown is a useful representation of the effect of combining several trains. Although the effect below 40 Hz is not necessarily consistent, it cannot be overlooked when one attempts to identify the causes of the frequency shift of the myoelectric signal during sustained contractions, for it is known that the firing rates and other statistical properties of MUAPTs are time dependent during sustained contractions (De Luca & Forrest, 1973).

Acknowledgment

This work was supported by Liberty Mutual Insurance Company.

References

BROMAN, H. (1973). *An investigation on the influence of a sustained contraction on the succession of action potentials from a single motor unit*. Unpublished doctoral dissertation, Chalmers University of Technology, Gotenberg, Sweden.

DE LUCA, C.J., & Forrest, W.J. (1973). Some properties of motor unit action potential trains recorded during constant-force isometric contractions in man. *Kybernetic*, **12**, 160-168.

LeFEVER, R., & De Luca, C.J. (1982). A procedure for decomposing the myoelectric signal into its constituent action potentials: Part I—Technique, theory and implementation. *IEEE Transactions on Biomedical Engineering*, **29**, 149-157.

LINDSTROM, L. (1970). *On the frequency spectrum of EMG signals*. Unpublished doctoral dissertation, Chalmers University of Technology, Gotenberg, Sweden.

MAMBRITO, B., & De Luca, C.J. (1983). Acquisition and decomposition of the EMG signal. In J.E. Desmedt (Ed.), *New Developments in EMG and Clinical Neurophysiology*, **10**, 52-72.

On Adaptability of the Motor Program Model for Human Locomotion

A.E. Patla
University of Waterloo, Waterloo, Ontario, Canada

T.W. Calvert
Simon Fraser University, Burnaby, British Columbia, Canada

Human locomotion, which involves selective or combined muscular recruitment, control of equilibrium, and adaptation to the external environmental conditions (Nashner, 1980), is characterized by a stable, stride-to-stride repeatable pattern of muscular activation (Arsenault, 1982), joint torque patterns (Herman, Wirta, Bampton, & Finley, 1976; Winter, 1982), and joint kinematic patterns (Herman et al., 1976; Shapiro et al., 1981; Winter, 1982). The evidence from these and other studies strongly suggests the presence of a motor program for locomotion in humans. The relative invariance of the support moment when compared to the individual joint moments (Winter, 1982) and the phase dependent response in which the movement of the whole limb alters in character in response to an external stimulus (Patla & Belanger, 1983) suggests that the limb is controlled as an independent unit. This intralimb motor program is similar to the autonomous spinal locomotor generator for a limb in mammals such as cats (e.g., Grillner, 1975).

For it to be useful, the basic pattern for propulsion must be adaptable to different speeds of locomotion. Many studies have attempted to characterize changes in average EMG activity (e.g., Brandell, 1977), joint torque patterns (e.g., Winter, 1982), and joint kinematic patterns (e.g., Shapiro et al., 1981) with speed of locomotion. These studies, along with the evidence for the motor program for human locomotion, have been descriptive in nature. Besides, as noted by many researchers (e.g., Bekey, Chang, Perry, & Hoffer, 1977), the motor program is characterized not only by the individual EMG pattern, or joint torque and kinematic pattern, but also the interaction between these patterns.

The focus of this paper is to use an analytical model for the mammalian locomotor pattern generators proposed by Patla (1982) to model the locomotor motor program in humans and study its adaptability to varying speeds of locomotion. In this study, the joint kinematic patterns are used as a measure of the locomotor motor program; these can be viewed as the locomotor motor program output after it has been passed through the musculoskeletal transfer function. Since the model can be used with the

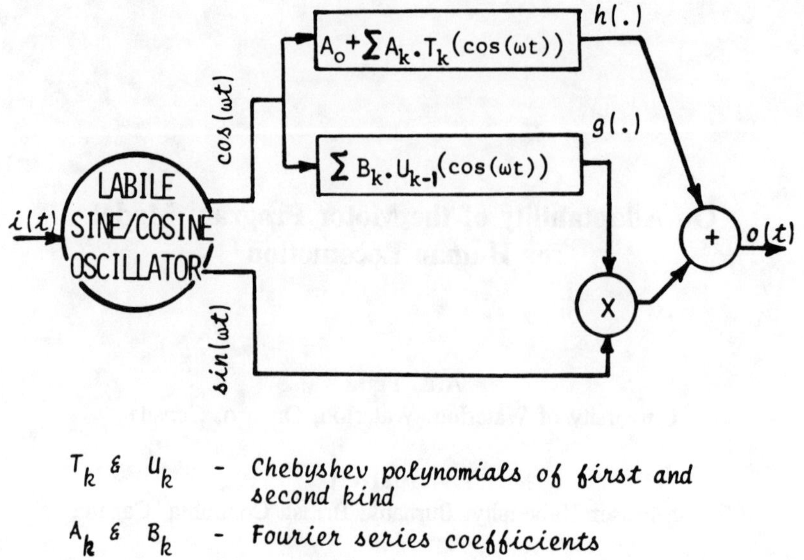

Figure 1—Labile synthesized relaxation oscillator model.

other measures of the motor program (e.g., the torque and EMG patterns), the use of kinematic patterns is not restrictive but represents the first stage in the study of an analytical model for the control of human locomotion.

Locomotor Motor Program Model

Unlike breathing, the locomotor activity is episodic and needs to be initiated. For a given speed of locomotion, the initiation signal is constant, but the outputs of the motor program are periodic. Thus the locomotor motor program (LMP) can be modeled as an oscillator community which alters its response to the tonic input which is the initiation signal. To study the behavior of the LMP, it is necessary to model an isolated oscillator. This pattern generator is modeled as a labile synthesized relaxation oscillator capable of producing any prescribed periodic output (Bardakjian, El-Sharkawy, & Damiant, 1981), and is characterized by the fundamental angular frequency ω and two nonlinear functions g(.) and h(.) (Bass, 1975; Chua & Green, 1974). The oscillator model is described in Figure 1.

Experimental Protocol

The CARS-UBC electrogoniometer, used in this study, monitors joint angles with the help of three potentiometers placed orthogonally to each other at each of the three joints of the lower limb. The double parallelogram arrangement and the sliding telescopic bar absorbs the linear translations and allows the monitoring of the relative angular displacements of the joints. The goniometer outputs were analog-to-digitally converted and stored on a PDP-11 minicomputer via a 16 channel multiplexor unit. The subject, a healthy male university student with no gait abnormality, was asked to walk along a straight walkway at subnormal, normal, and fast walking speeds. The sagittal plane motion of the hip, knee, and ankle joints, normalized to the peak values, was analyzed to give the fundamental frequency ω, and the nonlinear functions g(.) and h(.) (see Figure 2).

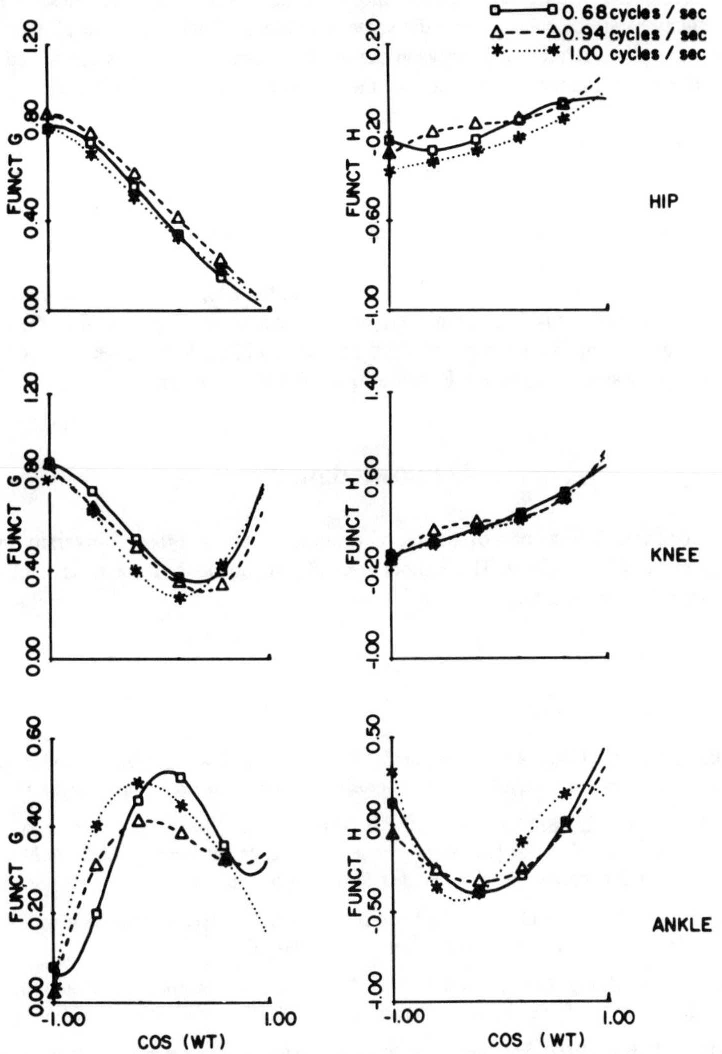

Figure 2—Shaping functions, g(.) and h(.) for the hip, knee, and ankle joint kinematics during walking at different speeds. Each curve is generated from 100 points fitted by a smooth line.

Results

The calculation of the oscillator parameters suggests that the locomotor program (LMP) is driven by one labile sine/cosine relaxation oscillator that produces the fundamental frequency oscillation of the outputs in response to the tonic input. The output of this oscillator is then fed into modules for each of the three joints used in the study to produce the output pattern measured during locomotion. The module for each joint consists of the two nonlinear functions g(.) and h(.). These functions depend only on the amplitude of the cosine functions and hence are time independent. Since these g(.) and h(.) functions are calculated from the same time block of angle data for all the three joints, the phase relationship between the joint kinematic patterns is accounted for.

Qualitatively, the model suggests that the LMP consists of a timing function (labile sine/cosine oscillator) with shaping functions [g(.) and h(.)] in cascade. With changes in speed of walking, analyses show (see Figure 2) that the shaping functions, g(.) and h(.) are invariant and that only the frequency and amplitude of the timing function change. Thus the locomotor program in humans controls the muscles of the lower limbs such that the basic shape of the final output, the kinematic patterns, remains the same with varying speeds of walking.

Unlike descriptive studies of human locomotion, the LMP model used in this paper is generative and amenable to simulation, and provides some insight into how the motor system can accomplish locomotion. Further studies will explore the use of this model with other measures of the LMP, the torque, and EMG patterns.

Acknowledgments

This work was done as part of a doctoral program at Simon Fraser University in the Department of Kinesiology. The authors gratefully acknowledge the assistance of John Chapman in programming.

References

ARSENAULT, B. (1982). A variability study of the EMG profiles in overground and treadmill walking in humans. Unpublished doctoral dissertation, University of Waterloo, Waterloo, Ontario.

BARDAKJIAN, B.L., El-Sharkawy, T.Y., & Damiant, N.E. (1981). On a population of excitable synthesized relaxation oscillators representing a labile bioelectric rhythm. *34th Annual Conference on Engineering in Medicine and Biology.* Houston.

BASS, S.C. (1975). The mathematical and laboratory generation of prescribed periodic waveforms. *IEEE Transactions on Circuits and Systems,* **22**(7), 603-610.

BEKEY, G.A., Chang, C.W., Perry, J., & Hoffer, M. (1977). Pattern recognition of multiple EMG signals applied to description of human gait. *IEEE Proceedings,* **65**(5), 674-681.

BRANDELL, B.R. (1977). Functional roles of the calf and vastus muscles in locomotion. *American Journal of Physical Medicine,* **56**, 59-74.

CHUA, L.O., & Green, D.N. (1974). Synthesis of nonlinear periodic systems. *IEEE Transactions on Circuits and Systems,* **21**, 286-294.

GRILLNER, S. (1975). Locomotion in vertebrates: Central mechanisms and reflex interaction. *Physiological Reviews,* **55**(2), 247-299.

HERMAN, R., Wirta, R., Bampton, S., & Finley, F.R. (1976). Human solution for locomotion: Single limb analysis. *Adv. in Behav. Biology,* **18**, 13-49.

NASHNER, L. (1980). Balance adjustments of humans perturbed while walking. *Journal of Neurophysiology,* **44**, 650-664.

PATLA, A.E. (1982). On pattern generators for locomotion in mammals. Unpublished doctoral dissertation, Simon Fraser University, British Columbia.

PATLA, A.E., & Belanger, M. (1983). Study of the response to nociceptive stimulus applied during walking in humans. *IX International Congress of Biomechanics.* Waterloo, Ontario.

SHAPIRO, D., Zernicke, R.F., Gregor, R.J., & Diestel, J.D. (1981). Evidence for generalized motor programs using gait pattern analysis. *Journal of Motor Behavior,* **13**(1), 33-47.

WINTER, D.A. (1982). Motor patterns in normal gait—Are we robots? Proceedings of the Second Conference of the CSB on Human Locomotion. *Locomotion II* (pp. 50-51). Kingston, Ontario.

Reflex Activity Changes with Human Muscular Strength Development

G.A. Wood, R.J. Lockwood, and F.L. Mastaglia
University of Western Australia,
Nedlands, Australia

In recent years neural factors have been found to be at least partly responsible for strength gains which accrue following relatively short periods of training (Komi, Viitasalo, Raurmamaa, & Vihko, 1978; Moritani & deVries, 1979; Sale, MacDougall, Upton, & McComas, 1983). Altered electromyographic patterns have been found to accompany strength increases (Wood, Lockwood, Cresswell, & Henstridge, 1983), and to be characteristic of the inherently strong (Stepanov & Burlakov, 1961). Milner-Brown, Stein, & Lee (1975) have demonstrated a higher incidence of motor-unit synchronization in weight lifters and have also shown that these athletes have more prominent late-reflexes ('long-loop' or 'supraspinal') evoked by electrical stimulation to the peripheral nerve of their trained muscles.

However, to date it has not been demonstrated that strength training brings about an alteration in reflex activity, nor that changes in reflex activity underlie observed changes in motor unit firing patterns. This study was directed toward these questions.

Method

Five female subjects (mean age 22 years) underwent a 6 week isometric strength training program for the ankle plantar flexors. Prior to and again following this training period reflex responses (tendon and Hoffmann) were elicited from the right triceps surae muscle group.

Reflex testing was undertaken in a quiet laboratory environment. The subject lay supine and tendon reflexes were obtained by the application of a controlled blow to the Achilles tendon by means of a solenoid-driven percussion hammer. Hoffmann reflexes were obtained by delivering a brief (1 ms) electrical stimulus to the posterior tribial nerve in the region of the popliteal fossa using a Grass S88 nerve stimulator. Once the intensity necessary to evoke a maximum H-response in the absence of a direct muscular response was determined, the intensity necessary to obtain a 50% response as well as the threshold level was established.

The H-reflex recovery profile was examined using paired electrical stimuli delivered to the posterior tibial nerve at interpulse intervals of 15 to 240 ms in steps of 25 ms. The intensity of the first (conditioning) stimulus was just below threshold, thereby ensuring that responses evoked by the second (test) stimulus would not be affected by a previous muscular response. The second stimulus was at an intensity which would have produced a one-half amplitude H-reflex had it not been preceded by a conditioning stimulus.

Electromyographic responses to each stimulus were detected through bipolar surface electrodes positioned immediately distal to the heads of the gastrocnemius. These responses were amplified (Grass P15 a.c. pre-amplifier; 3dB down at 10 hz and 10 Khz) and digitized on-line using a PDP-11/23 computer (sampling rate 1 Khz). Recordings were subsequently displayed on a graphics terminal and latencies and peak-to-peak amplitudes obtained by operator manipulation of the terminal graphics cursor. This data acquisition process is shown schematically in Figure 1.

Significant effects resulting from strength training were examined using a *t*-test for correlated samples.

Results and Discussion

Following the 6 weeks of specific isometric strength training, which produced mean gains of almost 20% in ankle plantar flexion strength, Achilles tendon and Hoffmann reflexes were found to be augmented ($t > 2.77$; $p < .05$). H-reflex recovery values were therefore normalized (test H-reflex/normal H-reflex) in order to make meaningful comparisons.

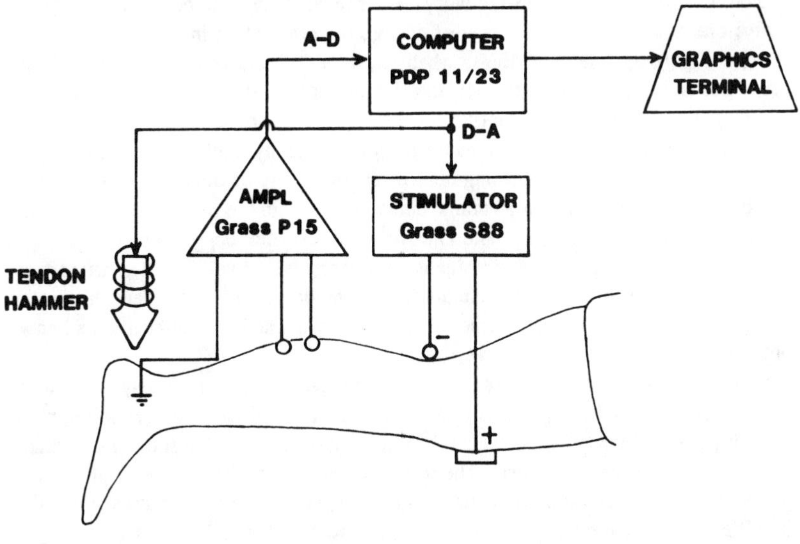

Figure 1—Schematic of data acquisition system.

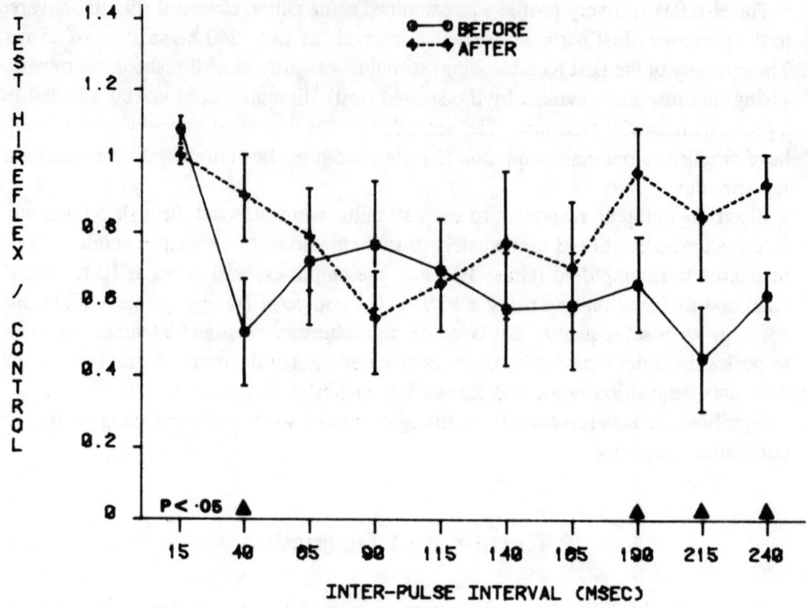

Figure 2—Normalized H-reflex recovery curves before and after six weeks of strength training (X ± SD).

Examination of the H-reflex recovery curves revealed two periods of heightened activity; one at an interpulse interval of 40 ms, the other at an interval of 190 to 240 ms (see Figure 2). These findings, when considered in the context of the postulated processes and pathways that underlie the H-reflex recovery curve (Taborikova & Sax, 1969), suggest that certain adaptive processes have occurred at a spinal level.

First, it would appear that spinal motor neuron excitability levels were elevated following strength training, although augmented tendon and individual H-reflex responses could be attributed to a greater volume conduction within the responding muscle arising from muscle hypertrophy. The later onset of H-reflex depression (90 ms cf. 40 ms) following a conditioning stimulus is, however, indicative of altered aftereffects. Whether these are due to local excitatory afterpotentials of the motor neuron's surface membrane (Eccles, 1946), or interneuron facilitatory/disinhibitory effects is not known, but certainly warrants further investigation.

Second, the results suggest that the recovery process following monosynaptic activity is accelerated. Whether this is due to 'long-loop' facilitation of the spinal motor neuron pool (Milner-Brown et al., 1975) or a more rapid replenishment of synaptic transmitter substance is not clear. The relatively late onset of the recovery process and previous findings that the early facilitation and longlasting depression are also obtained in patients with complete spinal cord section (Katz, Morin, Pierrot-Deseilligny, & Hibino, 1977) would suggest that the latter was the more plausible explanation.

Acknowledgment

This research was supported by grants from the Sir Robert Menzies Foundation for Health, Fitness and Physical Achievement.

References

ECCLES, J.C. (1946). An electrical hypothesis of synaptic and neuromuscular transmission. *Annals of the New York Academy of Science*, **47**, 429-455.

KATZ, R., Morin, C., Pierrot-Deseilligny, E., & Hibino, R. (1977). Conditioning of H-reflex by a preceding subthreshold tendon stimulus. *Journal of Neurology, Neurosurgery, and Psychiatry*, **40**, 574-580.

KOMI, P.V., Viitasalo, J.T., Raurmamaa, R., & Vihko, V. (1978). Effect of strength training on mechanical, electrical and metabolic aspects of muscle function. *European Journal of Applied Physiology*, **40**, 45-55.

MILNER-BROWN, H.S., Stein, R.B., & Lee, R.G. (1975). Synchronisation of human motor units: Possible roles of exercise and supraspinal reflexes. *Electroencephalography and Clinical Neurophysiology*, **38**, 245-254.

MORITANI, T., & deVries, H.A. (1979). Neural factors versus hypertrophy in the time course of muscle strength gain. *American Journal of Physical Medicine*, **58**, 115-130.

SALE, D.G., MacDougall, J.D., Upton, A.R.M., & McComas, A.J. (1983). Effect of strength training upon motorneuron excitability in man. *Medicine and Science in Sports and Exercise*, **15**, 57-62.

STEPANOV, A.S., & Burlakov, M.L. (1961). Electrophysiological investigation of fatigue in muscular activity. *Sechenov Physiological Journal of USSR*, **47**, 43-47.

TABORIKOVA, H., & Sax, D.S. (1969). Conditioning of H-reflexes by a preceding subthreshold H-reflex stimulus. *Brain*, **92**, 203-212.

WOOD, G.A., Lockwood, R.J., Cresswell, A.G., & Henstridge, J. (1983). Motor unit activity and muscular strength development. *Australian Phy. Eng. Sci. in Med.*, **6**, 71-75.

The Dependence of Reaction Times on Movement Patterns in Unilateral and Bilateral Upper Limbs in Trained Athletes

T. Kasai
Kokushikan University, Japan

There is much evidence on the differential role of the two hemispheres of the brain in human behavior, despite their anatomical symmetry. Investigations of the laterality of functions by reaction time (RT) experiments are usually associated with perceptual and cognitive functions (Dimond & Beaumont, 1974). Recent studies attempting to analyze the central mechanism for the motor programming and the execution of movements in a RT situation have demonstrated several differences in behavior between the preferred and nonpreferred hands (Hongo, Nakamura, Narabayashi, & Oshima, 1976; Nakamura, Taniguchi, & Oshima, 1976). Laterality variance, that is, right-left differences, have long been significant considerations in human movement research due primarily to the fact that these differences have been thought to influence the development of motor-integration patterns. However, Dimond's (1970) proposed hypothesis that each limb is exercised by a motor-output system and if one motor-output system is more sophisticated than another—for example, that relating to the right hand is more sophisticated than that which relates to the left—then it may be expected that the performance of the responding limb will be characterized by smoothness and efficiency. Not only will its performance be better than the other, but its use will be preferred over the other. Based upon this hypothesis, there must be some differences between the left and right hands in the motor functions which determine the hand preference. Thus, laterality research can also be interpreted with regard to differences in the effects of training on human motor function.

Few investigations that have studied laterality, however, have taken the RT paradigm and looked at the differences of motor control systems between bilateral athletes, that is, track runners (relatively equal participation of both arms), and unilateral athletes, that is, tennis players (demanding the use of predominantly one arm). In normal subjects, premotor time or electromyographic reaction time (EMG-RT) of the biceps brachii was definitely different in flexion (F) and supination (S) of the forearms. Accordingly, while the same muscle performs both functions, these two voluntary movements would be separately organized in the brain (Nakamura & Saito, 1974). Thus the aim of the present study is to compare EMG-RTs of bilateral athletes with those of unilateral athletes and to collect information with respect to training influences on motor control systems in the brain.

Methods

Subjects for this study were selected on the basis of proficiency in an activity which demanded use of predominantly one upper limb versus an activity which utilized equal participation of both upper limbs. Then, experiments were performed on 33 soccer players (all male students), 29 track runners (18 male and 11 female students), 21 tennis players (18 male and 3 female students) and 21 kendo (Japanese fencing) players (all male students), all without neurological disorders. The mean age was 19.8 years, SD = 1.0 year. They were all right-handed by self report. All subjects had considerable experience in their sporting event (5 years and over) and were continuing them at the time of the experiment.

The experimental procedures have already been reported in detail (Nakamura & Saito, 1974). In short, the subject sat in a quiet room on a chair and closed his eyes. The subject's arms were held in arm supporters in the following position: the shoulders were at 0°, the elbows at 60° flexion (full extension 0°) and the forearms in the pronated position. The tasks of the subject were flexion and supination of both arms simultaneously, responding to a sound signal (1000 Hz, 50 ms durations) as fast as possible. EMG-RTs were measured with the surface EMG of the biceps brachii muscle. The tasks were changed every six trials, and the EMG-RT was measured on 24 trials for each task. Data from trials just after the change in the task were discarded, so 20 trials for each movement pattern were subjected to analysis.

Results

Means and standard deviations for EMG-RTs are shown in Table 1. An analysis of the variance indicated that the EMG-RTs of the two movement patterns were not significant between the four sporting events' athletes but the EMG-RT differences obtained by subtracting EMG-RTs of supination (S) from those of flexion (F) were significant only on the right hand ($F = 5.66$, $df = 3/100$, $p < 0.01$). Then, the dependence of the EMG-RT difference to one's EMG-RT of F on both hands was examined. Figure 1 shows the scattergrams and the regression lines for the EMG-RT difference to one's

Table 1

Means and Standard Deviations of EMG-RTs of Flexion and Supination

		Left-Hand		Right-Hand	
	N	Flexion Mean (SD)	Supination Mean (SD)	Flexion Mean (SD)	Supination Mean (SD)
So	(33)	100.6 (13.5)	99.4 (10.8)	104.9 (12.9)	96.1 (11.1)
Tr	(29)	101.9 (10.7)	99.4 (9.9)	101.7 (12.3)	98.4 (11.3)
Te	(21)	102.4 (14.3)	97.5 (15.6)	106.3 (15.0)	99.4 (15.6)
Ke	(21)	95.3 (10.3)	96.0 (13.5)	99.0 (10.8)	98.1 (14.1)

Note: Values are in milliseconds. So = soccer, Tr = track, Te = tennis, Ke = kendo (Japanese fencing)

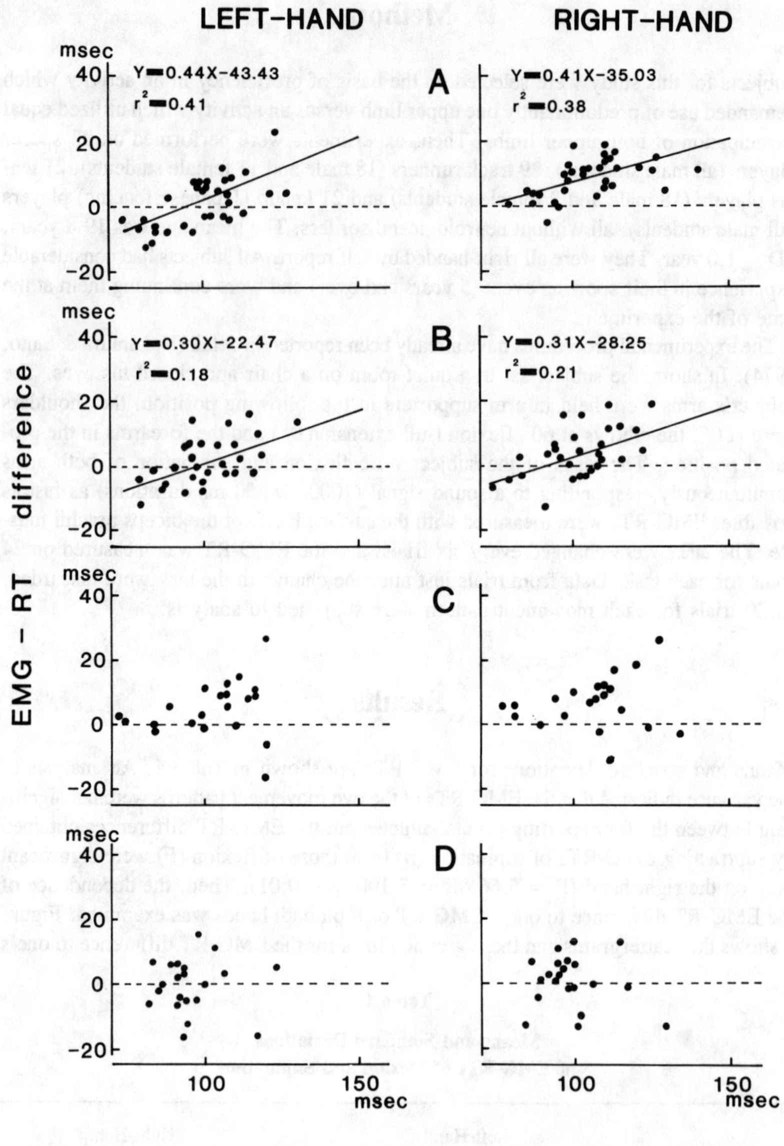

Figure 1—The relation of EMG-RT difference, subtracting EMG-RTs of S from those of F, to one's EMG-RT of flexion. A = soccer, B = track, C = tennis and D = kendo.

EMG-RT of F in each sporting event on both sides. The regression of the EMG-RT difference to one's EMG-RT of F was significant in soccer and track athletes on both hands but in tennis and kendo players was not significant. This means that there are different motor functions of arising movement patterns between the four sporting events' athletes on the preferred hand. Such treatment of the data would reveal the different motor output systems in the brain with respect to training effects of unilateral and bilateral athletes.

Discussion and Conclusions

Considerable attention has been given to the idea that many qualities of the athletic competitor may be explained by the principle of specificity of practice. Perhaps one of the most useful paradigms available to assess the quality of the neuromuscular response in an individual is the RT. Especially, EMG-RT is defined as the interval between stimulus presentation and the appearance of the first action potential in the target muscle group (Weiss, 1965) and also as the central information processing time.

Nakamura and Saito (1974) analyzed the difference in EMG-RT for right and left biceps, acting on the forearm in two different movement patterns, flexion and supination. Since the left-right difference of EMG-RT, the F of the nonpreferred hand was faster than that of the preferred hand and the S of the preferred hand was faster than that of the nonpreferred hand, they indicated that even in a simple movement there were differences in EMG-RTs for the right and left hands which did not depend on the muscles but on the movement patterns. It is suggested that lateral preference is not the same thing as cerebral dominance as has been stated by Dimond (1970) but rather is a set phenomenon in the motor-output system (Kasai, Nakamura, & Taniguchi, 1982; Wakabayashi, Nakamura, & Taniguchi, 1981). Moreover, the EMG-RT differences are dependent on one's EMG-RT of F and would be generated in the motor control system. This is consistent with the suggestion from clinical studies of patients (Nakamura, Taniguchi, & Yokochi, 1978; Nakamura & Taniguchi, 1980).

The main finding in the present study was that, despite unobserved differences of EMG-RT of F and S between the four sporting events' athletes, the dependence of the EMG-RT difference on one's EMG-RT of F disappeared in unilateral trained athletes. This clearly shows that there are different effects on motor control functions as a result of the unilateral upper limb exercised.

According to Nakamura and Taniguchi (1980) and Nakamura, Taniguchi, and Yokochi (1978), both the cerebral cortex and the basal ganglia are at least involved in the programming of movement patterns. The loss of dependence of EMG-RT difference on one's EMG-RT of F would correspond to a disorder of "programs" of movements for which the right cerebral cortex and the basal ganglia, but not the cerebellum, execute a crucial role.

In the light of additive models for RT (Rabbit, 1971), or a conduction model of neural processing, with a sensory projection system, motor control system and conduction pathway between them, the basal ganglia and the cerebellum should have quite different functions. Rather, the basal ganglia, sending information to the motor cortex through VL thalamus (Glickstein, 1972), would be functioning for the selection of movement patterns at the initiation of movements.

In view of the original questions, there was not a differential change in EMG-RT performance of F and S, but the dependence of the EMG-RT difference on one's EMG-RT of F disappeared in two unilateral sporting events' athletes on both hands. Since the EMG-RT is a reflection of the speed of processing in the central nervous system, the disappearance of dependence of the EMG-RT difference on one's EMG-RT of F indicates the change of the motor control functions. The results of the present study, which were obtained not from patients but from normal subjects, would indicate that the unilateral trained athletes detect these proposed changes to implicate specific functions that may directly alternate motor control generators both involved basal ganglia and motor cortex. Those facts obtained from the present study indicate the neural bases of the motor learning (Fleury & Lagasse, 1979).

References

DIMOND, S.J. (1970). Cerebral dominance or lateral preference in motor control. *Acta Psychologica,* **32**, 196-198.

DIMOND, S.J., & Beaumont, J.G. (1974). *Hemisphere function in the human brain.* London: Elek Science.

FLEURY, M., & Lagasse, P. (1979). Influence of functional electrical stimulation training on premotor and motor reaction time. *Perceptual and Motor Skills,* **48**, 387-393.

GLICKSTEIN, M. (1972). Brain mechanisms in reaction time. *Brain Research,* **40**, 33-37.

HONGO, T., Nakamura, R., Narabayashi, H., & Oshima, T. (1976). Reaction time and their left-right differences in bilateral symmetrical movements. *Physiology and Behavior,* **16**, 477-482.

KASAI, T., Nakamura, R., & Taniguchi, R. (1982). Effect of warning signal on reaction time of elbow flexion and supination. *Perceptual and Motor Skills,* **55**, 675-677.

NAKAMURA, R., & Saito, H. (1974). Preferred hand and reaction time in different movement patterns. *Perceptual and Motor Skills,* **39**, 1275-1281.

NAKAMURA, R., Taniguchi, R., & Oshima, Y. (1976). Preferred hand and steadiness of reaction time. *Perceptual and Motor Skills,* **42**, 983-988.

NAKAMURA, R., Taniguchi, R., & Yokochi, F. (1978). Dependence of reaction times on movement patterns in patients with cerebral hemiparesis. *Neuropsychologia,* **16**, 121-124.

NAKAMURA, R., & Taniguchi, R. (1980). Dependence of reaction times on movement patterns in patients with Parkinson's disease and those with cerebellar degeneration. *Tohoku Journal of Experimental Medicine,* **132**, 153-158.

RABBIT, P.M.A. (1971). Times for the analysis of stimuli and for the selection of responses. *British Medical Bulletin,* **27**, 259-265.

WAKABAYASHI, S., Nakamura, R., & Taniguchi, R. (1981). Movement patterns as an output variable in reaction time. *Perceptual and Motor Skills,* **53**, 832-834.

WEISS, A.D. (1965). The locus of reaction time changes with set, motivation and age. Journal of Gerontology, **20**, 60-64.

Effect of Preliminary Conditions (NO-EMG, EMG, TVR) on Voluntary Response

T. Kinugasa
University of Texas,
Austin, Texas, USA

It has been known that voluntary response latency is shorter with a preliminary contraction than without a preliminary contraction. Three aspects which might contribute to the explanation of this phenomenon are mechanical, psychological, and neurophysiological. From the mechanical standpoint, static tensing should "take the slack out of the muscle", which, as indicated by Ramsey and Street (1940), would increase the rate of tension development (Schmidt, 1967). From the psychological standpoint, a possibility for central locus of reaction time change with preliminary contraction is related to the "memory drum theory" proposed by Henry and Rogers (1960). They found that as the complexity of the intended movement increased, the reaction time increased, and the interpretation was that with complex movement the "motor program" was more complex and greater time was required to call the program from memory. Schmidt (1970) found premotor time being decreased with increased pre-tension, suggesting that partial contraction was equivalent to partially initiating a motor program to contract; further increases in pre-tension were thereby simpler, which shortened the central reaction time. From the neurophysiological standpoint, Yensen (1966) suggested that muscle spindle discharges, activated by tonic contraction, intensified both cortically and reflexly induced contraction. According to Vallbo (1970), primary muscle spindle discharges increased during voluntary contraction in man.

Therefore, the present study hypothesized that afferent discharges, especially primary afferent discharges, should contribute to the facilitation of a voluntary response. Studies in deafferentation have shown that a monkey with a deafferented limb can perform a variety of movements (Taub, 1975). But Vaughn, Gross, and Bossom, (1970) showed that deafferentation led to substantial delay in the time from the appearance of motor potentials evoked in the cortex related to the specific musculature involved in a given movement, and the onset of EMG, suggesting that afferent discharges might facilitate the voluntary response. It is well known that vibratory stimulation causes a powerful and rather specific activation of primary muscle spindles (Matthews, 1966) which induces muscle contraction through spinal reflex circuits; this phenomenon of a "tonic vibration reflex" (TVR) was first reported by Hagbarth and Eklund (1966). The present study was designed to investigate the effect of three preliminary muscle activities on voluntary response, using two different muscles (biceps brachii, gastrocnemius) to compare functional differences between the upper limb and the lower limb.

Methods

Eighteen healthy males (ages 20 to 34 years) were subjects for this study. Subjects were asked to react with contraction of the biceps brachii and the gastrocnemius separately. Six subjects performed both the biceps brachii and the gastrocnemius experiment.

Each of 12 subjects was asked to sit on a chair with the elbow of his preferred arm flexed 70° (0° at anatomical position), with the shoulder fixed 50°, and with the supinated forearm supported (shown in Figure 1). Each subject was instructed to relax his arm, in order to eliminate the activity of the antagonist (triceps brachii) during a trial, and to keep the elbow angle (70°) constant. Each subject was asked to respond with isometric elbow flexion as quickly as possible to a visual stimulus, which followed a 2 to 4 s warning signal. For the gastrocnemius experiment, each of twelve subjects was asked to lie prone with his ankles extending beyond the bed. Subjects were asked to respond with isometric contraction of the gastrocnemius to the visual stimulus as in the biceps brachii experiment.

Three experimental conditions were employed: NO-EMG, EMG, and TVR. In the NO-EMG condition, subjects were asked to maintain electrical silence during the warning period. In the EMG condition, subjects were asked to maintain a tonic level of contraction (20% of maximum EMG amplitude), during a warning period. The tonic level of contraction was practiced several times before the test session. In the TVR condition, a vibrator strapped to the limb so that it rested against the distal tendon of the given muscle was used to elicit a TVR during the warning period. The oscillatory frequency of the vibrator (Heiwa Electric Co. HV-13D) was 100 Hz, eliciting moderate tonic activity in the muscle. EMG signals were led off from a set of bipolar surface electrodes through a differential amplifier (Nihon Koden Co., RB-4) with filter settings of 10 Hz to 10 KHz and stored on an FM tape recorder (TEAC, R-410). Premotor time was measured on a digital oscilloscope (Nicolet Instrument Co., 201) at the level of 1 ms, in order to compare three experimental conditions.

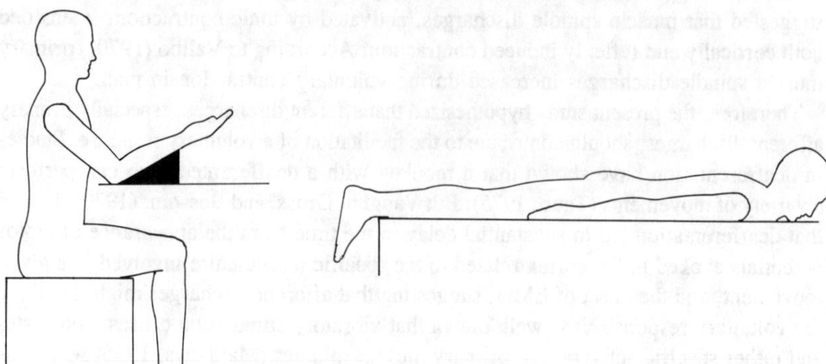

Figure 1—Subject's posture for the upper limb and the lower limb experiment.

Premotor time was defined as the time interval from the visual stimulus to the onset activity of the given muscle, so that premotor time represented delays involved in central processing. In the EMG condition, premotor time was defined as the time interval from the stimulus to initial activity above the level of the tonic contraction in the warning period or to initial activity after a silent period (shown frequently just prior to the reaction). Premotor time was obtained under all three conditions (NO-EMG, EMG, and TVR). Each subject performed ten trials for each condition. The order of conditions was randomized for subjects. Intertrial rest periods were provided (5 to 10 s) and a few minutes were allowed between trials and between conditions. Each subject practiced several times before the first condition.

Analysis of variance was used to compare mean premotor time across three conditions (NO-EMG, EMG, and TVR) within each muscle. A post hoc (Turkey) test was used when necessary to examine individual differences. Comparisons between muscles were conducted only on data from the six subjects who were tested in both limb experiments.

Results

Premotor times of twelve subjects for the biceps brachii experiment and for the gastrocnemius experiment were obtained under three conditions (See Figure 2). For the biceps brachii, the ANOVA revealed that the effect for condition was significant ($F[2,324] = 17.95, p < .001$), and a post hoc (Tukey) test at the .01 level indicated that mean premotor time for the EMG condition was shortest. There was no difference between mean premotor time for the NO-EMG and for the TVR condition. For the gastrocnemius, the ANOVA revealed that the effect for condition was significant ($F[2,324] = 33.43, p < .001$), and a Tukey test at the .01 level indicated that mean

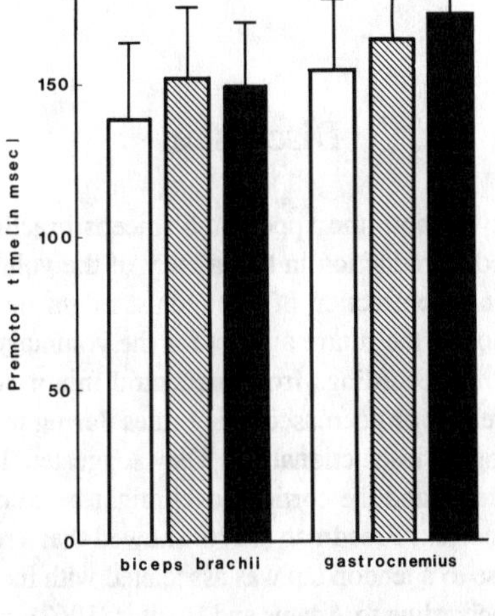

Figure 2—(Left) Mean premotor times for biceps brachii and gastrocnemius under three conditions (12 subjects). Error bars show one standard deviation.

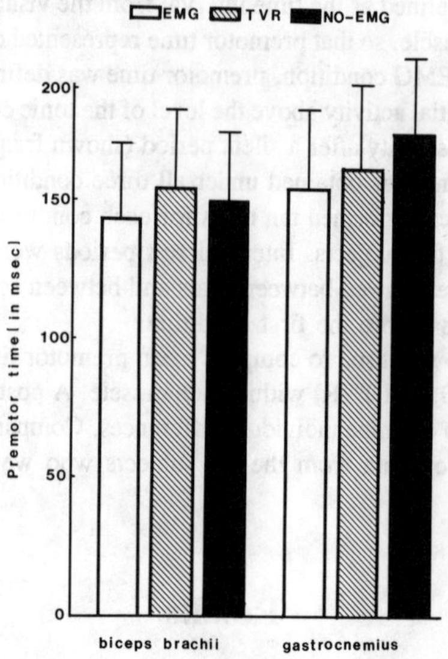

Figure 3—(Right) Mean premotor times for biceps brachii and gastrocnemius under three conditions by the six subjects who were tested in both experiments. Error bars show one standard deviation.

premotor time for the TVR condition was shorter than mean premotor time for the NO-EMG, but longer than mean premotor time for the EMG.

Figure 3 shows the premotor times for the six subjects who participated in both the biceps brachii experiment and the gastrocnemius experiment. The results of these six subjects were almost the same as the twelve subjects. The difference between mean premotor time for the EMG and mean premotor time for the NO-EMG was less in the biceps brachii than in the gastrocnemius (6 ms in biceps brachii; 20 ms in gastrocnemius).

Discussion

Increased initial activity in both the upper limb (biceps brachii) and the lower limb (gastrocnemius) resulted in reduction in the latency of the voluntary response, showing that preliminary tonic contraction of the agonist might result in "tuning," such as preliminary facilitation in the future agonist for the voluntary response. Evarts and Tanji (1976) reported that recordings from precentral motor cortex cells showed individual units with increased or decreased firing rates during the foreperiod of a reaction time task, depending on instructional set. They suggested that these automatic adjustments at the spinal level and the cortical level might be considered as preparatory "tuning" of the motor cortex. Freedman (1976) showed that a reduction in the latency of the voluntary response to a tendon tap was associated with increased values of initial tension. Furthermore, according to Adams and Boulter (1962), a subject's arousal state was a strong determinant of reaction time in vigilance situations. The difference between premotor time for the EMG and premotor time for the TVR might result from insufficient primary afferent discharges to raise the arousal level to the same level in the EMG condition for the response movement.

The difference which existed between the results of the upper limb experiment (premotor time for the TVR being the same as premotor time for the NO-EMG) and the results of the lower limb experiment (premotor time for the TVR being shorter than premotor time for the NO-EMG), suggests a difference in the neural connection between supraspinal and muscle spindle discharges for upper and lower limbs. Since primary spindle discharges from the biceps brachii might increase the excitability of the homonymous motorneurons, they might have a facilitatory effect on the voluntary response. However, if they had a facilitatory effect on the voluntary response, premotor time for the TVR should have been shorter than premotor time for the NO-EMG in the upper limb experiment.

The difference between mean premotor time for the EMG and for the NO-EMG in the six subjects was longer in the lower limb than in the upper limb, suggesting that the motor system controlling the lower limb might be more affected by the primary muscle spindle discharges than that of the upper limb. In other words, spinal facilitation from the primary spindle discharges might be stronger for the lower limb than for the upper limb.

References

ADAMS, J.A., & Boulter, L.R. (1962). An evaluation of the activationist hypothesis of human vigilance. *Journal of Experimental Psychology*, **64**, 495-504.

EVARTS, E.V., & Tanji, J. (1976). Reflex and intended responses in motor cortex pyramidal tract neurons of monkey. *Journal of Neurophysiology*, **39**, 1062-1068.

FREEDMAN, W., Minassian, S., & Herman, R. (1976). Functional stretch reflex (FSR)—A cortical reflex? In S. Homma (Ed.), *Progress in Brain Research*, **44**, 487-490.

HAGBARTH, K.E., & Eklund, G. (1966). Motor effect of vibratory stimulus in man. In R. Granit (Ed.), *Nobel Symposium I. Muscular afferents and motor control* (pp. 177-186).

HENRY, F.M., & Rogers, D.E. (1960). Increased response latency for complicated movements and a "memory drum" theory of neuromotor reaction. *Research Quarterly*, **31**, 448-458.

MATTHEWS, P.B.C. (1966). The reflex excitation of the soleus muscle of the decerebrate cat by vibration applied to its tendon. *Journal of Physiology (London)*, **184**, 450-472.

RAMSEY, R.W., & Street, S.F. (1940). The isometric length-tension diagram of isolated skeletal muscle fibers of the frog. *Journal of Cellular and Comparative Physiology*, **15**, 11-20.

SCHMIDT, R.A. (1967). Effect of positional tensioning and stretch on reaction latency and contraction speed of muscle. *Research Quarterly*, **38**, 494-501.

SCHMIDT, R.A., & Stull, G.A. (1970). Premotor and motor reaction time as a function of preliminary muscular tension. *Journal of Motor Behavior*, **11**, 96-110.

TAUB, E., Goldberg, I.A., & Taub, P. (1975). Deafferentation in monkeys: Pointing at a target without visual feedback. *Experimental Neurology*, **46**, 178-186.

VALLBO, A.B. (1970). Discharge patterns in human muscle spindle afferents during isometric voluntary contractions. *Acta Physiologica Scandinavica*, **80**, 552-566.

VAUGHN, H.G., Gross, E.G., & Bossom, J. (1970). Cortical motor potentials in monkey before and after upper limb deafferentation. *Experimental Neurology*, **26**, 253-262.

YENSEN, R. (1966). Neuromotor latency and take-up of musculo-tendious slack as components of RT. *Perceptual and Motor Skills*, **23**, 747-750.

Age-Related Changes in Postural Sway

K.C. Hayes, J.D. Spencer, C.L. Riach,
S.D. Lucy, and A.J. Kirshen
University of Western Ontario, London, Ontario, Canada

The periodic fluctuations in position of the body's center of pressure (CP) of ground reaction forces, i.e., postural sway, have been investigated in a large group of healthy children and adults. The primary purpose of this study was to generate a set of age-related normative data for the extent and power spectral density of postural sway, against which to evaluate the effects of various pathologies. In particular, consideration was given to recording the postural sway of young children (< 10 years) and aged individuals (> 65 years), with their eyes open (EO) and eyes closd (EC), i.e., to examine the Romberg effect.

Methods

Subjects (n = 102) ranged in age between 2 and 93 years old. All were in good health and free of neurological or orthopedic disabilities that would obviously affect their sway. The subjects stood upright, on a firmly secured steel force platform (AMTI OR-6 with resonant frequency Fz = 400 Hz), for 20 s. Their feet were positioned parallel and 6 cm apart. Subjects remained as still as possible, and visually fixated on an object 5 m away or completed the test with eyes closed. The order of administration of these trials was balanced across subjects.

The Fz (vertical force) and the Mx and My (moments of force about lateral [Lat] and antero-posterior [A-P] axes) signals from the force platform were amplified (DC-1 kHz) prior to A/D conversion at 51 Hz and storage in a digital oscilloscope. The digital output was processed in a Tek 4051 microcomputer to obtain estimates of the time varying excursions of CP (using an approximation of the form CPx = My/Fz) in the Lat and A-P directions. The digital coordinates for CP were then smoothed by passing them forward and backward in time through a second order Butterworth digital low pass (10 Hz) filter.

The smoothed digital coordinates were then processed through a FFT algorithm that computes the discrete Fourier transform expressed mathematically as:

$$Xd(k\Delta f) = \frac{1}{N\Delta t} \sum_{n=U}^{N-1} x(n\Delta t)e^{-j2\pi kn/N} \quad (1)$$

where t = time, T = time interval of sampling, Δt = time sample spacing, N = number of samples in T, n = time index (0,1,2,....,N−1), k = frequency index (0,1,2,...,N−1), Δf = frequency sample interval ($\Delta f = 1/T = 1/N\Delta t$), $Xd(k\Delta f)$ = the set of Fourier coefficients.

A rectangular windowing function was employed throughout. To assist in the identification of time truncation errors associated with the discontinuity of data at the window edges, some of the data were examined using a variety of modified cosine (Hanning) windows. The real and imaginary components of the FFT were then transformed into polar coordinates and subsequently to the power spectral density function.

Results

Young children (2 to 5 years) exhibited the greatest amount of sway in both the Lat and A-P directions. At 2 years old they were generally unable to stand with their eyes closed for the required period of time. For those older children who were able to stand unaided with eyes closed, the Romberg coefficients, based on RMS values, were small ($\leqslant$ 100%).

Adult males (> 20 years) exhibited a linear increase in sway with age in the Lat direction ($RMS_x = 1.37 + .027[Age]$: $p < .01$; $R^2 = .20$) with eyes open. This trend was paralleled in the eyes closed condition ($RMS_x = 1.64 + 0.025[Age]$: $p < .01$; $R^2 = .24$). Incorporation of the variables of height and weight into the multiple linear regression model (for eyes closed) increased the amount of explained variance to 36%. Romberg coefficients yielded overall means of $\bar{x}$ = 106% for Lat and $\bar{x}$ = 135% for the A-P direction.

Adult females did not exhibit an age-related increase in sway with their eyes open, but did reveal small but significant negative correlations between RMS (Lat) and both height ($r = -.34$) and weight ($r = -.26$) and between RMS (A-P) and height ($r = -.52$). With eyes closed the RMS (Lat) values were significantly correlated ($r = .43$) with age ($RMS_x = 1.45 + .025[Age]$: $p < .01$; $R^2 = .18$). Romberg coefficients for the RMS values were $\bar{x}$ = 124% for the Lat and 131% for A-P.

The power spectral density functions revealed the principal energy to be contained below 2 Hz with the peak power at < .2 Hz. In some older subjects there was increased energy in the 2 to 5 Hz bandwidth. This was accentuated in the EC condition. The power contained at this high frequency could not be simply attributed to time truncation artifacts as it was still present when various Hanning windows were employed. It appears this increased high frequency sway, which is also evident in cases of cerebellar dysfunction (Hayes, Spencer, Lucy, & Riach, 1983; Mauritz, Dichgans, & Hufschmidt, 1979; Silfverskiold, 1977) contributes at least in part to the exaggerated sway of the aging individual. The power contained at lower frequencies was also increased in these subjects. In children, the increased sway was generally associated with increased energy at frequencies < 2 Hz and infrequently was there evidence of energy > 3 Hz.

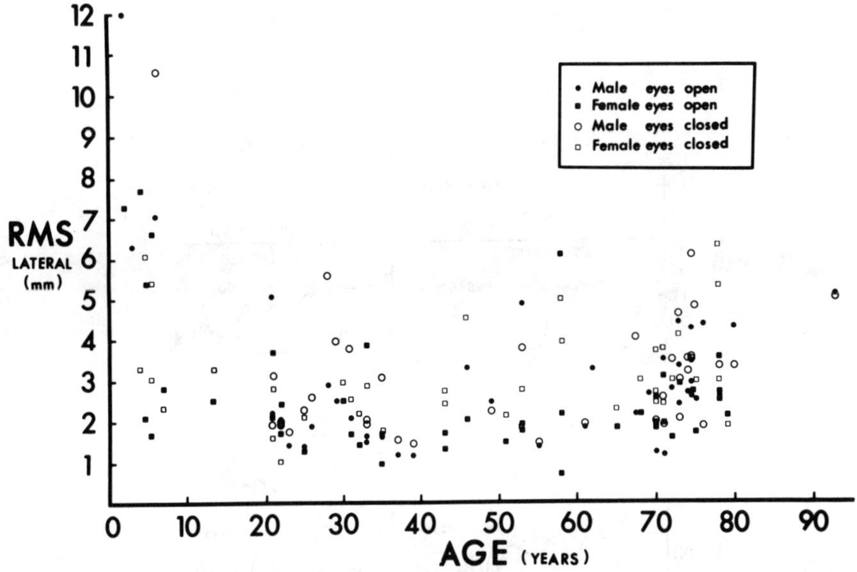

Figure 1—Changes in lateral postural sway with age.

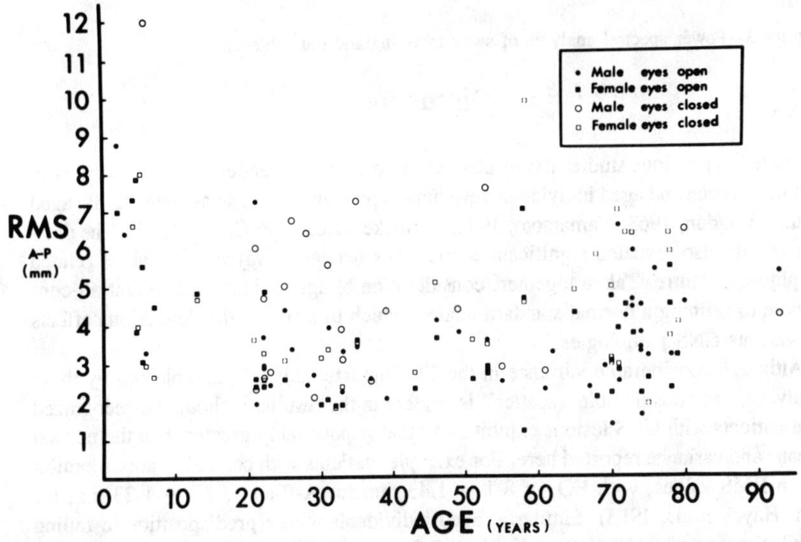

Figure 2—Change in antero-posterior sway with age.

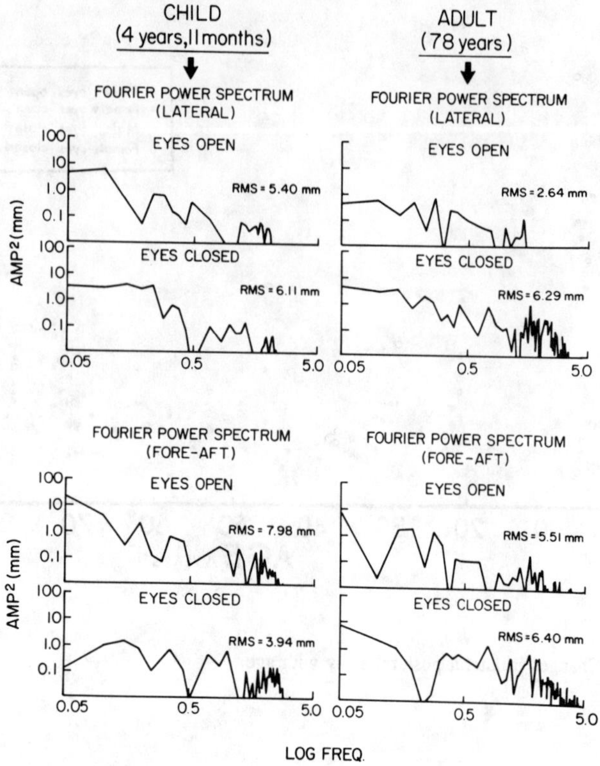

Figure 3—Power spectral analyses of sway in young and old subjects.

Discussion

As noted in previous studies the results indicated an age dependency in the sway, with young children and aged individuals revealing appreciably more sway than middle aged adults (Sheldon, 1963; Yamamoto, 1981; Zernicke, Gregor, & Cratty, 1977). The present results also revealed significant correlations between sway and variables related to physical stature. Taken together, consideration of age, height, and weight all contribute to refining a normal standard against which to evaluate the deleterious effects of various CNS pathologies.

Although considerable variance in the CP fluctuations is left unexplained by these analyses, and considerable "scatter" is evident in the results, it should be recognized that patients with CNS lesions exhibit sway that is noticeably greater than the normal means and variance reported here. For example, patients with cerebellar ataxia exhibit mean RMS values, with EO, of 8.1 $\pm$ 1.43 mm for A-P and 7.37 $\pm$ 1.23 mm for Lat (Hayes et al., 1983). Similarly, aged individuals with a predisposition to falling exhibit means of RMS (Lat) = 9.01 $\pm$ 2.8 mm and RMS (A-P) = 7.2 $\pm$ 1.1 mm (Kirshen, Hayes, Spencer, & Cape, 1983). The present results thus provide a useful data base by which to evaluate balance maturation or deterioration with aging or pathology.

References

HAYES, K.C., Spencer, J.D., Lucy, S.D., & Riach, C.L. (1983). Postural sway in subjects with cerebellar ataxia. *Society for Neuroscience Abstracts,* **13**, 21.15.

KIRSHEN, A.J., Hayes, K.C., Spencer, J.D., & Cape, R.D.T. (1983). Physiological factors in falls. *40th Annual Meeting of the American Geriatric Society,* New York.

MAURITZ, K.H., Dichgans, J., & Hufschmidt, A. (1979). Quantitative analysis of stance in late cortical cerebellar atrophy and other forms of cerebellar ataxia. *Brain,* **102**, 461-482.

SHELDON, J.H. (1963). The effect of age on the control of sway. *Gerontologica Clinica,* **5**, 129-138.

SILFVERSKIOLD, B. (1977). Cortical cerebellar degeneration associated with a specific disorder of standing and locomotion. *Acta Neurologica Scandinavica,* **55**, 257-272.

YAMAMOTO, T. (1981). Changes in postural stability with special reference to age. In A. Morecki, K. Fidelus, K. Kedzior, & A. Wit (Eds.), *Biomechanics VII-A* (pp. 169-173). Baltimore: University Park Press.

ZERNICKE, R.F., Gregor, R.J., & Cratty, B.J. (1978). Quantification of postural stability in normal children. In E. Asmussen & K. Jorgenson (Eds.), *Biomechanics VI-A* (pp. 130-134). Baltimore: University Park Press.

Effect of Extended Habituation on Variability Patterns of Within-Movement Forces Applied in Pedaling

J.D. Brooke and B.R. Goslin
University of Guelph, Ontario, Canada

The forces which result from movement at the whole person level reflect the final common path of the underlying neural activity (with standardization of muscle metabolic factors). To gain more insight into these forces, the authors and colleagues have developed a biomechanical technique to transduce continuously the force applied to the pedals when humans cycle on an ergometer. Extending from earlier work (Daly & Cavanagh, 1976; Hoes, Binkhorst, Smeekes-Kuyl, & Vissers, 1968) by using online computerization to mini- and main-frame, many cycles of the movement can be described. Force data can be obtained every 20 ms with system accuracy to ± 1 N (Brooke, Hoare, Rosenrot, & Triggs, 1981).

A clear pattern of force appears over the movement cycle, reflecting internal CNS pattern generation. For the power generation phase the mean curve is an inverted U. With repeated cycles, the standard deviations (*s*) around the mean forces are obtained and the coefficients of variation (C.V.) calculated. These also pattern over the cycle, and reflect control limits of the neural pattern generation. Subjects are very stable in response, with C.V.s around 5% at peak force, rising in a U curve to increase threefold at the extremes of the power phase. We have reported that there is increase in the variability of force applied and altered mean distribution of forces within a movement cycle, when the fasting blood glucose is depressed (Brooke, Chapman, Fischer, & Rosenrot, 1982a, 1982b)

The within-movement patterns of force variability are impressively stable within a trial and high replicability of mean forces has been reported between trials and days (Brooke et al., 1981). The degree to which there is plasticity in the variability, pattern or quantity, is important for our intended investigation of its neural components, as well as for experimental control.

Therefore the present study investigated what change occurs in the *s* and C.V. patterns of within-movement forces applied in pedaling, when habituation extends over several weeks.

Method

Two male informed-volunteer samples, age mean ($\bar{x}$) 28 years, s 6.5 years, exercised at 60 pedal rev/min at power output increasing in three minute steps without rest from 180 W by 30 W to maximum. The mean maximum power was 272 W, with a standard deviation of 28 W. One sample (control, $n = 7$) performed one further trial. The other group (experimental, $n = 5$) made between 10 and 17 further trials over the following 3 weeks. These subsequent (data) trials for both groups were at a power output of 50% of the individual's maximum power on the first trial, exercising continuously for 12 min. After 5 min of exercise, pedal force data from 20 complete pedal cycles were collected in each of the following 5 min. Intra-cycle means, and s and C.V. about them, were calculated for each subject in each sample and trial. These statistics were plotted against time, and negative-slope linear regressions were tested at $p < .05$. Paired comparisons of first and last trials for the experimental group were by t statistic at $p < .05$.

The apparatus has been reported (Brooke et al., 1981). A strain gauge, mounted on a rolled steel plate bolted to the pedal, continuously transduced an estimate of the force on that point of the pedal in that plane. The pedals, on a Monark ergometer clamped to the floor, had toe clips, straps, and blocks positioned under the feet. Using a pulse from the rotation of the chain wheel teeth, the force signal was digitized and stored in a PDP-8E minicomputer for 26 equidistant points in the pedal cycle.

Before and after each trial, the pedal was calibrated to $\pm$ 1 N by applying known weights at the ranges of force observed (21 N and 352 N). The sampling rate was well within the reported dynamic frequency response of the force pedal (Brooke et al., 1981).

Control of subject state and experimental setting were very strongly emphasized. Control subjects were asked to avoid drugs within 24 hr before, and vigorous activity 12 hr before trials. Experimental subjects were additionally instructed to standardize hours of sleep each night, three normal meals each day, eat 1.5 hr before trials and avoid intentional body weight change. For all subjects, pedal foot placement, saddle height, and clothing were standardized, body temperature and attitude scales were used to screen for ill-health and laboratory disturbances were avoided. All subjects were recreational cyclists. Fifty grams of d-glucose were ingested in water prior to each ride to ensure supra-fasting blood glucose levels. Each ride, subjects were instructed to "pedal as consistently as possible" using the cycle speedometer and also a digital display of mean pedal rate. Experimental subjects rode two to three times a week at the same time of day.

Results

From personal history questionnaires, subjects appeared to follow instructions. Figure 1 (top) shows that mean forces across the pedal cycle were similar for the two groups initially. By the last trial the experimental group mean forces were lower ($p < .05$). Figure 1 (middle) shows that force s over the pedal cycle was significantly higher in the control group than in the experimental group's first trial, which itself was higher than the last experimental trial. C.V.s of force over the cycle were also higher in the control group than in the first experimental trial but did not differ pre-post habituation, as is shown in Figure 1 (bottom).

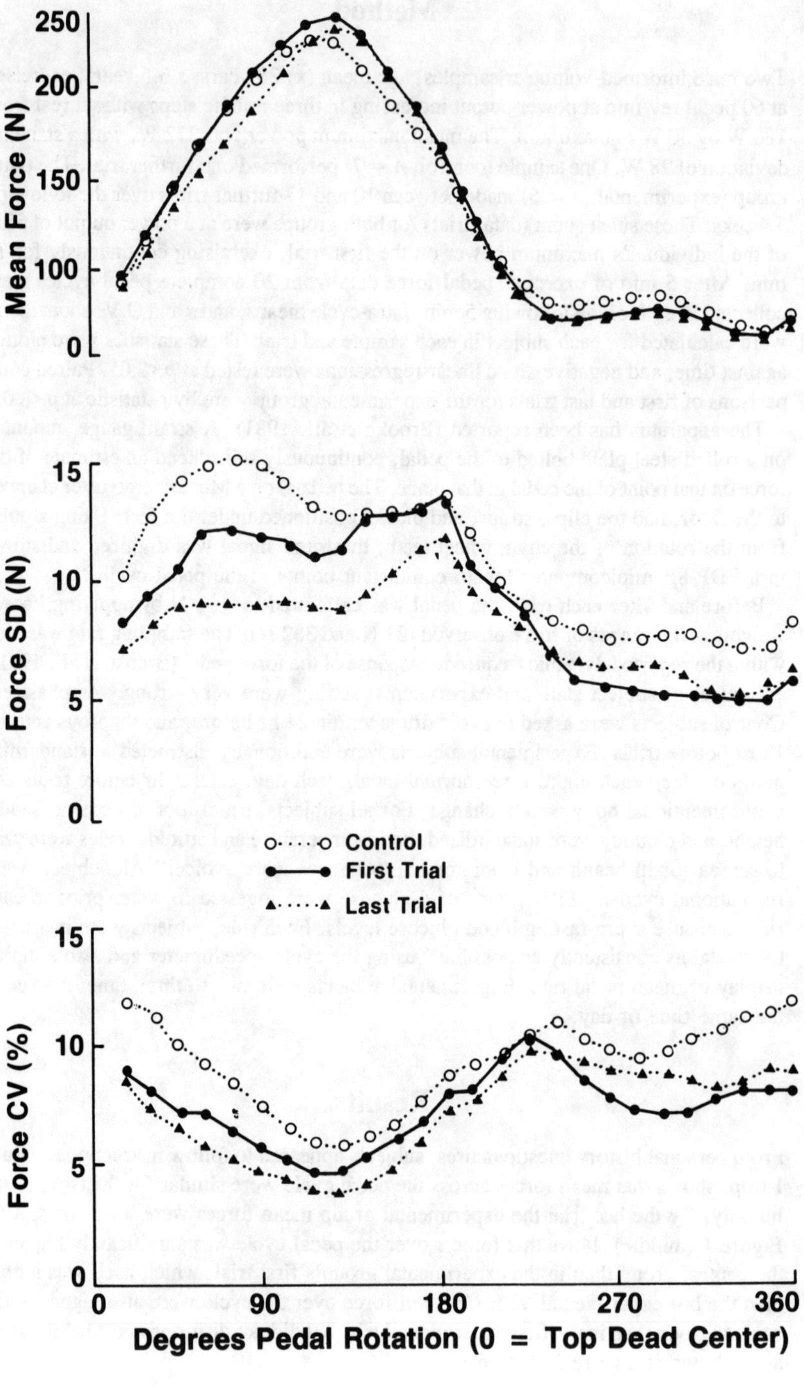

Figure 1—Force parameters over trials.

Table 1

Parameters for Linear Regressions of Individual Subjects' Coefficients of Variation of Pedal Force, Averaged Over the Pedal Cycle Phases of Power and Recovery, Against Trials Over the Habituation Period

Subject	Slope*	Intercept	% Variance Accounted
Power phase			
A	−0.01	5.66	0.2
B	−0.03	7.52	3.2
C	−0.09*	5.73	33.4
D	0.02	6.08	1.0
E	0.02	4.92	4.6
Recovery phase			
A	−0.05	11.14	1.85
B	0.01	9.59	0.20
C	0.09	4.97	12.45
D	0.06	7.02	6.00
E	0.08	8.30	11.44

*Slope significant $p < .05$.

For C.V.s averaged across the pedal cycle for each individual in the habituation trials, none of the slopes of the five subsequent regression lines were significant. With such data split into power and recovery phases, one of the power phase regressions showed significant negative slope, as Table 1 shows. Intercepts ranged from 5 to 11%. Within trials, five 1-min samples of within-cycle mean forces were highly correlated, with a range of 0.97 to 0.99.

Discussion

It is remarkable that no significant reduction in the coefficients of variation occurred over more than 3 weeks of practice. The present task is noteworthy in that initially the coefficients of variation were low, ranging from 4% (reported for peak force by Sargeant & Davies, 1977) to 12% at extremes of the power phase, suggesting that lack of experimental control was not why nonsignificance occurred. Nor should it be, for the procedures heavily emphasized such control. Subjects did perform the task with ease, as shown by the between sample inter-correlations, and the patterns of both mean force and its surrounding variability were characteristic, both for the group and the individual. On these grounds, the control of force applied in this activity may involve more automated neural subroutines (Grillner & Zangger, 1979; Tatton & Bruce, 1981) which are less trainable than the psychomotor tasks often studied.

The lower variability for the experimental vs. control group in trial one may result from subject differences. Alternatively it may be due to the additional constraints applied on experimental group behavior before testing started. A procedural conclusion is that this task does not demand extensive habituation, nor multiple samples averaged.

Acknowledgment

Supported by a grant from the Natural Sciences and Engineering Research Council of Canada.

References

BROOKE, J.D., Hoare, J., Rosenrot, P., & Triggs, R. (1981). Computerized system for measurement of force exerted within each pedal revolution during cycling. *Physiology and Behavior,* **26**, 139-143.

BROOKE, J.D., Chapman, A., Fischer, L., & Rosenrot, P. (1982a). Repetitive skill deterioration with fast and exercise-lowered blood glucose. *Physiology and Behavior,* **29**, 240-251.

BROOKE, J.D., Chapman, A., Fischer, L., & Rosenrot, P. (1982b). Between-minute stability in habituated human skill with modestly lowered blood glucose. *Physiology and Behavior,* **29**, 1111-1115,

DALY, D.J., & Cavanagh, P.R. (1976). Asymmetry in bicycle ergometer pedalling. *Medicine and Science in Sports,* **8**, 204-208.

GRILLNER, S., & Zangger, P. (1979). On the central generation of locomotion in the low spinal cat. *Experimental Brain Research,* **34**, 241-262.

HOES, M.J., Binkhorst, R.A., Smeekes-Kuyl, A.E., & Vissers, A.C.A. (1968). Measurement of forces exerted on pedal and crank during work on a bicycle ergometer at different loads. *Internationale Zeitschrift für Angewandte Physiologie,* **26**, 33-42.

SARGEANT, A.J., & Davies, C.T.M. (1977). Forces applied to cranks of a bicycle ergometer during one- and two-leg cycling. *Journal of Applied Physiology,* **42**, 514-518.

TATTON, W.G., & Bruce, I.C. (1981). Comment: A schema for the interactions between motor programs and sensory input. *Canadian Journal of Physiology and Pharmacology,* **59**, 691-699.

A Neuromuscular Approach in Post-Immobilization Clinical Evaluation

L.E. Tremblay, P.P. Lagasse, and M.C. Normand
Ecole de Réadaptation et Laboratoire Des Sciences De L'Activité Physique, Université Laval, Québec, Canada

Joint immobilization for several weeks is known to induce modifications in neuromuscular responses (Burke & Edgerton, 1975) which are difficult to assess clinically. At the present time, skin state, scar, edema, limb girth, and range of movement are all used to estimate the state of the limb after prolonged immobilization. Few kinesiological techniques are presently available to assess the functional quality of the neuromuscular system, at the time of cast removal.

There exists, however, an electrophysiological technique which is commonly used in neurological studies to assess neuromuscular disorders and which represents an interesting alternative to evaluate the neuromuscular system after immobilization. It is called the M response. The purpose of this study is to introduce a new post-immobilization clinical approach based on the electrophysiological M response for the evaluation of the neuromuscular system.

Material and Methods

Subjects

Twenty-two volunteers participated in this study. Nine healthy men ($\bar{x}$ age = 23 ± 1 years) were measured for the M response in both lower limbs. The other group was composed of 13 male patients whose average age was 27.6 ± 9 years. Their mean immobilization time was 36.1 ± 19 days following a lower limb musculoskeletal injury (fracture or severe sprain of ankle or knee). These injuries were treated by a cast which maintained the ankle at a constant angle during the immobilization period. They were all informed of the potential risks and hazards of their participation in this study and signed an informed consent form.

M Response

The M response is a muscle response evoked on the nerve by electrostimulation and recorded by EMG surface electrodes. The supramaximal stimulation recruits 100% of the available motor units (Hugon, 1973).

Apparatus

The M response was evoked in the soleus muscle by a 1 ms electrical square wave pulse (Grass Stimulator S88) applied to the posterior tibial nerve at the popliteal fossa, via a Stimulus Isolation unit (Grass SIU5A). The M response was amplified (Tektronix AM502) and recorded on an oscilloscope (Tektronix 5103N) and finally photographed (Camera Tektronix C5). The electrical stimulus was delivered by a surface electrode (stimulation electrode Siemens 155 847) placed on a moving frame with magnetic base, fixed to an iron plate close to the testing table. The reference electrode (3 × 4 cm) was placed just above the knee on the anterior surface of the thigh. A standardized technique (Hugon, 1973) was used for skin preparation and bipolar EMG surface electrode placement.

Experimental Procedure

During testing, the position of the subject remained constant (Mongia, 1972). The stimulation schedule was divided in two phases. Five stimuli were applied to the noninjured limb (control limb) and then to the injured limb. During testing, the subject was instructed to relax physiologically and mentally.

The mean interelectrode impedance was measured at 1200 Ω and never exceeded 2000 Ω. Patients showing edema in the injured leg at cast removal were not measured.

Results

Using the negative deflection of the M wave and comparing the mean of the maximal M responses in all nonimmobilized limbs in both healthy and injured patients with the mean of the maximal M responses of all immobilized limbs, the latter shows a 37.2% drop ($p < .05$). However, when the comparison is between immobilized and nonimmobilized limbs of the injured patients the drop in the M response is 22.4% ($p < .05$). These results are shown in Figure 1.

The comparison between the right and left limbs of the healthy subjects does not show any difference. The peak to peak amplitude of the M response of the immobilized limb decreased significantly ($p < .05$) by 39.1% when compared to the nonimmobilized control limb (see Figure 2).

Moreover, a significant correlation ($r = -.69$) was ($p < .01$) shown between the amplitude of the M peak to peak response and the duration of immobilization.

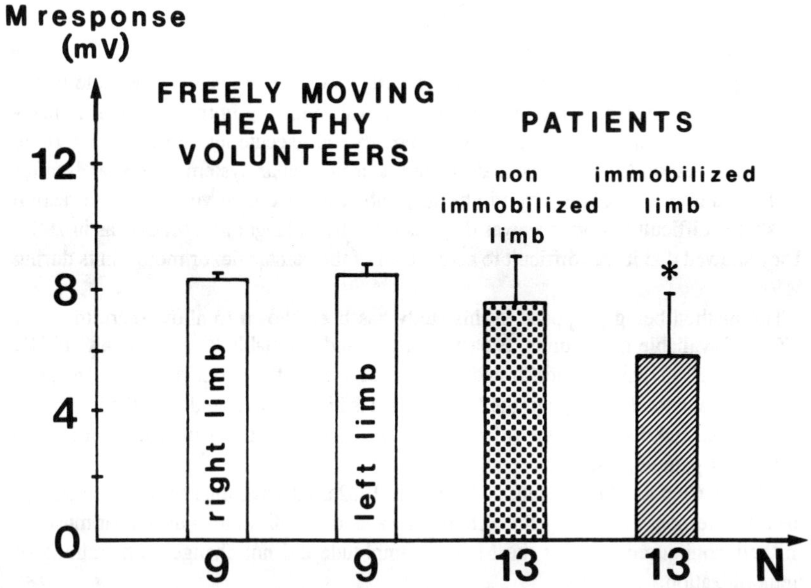

Figure 1—Mean (± S.D.) of the amplitude measured by the negative deflection wave of the M evoked response. The results of the immobilized limb are significantly ($p < .05$) different from the three other means.

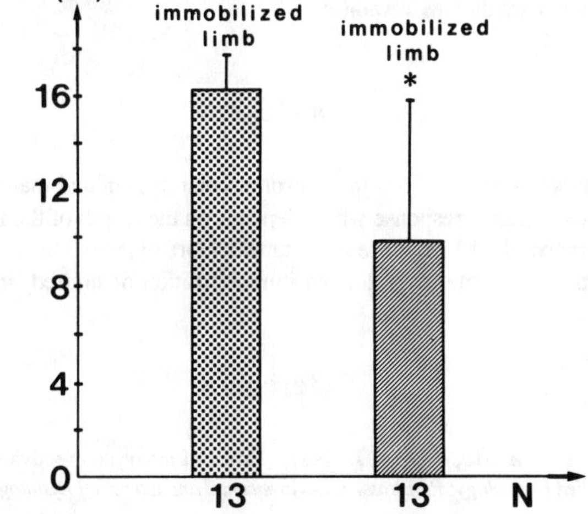

Figure 2—Means (± S.D.) of the amplitude measured by peak to peak for the M evoked response. The results of the immobilized limb are different from the nonimmobilized control limb significantly ($p < .05$).

Discussion

At the present time, few techniques are readily available to evaluate the state of the neuromuscular system at the time of cast removal following immobilization. A maximal voluntary contraction is not always possible to execute and does not constitute a reliable method to assess the state of the neuromuscular system immediately after prolonged immobilization. Even in healthy subjects a maximal voluntary contraction (MVC) is difficult to obtain as was demonstrated by Bélanger and McComas in 1981. They showed that it was difficult to activate all of the plantar flexor motor units during MVC.

The method being proposed in this study has been shown to allow recruitment of 100% of available motor units, is easy to apply, and is reliable (Gans & Kraft, 1981). Gans and Kraft have demonstrated that the M index (the product of peak to peak amplitude and total duration of the M response) is highly correlated with the integrated M response and that it can be utilized to assess the state of the neuromuscular system in various pathologies.

Sale, McComas, MacDougall, and Upton in 1982 used a maximal M wave to measure the reflex response following strength training and immobilization of the thenar muscles. In their noninjured subjects the M wave amplitude did not change with training or immobilization.

The M response represents a synchronized discharge of the motor axons which are excited by a supramaximal electrical discharge. When its value is assessed by measuring the peak to peak amplitude of the EMG signal, the M response decreases by 39.1% after prolonged immobilization. Edema or skin condition cannot explain these modifications following immobilization. Perhaps the injury is an important variable in these results. Since there is a significant correlation ($r = -.69$) between the length of the immobilization period and the M response, the length of the immobilization period tends to induce a smaller M response.

Conclusion

Prolonged immobilization affects the neuromuscular system and manifests itself by a decreased evoked motor response which depends on the length of the immobilization. It thus appears that the M response constitutes a more objective technique to evaluate motor deficits attributable to prolonged immobilization of injured limbs.

References

BÉLANGER, A.Y., & McComas, A.J. (1981). Extent of motor units activation during effort. *Journal of Applied Physiology: Respiratory, Environmental and Exercise Physiology*, **51**, 1131-1135.

BURKE, R.E., & Edgerton, V.R. (1975). Motor units properties and selective involvement in movement. In J. Keogh & R.S. Hutton (Eds.), *Exercise and Sport Sciences Reviews*, **3**, 31-81.

GANS, B.M., & Kraft, G.H. (1981). M-response quantification: A technique. *Archives of Physical Medicine and Rehabilitation*, **62**, 376-380.

HUGON, M. (1973). Methodology of the Hoffmann reflex in man. In J.E. Desmedt (Ed.), *New developments in electromyography and clinical neurophysiology* (Vol. 3, pp. 277-293). Basel: Karger.

MONGIA, S.K. (1972). H reflex from quadriceps and gastrocnemius muscle. *Electromyography,* **12**, 179-190.

SALE, D.G., McComas, A.J., MacDougall, J.D., & Upton, A.R.M. (1982). Neuromuscular adaptation in human thenar muscles following strength training and immobilization. *Journal of Applied Physiology: Respiratory, Environmental and Exercise Physiology,* **53**, 419-424.

V.
NORMAL GAIT

Ground Reaction Patterns of Normal Human Gait

S.S. Nakhla and A.I. King
Wayne State University, Detroit, Michigan, USA

Ground reaction forces are routinely measured on instrumented walkways. Several attempts have been made to use an instrumented shoe in place of the walkway force plate. Spolek and Lippert (1976) developed triaxial load cells for use under the heel and forefoot. This two-cell system was used by Harrington, Lippert, and Maritz (1978) to measure ground reaction force during running. The use of multiple shoe-borne load cells was first described by Ranu, Cheng, Denton, Levine, and King (1980), who validated their results against a walkway force plate. They were used by Cheng et al. (1980) to study the gait of below-knee amputees and by White, Winter, King, and Nakhla (1982) to demonstrate the difference in joint moments between overground and treadmill gait. Nakhla, King, and Begeman (1983) obtained ground reaction data for treadmill and overground normal gait and showed that the fore-aft toe-off force for treadmill gait was much smaller than that for overground gait. The objective of this paper is to quantify the observed differences in overground and treadmill gait and to compare the overground data with those reported previously.

Method

Six miniature load cells were used on the instrumented shoe. The cells were attached to the sole by Velcro tape which was glued to the back of each cell and over the entire surface of the sole, facilitating positioning of the cells. Each transducer was calibrated statically and dynamically with the Velcro tape in place. Sinusoidal and triangular wave shapes of up to 30 Hz were used in the dynamic calibration. Three of the six load cells were placed under the heel and three under the forefoot. There were two lateral heel cells and one medial cell. The rearmost lateral cell was placed over the site of maximum wear of the heel. The other cells were positioned under the first and fifth metatarsal heads and the big toe. Cork elevators about 20 mm thick were placed under the contralateral shoe with Velcro tape. The effect of elevating the shoe was studied by comparing ground reaction data from a walkway force plate. Six normal male subjects were asked to walk at three different speeds with and without the miniature load cells under their shoes. Each subject made 60 runs, 30 on each foot.

For the main experiment, a total of 7 normal male right-handed subjects were used. Their age and weight range were 20 to 48 years and 60 to 85 kg, respectively. Each subject was asked to walk over ground and on the treadmill at four different speeds on the same day. The left and right shoes were instrumented alternately. The walking speed was determined by the subject during overground gait. He was asked to first walk at his normal speed, followed by a run at a slower speed and a run at a faster speed. The last run was again at his "normal" speed which was in fact different from the first run. This protocol resulted in 16 runs per subject. Vertical (Z-axis) and fore-aft (X-axis) forces were monitored during each run. A portable data acquisition system was used by the subject. After each run, the data were transferred to a digital cassette tape for processing on a main frame computer. The analog signals were also recorded on tape via trailing cables as a back-up. The data were prefiltered at 250 Hz prior to digitization at 500 Hz. Over-ground walking speed was determined by two light sources placed 6.6 m apart and a photo diode mounted on the shoulder of the subject. Treadmill speed was usually set slightly lower than the equivalent overground speed and was measured by a speedometer on the treadmill.

During each run, up to eight consecutive steps of data were collected. For each step, time was normalized as a percentage of stance time and the forces were normalized as a percentage of body weight. The vertical ground reaction force was identified by three parameters, the two peaks and the intervening valley denoted by Z1, Z3, and Z2 in Figure 1. The fore-aft force was characterized by its positive and negative peaks, X1 and X2, as shown in Figure 2. The means and standard deviations were computed for each subject and for the entire data set. In Figure 1, the ensemble average of the vertical ground reaction forces is shown as a solid line. Each dotted line represents one standard deviation. The same format is used in Figure 2 for the fore-aft forces. A linear regression analysis was performed using walking speed as the independent variable and the peak forces as dependent variables. Four groups of data were analyzed. They are data from the right and left foot for overground and treadmill gait. A Chi-square test was performed to look for differences between elevated and normal shoe gait, overground and treadmill gait, and the right and left foot.

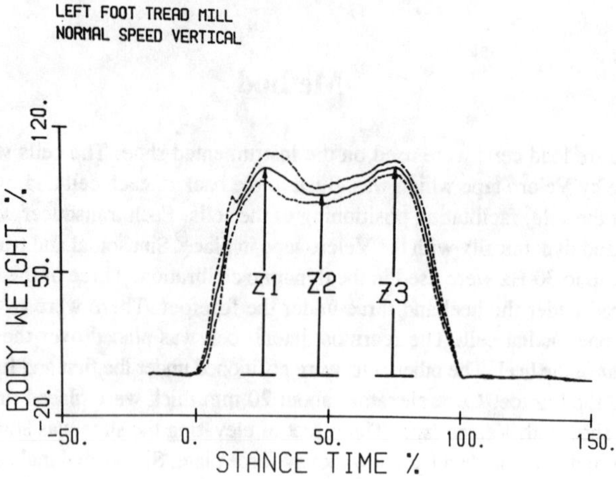

Figure 1—Vertical treadmill reaction force (average ± 1 S.D.).

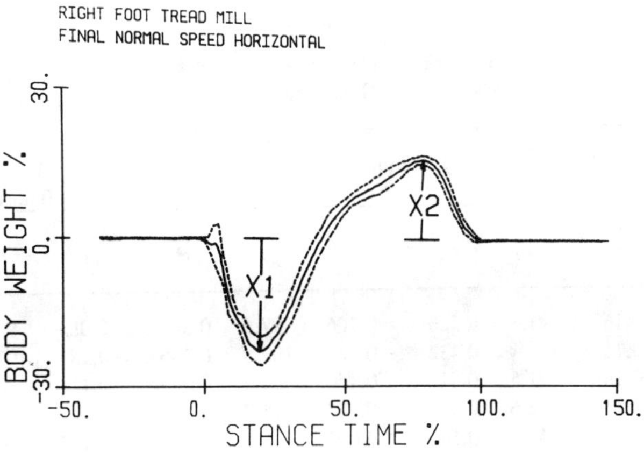

Figure 2—Fore-aft treadmill reaction force (average ± 1 S.D.).

Results

The effect of elevating the shoes by 20 mm was analyzed first. There was no significant difference between the regression lines for the two feet wearing elevated or normal shoes, based on data taken from the six subjects used in this test. The data from the two feet were combined into a single group and a comparison was made between the elevated and normal groups. There was again no significant difference between the two lines for all parameters. The detailed results are shown in Table 1.

There was no significant difference between the left and right foot for over-ground gait, based on the data from the seven subjects used for the main experiment. Thus, over-ground data from the two feet were combined into a single regression line. The regression lines for the two feet in treadmill gait were significantly different at the 99% level except for X2, which is significant at the 95% level. Treadmill and over-ground gait were significantly different for all parameters at the 99% level. This is true for the left treadmill, right treadmill, or combined treadmill data compared with combined ground data. These results are shown in Table 2.

Table 1

Peak Force-Velocity Regression Lines for Elevated Shoe and Normal Shoe Gait (y ⁵ mx ⁵ b)

Force	Normal				Elevated			
Amplitude	b	m	r^2	%	b	m	r^2	%
Z1	93.6	0.141	0.510	± 8.1	94.2	0.133	0.483	± 8.3
Z2	109.2	−0.160	0.433	± 5.3	107.1	−0.179	0.496	± 5.8
Z3	91.7	0.137	0.617	±11.6	92.3	0.126	0.582	±13.0
X1	19.3	0.091	0.561	± 7.1	16.4	0.083	0.570	± 7.7
X2	4.0	0.104	0.469	± 6.2	5.8	0.109	0.519	± 6.1

Table 2

Peak Force-Velocity Regression Lines for Treadmill and Over-Ground Gait ($y\ ^{5}\ mx\ ^{\$}\ b$)

Peak Value	Treadmill Right Foot m	b	r^2	Treadmill Left Foot m	b	r^2	Ground Both Feet m	b	r^2
Z1	0.167	91.3	0.526	0.106	97.6	0.308	0.082	96.7	0.427
Z2	−0.152	102.1	0.432	−0.126	103.8	0.336	−0.155	107.3	0.530
Z3	0.166	91.5	0.512	0.150	91.3	0.569	0.117	94.5	0.464
X1	0.151	12.6	0.473	0.115	15.7	0.412	0.095	13.9	0.500
X2	0.119	4.04	0.542	0.092	6.1	0.526	0.117	4.5	0.464

Discussion and Conclusions

Table 3 summarizes the regression data for over-ground and treadmill gait obtained by means of this instrumented shoe. It also shows data reported by Andriacchi et al. (1977). The sensitivity of these peaks to walking speed is enhanced on the treadmill, particularly Z1, X1, and X2 of the right foot. The differences between the left foot treadmill gait and over-ground gait are not consistent. If the data from the instrumented shoe are compared with those reported by Andriacchi, Ogle, and Galante (1977) for overground gait, it is seen that there is a large difference in the slopes of the regression lines. For each of the five parameters, data from the instrumented shoe indicate that their sensitivity to walking speed is not as high as that reported previously. No ready explanation of this discrepancy can be made.

The conclusions of this study are that

1. Elevation of the shoe did not seem to affect normal gait.
2. Treadmill and overground gait are significantly different.

Table 3

Comparison of Regression Line Slopes Between Data Obtained From Instrumented Shoe and From Walkway

Force Amplitude	Treadmill Right m	%	Treadmill Left m	%	Ground 7 Subjects m	%	Ground Andriacchi (77) m	%
Z1	0.167	± 6.3	0.106	± 6.7	0.082	± 7.4	0.257	± 7.8
Z2	−0.152	± 4.8	−0.126	± 4.8	−0.155	+ 5.3	−0.330	± 5.1
Z3	0.166	± 7.1	0.150	± 7.4	0.177	±10.1	0.193	±12.4
X1	0.151	±13.0	0.115	±12.6	0.095	± 7.8	0.125	± 7.5
X2	0.199	±12.4	0.095	±16.8	0.117	± 5.9	0.118	± 7.6

3. During treadmill gait, there is an observed difference between the right and left foot. This is not significant for overground gait.

References

ANDRIACCHI, T.P., Ogle, J.A., & Galante, J.O. (1977). Walking speed as a basis for normal and abnormal gait measurements. *Journal of Biomechanics,* **10**, 261-268.

CHENG, R., Ranu, H.S., Ross, R., Nakhla, S.S., Levine, R.S., & King, A.I. (1980). Gait analysis of an amputee with miniature triaxial shoe-borne load cells. *Proceedings of International Conference on Rehabilitation Engineering* (pp. 236-239).

HARRINGTON, R.M., Lippert, F.G. III, & Maritz, W.E. (1978). A shoe for measuring foot-to-ground forces while running. In D. Jaron (Ed.), *6th Annual New England Bioengineering Conference* (pp. 158-161).

NAKHLA, S.S., King, A.I., & Begeman, P.C. (1983). Ground reaction forces in treadmill and overground gait. *Proceedings of the 29th Annual Meeting of the Orthopedic Research Society* (p. 153).

RANU, H.S., Cheng, R., Denton, R.A., Levine, R.S., & King, A.I. (1980). A study of normal and abnormal human gait with miniature tri-axial show-borne load cells. *26th Annual Meeting of the Orthopedic Research Society* (p. 136).

SPOLEK, G.A., & Lippert, F.G. (1976). An instrumented shoe. A portable force measuring device. *Journal of Biomechanics,* **9**, 779-783.

WHITE, S.C., Winter, D.A., King, A.I., & Nakhla, S.S. (1982). Kinetic analysis of gait using an instrumented force-shoe. In *Human Locomotion II.* Proceedings of a Special Conference of Canadian Society of Biomechanics (pp. 62-63).

Stride Length and Cadence: Their Influence on Ground Reaction Forces During Gait

R.W. Soames and R.P.S. Richardson
King's College London,
Strand, London, England.

Ground reaction forces have been studied by many workers investigating the gait of patients with specific pathologies and following joint replacement therapy. When assessing the behavior of joint prostheses, and/or calculating joint transmission forces from these ground reaction forces, one needs to know whether any differences observed are due to the pathology or to the characteristics of the prostheses, or are merely due to changes in the velocity of walking: more specifically to changes in stride length or cadence. Paul (1970) has shown that the average joint forces, for both the hip and knee joints, increase with increasing stride length, while Stauffer, Chao, and Brewster (1977) have observed that an increase in cadence during 'free walking' in normal subjects does not significantly change the magnitude of peak compressive forces across the ankle joint. Demotazz, Mazur, Thomas, Sledge, and Simon (1979), in comparing patients who had undergone total ankle replacement (T.A.R.) therapy with normal subjects, observed differences in walking speed, stride length, cadence and single-limb stance times, as well as noticeable differences in the vertical ground reaction forces, both at heel-strike and toe-off, between the two groups.

The velocity of walking is the product of stride length and cadence, consequently an increase in velocity may be achieved by increasing either or both of these components. It might be expected that ground reaction forces would increase with increasing velocities of gait due to the greater propulsive forces associated with these higher velocities. Indeed Jansen and Jansen (1978) have observed a general increase in vertical ground reaction forces at both heel-strike and toe-off with increasing speed, together with a marked decrease in force during the midstance period. The antero-posterior forces were also observed to increase proportionally with increasing speed, but there was no relationship between speed of walking and the medio-lateral reaction forces. Although the velocity of gait may influence ground reaction forces, whether it is stride length or cadence which has the major influence is not known. This may be an important issue since many investigators are now allowing their subjects to "free walk" when observing gait and its characteristics.

Methods

Recordings of the three orthogonal ground reaction forces during gait were taken from 12 subjects (six male, six female), aged 19 to 23, who had no known or apparent disorders of their gait. Subjects walked across a multicomponent biomechanical force platform (AMTI, model OR6-3) with each of nine randomly presented patterns of gait, which were all constrained for both stride length and cadence. Stride length was standardized as either 50%, 75%, or 100% of the subject's leg length, which was taken as standing height to the greater trochanter. Both Stauffer et al. (1977) and Demottaz et al. (1979) have observed that step frequencies of free walking normal subjects vary between 35 and 75 steps/min. Consequently step frequency in the present study was constrained to either 42, 52, or 62 steps/min.

So that footwear should not modify gait all subjects wore training shoes during each trial. From each of the three ground reaction records the peak forces associated with heel-strike and toe-off were determined and expressed as a percentage of the subject's body weight. Three-way analyses of variance were conducted on these data to determine a) whether there is any difference between the sexes in the magnitude of the ground reaction forces at heel-strike and toe-off, in each of the directions of measurement, and b) whether stride length and/or cadence influence these ground reaction forces.

Results and Discussion

The vertical (Fz), and antero-posterior (Fy), and medio-lateral (Fx) patterns of the ground reaction forces and their magnitude during gait were similar to those reported by other workers (see Figure 1). However, slight differences were observed within these patterns with increasing velocity of walking. The pattern of vertical loading showed the greatest variation among subjects, particularly with respect to the magnitude of the peak forces at heel-strike and toe-off. These peak forces did, however, increase with increasing velocity, while the midstance minimum decreased (see Figure 1a). In the antero-posterior direction the posteriorly directed peak forces were generally of greatest magnitude. In most of these gaits, however, there was not a smooth, constant application of load at heel-strike, shown by a flattening of the initial peak and the presence of several smaller peaks during the initial loading of the foot (see Figure 1b). For the medio-lateral direction there was a plateau between the two medially directed peaks, while at higher velocities smaller peaks appear in this region (see Figure 1c).

No significant differences were observed between the sexes in the magnitudes of the peak forces associated with heel-strike and toe-off in any of the three directions of measurement. Thus the male and female data were combined in all subsequent analyses.

Three-way analysis of variance showed there to be significant differences ($P < 0.01$) in peak reaction forces at heel-strike between cadences in all three orthogonal directions. There were also significant differences ($P < 0.01$) at heel-strike in peak reaction forces between stride lengths in all three orthogonal directions. However, there was no interaction between cadence and stride length in the magnitude of these peak forces at heel-strike, suggesting that independent mechanisms are operating. The only significant

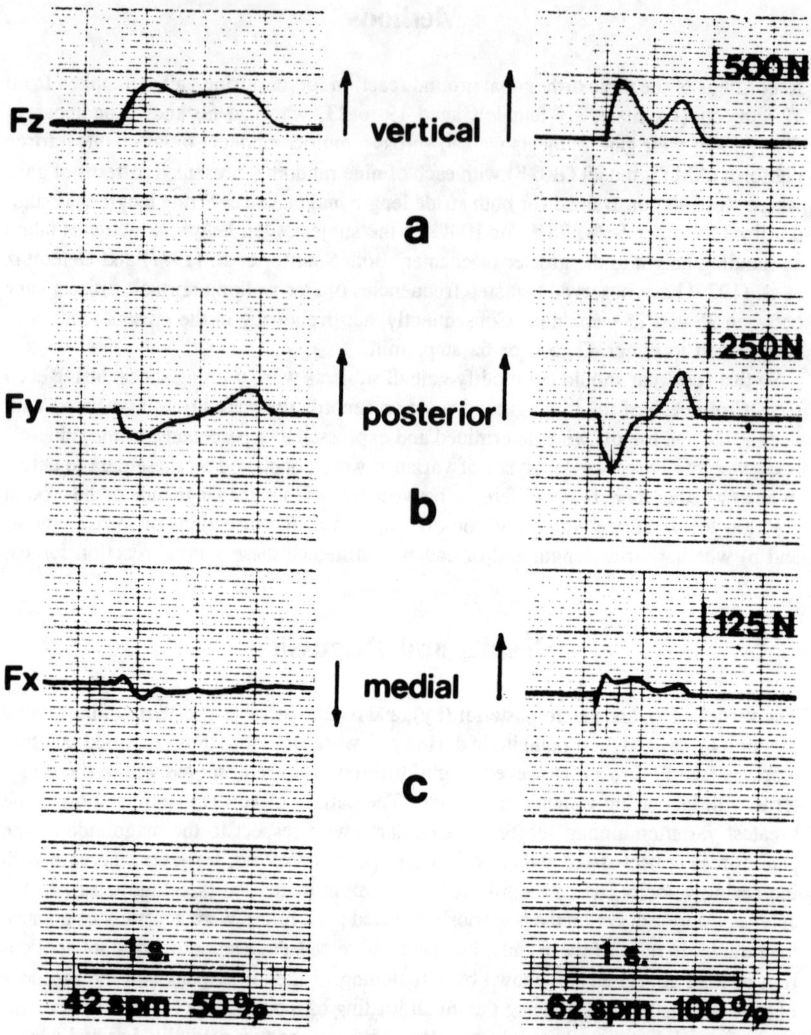

Figure 1—Examples of the three orthogonal ground reaction forces (Fx, Fy and Fz) at low (42 steps per minute, 50% leg length stride) and high (62 steps per minute, 100% leg length stride) velocities of walking.

differences in peak forces at toe-off were between cadences in the antero-posterior directions. The means and standard deviations of the peak forces at heel-strike in the medio-lateral, antero-posterior and vertical directions, together with those for toe-off in the antero-posterior direction, are presented in Figure 2.

In the medio-lateral direction peak forces associated with the high and low cadences only were significantly different. However, in the antero-posterior direction peak forces at heel-strike differed at all three cadences, while at toe-off the peak forces at the lowest cadence differed from those at the two higher cadences. In the vertical direction the

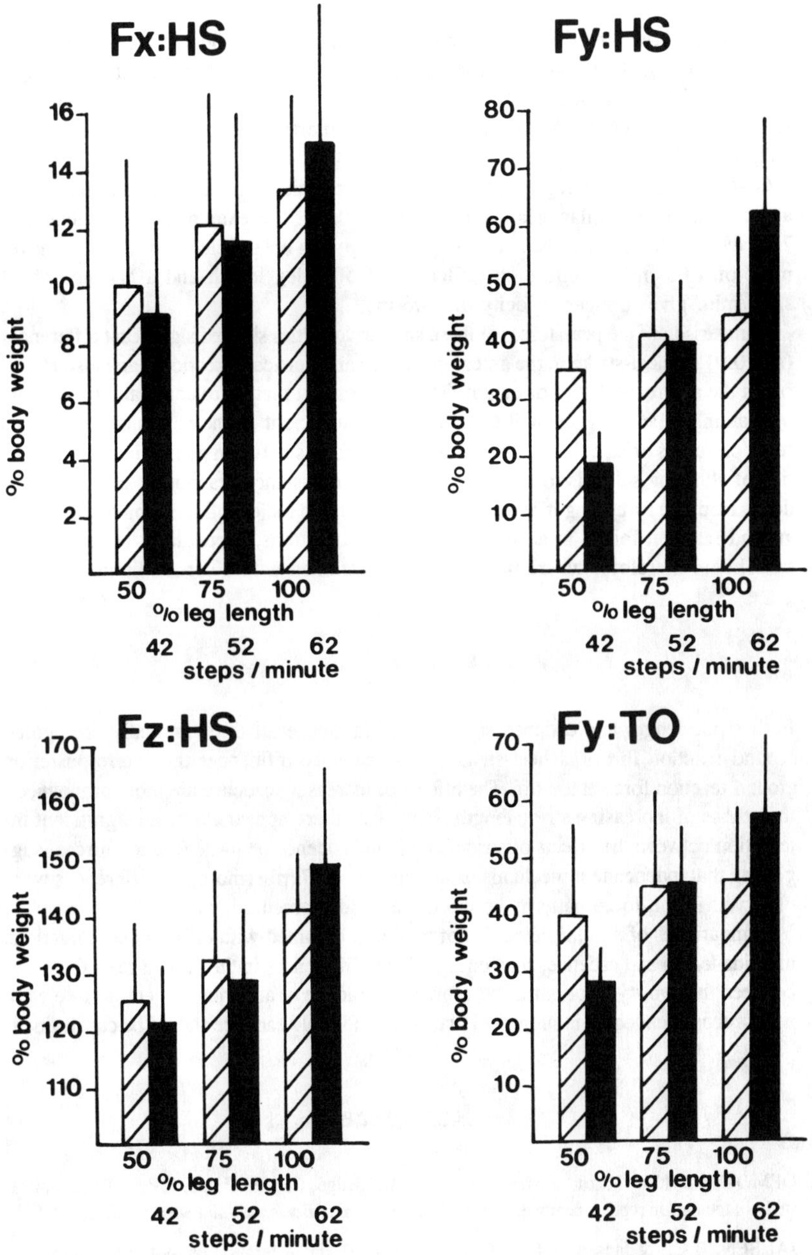

Figure 2—The relationship between ground reaction forces at heel-strike (Fx, Fy, Fz: HS) and toe-off (Fy: TO), with leg length and cadence.

peak forces associated with the highest cadence differed from those at the two lower cadences. The results of similar t-tests conducted using the stride length data revealed that, for all three directions of measurement, peak forces at heel strike differed only between the longest and shortest stride lengths. Although both cadence and stride length have a significant influence on the observed peak reaction forces, particularly at heel-strike, the effects of changes in cadence are much more pronounced than those of stride length (see Figure 2).

Given that the mean leg length was 0.90 m, certain combinations of stride length and cadence give similar mean velocities of walking. For example, a stride length of 75% leg length and a cadence of 42 steps/min. gives a mean velocity of forward movement of 1.05 ms^{-1}, while a stride length of 50% leg length and a cadence of 62 steps/min. gives a mean velocity of 1.06 ms^{-1}.

Comparison of the peak forces at these similar velocities showed significant differences ($P < 0.01$) at heel-strike in the antero-posterior and vertical directions, and also at toe-off in the antero-posterior direction. This implies that merely to constrain the velocity of walking in an experimental program is not sufficient if one is comparing ground reaction forces or calculating joint transmission forces between populations. The practice of allowing subjects to free walk in such circumstances cannot be supported. Indeed, from the foregoing it appears that cadence has a major influence on the observed ground reaction forces in all three orthogonal directions, particularly at heel-strike, and should therefore, where possible, be constrained within certain limits.

Conclusions

Both stride length and cadence significantly influence all three mutually orthogonal ground reaction forces at heel-strike, while cadence influences the antero-posterior ground reaction force at toe-off. The effects of increasing cadence are more pronounced than those of increasing stride length. However, there appears to be no significant interaction between the effects of stride length and cadence on these reaction forces, suggesting that independent mechanisms are operating. Furthermore, no differences were observed in the force patterns between men and women.

Comparisons of similar velocities of walking, obtained with different combinations of stride length and cadence, showed significant differences in the ground reaction forces between the two styles of gait. Therefore, the practice of allowing subjects to free walk is not recommended. It is suggested that where possible cadence should be constrained.

References

DEMOTTAZ, J.D., Mazur, J.M., Thomas, W.H., Sledge, C.B. & Simon, S.R. (1979). Clinical study of total ankle replacement with gait analysis. *Journal of Bone and Joint Surgery*, **61A,** 976-988.

JANSEN, E.C., & Jansen, K.F. (1978). Vis-velocitas-via: Alteration of foot-to-ground forces during increasing speed of gait. In E. Asmussen & K. Jorgensen (Eds.), *Biomechanics VI-A* (pp. 267-271). Baltimore: University Park Press.

PAUL, J.P. (1970). The effect of walking speed on the force actions transmitted at the hip and knee joints. *Proceedings of the Royal Society of Medicine*, **63**, 200-202.

STAUFFER, R.N., Chao, E.Y.S., & Brewster, R.C. (1977). Force and motion analysis of the normal, diseased, and prosthetic ankle joint. *Clinical Orthopaedics and Related Research*, **127**, 189-196.

Congruity of Normal Gait Patterns Over a Three-Year Period

J.S. McIlwain and R.K. Jensen
Laurentian University, Sudbury, Ontario, Canada

A structural pattern recognition procedure has recently been applied to quantitatively assess the congruity of various lower limb segmental gait patterns for treadmill walking (Whiting & Zernicke, 1982). An estimate of congruity or similarity in shape between any two XY patterns was obtained by chain encoding each pattern and then determining the cross-relation function from the two generated chains. The peak value of the recognition coefficient (R) served as the criteria measure for intercurve comparisons. This coefficient could vary from -1.0 to 1.0, with a value of 1.0 indicating perfect congruity and a value of 0.0 indicating absence of congruity between patterns. The present investigation reports comparative and additional R values for lower limb movement patterns for normal speed, level floor walking at normal speed.

Methods

Angular kinematic data of the right lower limb segments were obtained from 80 frames/s, 16 mm Locam film records. Two complete normal speed walking cycles were recorded each year for each subject (n =3) over a period of 3 years, providing 18 trials of data for each correlation analysis. The normal male subjects were initially 6, 9, and 12 years of age.

Digitized raw displacement film data were filtered using a low pass Butterworth filter at 5.0 Hz cutoff and differentiated using finite difference equations to obtain estimates of angular velocity. Angular displacement conjoint pairs chosen for individual cross correlation analysis included: (a) Shank Absolute vs. Time, (b) Thigh Absolute vs. Time, (c) Knee Relative vs. Time, (d) Hip Relative vs. Time, (e) Knee Relative vs. Thigh Absolute, and (f) Knee Relative vs. Hip Relative (e.g., see Figures 1 and 2). Six similar angular velocity pairs were also investigated (e.g., see Figures 3 and 4).

Each pattern of a given conjoint pair was chain encoded to a curve of 25 chain elements according to the grid intersect quantization procedures outlined by Freeman (1961). In brief, a sequence of elements was constructed to approximate each curve by superimposing a grid overlay onto the square aspect ratio XY curve and connecting successive

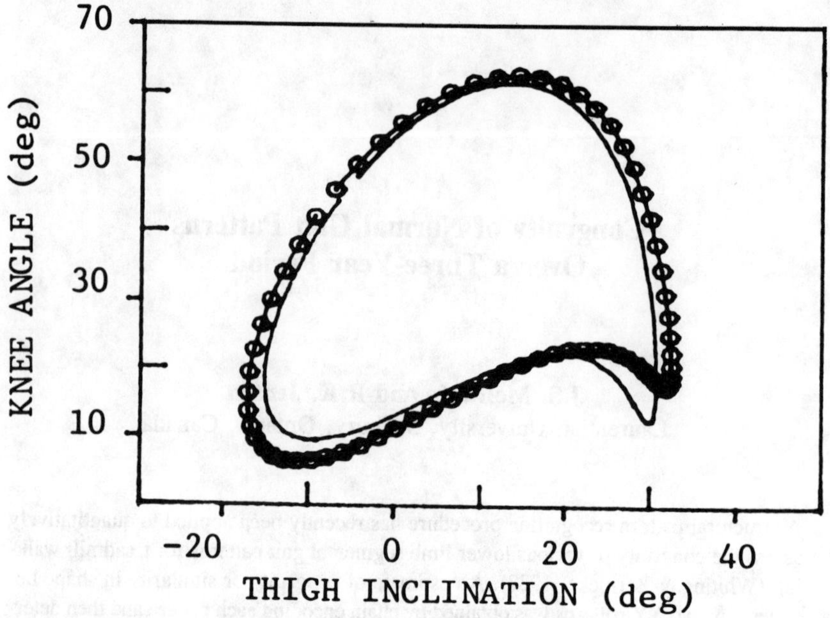

Figure 1—Exemplar angle-angle curves of normal level floor walking of two consecutive trials (Subject 1, Year 1, Trial 1 [——] and Trial 2 [0——0]. Recognition coefficient between curves is *R* = .83.

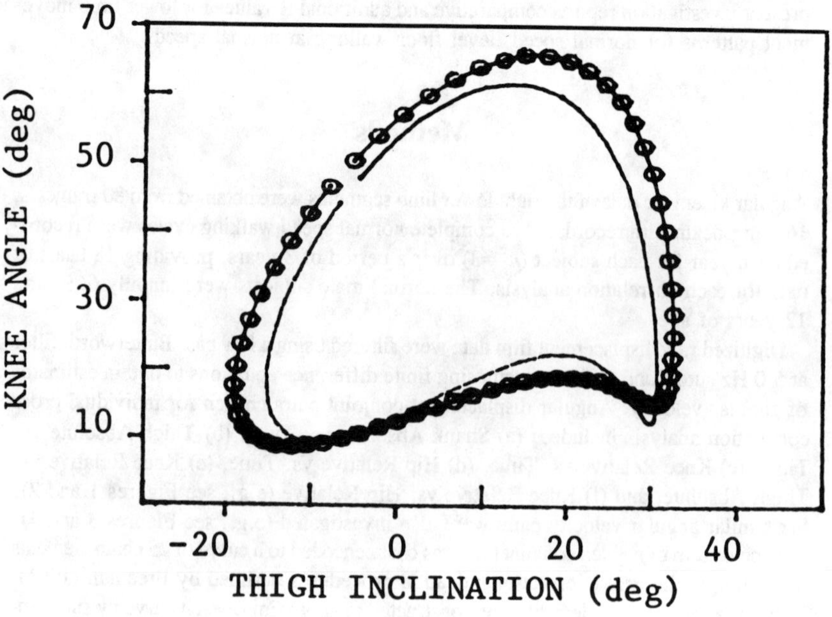

Figure 2—Exemplar angle-angle curves of normal level floor walking (Subject 1, Year 1, Trial 1 [——] and Year 3, Trial 1 [0——0]). Recognition coefficient between curves is *R* = .80.

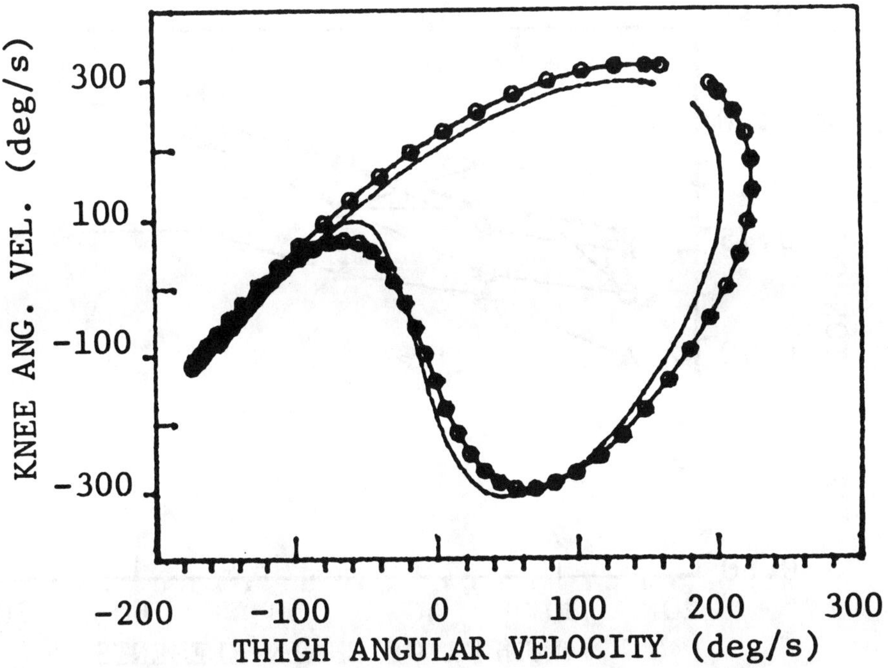

Figure 3—Exemplar angular velocity-velocity curves of normal level floor walking of two successive trials (Subject 1, Year 1, Trial 1 [——], and Trial 2 [0——0], *R* = .77).

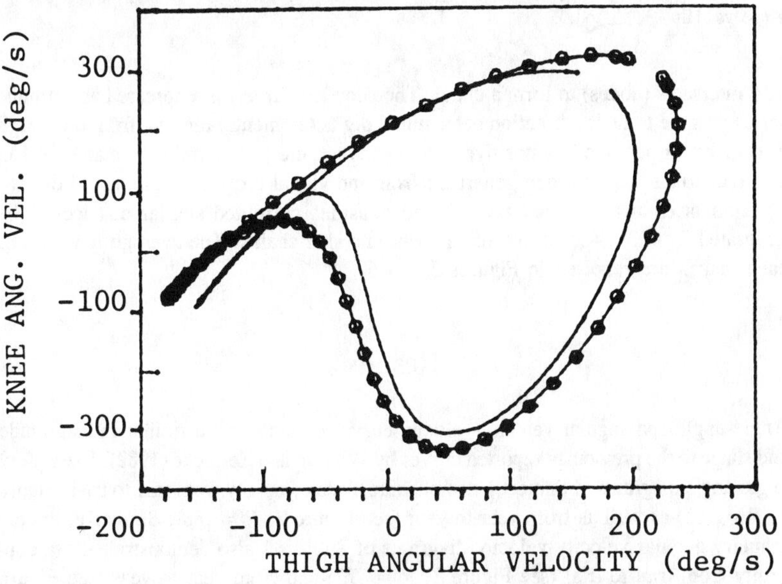

Figure 4—Exemplar angular velocity-velocity curves of normal level floor walking (Subject 1, Year 1, Trial 1 [——] and Year 3, Trial 1 [0——0], *R* = .83).

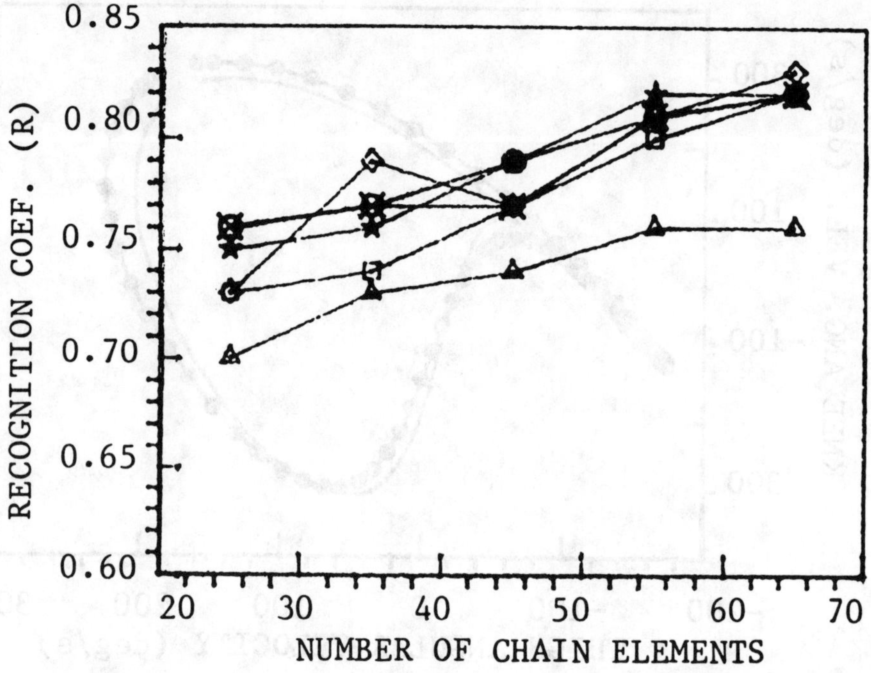

Figure 5—Average recognition coefficients of lower limb angular displacement patterns of normal level floor gait. Angular kinematic patterns include Shank vs. Time [0——0], Thigh vs. time [x——x], Knee vs. Time [Δ——Δ], Hip vs. Time [□——□], Knee vs. Thigh [*——*], and Knee vs. Hip [◇——◇].

grid intersects (nodes) to form a chain. The encoded curve was expressed as a unique series of node to node, direction determined digits (elements) ranging from 0 through 7 inclusive (x positive 1, y positive 2, x negative 4, etc.). A correlation matrix of 153 non-redundant R's was then generated from individual cross correlations on the 18, 25 element chains. Manipulation of grid densities permitted similar matrices to be generated from 35, 45, 55 and 65 element encoded chains. The average R values of each matrix are reported in Figures 5 and 6.

Results

Angle-angle and angular velocity-velocity curves obtained were similar in magnitude and shape to the previously reported curves by Whiting and Zernicke (1982). Exemplary angle-angle diagrams of Subject 1 demonstrate close congruity from trial to trial (Figure 1, R = .83) as well as from year to year (see Figure 2, R = .80). Similarly, the exemplary angular velocity-velocity diagrams of Subject 1 also demonstrate close congruity from trial to trial (see Figure 3, R = .77) and from year to year (see Figure 4, R = .83).

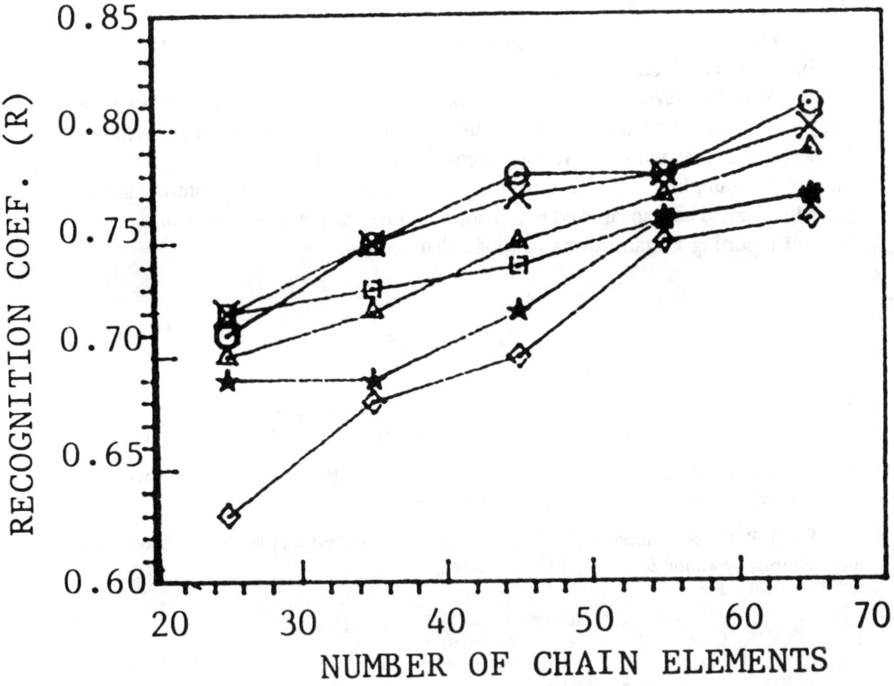

Figure 6—Average recognition coefficients of lower limb angular velocity patterns. The legend details are similar to those in Figure 5.

The average R values for each correlation matrix are summarized in Figure 5 (displacement data) and Figure 6 (velocity data). Overall displacement coefficients (R = .78) were slightly greater than velocity coefficients (R = .74) and demonstrated less spread between the type of conjoint pair investigated. Coefficient magnitudes tended to increase as the number of chain elements increased for displacement and velocity data.

The average intersubject values for Knee Relative vs. Thigh Absolute displacement (R = .78) and velocity (R = .72) patterns were similar to the comparative treadmill values (R =.80, R =.66) of Whiting and Zernicke (1982). Overall average within subject displacement and velocity coefficients (R = .79 and R = .75) were also similar to comparative average between subject coefficients (R = .77, R = .74). Furthermore, mature gait patterns were indicated by stable year-to-year-within-subject coefficients (Sutherland, Olshen, Cooper, & Woo, 1980).

Conclusions

Based upon the limited sample and kinematic patterns investigated, the results indicate that the same level of congruity exists for treadmill and level floor walking patterns at normal speed. This degree of congruity, however, appears to be dependent on the type of kinematic pattern investigated and the encoding resolution used. Furthermore, for a given kinematic pattern, the degree of congruity remains relatively constant within and between individuals over time.

Along with the previous data, the results also indicate the chain encoding cross-relation analysis method to be a useful procedure for certain comparative and longitudinal biomechanics investigations. Although sensitive to certain data processing procedures such as data sampling rate, digital filtering, and method of encoding interpolation, the procedure should be standardized to provide a useful quantitative tool for storing, analyzing and reporting certain forms of biomechanical data.

References

FREEMAN, H.A. (1961). A technique for classification and recognition of geometric patterns. *Proceedings of the 3rd International Congress of Cybernetics.* (pp. 348-369). Namur, Belgium.

SUTHERLAND, D.H., Olshen, R., Cooper, L., & Woo, S. (1980). The development of mature gait. *Journal of Bone and Joint Surgery,* **62A,** 336-353.

WHITING, W.C., & Zernicke, R.F., (1982). Correlation of movement patterns via pattern recognition. *Journal of Motor Behavior,* **14(2),** 135-142.

Dynamic Changes of the Arches of the Foot During Walking

S.M. Yang, J. Kayamo, T. Norimatsu, M. Fujita, N. Matsusaka, and R. Suzuki
Nagasaki University School of Medicine, Nagasaki, Japan

H. Okumura
Nagasaki University, Nagasaki, Japan

When observing the use of the arch support in cases of flat foot and calcaneous fracture, and so forth, clinically, we understand that the arches of the foot are very important for a human being for the activities of standing and walking.

The purpose of this paper is to analyze the dynamic changes of the arches of the foot during walking, especially for the purpose of clinical evaluation of foot diseases.

Materials and Methods

As shown in Figure 1, we recorded the ground reaction force (G.R.F.), electric arch gauge (E.A.G.), metatarsophalangeal joint (MP—joint), electromyogram (E.M.G.) and optical stick picture camera (S.P.C.) simultaneously.

For the MP-joint and E.A.G. we made an original electrogoniometer which consisted of a piece of thin elastic steel on which a strain gauge was fixed. To cancel The natural vibration of the steel, a 2-cm-round axis was formed especially for the medial arch. The E.M.G., flexor hallucis longus (F.H.L.), peroneous longus (P.L.), tibialis anterior (T.A.) and gastrocnemius were examined by surface electrodes.

Five normal adult males aged 25 to 35 were used, and asked to walk 20 times each on a floor on which Matake's Force Plate was fixed. In addition, the optical S.P.C. was used for confirming changes in the pattern of the medial arch.

Before measuring the length and height of the anterior arch, the medial arch and the MP-joint, the instruments were calibrated and their linearity confirmed. All of the data were calculated and displayed by a HITACHI-PC-8100 computer.

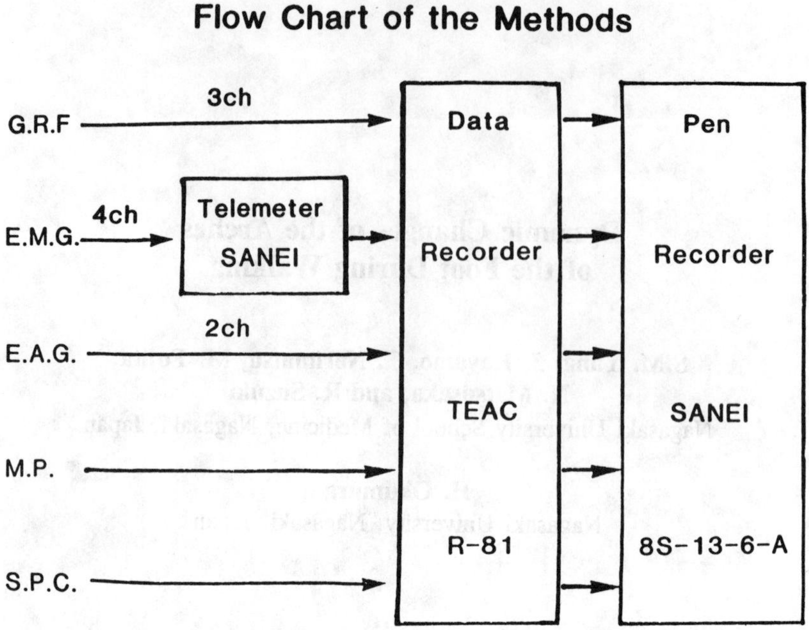

Figure 1—Ground reaction force (G.R.F.), E.M.G., electric arch gauge (E.A.G.), metatarso-pharyngeal joint (MP-jont) and stick picture camea (S.P.C.) were recorded simultaneously.

Results

First, the mean pattern of the anterior arch (see Figure 2) showed a one-phase pattern with a small notch in the area of 5%, a large notch in the area of 95 to 100%, and a flattening in the area of 40 to 50% when the whole stance phase walking cycle equaled 100%. Since the positive value of the Y-axis represents the length of the arch, the shortest length (and greatest height) of the anterior arch will occur in the area of 95 to 100% and the longest and most shallow arch will be in the area of 40 to 50% of the stance phase. The mobility of the anterior arch will be about 16 mm during the whole stance phase.

Second, the mean pattern of the medial arch (see Figure 3) shows a two-phase pattern which is different from the mean pattern of the anterior arch, but with the two same special notches in the areas of 5% and 95%, and a flattening in the area of 30%. It seems that, therefore, the shortest length of the medial arch will be also in the area of 95% of the stance phase, but the longest length will be in the area of 30%. The mobility of the medial arch will be about 4 mm during the whole stance phase.

Third, the mean pattern of the MP-joint (see Figure 4) shows the two same notches in the areas of 5% and 95% of the stance phase, which indicates that most dorsiflexion of the MP-joint will be in the area of 95% of the stance phase while the angular change of the MP-joint will be at 40 to 50 degrees.

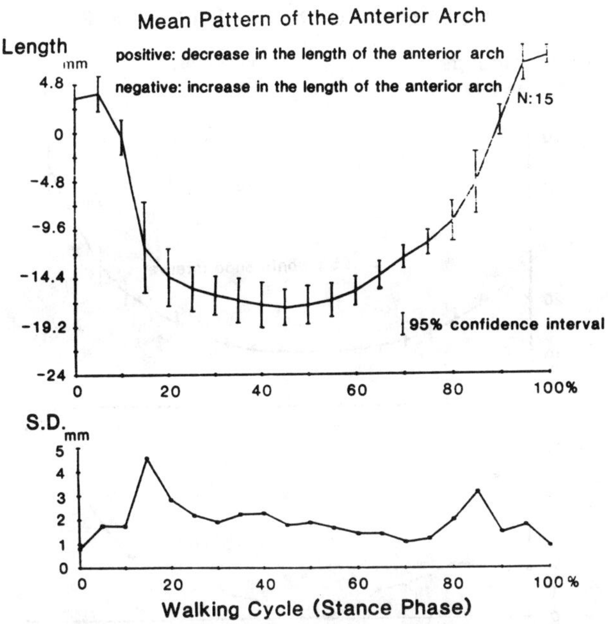

Figure 2—Mean pattern of the anterior arch, which is one phase, showing two notches in the areas of 5% and 95 to 100% of the stance phase.

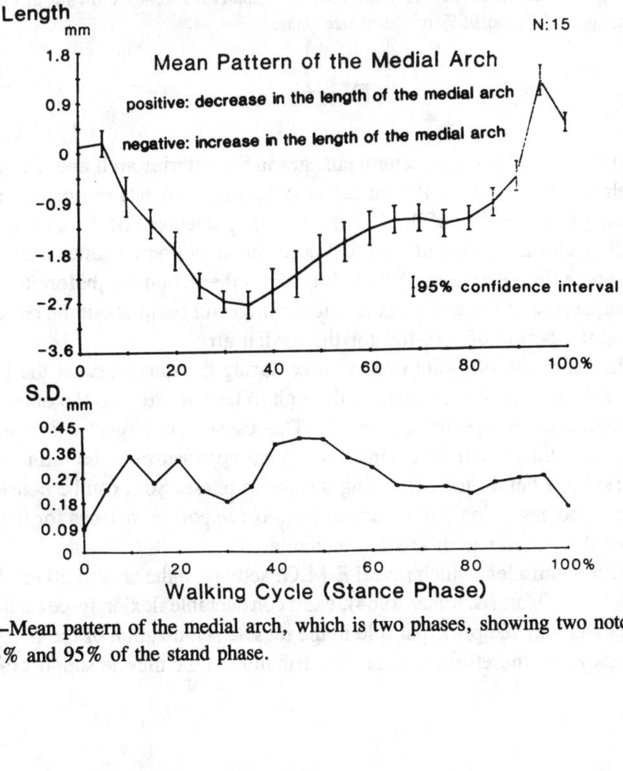

Figure 3—Mean pattern of the medial arch, which is two phases, showing two notches in the areas of 5% and 95% of the stand phase.

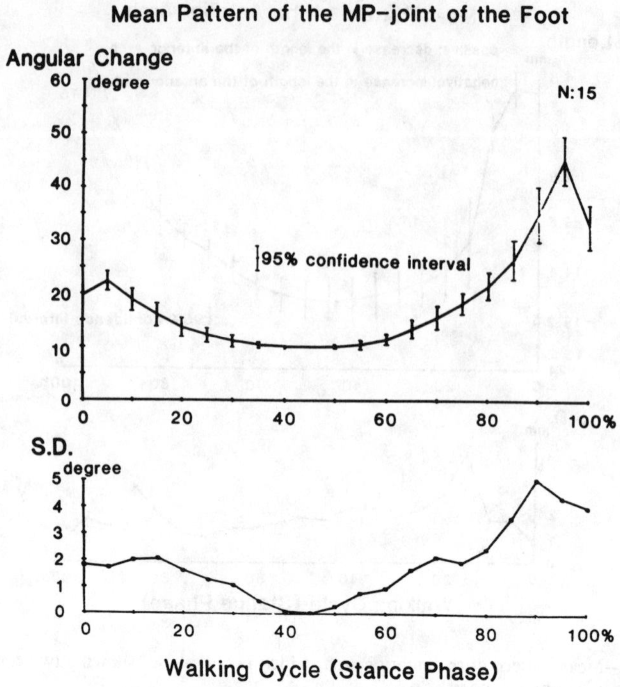

Figure 4—Mean pattern of the MP-joint with the same two notches showing in Figures 2 and 3 in the areas of 5% and 95% of the stance phase.

Discussion

The relationships between the length changes in the anterior arch and the medial arch and angular charges in the MP-joint can be observed from data presented in Figure 5.

According to the period of the G.R.F. and the percentage of the stance phase, the first small notch (in the area of 5%) will occur in the period just after heel-strike, and the large notch (in the area of 95%) will occur in the period just before toe-off. Then the greatest degree of flattening of the anterior arch will occur about the period of heel-off, and at the period of foot-flat for the medial arch.

Dorsiflexion of the MP-joint occurs twice during the stance phase: the first begins just before the heel-strike, continuing through to heel-strike, and the other starts just after heel-off continuing through toe-off. This causes the largest windlass action of the soft tissue—the plantar ligaments, the plantar apponeurosis and retinacula cutis—in the period just before toe-off during the stance phase. So it can be considered that the largest windlass action will be one of the most important reasons for the shortness (height) of the anterior arch and the medial arch.

The intrinsic muscles, which reveal E.M.G. activity in the area of 60 to 95% during the stance phase (Mann & Inman 1964), exert considerable flexion force on the forepart of the foot and play the principal role in the muscle stabilization of the transverse tarsal joint; they are, therefore, the main contributors to the muscle support of the arch.

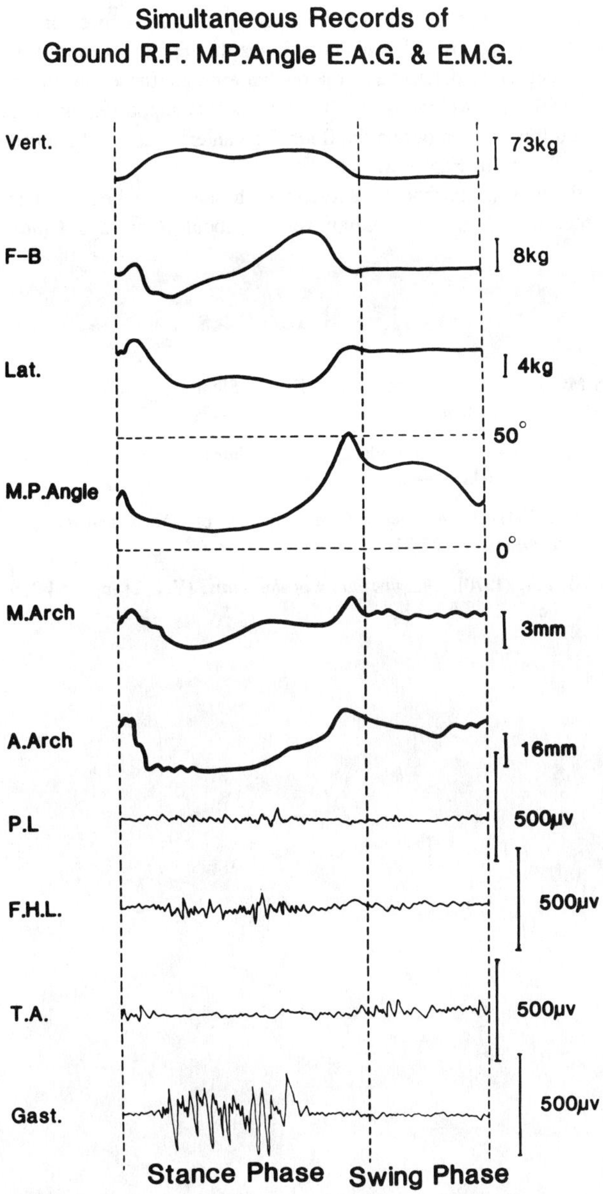

Figure 5—Simultaneous records of G.R.F., MP-joint, E.A.G., and E.M.G. showing the relationships between them.

Thus, it can be stated that there is also a very close relationships between the activity of the intrinsic muscles and the arches: that is, the activity of these muscles make the arches shorter. In other words, the ligaments and muscles act as the tightners of the arches.

In the period of foot-flat during the stance phase, the entire surface of the sole of the foot rests on the ground. This constitutes the maximum foot print, which causes the highest degree of flattening of the medial arch confirmed in Figure 3. The period just after heel-off is the moment when the body is supported on the anterior part of the foot and it is splayed out on the floor. The anterior arch is flattened in turn, which can be confirmed in Figure 2.

The entire movements from the shortest, highest arch, to the longest and most shallow of the anterior arch and the medial arch are about 16mm and 4 mm.

References

BOJSEN-MOLLER, F., & Lamoreux, L. (1979). Significance of free dorsiflexion of the toes in walking. *Acta Orthopedica Scandinavia,* **50,** 471-479.

MANN, R., & Inman, T. (1964). Phase activity of intrinsic muscles of the foot. *Journal of Bone and Joint Surgery,* **46A,** 469-481.

HICKS, J.H. (1954). The mechanics of the foot II: The plantar aponeurosis and the arch. *Journal of Anatomy (London),* **99,** 25-31.

KAPANDJI, I.A. (1970). *The physiology of the joints.* (Vol. 2) (pp. 154-219). Edinburgh: E & S Livingston.

A Speed-Related Kinematic Analysis of Overground and Treadmill Walking

C.L. Taves, J. Charteris, and J.C. Wall
University of Guelph, Ontario, Canada

In the past, the convenience of the treadmill resulted in its widespread use in research with several investigators considering treadmill locomotion to be biomechanically equivalent to overground locomotion (Astrand & Rodahl, 1970; Durnin & Passmore, 1967; Margaria, 1976). More recently kinematic and energy cost comparisons of treadmill and overground running have resulted in a controversy regarding the similarity of the two conditions (Elliott & Blanksby, 1976; Nelson, Dilman, Lagasse, & Bickett, 1972). The discussion has not been limited to running, however, and examination of studies focusing on treadmill and overground walking indicates a similar state of disagreement (Parker, 1979; Rozendal, 1966; Whittle, 1980). Locomotion studies have been plagued by a number of factors that may have had a confounding effect on the results. With allometric differences accounted force, recent studies have established speed dependent trends in several gait parameters (Grieve & Gear, 1966; Nottrodt, Charteris, & Wall, 1982). It follows, then, that additional variability between subjects linked with fixed absolute velocities may have tended to obscure the importance of differences due to condition (Brandell & Williams, 1974). During treadmill locomotion, although an infinite number of combinations of stride length and stride frequency can be chosen, after an adequate habituation period this variability is significantly reduced (Charteris & Taves, 1978; Wall & Charteris, 1981) and more closely approximates the ranges observed in overground conditions (Taves, 1982). The variation of stride length has been demonstrated to be a significant factor in energy cost (Zarrugh, 1981). These points, then, require use of a treadmill that allows a habituated subject the freedom to choose his optimal stride length; they also indicate how the use of compact treadmills and non-habituated subjects for the comparative assessment of the conditions may have greatly contributed to the continuation of the debate. It was felt a kinematic study incorporating controls on these factors, in the experimental procedures and equipment, with sufficient numbers of habituated subjects, would be more effective in examining locomotor pattern alteration due to the treadmill condition.

Method

Prior to data collection the 13 female subjects had all habituated to treadmill walking at the walking speeds to be analyzed in addition to a jogging speed, which is in agreement with the recommendations of Wall and Charteris (1981). The locomotor patterns were recorded on film at 70 frames/s. as each subject walked at randomly-ordered speeds representing 0.5, 0.9 and 1.3 stature/s for both the treadmill (Quinton 24-72) and the overground condition. Speeds were considered equivalent when they did not vary more often than 2% from the required speed over a 6-meter sample for each condition.

The analytical technology consisted of an L-W Stop Action Motion Analyzer projector model 800 set to project normally onto a VW01 Sonic Digitizer Plate (DEC) interfaced with a PDP-Lab 8E computer (DEC). The image of each subject (scale = 1/8 x actual) was projected onto a 30-cm^2 digitizer screen. Temporal distance and angular data were analyzed statistically for the three speeds over both conditions of treadmill and overground walking.

An ANOVA (with equal n's) with a *post-hoc* Tukey multiple comparison analysis (both at $p < .01$) focused initially on the more general kinematic data of distance and temporal measures between the two conditions for each of the three speeds. Since the functional significance of portions of the cycle have been identified, the occurrence and duration of selected subphase, determined by the observation of key events (e.g., heel-rise to toe-off) in terms of raw time and percentages of cycle time were examined. Statistical analysis of the angular patterns included the relative angular positions of the lower limb segments at 10% intervals throughout the stride cycle as well as at the occurrence of selected key events.

Results

The stride length and cadence parameters were found to be significantly different statistically due to the speed variable. The transition in the relative position graphical-

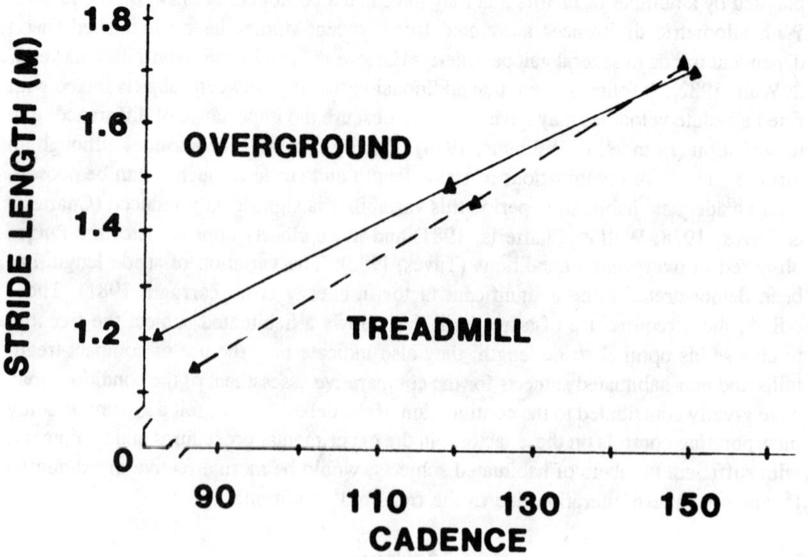

Figure 1—Stride length versus cadence relationship. This represents the group averages for the three speeds of walking overground and on the treadmill. Stride lengths are in meters and cadence is the number of steps per minute.

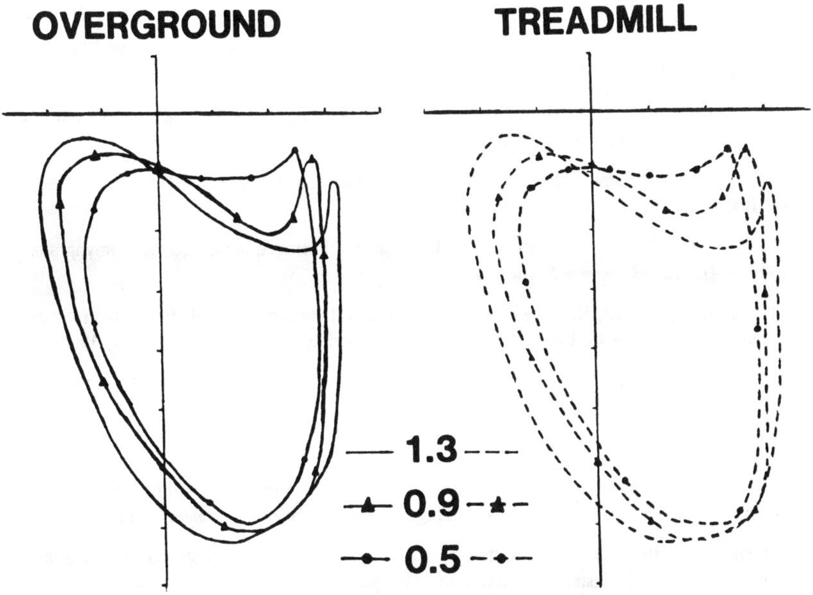

Figure 2—Thigh-knee cyclograms. Vertical axes (knee) and horizonal axes (thigh) are both marked in 10° intervals with the intercept representing 0° on both axes. Points indicate 10% of cycle.

ly of the condition at 1.3 statures/s velocity (see Figure 1), although not significant, does illustrate the same relationship reported by Nelson et.al. (1972) who found significant differences between treadmill and overground conditions at fast running speeds.

When calculated in terms of the percentage of cycle, all temporal measures except time to foot-flat had responded to speed changes (heel-strike to ball-on remained at 8% of cycle time) but none had demonstrated significant differences due to condition. Angular measures were generally observed to be significantly different due to the speed (see Figure 2) but when the effect of condition was analyzed significant differences were observed only during the double support and early and late swing phases of the walking cycle.

Conclusion

Results of this study indicate precise kinetic analysis of treadmill and overground walking can be narrowed to specifically examine double support and limited portions of the swing cycle in speed-matched efforts by thoroughly habituated subjects on suitably-sized treadmills since it is only in these portions that kinematic differences exist.

References

ASTRAND, P.O., & Rodahl, K. (1970). *Textbook of work physiology*. New York: McGraw-Hill.

BRANDELL, B.R., & Williams, K. (1974). The interrelationships of lower limb movements and muscle contractions in normal human locomotion. *5th Canadian Medical and Biological Engineering Conference.* Montreal.

CHARTERIS, J., & Taves, C. (1978). The process of habituation to treadmill walking: A kinematic analysis. *Perceptual and Motor Skills,* **47**, 659-666.

DURNIN, J.V., & Passmore, R. (1967). *Energy, work and leisure.* London: Heinemann Educational Books.

ELLIOTT, B.C., & Blanksby, B.A. (1976). Reliability of averaged integrated electromyograms. *Journal of Human Movement Studies,* **2**, 28-35.

GRIEVE, D.W., & Gear, R.J. (1966). The relationships between length of stride, step frequency, time of swing and speed of walking for children and adults. *Ergonomics,* **5**(9), 379-399.

MARGARIA, R. (1976). *Biomechanics and energetics of muscular work.* Oxford: Claredon Press.

NELSON, B.C., Dillman, C.J., Lagasse, P., & Bickett, P. (1972). Biomechanics of overground versus treadmill running. *Medicine and Science in Sports,* **4**, 233-240.

NOTTRODT, J.W., Charteris, J., & Wall, J.C. (1982). The effects of speed on pelvic oscillations in the horizontal plane during level walking. *Journal of Human Movement Studies,* **8**, 27-40.

PARKER, G.H. (1979). Locomotor and spatial differences between overground and treadmill walking. Unpublished Master's Thesis, Trent University.

ROZENDAL, R.H. (1966). Comparison between human gait on a fixed track and on a motor driven treadmill. *Acta Morphologica Neerlands—Scandinavia,* **6**(3), 319-320.

TAVES, C.L. (1982). A speed-related kinematic analysis of overground and treadmill walking. Unpublished Master's thesis, University of Guelph, Ontario.

WALL, J.C., & Charteris, J. (1981). A kinematic study of long-term habituation to treadmill walking. *Ergonomics,* **24**(7), 531-542.

WHITTLE, W.M. (1980). Kinematics of treadmill walking. *Proceedings of the Conference of Human Locomotion I* (pp. 30-31). London, Ontario: Canadian Society for Biomechanics.

ZARRUGH, M.Y. (1981). Power requirements and mechanical efficiency of treadmill walking. *Journal of Biomechanics,* **14**(3), 157-165.

Relationship Between Right and Left Legs in Human Gait, From a Viewpoint of Balance Control

N. Matsusaka, M. Fujita, A. Hamamura, T. Norimatsu, and R. Suzuki
Nagasaki University, Nagasaki, Japan

The human gait is characterized by the smoothness of the displacement of the body center of gravity along the progression path. In bipedal walking it is necessary to manage a smooth forward movement while at the same time controlling the medial-lateral balance with each leg; the leg which steps forward must control the medial-lateral oscillation of the body caused by the thrust of the posterior leg (Ducroquet, 1968). The authors previously reported that the lower leg msucles and the subtalar joint played a very important role in controlling medial-lateral balance in walking.

Does a dominant side exist in the leg just as a dominant arm exists? What function does a ''dominant leg'' have if it is existent? The purpose of this paper is to examine the question of which leg is more dominant in performing a smooth forward movement and controlling the medial-lateral balance.

Subjects and Methods

Thirty-three normal adults, whose ages ranged from 22 to 37 years, were enlisted for this study. Twenty-six of the subjects were male and seven were female. Twenty-eight were right-handed and five were left-handed.

The subjects were asked to walk 15 to 30 trials with their customary gait on a walkway, in the center of which 2 large sized force plates (40 cm × 250 cm, made by Amima Co.) were embedded side by side for each leg. The three components (vertical, forward-backward, and lateral) of the ground reaction forces of three to five serial steps were recorded for each trial using the force plates.

The maximum amplitudes of the lateral component in the phase of deceleration and that of acceleration were termed FL1 and FL2 respectively (see Figure 1). The relationship between FL1 and FL2 of the same side, and between FL2 of the right (left) step and the next FL1 of the left (right) step were investigated.

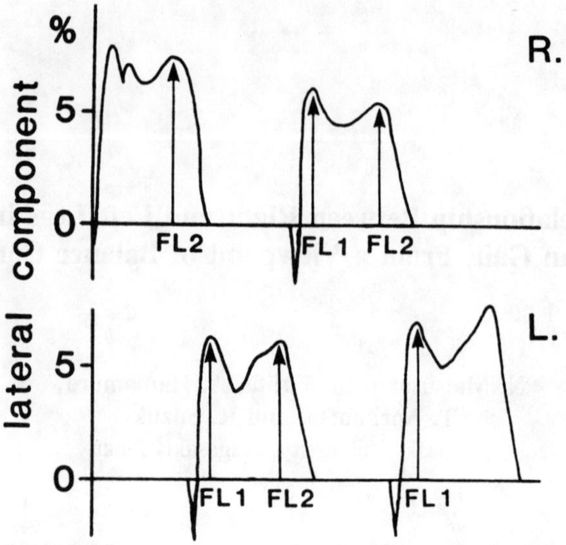

Figure 1—Points of measurement in lateral component of ground reaction force.

Results

A *t*-test was applied to the correlation coefficients. FL1 was directly proportional to FL2 on each side in all subjects. In 23 subjects (70%) FL2 of the left step was directly proportional to the next FL1 of the right step (see Figure 2A). In 5 subjects (15%) there was a positive correlation between FL2 of the right and the next FL1 of the left (see Figure 2B). In the others (15%) there was no significance between FL2 of the left and the next FL1 of the right, nor between FL2 of the right and the next FL1 of the left (see Table 1).

In 22 of 29 right-handed subjects, there was a positive correlation between FL2 of the left and the next FL1 of the right, but in 1 subject FL2 of the right was directly proportional to the next FL1 of the left, in 6 subjects there was no significance between FL2 and the next FL1 of both steps (see Table 2).

In 4 of 5 left-handed subjects there was a positive correlation between FL2 of the right and the next FL1 of the left, and in 1 subject FL2 of the left was directly proportional to the next FL1 of the right (see Table 2).

Discussion

In human gait (bipedal walking) the center of gravity is displaced to the supporting side, and the leg which steps forward must control the medial-lateral oscillation of the body caused by the thrust of the posterior leg. In this study FL2 (lateral braking force) of the left step was directly proportional to the next FL1 (lateral thrust) of the right step, and there was no significant difference in the FL2 of the right step and the next

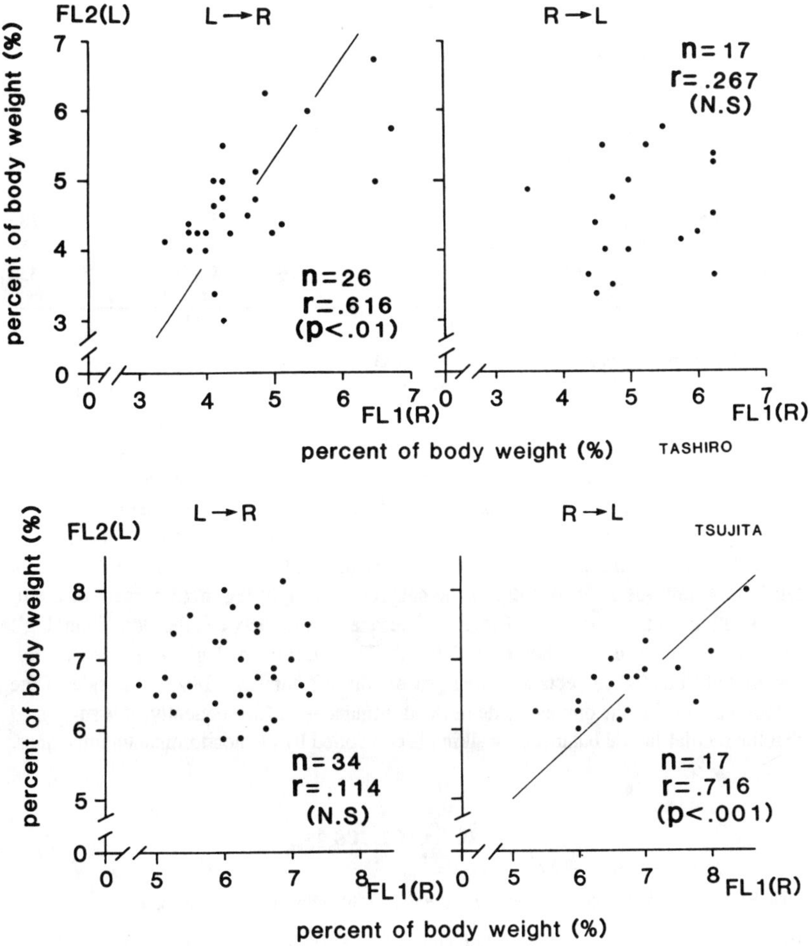

Figure 2—Relationship between FL2 of the left (right) step and the next FL1 of the right (left) step. A: right-handed male. B: left-handed male.

Table 1

Results of t-Test When Applied to Correlation Coefficients

L → R : p < .05 R → L : N.S. 23 (69.8%)	L → R : N.S. R → L : p < .05 5 (15.1 %)	L → R : N.S. R → L : N.S. 5 (15.1%)

FL1 of the left step in 70 percent of the subjects. The lateral braking force of the right step, in other words, is influenced by the lateral thrust of the left step and the lateral braking force of the left step is independent of the lateral thrust of the right step. Based

Table 2

Relationship Between Dominant Arm and "Dominant Leg"

	"Dominant Leg"			
	Left	Right	Left & Right	Total
Right-handed	22	1	5	28
Left-handed	1	4	0	5
Total	23	5	5	33

on these findings, it is suggested that the medial-lateral balance in walking is controlled by the left leg in 70 percent of the subjects, in 15 percent the right leg controls the balance, and in the remaining 15 percent both legs control the balance.

The concept of a dominant arm is characterized by skill movement but that of "dominant leg" is under discussion. Hirasawa (1979) suggested that "dominant leg" was the left leg which has a stabilizing function from the viewpoint of stasiology. In the human gait the function of the "dominant leg" is regarded as controlling the medial-lateral balance. In this study, therefore, it is reasonable to presume that the "dominant leg" is the left one in 70 percent of the subjects, the right leg in 15 percent of them, and both legs in the others (see Table 1). The relationship between the "dominant leg" and the dominant arm is shown in Table 2. The "dominant leg" is mainly the left one in right-handed subjects and the right one in left-handed subjects; the side of the "dominant leg" is the opposite side of the dominant arm. Consequently, it is suggested that the medial-lateral balance in walking is controlled by the nondominant hemisphere.

References

DUCROQUET, R. (1968). *Walking and limping*. Philadelphia: Lippincott Co.

HIRASAWA, Y. (1979). An observation on standing ability of Japanese males and females. *J. Anthrop. Soc. Nippon*, **87**, 81-92.

Quantitative Evaluation of Cocontraction of Knee and Ankle Muscles in Normal Walking

S.J. Olney
Queen's University, Kingston, Ontario, Canada

Cocontraction is defined as the simultaneous activity of agonist and antagonist muscles crossing the same joint and acting in the same plane. Although abnormally high levels of cocontraction represent wasted force generated by the muscles, some cocontraction is seen as part of normal movement patterns. It has been suggested that cocontraction varies in normal movement with the stage of development (Gatev, 1972) and skill level (Payton & Kelley, 1972) of the individual and with the levels of contraction of the muscle groups (Patton & Mortensen, 1971). Quantification of cocontraction would be useful in studying these factors in normal movements, in assessing the importance of cocontraction in abnormal movements and in recommending treatment in pathological conditions. Few attempts have been made to quantify cocontraction (Falconer, 1978) and these have relied on monitored EMG activity alone without considering its relationship to moment production. The purpose of this study was to quantify cocontraction, in terms of moment production, that was present at the knee and ankle during normal walking.

Method

All data were collected at the Gait Laboratory of the University of Waterloo and processed at the University of Waterloo and Queen's University. Four normal subjects participated in calibration and walking trials. For the knee, two gait cycles for each of 3 subjects were analyzed; for the ankle, two gait cycles for each of 4 subjects were included.

The full model (Model 2) used for prediction of moments for each muscle group during walking was: $M(t) = K_1E_1(t)[1 + K_2(\theta(t) - \theta_1) - K_3\omega(t)]$ when $M(t)$ was the instantaneous moment; $E_1(t)$ was the instantaneous amplitude of EMG from the representative muscle, processed to yield optimal correlation between it and force; θ_1 was the joint angle at which isometric calibration contractions were conducted to derive the moment-EMG relationship; $\omega(t)$ was the instantaneous angular velocity; K_1 was an experimentally derived constant relating EMG to moment, while K_2 and K_3 were constants iterated to yield minimal root mean square errors between predicted resultant moments about the joints and those obtained from standard link segment analyses (Bresler & Frankel, 1950) of cine film and force plate data as described by Winter (1980). In Model 1, K_2 and K_3 were set to zero. In summary, Model 1 predicted moments from EMG using only the static moment-EMG relationship and the velocity effects which were allowed to occur with a fixed tendon-to-tendon length, whereas Model 2 predicted moments from EMG including also the effects of changes in muscle length (as joint angle) and contraction velocity (as joint angular velocity).

Calibration contractions consisted of cyclical isometric contractions performed in bursts closely approximating the amplitude and duration of that seen in the subject's gait. EMG and force signals were sampled, digitally stored and used to determine the cut-off frequency of the low pass filter for EMG processing and the K_1 constant, both of which were needed to use the model with the EMG derived from walking trials.

All positive and negative moments predicted from the full model were digitally stored for each trial. The gait cycles were divided into segments: stance, mid-stance, heel-strike transition and toe-off transition periods (see Figure 1). The instantaneous cocontraction value was the lower of the two opposing moments. The index of cocontraction, I, was defined as $I = 2C/(P + N)$ (100%) where C was the integral of cocontraction over the time period, and P and N were, respectively, the time integrals of the positive and negative moments (Falconer, 1978). Mean positive, negative, and cocontraction moments for each segment were expressed as a percentage of the maximum moment of either polarity occurring during the gait cycle. Integrals and cocontraction indices were calculated for each segment.

Results

Examples of use of the predictive model are shown in Figure 2. In each case the basic model (Model 1) demonstrates the contrast between predictions neglecting length and velocity factors ($K_2 = 0$, $K_3 = 0$), that is, the errors that are made if processed EMG

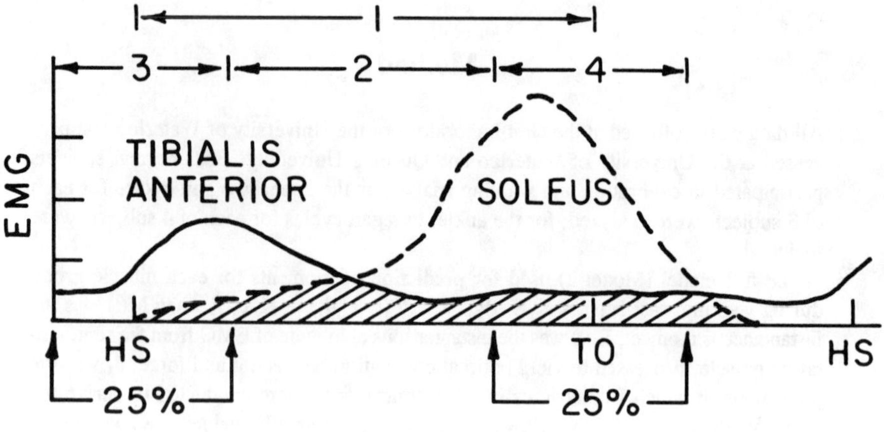

Figure 1—Segments of gait cycle: 1) stance, 2) mid-stance, 3) heel-strike transition, 4) toe-off transition. Cocontraction area is shown hatched. HS = heel-strike; TO = toe-off.

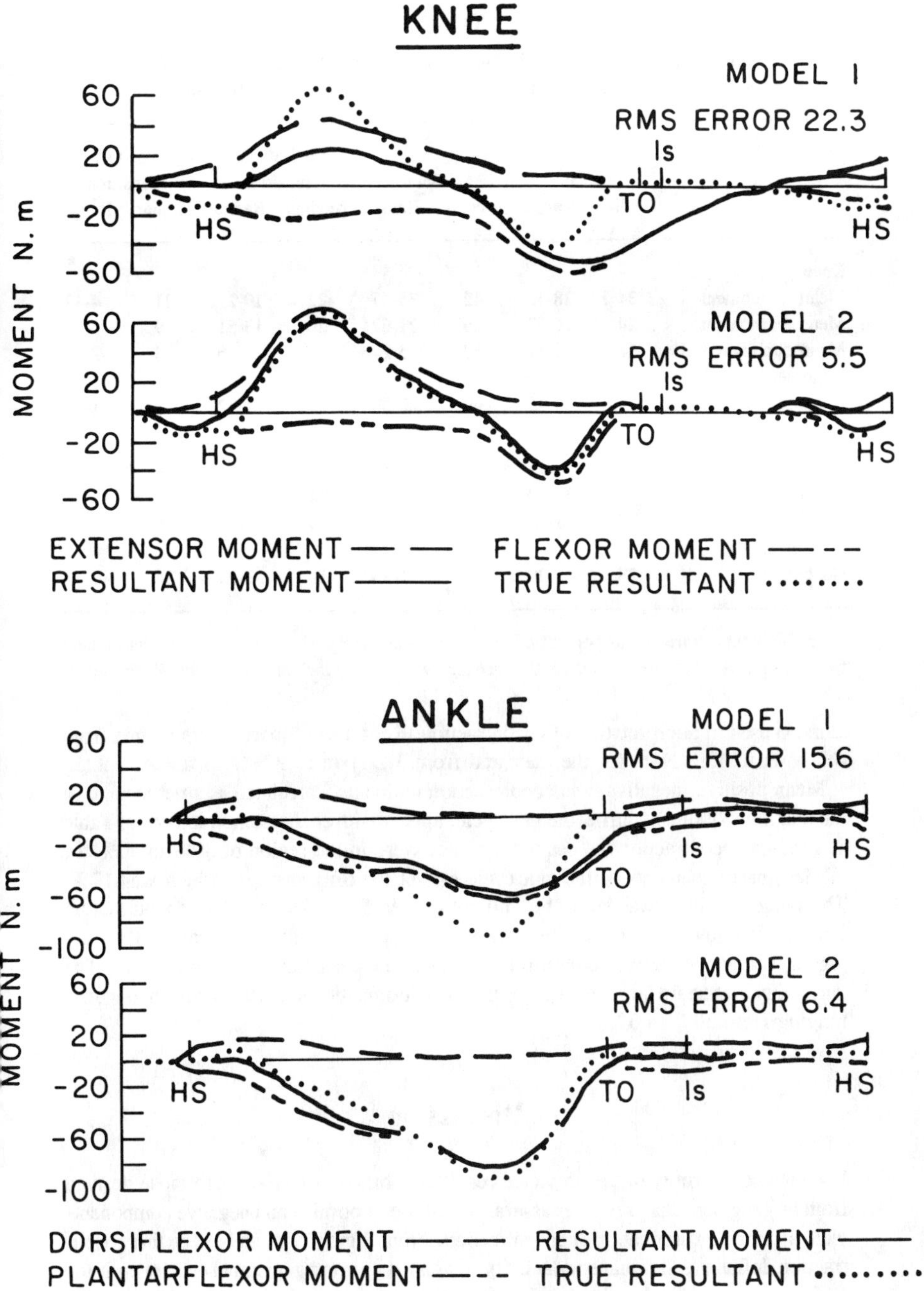

Figure 2—Moments predicted about the knee and ankle during one gait cycle. Model 1, angle and velocity constants set to 0; Model 2, full model. HS = heel-strike, TO = toe-off; s = second.

Table 1

Cocontraction Indices and Median and Range of Average Moments About Knee and Ankle Occurring During Four Segments of the Gait Cycle

	Stance		Mid-Stance		Heel-Strike Transition		Toe-Off Transition	
	Median	Range	Median	Range	Median	Range	Median	Range
Knee								
Mean + moment	34	28-42	42	35-47	22	19-29	21	4-43
Mean − moment	24	16-62	25	21-67	20	13-51	9	6-34
Mean cocontraction moment	12	5-21	12	5-27	6	3-16	3	2-9
Cocontraction index	37	9-45	40	20-50	34	21-52	17	1-40
Ankle								
Mean + moment	7	2-12	5	1-11	10	6-18	6	5-12
Mean − moment	41	37-60	53	40-72	4	2-18	36	23-54
Mean cocontraction moment	5	1-11	5	1-10	4	3-15	6	4-11
Cocontraction index	23	2-30	18	1-23	59	33-93	34	30-37

Note. Moments expressed as percent of maximum occurring during the cycle. Results include two trials for each of three subjects for the knee and two trials for each of four subjects for the ankle.

alone is used in interpretation of cocontraction. Root mean square errors ranged from 4.7 N•m to 13.0 N•m for the knee and from 3.2 N•m to 9.5 N•m for the ankle.

Mean positive, negative, and cocontraction moments, calculated as a percentage of the maximum moment during the gait cycle, appear with cocontraction indices in Table 1. Median cocontraction indices for the knee were in the region of 30% to 40% for all designated segments of the gait cycle except toe-off transition, which was 17%. The range of values was large, but did not exceed 52%. At the ankle, cocontraction indices were lower during stance than comparable knee values (23% and 18%), but higher median values were obtained for both transition segments (59% and 34%). The variability was again large, except for the toe-off transition segment, which had values ranging from 30% to 37%.

Discussion

The close conformity of the predicted resultant moments of Model 2 to those derived from link segment analyses gives assurance that use of positive and negative components yield accurate quantification of cocontraction. From Figure 2, it can be seen that cocontraction defined as simultaneous activity of agonist and antagonist muscles, even if the EMG is processed and scaled to relate to static moment production (Model 1), is quite different from cocontraction defined in terms of true moment production (Model 2). This is particularly evident in the example of the knee. It can be seen that, for the first two-thirds of stance, using Model 1 the predicted extensor moment was only slightly larger than the predicted flexor moment, which would result in a very high cocontraction index. However, when the full model was used (Model 2), the extensor moment was increased and the flexor moment decreased, producing a much smaller cocontraction index. Particularly in pathological movement, the implications are clear: EMG used to assess cocontraction must be interpreted using muscle length and contraction velocity changes. Although not apparent from the results presented here, muscle length effects were considerably more important than velocity differences. This is likely attributable to the inclusion in Model 1 of the velocity effects which could occur within a fixed tendon-to-tendon length, a feature determined by the EMG processing.

The extent of cocontraction in terms of moment production in normal walking is considerable. This is particularly true of muscles surrounding the knee, which may be attributable to the fact that all knee flexors and one knee extensor perform functions at one other joint in addition to the knee. The extent of variability present was not altogether unexpected; the variations in resultant knee moment reported by Winter (1982) might be expected to involve varying amounts of cocontraction.

Cocontraction indices at the ankle were substantially lower during stance and mid-stance segments than were corresponding indices at the knee. It appears that values exceeding 30% would be suspected of being abnormal. With values ranging between 30% and 37%, the toe-off transition period was the only segment displaying a high degree of consistency, demonstrating potential for application in quantitative assessment. Recommendations for further research to improve the sensitivity of the measure include the development of a less cumbersome method of determining opposing moments generated about the joints, the extension of the study to include more trials and more subjects, and the assessment of factors thought to affect the extent of cocontraction in normal movement.

Acknowledgments

The author acknowledges the assistance of Dr. D.A. Winter, under whose supervision part of this study was undertaken, the technical assistance of Mr. P. Guy, and financial support of the Medical Research Council of Canada (MT-4343) and Queen's University Faculty of Medicine.

References

BRESLER, B., & Frankel, J.P. (1950). The forces and moments in the leg during walking. *Transactions of the American Society of Mechanical Engineers*, **72**, 27-36.

FALCONER, K. (1978). *A method of quantification of the amount of co-contraction at the ankle joint in human gait*. Unpublished master's thesis, University of Waterloo, Waterloo.

GATEV, V. (1972). Role of inhibition in the development of motor co-ordination in early childhood. *Developmental Medicine and Child Neurology*, **14**, 336-341.

PATTON, N.J., & Mortensen, O.A. (1971). An electromyographic study of reciprocal activity of muscles. *Anatomical Record*, **170**, 255-268.

PAYTON, O.D., & Kelley, D.L. (1972). Electromyographic evidence of the acquisition of a motor skill. *Physical Therapy*, **52**, 261-266.

WINTER, D.A. (1982). Motor patterns in normal gait—Are we robots? *Proceedings of the 2nd Biannual Conference of the Canadian Society for Biomechanics* (pp. 50, 51). Kingston, Canada.

WINTER, D.A. (1980). Overall principle of lower limb support during stance phase of gait. *Journal of Biomechanics*, **13**, 923-927.

Calf Muscle Work and Trunk Energy Changes in Human Walking

A.L. Hof, M.A.A. Schallig, and Jw. van den Berg
University of Groningen, Groningen, The Netherlands

The calf muscles are important in walking, but their function is still unclear. Several authors have proposed various possible functions of the calf muscles (cf. Sutherland, Cooper, & Daniel, 1980, for a review). These can be summarized as follows. The negative work of the calf muscles, that is, the work done in the eccentric phase of the contraction, is supposed to restrain the forward rotation of the tibia around the ankle. It remains unclear whether this restraining will also manifest itself at the center of gravity of the head-arms-trunk (HAT) body segment. Concerning the positive work, that is, the concentric phase of the contraction, two hypotheses have been put forward. One suggests that the calf muscles propel the body forward (push-off). The alternate view is that it initiates the swing of the leg.

If the calf muscles are indeed engaged in accelerating or decelerating the HAT, and if contribution of other muscle groups is minimal, a simple relationship between the production of the calf muscle work and the increase of the HAT energy, both in time and in magnitude is to be expected.

Methods

We investigated the relationship between the work done by calf muscles and total HAT energy. The total HAT energy consists of two components: potential and kinetic energy. Several approximations have been made in measuring the total HAT energy. We considered only the movements in the sagittal plane, because the rotational energy is small (Winter, Quanbury, & Reimer, 1976; Zarrugh, 1981). The kinetic energy due to vertical movements is also negligible, as confirmed by our own measurements. In the introduction it is suggested that a significant part of the variations of the total HAT energy is contributed by the calf muscles. Simultaneous measurements of the calf muscle work and the HAT energy should reveal this supposed contribution.

The subjects walked on a treadmill. At their back two strings were attached with tape, at the T10 level, the approximate center of gravity of the HAT. One string (see Figure 1), running horizontally, actuated a tachogenerator which measured the HAT's

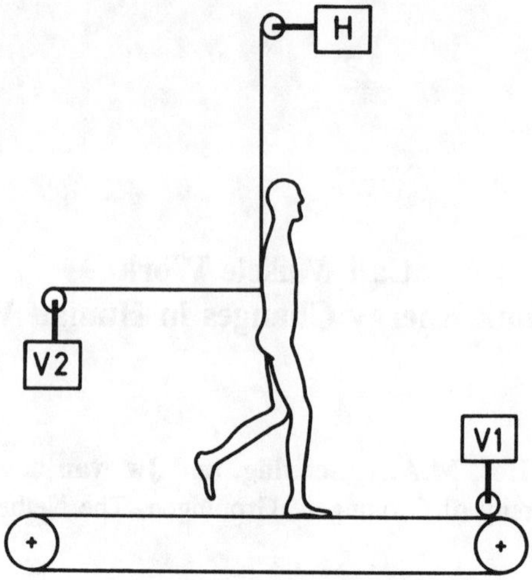

Figure 1—The set-up.

horizontal velocity (V_2). The second string was used to measure the vertical displacements of the HAT via a potentiometer and pulley assembly in order to calculate the potential energy (mgh). Care was taken to make the set-up as light and stiff as possible. The set-up is similar to those used by various other authors (Ohmichi & Miyashita, 1981; Ralston & Lukin, 1969; Zarrugh, 1981). It was necessary to measure the instantaneous treadmill speed (V_1) with a tachogenerator because the variations of the belt speed showed peaks of 5% during a stepcycle. The signals from both tachogenerators were filtered by 23 Hz-12 dB/oct. Butterworth filters. An analog summing and squaring circuit in cascade with the filter allowed the calculation of the kinetic energy ($\frac{1}{2}$ m $[V_1 + V_2]^2$). Neglecting the belt speed variations would result in a 10% error in the kinetic energy, and even more in the total HAT energy. The total time delay between movement and signal from the circuits was below 20 ms. The amplification of the circuits of the kinetic and the potential energy was adjusted to match the mass of the HAT, estimated as 0.678 times the body mass (Winter, 1979).

The calf muscle work of the right leg was measured by EMG to force processing (Hof & van den Berg, 1981). The work of the left calf was estimated by shifting the signals over half a stride.

The various signals from the EMG to force processor and the analog circuit were recorded on a 12-channel paper recorder. No averaging was done in order to preserve the correct time relationship between the various signals. Recordings were made on 3 male individuals. They walked at various speeds from 0.5 to 2.0 m/s. At 1 and 1.5 m/s they also walked at forced step lengths, ranging from 0.5 to 1.1 meter.

Results

Representative recordings of one subject are given in Figure 2. The amplitude of the energy curves (cf, Cavagna & Margaria, 1966; Ohmichi & Miyashita, 1981; Zarrugh,

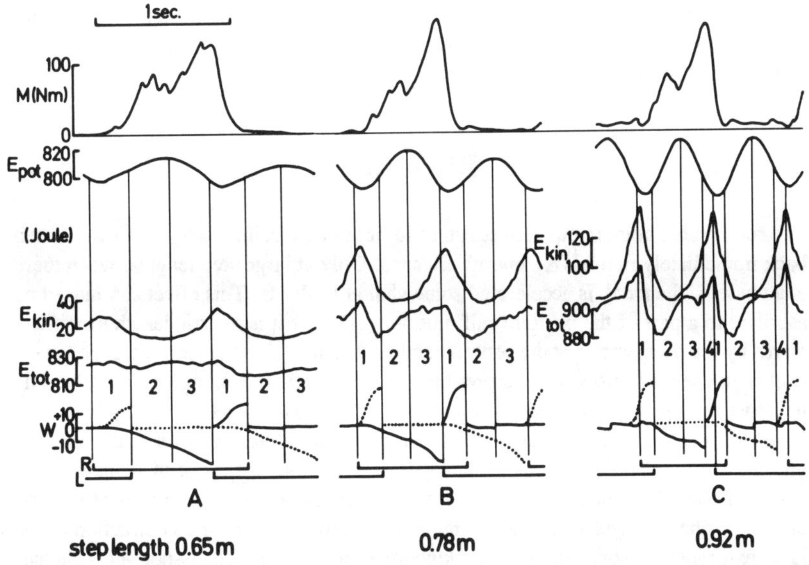

Figure 2—Recordings for one of the subjects (36 years, 1.92 m, 89.5 Kg). A: slow speed, 0.75 m/s, step length 0.65 m; B: moderate speed, 1.25 m/, step length 0.78 m; C: fast speed, 1.75 m/s, step length 0.92 m. From top to bottom, right calf muscle moment M, obtained by EMG processing; potential HAT energy, kinetic HAT energy, total HAT energy; right calf muscle work $W = \int M d\phi$ (drawn line), estimated W for the left calf muscle (dotted line), footcontact. The energy and work signals are at the same scale. Division in parts 1 to 3, 4 starts at right heel-strike.

1981), as well as the calf muscle work (Hof, Geelen, & van den Berg, 1983) increases with step length.

We can discern three, sometimes four, phases in a step (half a stride). These are discussed beginning with the right heel contact, and they can best be seen at average step length (see Figure 2B).

1. Double stance: The left calf muscles perform a considerable amount of work, while the right calf muscles are hardly active. The positive work cannot be seen back in the total HAT energy: actually the kinetic energy decreases markedly while the potential energy remains almost constant.

2. From left toe-off to midstance: Negative work is done (absorbed) by the right calf muscles. However, the HAT energy increases due to an increase in potential energy.

3. From midstance to push-off: In this phase potential energy of the HAT is transformed into kinetic energy, and the total energy change is almost zero. The negative work of the right calf muscle is not reflected in a HAT energy decrease.

4. Usually the start of the push-off (plantarflexion) coincides with the contralateral heel contact, and phase 3 is followed by a next phase 1. At the largest step length, however, the push-off begins earlier (see Figure 2C). In that case there is a sharp increase in total and kinetic HAT energy which approximately corresponds to the first half of positive calf muscle work. As soon as the contralateral foot is flat the sharp energy decrease of phase 1 follows.

At small step lengths (see Figure 2A) the variations in total HAT energy are very small, and the exchange between potential and kinetic energy is nearly perfect. The conclusion, however, that the calf muscle work is not seen back in the total HAT energy changes, still applies.

Discussion

The remarkable point in our results is that so little of the calf muscle work can be seen back immediately in the HAT energy changes. Only at large step lengths, when there is a phase 4, the trunk is accelerated somewhat at push-off. This effect can indeed be ascribed to a part of the positive calf muscle work, but it is overridden shortly afterwards by the deceleration at the contralateral heel-strike (phase 1). In general however, it does not seem justified to interpret the term push-off in walking too literally. The negative calf muscle work cannot be seen back at all in the HAT energy level.

Several explanations are possible for why no simple relationship between the calf muscle work and the HAT energy changes exists. First, the work of other muscles may complicate the picture. For instance, the decrease and increase of the HAT total energy in phases 1 and 2 may be related to an eccentric-concentric contraction of the knee extensors. Second, the (mainly kinetic) energy of the legs is not yet included. Experiments in which the kinetic energy of the legs is measured as well are in preparation.

References

CAVAGNA, G.A., & Margaria, R. (1966). Mechanics of walking. *Journal of Applied Physiology*, **21**, 271-278.

HOF, A.L., & van den Berg, Jw. (1981). EMG to force processing I-IV. *Journal of Biomechanics*, **14**, 747-792.

HOF, A.L., Geelen, B.A. & van den Berg, Jw. (1984). Calf muscle moment, work and efficiency in level walking. *Journal of Biomechanics*, **16**, 523-537.

OHMICHI, H., Miyashita, M. (1981). Analysis of the external work derived from the kinematics of human walking. In A. Morecki, K. Fidelus, K. Kedzior, & A. Wit (Eds.), Biomechanics VII-B (pp. 184-189). Baltimore: University Park Press.

RALSTON, H.J., & Lukin, L. (1969). Energy levels of human body segments during level walking. *Ergonomics*, **12**, 39-46.

SUTHERLAND, D.H., Cooper, L., & Daniel, D. 1980). The role of the ankle plantar flexors in normal walking. *Journal of Bone and Joint Surgery*, **62**, 184-189.

WINTER, D.A., Quanbury, A.O., & Reimer, G.D. (1976). Analysis of the instantaneous energy of normal gait. *Journal of Biomechanics*, **9**, 253-257.

WINTER, D.A. (1979). *Biomechanics of human movement*. Toronto: John Wiley & Sons.

ZARRUGH, M.Y. (1981). Power requirements and mechanical efficiency in treadmill walking. *Journal of Biomechanics*, **14**, 157-165.

Intracompartmental Pressure Changes During the Walking Cycle

J.C. Wall and A.G.P. McDermott
Dalhousie University, Halifax, Nova Scotia

The recognition of the exercise-induced compartment syndrome has led to an increased need to measure dynamic intracompartmental pressures. Since this syndrome is caused by increased intracompartmental pressure, study of the factors which cause these pressures to rise, for example, muscular contraction, is essential to an understanding of the associated pathology and physiology. In the past, electromyography (EMG) has been the principal method used to determine muscle activity, particularly in determining the phasic activity during dynamic movements. However, Baumann, Sutherland, and Hanggi (1979) have pointed out the weaknesses in using EMG to determine muscle tension in dynamic situations.

The relationship between muscle tension and intramuscular pressure has been investigated by several research groups (Owen et al., 1977; Sylvest & Hvid, 1959). The general consensus is that under static conditions the relationship is linear. However, when a muscle is stretched, it has an elastic component which contributes, along with the active component, to the total force developed by the muscle during contraction. Therefore under dynamic conditions one would question the linear relationship between muscle tension and pressure.

Baumann et al. (1979) investigated the dynamic changes of intramuscular pressure in tibialis anterior during walking. Their results showed that the complex changes of intramuscular pressure were related not only to muscle activity, but also to changing muscle length and variation in passive loading. However, the wick catheter they used had too low a frequency response which resulted in dampening in pressure changes over time. This paper reports the use of a catheter with a high frequency response capability (20,000 Hz) designed to monitor intramuscular pressure of the anterior compartment of the leg during normal walking.

Materials and Methods

Five male subjects volunteered to participate in this study. Details of the age, weight, and height of each subject are given in Table 1. The subjects were asymptomatic.

Table 1

Results of Maximal Dorsiflexion and Plantarflexion Tests

		Subject Number					
		1	2	3	4	5	Group
Age (years)		30	23	45	19	34	30.2
Weight (kg)		72.0	69.3	74.7	58.9	70	68.98
Height (m)		1.93	1.78	1.73	1.68	1.63	1.73
PRE-EXERCISE PRESSURES (mm•Hg)							
Supine							
Rest		8	5	10	8	4	7
Max D. flex		140	175	190	210	125	168
Max P. flex		37	60	75	100	102	74.8
Standing							
Tip toes		36	22	22	32	22	26.8
On heels		105	97	100	100	68	94
One leg		42	43	50	65	46	49.2
GAIT STUDY							
Speed							
(Stat/s)		0.73	0.75	0.78	0.80	0.82	0.78
Stride	$\bar{X}$	1.02	1.18	1.12	1.08	1.00	1.08
Time (s)	S.D.	0.01	0.03	0.03	0.03	0.03	0.07
Support	$\bar{X}$	63.8	64.4	65.6	65.6	63.1	64.4
Time (%)	S.D.	1.76	2.32	1.17	2.39	0.82	2.02
PEAK A							
Time	$\bar{X}$	3.7	5.8	5.2	3.5	2.0	4.47
(%)	S.D.	1.0	1.19	1.15	1.1	1.6	1.87
Pressure	$\bar{X}$	53.9	66.3	84.6	89.1	128.7	84.5
(mm•Hg)	S.D.	5.28	8.78	8.10	5.36	15.03	27.2
PEAK B							
Time	$\bar{X}$	50.1	47.2	47.8	45.9	41.4	46.5
(%)	S.D.	1.9	1.81	1.56	1.60	1.65	3.36
Pressure	$\bar{X}$	49.3	83.9	76.7	84.1	112.7	81.3
(mm•Hg)	S.D.	2.26	5.28	10.27	2.77	7.12	21.3

The technique for insertion of the STIC catheter was similar to that used by McDermott, Marble, Yabsley, and Phillips (1982). Following insertion, the leg was repeatedly squeezed to insure transducer response. A test sequence of supine and standing maneuvers was performed.

Footswitches were attached to the heel and toe regions of the right shoe. The signals from the footswitches were recorded along with those from the catheter on a Hewlett Packard 3964A strip chart recorder.

Prior to inserting the catheter, the subjects were walked on a level motorized treadmill at 3 miles/hr for several minutes until they felt comfortable. With the catheter in place the subjects walked on the treadmill at the same speed for 1 min, during which time data was collected. From the footswitch data, stride time and support time were measured, and the times and magnitudes of the peaks were obtained from the tracing.

Results

Ten strides were analyzed for each subject and the results obtained are shown in Table 1 together with the results of the static tests. This table also shows the mean and standard deviations for the group as a whole obtained from 50 strides (10 per subject). The data has been plotted in Figure 1 which also includes a representative plot of the pressure curve.

Discussion

Since it is the purpose of this paper to look for possible causes for increases in intracompartmental pressure during walking only the peak pressures that occur during the walking cycle are here considered. There are two such peaks and these occur during the stance phase as shown in Figure 1.

Sylvest and Hvid (1959) have shown that isometric muscle activity leads to an increased pressure. This is confirmed by the results of the maximal dorsiflexion and plantarflexion tests shown in Table 1. These suggest that both pre-tibial and calf muscle activity increase the pressure in the anterior compartment.

Unfortunately, no EMG data was collected during this study, and consequently no direct correlation between muscle activity and compartmental pressure could be made. The EMG data used in Figure 1 is representative of that found in previous studies and is adapted from Inman, Ralston, and Todd (1981).

Peak A occurs just after heel-strike during the double braking support phase. This is the time at which the tibialis anterior is most active. From the goniogram it can be seen that the ankle is plantar flexing at this time and therefore the pre-tibials are contracting eccentrically, preventing foot slap. Dubo et al. (1976) showed that the peak amplitude of tibialis anterior activity during walking was 73% of that found under maximal isometric conditions. The pressure developed, however, is approximately half of that resulting from the maximal resisted dorsiflexion test.

The extent of the ankle excursion coincident with Peak A is only about 15° from the anatomical position. The resulting stretch in the tibialis anterior muscle from this position of the foot is therefore small and unlikely to contribute greatly to the increased pressure in the anterior compartment.

Peak B occurs at approximately 46% of the gait cycle during the latter portion of the single support phase. At this time the triceps surae are most active. During the initial activity the foot is dorsiflexing, implying an eccentric contraction. From approximately 40% of the gait cycle the foot is plantar flexing, and therefore the calf muscles are undergoing concentric contraction. Dubo et al. (1976) show the maximum amplitude of gastrocnemius EMG activity during walking to be 63% of that found under maximal isometric conditions. However, the peak activity of soleus during normal gait has been shown by Arsenault and Winter (1980) to be from 114 and 167% of maximum isometric activity. The results of the present study show that the peak pressure at this time is likewise greater than that achieved during maximal resisted plantarflexion in the supine position and about three times that developed while standing on tip toes.

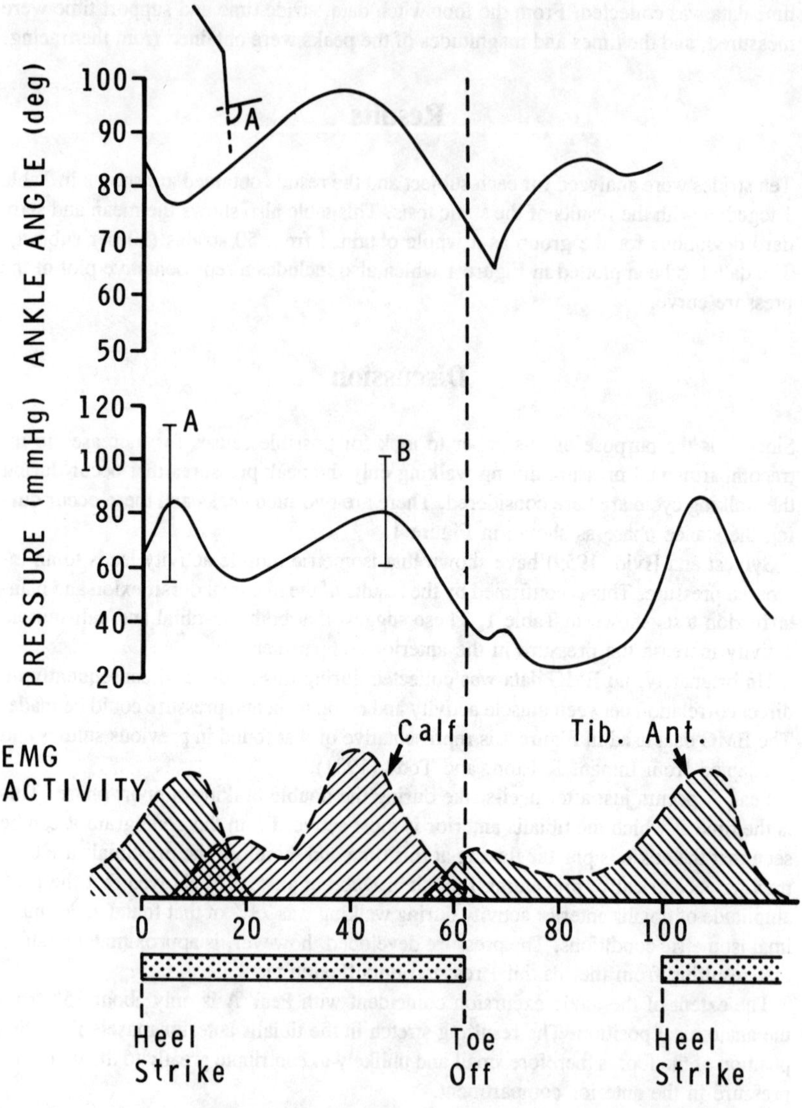

Figure 1—Changes in ankle angle, intracompartmental pressure and EMG activity through the gait cycle.

Grieve, Pheasant, and Cavanagh (1978) measured the change in length of the gastrocnemius resulting from changes in angle of the knee and ankle joints. Their results predict that throughout the gait cycle the muscle is always less than the length encountered in perpendicular standing. The length given for the standing position is 6.5% greater than their reference length, which was taken with knee and ankle angles of 90°. Between 40% and 50% of cycle time they predict that at a normal walking speed the gastrocnemius is about 6% longer than the reference length, that is, close to the length in perpendicular standing. Again, unless intracompartmental pressure increases can be caused by very small changes in muscle length, one must conclude that the primary cause for the increased pressure represented by Peak B is the activity of the calf muscle group. This finding suggests that decompression of symptomatic patients with exercise induced compartment syndrome would be enhanced by fasciotomies of both the involved compartment and the adjoining compartment.

References

ARSENAULT, B., & Winter, D.A. (1980). Repeatability of electromyographic activity during gait. *Proceedings of the Canadian Society for Biomechanics Conference: Human Locomotion I* (pp. 22-23). London, Ontario.

BAUMANN, J.U., Sutherland, D.H., & Hanggi, A. (1979). Intramuscular pressure during walking. *Clinical Orthopedics and Related Research,* **145**, 292-299.

DUBO, H.I.C., Peat, M., Winter, D.A., Quanbury, A.O., Hobson, D.A., Steinke, T., & Reimer, G. (1976). Electromyographic temporal analysis of gait: Normal human locomotion. *Archives of Physical Medicine and Rehabilitation,* **57**, 415-420.

GRIEVE, D.W., Pheasant, S., & Cavanagh, P.R. (1978). Prediction of gastrocnemious length from knee and angle joint posture. In E. Asmussen & K. Jorgensen (Eds.), *Biomechanics VI-A* (pp. 405-412). Baltimore: University Park Press.

INMAN, V.T., Ralston, H.J., & Todd, F. (1981). *Human walking*. Baltimore: Williams and Wilkins.

McDERMOTT, A.G.P., Marble, A.E., Yabsley, R.H., & Phillips, B. (1982). Monitoring dynamic anterior compartment pressures during exercise. *American Journal of Sports Medicine,* **10**, 83-89.

OWEN, C.A., Garetto, L.P., Hargens, A.R., Schmidt, D.A., Mubarak, S.J., & Akeson, W.H. (1977). Relationship of intramuscular pressure to strength of muscular contraction. *Trans. Orthop. Res. Soc.,* **2**, T246.

SYLVEST, O., & Hvid, N. (1959). Pressure measurements in human striated muscle during contraction. *Acta Rheumatologica Scandinavica,* **5**, 216-222.

Heel Height Induced Changes in Metatarsal Loading Patterns During Gait

R.W. Soames and C. Clark
King's College London,
Strand, London, England

Schwartz, Heath, Morgan, and Towns (1964) observed that wearing a heel leads to an increase in the proportion of weight borne by the forefoot. In addition there are changes in the pressure patterns under the metatarsals, with an increase under the first and a decrease under the fifth metatarsal. These changes they attribute to an increased pronation of the foot due to the instability of wearing high heels. In contrast Godfrey, Lawson, and Stewart (1967) reported a reduction in pressures under both the first and fifth metatarso-phalangeal joints when wearing heels: there is also a significant increase in the time for which the first metatarso-phalangeal joint is weight bearing.

If relationships between changes in heel height and pressures under the individual metatarsal heads can be established, it should be possible to gain an insight into the mechanical behavior of the metatarsals. Such information could have direct implications with respect to foot pathologies where either anatomical anomalies or abnormalities of gait, may lead to forefoot loading patterns which create local areas of high stress, with the possibility of subsequent stress fracture. Being able to recognize such potentially damaging patterns of loading will reduce the likelihood of further pathology, for example, by redistributing the load within the forefoot to a less harmful pattern.

Method

Pressures under the posterior heel, the five metatarsal heads and the great toe were recorded from 26 female subjects, aged 20 to 32, while walking in each of three pairs of shoes with different heel heights—low, medium, and high. The order of heel heights was randomized to minimize fatigue effects. Absolute measures of each heel height for each subject were taken as the vertical distance between the sole-heel junction and the ground. The means and standard deviations of each of the heel heights are given in Table 1, from which it can be seen that the high heel height was twice as high as the medium heel height. The low heel height was taken as zero even though the subjects were wearing training shoes, as it was deemed that such shoes had a constant sole thickness.

Table 1

Means and Standard Deviations of Each of the Heel Heights and Metatarsal Ray Lengths

	Heel Height (mm)			Ray Length (mm)				
	Low	Medium	High	1	2	3	4	5
X	0	34.3	68.8	139.8	142.3	139.0	133.0	124.8
S.D.	—	13.1	16.5	11.2	11.1	8.6	8.9	9.4

Prior to recordings being taken, and after the attachment of the transducers, the length of each metatarsal ray was noted as the distance from the posterior heel to the metatarsal head transducer. These lengths were recorded so that the change in ground angle of the metatarsals with increasing height could be determined. From Table 1 it can be seen that the second ray is the longest, although there is no significant difference in length between the medial four rays. However, the fifth ray is significantly ($P = .01$) shorter than the rest.

The pedal pressures were recorded using small semiconductor strain gauge transducers, the output of which are led to a PDP 11/10 computer, digitized online at 200 Hz and written to magnetic tape. Further details of the transducer system can be found in Soames, Stott, Goodbody, Blake, and Brewerton (1982).

Results

The mean peak pressures under the heel, metatarsal heads, and great toe for all three heel heights are shown in Figure 1. For each heel height the lowest pressures are found under the lateral three metatarsals. However, as heel height increases the differences between medial and lateral pressures increases, particularly for the first metatarsal. With increasing heel height the pressure under the heel decreases slightly, while that under the great toe increases, so that when wearing medium or high heels pressures under the great toe are larger than those under the heel. From Figure 1 it can also be seen that the gradient of the increase in pressure between the second and third metatarsals increases with heel height.

Subjects' assessments of medium and high heels differed to such an extent that what was deemed high by one subject would be considered medium by another. Consequently, to determine the relationship between absolute heel height and peak pressure, polynomial regressions were conducted on the peak pressures with either heel height or the heel height-ray length ratio as the independent variable.

The correlation coefficients from these regressions are shown in Table 2. In all but one case linear regression provided the best fit to the data, the exception being M5 with the ratio data as the independent variable. As can be seen from Table 2, there are significant negative correlations between heel height and peak pressure under the heel and lateral three metatarsals, and significant positive correlations between heel height and peak pressures under the two medial metatarsals.

It appears, therefore, that a direct relationship exists between heel height and the pressures under the forefoot, with the pattern of loading changing as heel height increases.

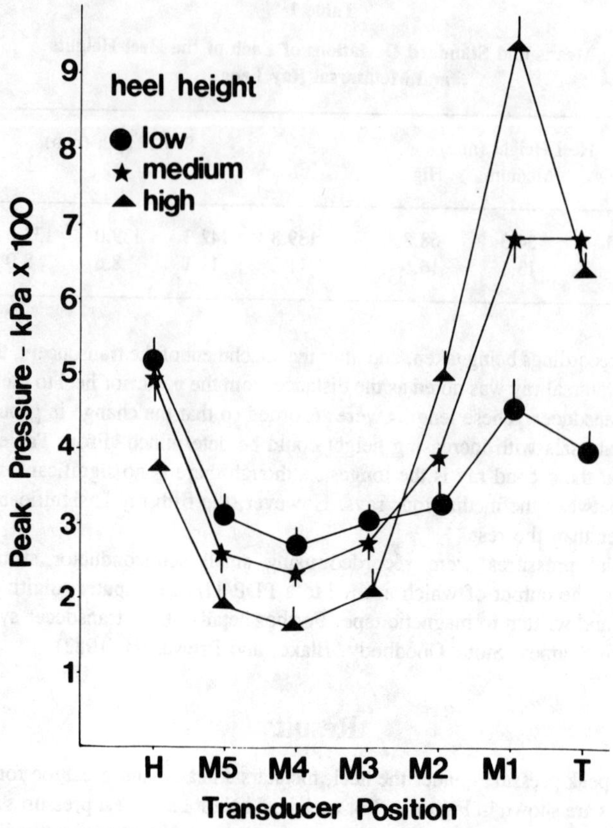

Figure 1—Mean peak pressure (kPa X 100) under the heel (H), metatarsal heads (M5 to M1) and first toe (T) during gait when wearing low, medium, and high-heeled shoes.

Table 2
Correlation Coefficients Obtained between Peak Pressures Observed with Either Absolute Heel Height or the Heel Height-Ray Length Ratio as the Independent Variable

Peak Pressure under . . .	Absolute Heel Height	Heel Height-Ray Length Ratio
Heel	−0.51*	—
M5	−0.54*	−0.58*
M4	−0.69*	−0.76*
M3	−0.50*	−0.56*
M2	0.53*	0.56*
M1	0.48	0.38
Great toe	0.31	—

Discussion

There is no doubt that wearing heels leads to changes in the pattern of pressure distribution across the metatarsal heads. Indeed, the data presented supports the earlier findings of Schwartz et al. (1964). They propose that the changes in loading are due to increased pronation of the foot in response to decreased stability induced by the high heels. However, both Adrian and Karpovich (1966) and Merrifield (1971) found no evidence of instability with high heels. It is suggested that these changes are brought about by changes in load transmission by the metatarsals due to changes in their orientation with increases in heel height.

The question arises as to how the metatarsals are transmitting the loads applied to the forefoot. Stokes, Hutton, and Stott (1979) have suggested that the foot acts partly as an arch structure and partly as a beam or lever. If the metatarsals are conforming solely to the role of peripheral elements of an arch, then the compressive forces they experience at the metatarsal head can be resolved into horizontal and vertical components. The vertical component (R) is given by

$R = K.\sin \phi$ where K is the compressive force acting along the metatarsal, and ϕ is the angle subtended by the long axis of the metatarsal with the ground.

As heel height is raised, the angle ϕ increases and consequently R increases. The medial metatarsal head pressures do increase but the lateral metatarsal head pressures decrease. However, the proportional increase in R in going from one heel height to another is independent of the change in angle ϕ. Consequently it must be concluded that the metatarsals do not act as arch elements. That the angle ϕ increases with increasing heel height is evidenced by the decrease in pressure beneath the great toe, since increases in angle ϕ results in lower horizontal reaction forces transmitted to the great toe.

If the metatarsals act purely as beams, pressure recorded under the heads will be a function of shear forces, axial forces, and bending moments. Axial load (A) $\propto F.\sin \phi$, shear force (S) $\propto F.\cos \phi$ and bending moment at the metatarsal base (M) $\propto F.h/\tan \phi$; where F is the pressure at the metatarsal head, ϕ is the angle subtended by the long axis of the metatarsal and h is the length of the metatarsal.

As heel height is raised, and assuming that metatarsal head pressures remain constant, then axial loading of the metatarsal would increase, and shear forces and bending moments decrease. However, pressures under the third, fourth and fifth metatarsal heads actually decrease with increasing heel height, thus those components which increase with increases in ϕ must be more than offset by those which decrease under the same conditions. The conclusions thus drawn are that, as the heel is raised, shear forces and bending moments increase in the first and second metatarsals, and decrease in the third, fourth, and fifth metatarsals.

In a structurally homogeneous material with uniform cross-sectional area, the longer the specimen length the shorter the fatigue life because of the increased bending moments at the fixed end. March fractures have been found to be more common in the second and third metatarsals with the center of the bone more commonly being the site of fracture (Devas, 1975), suggesting that the bones have yielded under bending stresses. According to Levy (1978), the first metatarsal is more likely to be fractured at its head, with sclerosis being apparent perpendicular to the direction of the stress, suggesting that shear forces are responsible. Therefore loading configurations which increase shear forces in the forefoot are likely to give rise to fracture of the first metatarsal, since this appears to undergo a greater change in shear force for a given changed loading configuration than any other metatarsal.

Conclusions

Mean peak pressures under the third, fourth, and fifth metatarsal heads are consistently lower than those under the first and second metatarsal heads irrespective of heel height. However, with increasing heel height the magnitude of this difference increases. Raising the heel also leads to a reduction in mean peak pressures under the heel and an increase under the great toe.

When considered with respect to either absolute heel height or to the heel height-ray length ratio, peak pressures under the third, fourth, and fifth metatarsal heads decrease, while peak pressures under the first and second metatarsal heads increase with increasing heel height. From the data it is suggested that, with the changed loading conditions associated with increased heel height, greater shear stresses and bending moments are developed in the first and second metatarsals and lower shear stresses and bending moments are developed in the third, fourth, and fifth metatarsals.

References

ADRIAN, M.J., & Karpovich, P.V. (1966). Foot instability during walking in shoes with high heels. *Research Quarterly,* **37**, 168-175.

DEVAS, L.M. (1975). *Stress fractures.* Edinburgh: Churchill-Livingstone.

GODFREY, C.M., Lawson, G.P., & Stewart, W.A. (1976). A method for determination of pedal pressure changes during weight-bearing: Preliminary observations in normal and arthritic feet. *Arthritis and Rheumatism,* **10**, 135-140.

LEVY, J.M. (1978). Stress fractures of the first metatarsal. *American Journal of Roentgenology,* **130**, 679-681.

MERRIFIELD, H.H. (1971). Female gait patterns in shoes with different heel heights. *Ergonomics* **14**, 411-417.

SCHWARTZ, R.P., Heath, A.I., Morgan, D.W., & Towns, R.C. (1964). A quantitative analysis of recorded variables in the walking pattern of normal adults. *Journal of Bone and Joint Surgery,* **46A**, 324-334.

SOAMES, R.W., Stott, J.R.R., Goodbody, A., Blake, C.D., & Brewerton, D.A. (1982). Measurement of pressure under the foot during function. *Medical and Biological Engineering Computers,* **20**, 489-495.

STOKES, I.A.F., Hutton, W.C., & Stott, J.R.R. (1979). Forces acting on the metatarsals during normal walking. *Journal of Anatomy,* **129**, 579-590.

Foot Angles During Walking and Running

J.P. Holden and P.R. Cavanagh
The Pennsylvania State University, University Park, Pennsylvania, USA

K.R. Williams
University of California-Davis, Davis, California, USA

K.N. Bednarski
Converse Inc., Wilmington, Massachusetts, USA

The term "angle of gait" is frequently used to describe the position of the feet relative to the direction of forward movement during locomotion. Because this in-toeing and out-toeing is governed by factors such as hip joint motion, the amount of tibial and malleolar torsion present, and the relative adduction and abduction of the whole foot to the body of the talus, deviations from normal values are assumed to indicate structural and/or functional abnormalities in the lower limb. Unfortunately, research studies to determine a normal range for various populations and conditions have been limited in number and inconsistent in results.

Most of the major studies examining angle of gait during walking have made measurements with respect to a longitudinal line through the middle of a walkway (Dougan, 1924; Morton, 1932; Murray, Drought, & Kory, 1964; Padek, 1926). However, Chodera and Levell (1973) showed that straight line walking does not exist, and so they chose the mean direction of the two line segments meeting at the centroid of successive left and right footprints to represent the instantaneous direction appropriate for that footprint pair. This method has been adopted by the present authors for a larger study of stride parameters in walking and running, and the results for foot angle (FA) are presented here.

Before discussing the measurement of foot angles, it is important to note that this and all previous studies have assumed that there is a single value that represents the position of the foot during contact with the ground. Therefore, a preliminary investigation was conducted to examine changes in FA during foot contact. Five male subjects ran at 3.8 m/s for 2 trials each over waxed and high friction surfaces which were intended to represent extremes of frictional conditions. The subjects wore a shoe with a target attached to the midsole at the front of the shoe, and film taken at 400 frames/s enabled measurement to be made of shoe movement relative to the ground during foot contact.

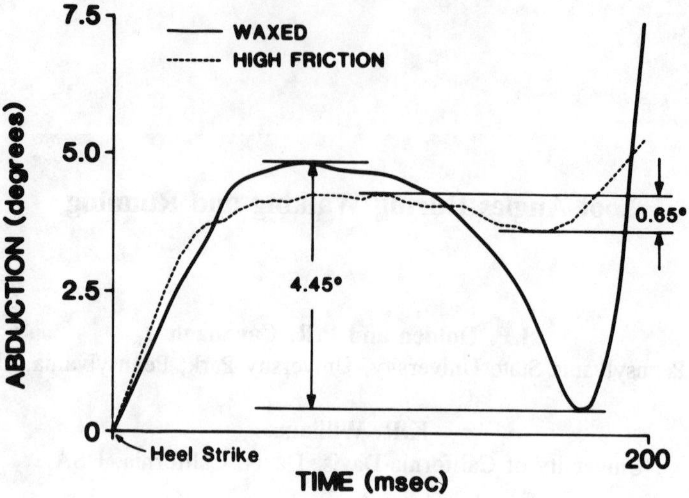

Figure 1—Mean changes in FA of 5 subjects when running over waxed and high friction surfaces. Changes are with respect to foot orientation at time of heel strike. The angle at initial contact is set to zero and abduction is indicated as positive.

The results are presented in Figure 1 as time-normalized plots of the mean changes in FA during foot contact, relative to foot orientation at heel strike. The curves illustrate that during foot-flat, FA remained within a fairly narrow range (.6°) when friction was high, but when friction was low, the foot tended to adduct an average of 4.4° during the time after heel-off. These results indicate that unless investigators provide a high friction surface capable of resisting the moments that cause these rotations, FA should not be represented by a single value for a given foot contact.

Procedures

Sixty-four symptom-free subjects (32 males, 32 females, mean age 26.2, height 172.7 cm, weight 67.2 kg) completed 6 trials each, 3 walking and 3 running. The 3 trials in each mode were at self-chosen speeds (slow, intermediate, fast) intended to represent a large part of the subject's available range. Ink marks left by the footwear on a 12-m section of a wooden walkway were measured to determine the foot position for each contact, and FA was calculated as an average of between 6 and 15 foot contacts for each trial. Because the ink marks were not smudged, it was assumed that the combination of the wooden walkway and the rubber-soled shoes used in this study did provide the high friction condition necessary, as discussed earlier, to characterize foot position by a single value. FA was defined as the angle formed by the midline of the shoe (i.e., the line connecting ink marks) and the corresponding direction of travel (see Figure 2). Analysis of variance was used to determine significant ($p < .05$) differences in FA due to gender, right vs. left feet, and speed.

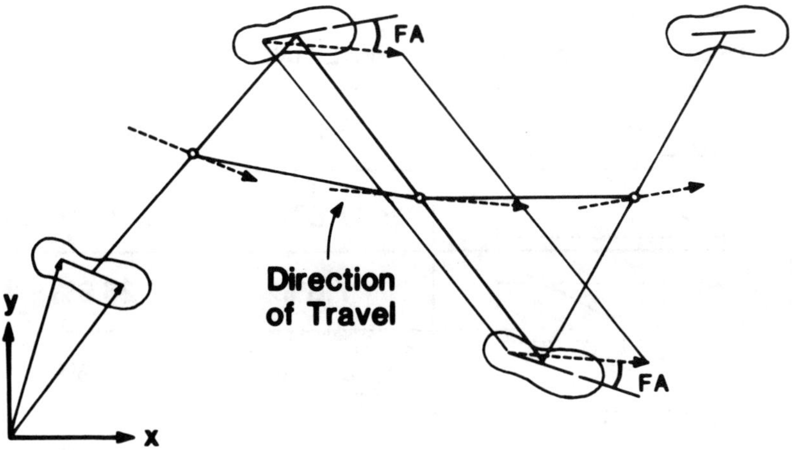

Figure 2—Method of calculation of foot angle, according to Chodera and Levell (1973).

Results

The means, SD, and ranges of right and left FA values over all subjects and speeds are shown in Table 1. In both walking and running, the right FA was significantly larger (more abduction or "toe out") than the left. FA trends with speed showed no significant differences between right and left feet, and are therefore described for the averages of both feet.

An initial examination of the data revealed that FA changes with speed were highly variable from subject to subject. Therefore, subjects were grouped within the walking and running categories according to the observed change in FA with speed, relative to the FA measured at the slow speed. The trends are shown schematically in Figure 3, and the distribution of subjects over trends reveals that in walking and in running, only just over half of the subjects showed monotonically changing (in either direction)

Table 1

Means, SD, and Ranges for Right and Left Foot Angles During Walking and Running for All Subjects and All Speeds (Abduction Positive)

		Mean	SD (Rt-Lt)	Range
Walking	Rt	+5.4	5.0	−15.9 to +18.5
	Lt	+4.1		−10.0 to +18.2
Running	Rt	+6.9	5.4	−9.2 to +39.9
	Lt	+5.3		−5.9 to +33.1

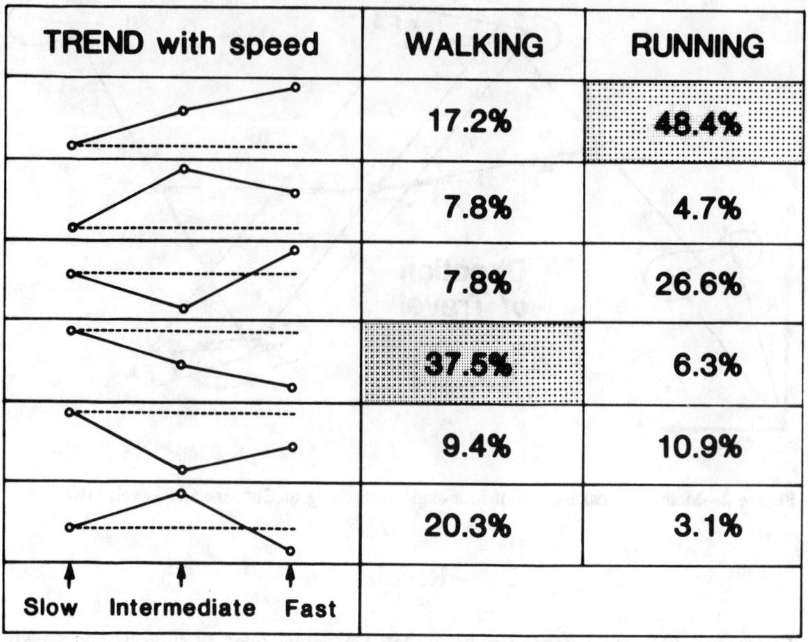

TREND with speed
WALKING
RUNNING
17.2%
48.4%
7.8%
4.7%
7.8%
26.6%
37.5%
6.3%
9.4%
10.9%
20.3%
3.1%
Slow
Intermediate
Fast

Figure 3—The distribution of subjects according to changes in FA with increases in speed. The largest groups for walking and running are shaded.

FA as speed increased. The most common responses were for the foot to be progressively less abducted ("toed out") as speed of walking increased (38% of subjects) and progressively more abducted as speed of running increased (48% of subjects). When the combination of walking and running trends were considered, the largest group contained only 19% of subjects. These subjects were those who fell into the most common groups in both walking and running.

Because speeds were self-chosen by the subjects, statistical analysis was performed on the data of those subjects whose slow, intermediate, and fast speeds (normalized by dividing by leg length) fell within specified ranges. This served to eliminate any effects that might otherwise have been due to the overlapping speed ranges chosen by the subjects. The analyses included 40 subjects in walking and 44 subjects in running, with both groups evenly divided between males and females. In walking, the mean values showed a trend of progressively less abduction as speed increased, but no mean differences were significant. In running, the mean trend was in the opposite direction to that in walking, and mean FA in the "fast" speed range was significantly greater (more abduction) than in either of the two slower running speed ranges. However, the mean FA values for the two slower running speed ranges were within the range of mean values obtained throughout the walking speed ranges, and it was only at the fast running speed range that mean FA differed from any other mean value in walking or running by more than 1 degree. There were no effects in either walking or running due to gender (see Table 2).

Table 2

Mean Foot Angles in the 3 Speed Ranges

	Slow	Intermediate	Fast
Walking (n = 40)			
Speed range (*)	0.84 - 1.60	1.60 - 2.00	2.00 - 2.70
Foot angle (**)	5.2	5.0	4.6
Running (n = 44)			
Speed range	2.34 - 3.61	3.61 - 5.13	5.13 - 8.61
Foot angle	4.7	5.1	9.0

(*) leg lengths/s; (**) degrees of abduction

Discussion

The significantly larger (more abducted) right FA may be related to the fact that over 90% of the subjects in this study indicated that they were right leg dominant, according to their chosen leg for kicking. Sgarlato (1965) noted that most reports on tibial torsion describe a higher average value in the right tibia, indicating a possible structural cause of the difference. Four earlier studies (Dougan, 1924; Morton, 1932; Murray et al., 1964; Padek, 1926) were evenly divided in reporting that the right or the left foot was more abducted. Their conflicting results may have been due to the methods they employed, which did not account for the bias produced by a changing direction of travel. While Chodera and Levell (1973) did account for this factor, they reported data for only one trial by a single healthy subject.

The results for FA changes with speed, which are based on within-subject comparisons in both walking and running, raise doubts about the accuracy of previous hypotheses, the most popular of which was put forward by Morton (1932). He suggested that 22.5° of abduction in standing, and 0° in running, represent functional extremes for a continuum of foot angles. However, only one of our subjects showed such monotonic changes in FA across the entire speed range, and this subject showed a trend in the opposite direction to that suggested by Morton (1932).

Perhaps the most important finding of this study is that FA, unlike stride length and many other gait parameters, shows changes with speed that are complex and highly variable between subjects. The reason for this variability must lie in structural factors which were not measured in this study. Therefore, future work must attempt to explain the new observations presented here on the basis of morphological and structural factors.

Acknowledgments

This study was supported in part by a grant from PUMA. The authors would like to acknowledge the assistance of Ann Bickford in data collection.

References

CHODERA, J.D., & Levell, R.W. (1973). Footprint patterns during walking. In R.M. Kenedi (Ed.), *Perspectives in biomedical engineering* (pp. 81-90). Baltimore, MD: University Park Press.

DOUGAN, S. (1924). The Angle of Gait. *American Journal of Physical Anthropology,* **7**, 275-279.

MORTON, D.J. (1932). The angle of gait: A study based upon examination of feet of central African natives. *Journal of Bone and Joint Surgery,* **14**, 741-754.

MURRAY, M.P., Drought, A.B., & Kory, R.C. (1964). Walking patterns of normal men. *Journal of Bone and Joint Surgery,* **46A**, 335-360.

PADEK, S.D. (1926). The angle of gait in women. *American Journal of Physical Anthropology,* **9**, 273-291.

SGARLATO, T.E. (1965). The angle of gait. *Journal of the American Podiatry Association,* **55**, 645-650.

Kinematics of the Female Upper Body During Ambulation at Various Speeds

F. Figura and M. Marchetti
Università di Roma, Italy

T. Leo
Università di Ancona, Italy

The upper body kinematics of males have been extensively studied both in normal and pathological gait and their relevance in gait evaluation stressed (Cappozzo, Figura, Leo, & Marchetti, 1978; Saunders, Inman, & Eberhart, 1953). However, only limited information concerning the locomotion of females is available. The most comprehensive work in this respect (Murray, Kory, & Sepic, 1970) does not take into account frontal and sagittal rotations.

The aim of our study was the description of those aspects of female ambulation that would permit the identification of stereotypic patterns. These patterns are needed as background information for the assessment of effective procedures of female pathological gait evaluation. In this paper the following rotations of the upper body have been considered: a) frontal and horizontal rotations of the transverse axes of shoulders and pelvis; b) frontal and sagittal rotations of the vertical axis of the trunk.

Methods

Five young Caucasian women performed a total of 20 tests. All of them wore low heeled shoes. The subjects walked both at free and high speed. Free speed ranged from 1.17 to 1.45 m/s; high speed varied from 1.59 to 2.2 m/s.

Subjects and test characteristics are summarized in Table 1. Due to the reduced sensitivity of the female locomotion pattern to the subjects' anthropometry (Murray et al., 1970), the tested sample was assumed to be statistically significant for a population with similar age and ethnical characteristics.

Shoulder and pelvis segments were defined by means of landmarks placed on the acromial and trocanteric processes. The trunk axis was defined as the segment joining the medial points of the shoulder and pelvis segments. The rotational displacements of the above segments were obtained by simple trigonometric processing of the landmark motion data. These latter were measured by a chronostereophotogrammetric tech-

Table 1

Subject and Test Data

Subject	Age (year)	Stature (m)	Body Mass (kg)	Speed (m/s)	Stride Period (s)
1	21	1.67	63	1.29	1.07
				1.59	0.98
				1.62	0.93
				2.08	0.83
2	16	1.58	54	1.25	1.06
				1.25	1.07
				1.71	0.92
				1.93	0.76
3	20	1.62	49	1.39	0.99
				1.45	0.98
				1.68	0.93
				2.20	0.78
4	19	1.65	54	1.17	1.19
				1.25	1.03
				1.62	0.97
				1.76	0.97
5	22	1.65	52	1.40	1.02
				1.72	0.91
				1.94	0.87
				2.07	0.83

nique described elsewhere (Cappozzo, 1981). Sampling frequency was fixed at 30 and 60 samples/s during free speed and high speed tests, respectively.

In a previous study (Cappozzo, Figura, Gazzani, Leo, & Marchetti, 1982) the frequency domain representation has been shown to be the most effective way to evidence the characteristics of the upper body rotations. Such a representation has been adopted in the present study as well. Each rotation was described as follows:

$$\Theta(t) = M_o + \sum_{i=1}^{N} M_i \sin(i\Omega t + \Phi_i) \qquad (1)$$

where M_i = amplitude of the ith harmonic; Φ_i = phase of the ith harmonic, with reference to a time origin coinciding with left heel strike; Ω = fundamental frequency, corresponding to the stride period; M_o = mean value of $\Theta(t)$ over the stride period; N = maximum number of significant harmonics.

Each harmonic is fully described by a vector in the polar plane having an amplitude equal to M_i and a phase equal to Φ_i. Based on the harmonic amplitude estimate error, harmonics having an amplitude less than 0.5° were considered not significant.

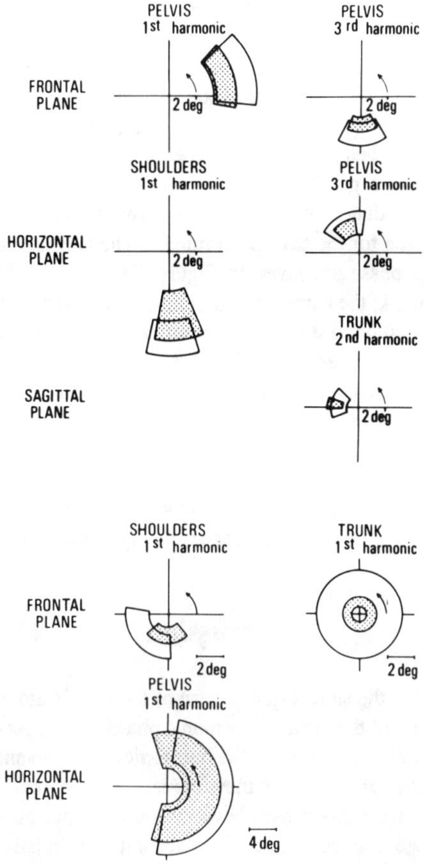

Figure 1—Vectorial representation of the significant harmonics composing the upper body rotations during female ambulation. Shaded sectors are relative to free speed tests.

Results

Three groups of results have been obtained:

1) Synthetic description of the upper body movement pattern by means of a small number of harmonics. These are: first and third harmonics of pelvic rotations both in the frontal and horizontal planes; first harmonic of shoulder rotations both in the frontal and horizontal planes; first harmonic of trunk rotation in the frontal plane; first and second harmonics (the former in a few cases only) of trunk rotation in the sagittal plane.

2) Assessment of the stereotypic pattern of female ambulation. This was based on the inter- and intra-individual repeatability of some of the intrinsic harmonics (Cappozzo et al., 1982), that is, the only ones which should exist under complete symmetry of

the locomotor act. These are the first and third harmonics for the pelvis and shoulder rotations both in the horizontal and frontal planes; the second and fourth for the trunk sagittal rotations; the first and third for the trunk frontal rotations. No significant extrinsic harmonics, that is, those due to asymmetries in the walking pattern, were found in our experiments except for the first harmonic of the trunk sagittal rotation. For each subject all harmonics showed a good phase repeatability irrespective of walking speed. The harmonics showing a good inter-individual repeatability are depicted in Figure 1A, where the envelope diagrams of the vectorial representations of the harmonics in the polar plane are given for the two speed ranges. The harmonics that resulted in being more scattered in phase are given in Figure 1B.

As shown in Figure 1A, the harmonic's amplitude increased with the walking speed, except the second harmonic of the trunk rotation in the sagittal plane. The above harmonics are apt to be a component of the female ambulation stereotypic pattern.

3) Comparison of the female and male ambulation stereotypic patterns. The masculine pattern was assessed in a previous investigation (Cappozzo et al., 1982). Two main differences are evident: a) the first harmonic of the trunk frontal rotation is part of the masculine stereotypic pattern (phase range from 30° to 100°), while in women's ambulation it is largely scattered in phase; b) in women the first harmonic of the pelvis frontal rotation showed an amplitude approximately double that in man.

Discussion

In the preceding section the stereotypic pattern of the female ambulation was assessed in terms of repeatability of the intrinsic harmonic phase values. As discussed elsewhere (Cappozzo et al., 1982), the phases of the harmonics are relevant as far as the shape of the time course of the variable which they represent is concerned. Variables described using harmonics with the same phases have the same shape; maxima and minima of the time course change in dependence of the absolute and relative values of the harmonic amplitudes. Therefore it can be said that the shape of the upper body angular displacements, to which the harmonics in Figure 1A refer, tends to be an invariant characteristic of level walking irrespective of walking speed.

What should, in this context, be the meaning of the large scatter of the phase values of the trunk frontal rotation first harmonic? A possible suggestion, based on anatomical evidence, is that women do not need to synchronize the upper body lateral bending with the step phases because the lateral displacement of the pelvis gives the prevailing contribution to the alignment of the center of gravity over the supporting foot.

In relation to the differences between female and male amplitude of pelvic tilt, the following can be stressed. It has been emphasized by Murray et al. (1970) that the vertical displacement of the head is smaller in women than in men. The same authors associated this female behavior with the reduced stride length and with the more pronounced knee flexion during stance. The larger frontal pelvic tilt shown by our findings is a cooperating mechanism in view of the vertical upper body displacement reduction. The anatomical findings of Brinkmann, Hoefert, and Jongen (1981) relative to the sex-dependency of the dimensions of the femoral head indicate that the hips are intrinsically more exposed to stress in women than in men. All three mentioned mechanisms seem to aim at compensating for the above female anatomical characteristic and to cooperate in enhancing the specific shock absorption capability of the musculoskeletal system.

The role of the upper body rotations with respect to the enhancement of the reliability and the safety of the musculo-skeletal system during ambulation (Cappozzo et al., 1978) seems to be confirmed by these results.

References

BRINKMANN, P., Hoefert, H., & Jongen, H.Th. (1981). Sex differences in the skeletal geometry of the human pelvis and hip joint. *Journal of Biomechanics,* **14**, 427-430.

CAPPOZZO, A., Figura, F., Leo, T., & Marchetti, M. (1978). Movements and mechanical energy changes in the upper part of the human body during walking. In E. Asmussen & K. Jorgensen (Eds.), *Biomechanics VI-A* (pp. 272-279). Baltimore: University Park Press.

CAPPOZZO, A. (1981). Analysis of the linear displacement of the head and trunk during walking at different speeds. *Journal of Biomechanics,* **14**, 411-425.

CAPPOZZO, A., Figura, F., Gazzani, F., Leo, T., & Marchetti, M. (1982). Angular displacements in the upper body of AK amputees during level walking. *Prosthetics and Orthotics International,* **6**, 131-138.

MURRAY, M.P., Kory, R.C., & Sepic, S.B. (1970). Walking pattern of normal women. *Archives of Physical Medicine and Rehabilitation,* **51**, 637-650.

SAUNDERS, J.B., Inman, V.T., & Eberhart, H.D. (1953). The major determinants in normal and pathological gait. *Journal of Bone and Joint Surgery,* **35A**, 543-558.

Development of Gait in Childhood

M. Noguchi and A. Hamamura
National Sanatorium Nagasaki Hospital, Nagasaki, Japan

N. Matsusaka, M. Fujita, T. Norimatsu, S. Ikeda, and R. Suzuki
Nagasaki University School of Medicine, Nagasaki, Japan

Generally children start bipedal walking at approximately the end of the first year of life, due to the maturation of the neuromuscular system. Their gait pattern, however, is quite different from that of an adult. Therefore it is important to understand the normal gait pattern of children in order to evaluate their pathological gait.

The purpose of this paper is to assess the characteristics of the gait pattern in childhood and to estimate the time of gait maturation, primarily using a force plate.

Subjects and Methods

The gaits of 57 normal children between the ages of 3 and 10 years were studied. For this study a newly developed computerized gait analysis system and force plate (Anima Corp., Tokyo, Japan) were used.

The subjects were asked to walk freely on a large force plate which recorded the three dimensional components of the ground reaction force of several steps (see Figure 1). Vertical, fore-aft, and lateral components of the ground reaction force were normalized by microcomputer and analyzed by making Lissajous figures. The locus of the center of pressure (C.O.P.) was also obtained. The data were plotted by X-Y plotter and compared with those of 30 normal adults (see Figure 2).

Results

For the lateral component, the peak of normalized lateral force was higher than that of adults (see Figure 6). Y1 became the adult pattern at 10 years old, Y2 at 8 years old and YO at 9 years old (see Figure 3).

In the fore and aft component, the peaks of the heel-strike (XI) were higher than the peaks of the toe-off (X2), particularly when the subjects were younger (see Figure

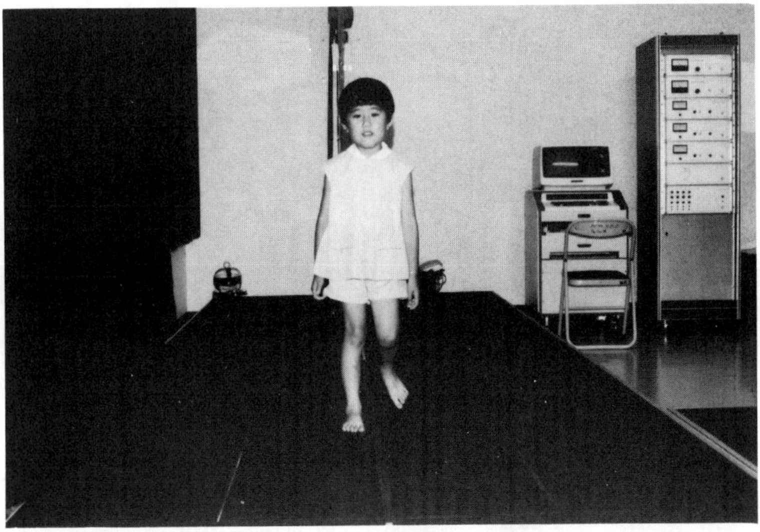

Figure 1—The subjects were asked to walk freely on a large force plate.

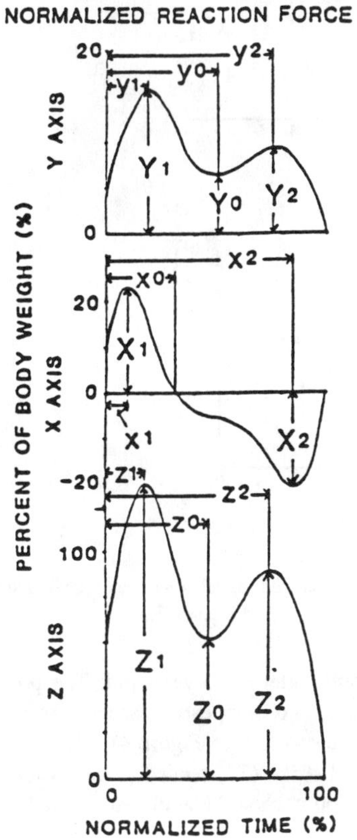

Figure 2.

Y	:	Lateral component
Y1	:	The peaks of the heel-strike
Y2	:	The peaks of the toe-off
YO	:	The valley of Y1 and Y2
X	:	Fore and aft component
X1	:	The peaks of the heel-strike
X2	:	The peaks of the toe-off
xO	:	Normalized time (N.T.) of the point at which the fore component changed to the aft component
Z	:	Vertical component
Z1	:	The peaks of the heel-strike
Z2	:	The peaks of the toe-off
ZO	:	The valley of Z1 and Z2

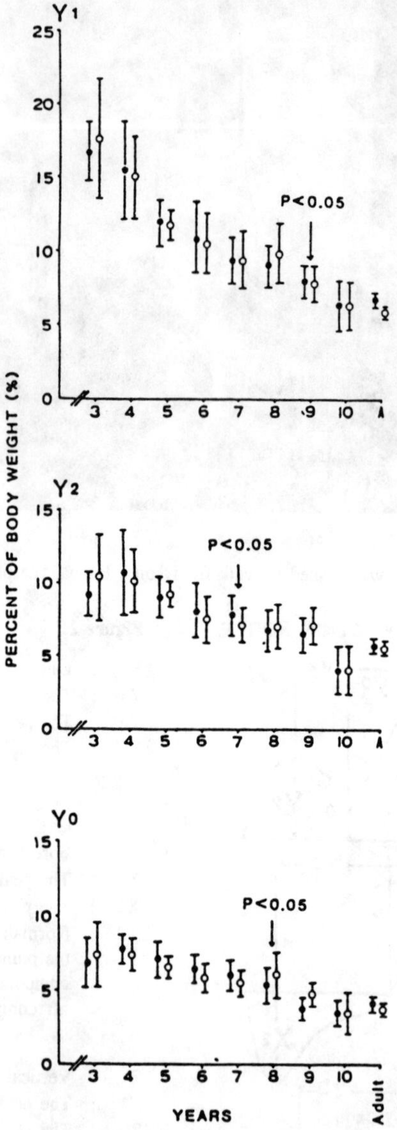

Figure 3—In the lateral component, the peak of normalized lateral force was higher than that of adults. Y1 became the adult pattern at 10 years old, Y2 at 8 years old.

6). The ratio of XI to X2 approached the adult pattern at 5 years old. The point of changing from fore component to aft component (xO) appeared earlier than that of adults but approached the adult pattern at 6 years old (see Figure 4).

In the vertical component, the peaks of heel-strike (Z1) tended to be higher than the peaks of toe-off (Z2) (see Figure 6). Z2 approached the adult pattern at 6 years old and ZO at 6 years old (see Figure 5).

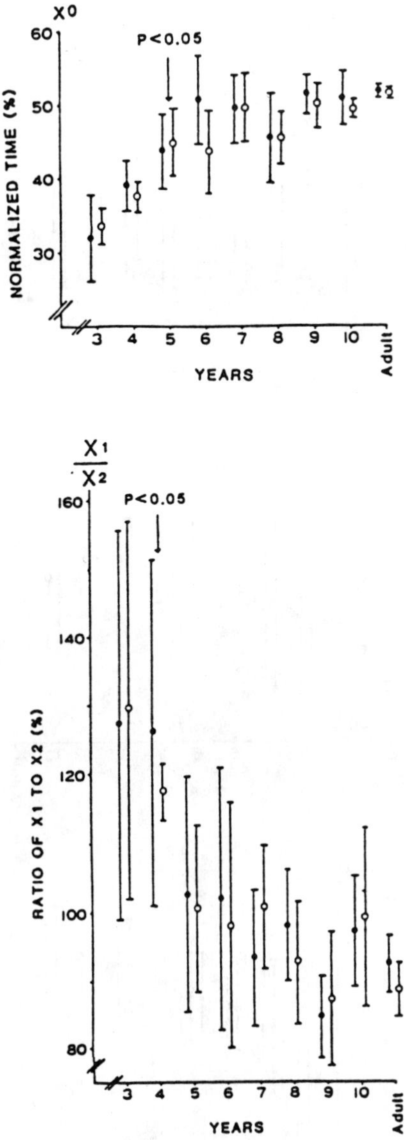

Figure 4—In the fore and aft component, xO became the adult pattern at 6 years old and X1/XO at 5 years old.

Discussion

The time of gait maturation depended on the parameter chosen. For example, a mature gait pattern as determined by duration of single-limb stance, walking velocity, cadence, step length, and the ratio of pelvic span to ankle spread was well established by the

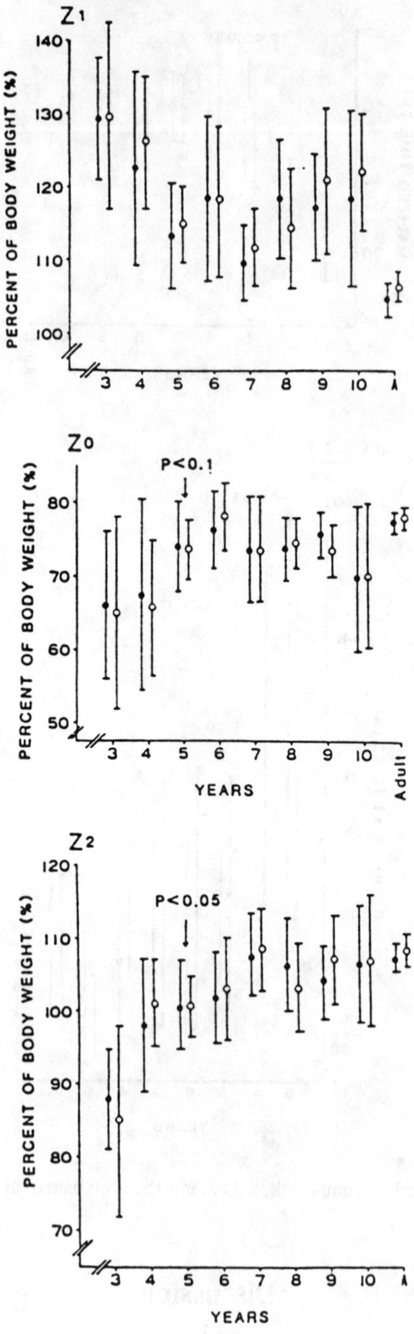

Figure 5—In the vertical component, the peaks of the heel-strike (Z1) tended to be higher than the peaks of the toe-off (Z2). Z2 became the adult pattern at 6 years old and ZO, the same.

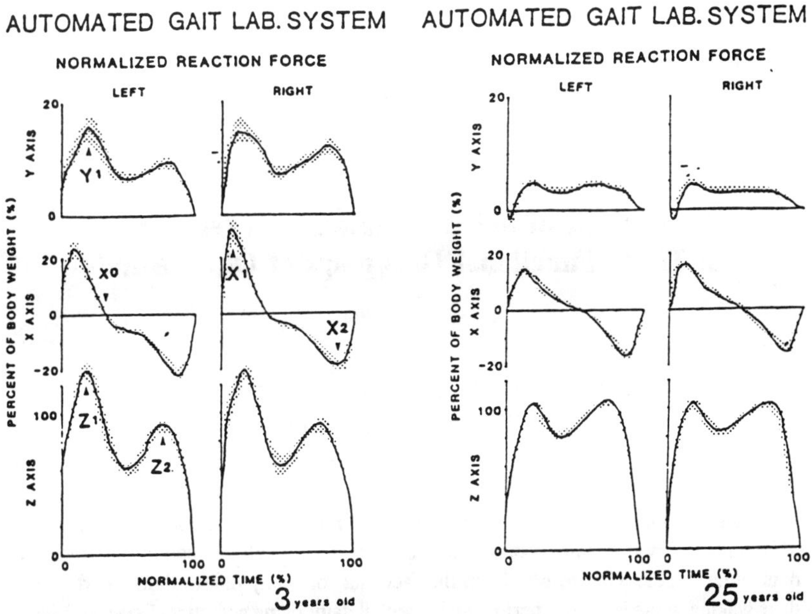

Figure 6—(a) A 3-year-old girl. **(b)** A 25-year old man. In the lateral component (Y), the peak of normalized lateral force was higher than that of adults.

In the fore and aft component (X), the peaks of the heel-strike (X1) were higher than the peaks of the toe-off (Z2) particularly when the subjects were younger. The point at which the fore component changed to the aft component (xO) appeared earlier than that of adults.

In the vertical component (Z), the peaks of the heel-strike (Z1) tended to be higher than the peaks of the toe-off (Z2).

age of three years. The primary fore and aft component approached the adult pattern at about 5 years old, the secondary vertical component at about 6 years old and finally the lateral component at about 9 years old. We propose that the development of the fore and aft component is necessary for the primary gait, while the development of the lateral component is not necessary for the primary gait but is necessary for a complete gait.

References

BECK, R.J. (1981). Changes in the gait pattern of growing children. *Journal of Bone and Joint Surgery,* **63A,** 1452-1457.

SUTHERLAND, D.H., Olshen, R., Cooper, L., & Woo, S.L.Y. (1980). The development of mature gait. *Journal of Bone and Joint Surgery,* **62A,** 336-353.

Analysis of Selected Gait Characteristics in Three Functional Groupings of Older Adults

G.W. Marino, C.A. Finch, and J.H. Duthie
University of Windsor, Windsor, Ontario, Canada

Since walking is at least a minor component of most daily living activities, the need to be ambulatory is a major requirement for noninstitutionalized, community independent older adults. This results from the fact that for many older adults nearly all of their waking hours are spent performing essential daily living activities (Cooper, Adrian, & Glassow, 1982).

Although the importance of walking as a basic skill for older adults is obvious, it appears that walking has not been of primary concern in studies of older adults. In contrast, most studies investigating the physical domain of the population in question have focused on either the physical fitness or the technical aspects of more task specific daily living activities (Faletti, 1982; Frekany & Leslie, 1975; Sidney & Shepard, 1977). In view of the necessity of walking in the performance of many essential living skills it was felt that a study of basic walking mechanics might provide: (1) information allowing an initial differentiation between functional groupings of older adults, and (2) the basis from which rehabilitative programs for high risk older adults could be developed.

The purpose of this study was to assess selected characteristics of the walking movement patterns of three functional groups of older adults. These groups included: a) those who are community independent, b) those with a high risk of institutionalization, and c) the institutionalized. Between-group differences, within-group differences and within-subject differences were used to differentiate between the functional capabilities of the three groups.

Methods

The normal walking patterns of 54 subjects were filmed with a Locam 16 mm camera set to operate at 100 frames/s. No attempt was made to pace the subjects and each subject was filmed during the latter stages of a 15 m walk to ensure the recording of a typical sequence of walking strides.

Film analysis of two strides per subject was facilitated through use of a Vanguard M-16 projector, a Numonics graphics calculator, and an Apple II microcomputer system. The variables selected for analysis included horizontal velocity (HV), stride length (SL), stride rate (SR), and support time percentages (SS, DS) and time of stride (TOS). In addition, vertical displacement of the center of gravity, range of motion of arm and forearm segments and hip, knee, and ankle angles at four strategic points during the stride were measured. Summary statistics, One Way Analysis of Variance, the Sheffe Multiple Comparisons test, and Pearson Product-Moment correlation analysis were used to determine statistically significant between- and within-group differences and relationships.

Results

Means of basic stride characteristics of each of the two strides analyzed for each group of subjects are summarized in Table 1. It was determined that community independent older adults walked at a higher velocity and with longer strides than high risk and institutionalized subjects. No statistically significant differences were found in stride rates in a between-group comparison. Although the differences were not statistically significant due to high variability, it was also noted that independent subjects tended to spend a greater percentage of the time of stride (T.O.S.) in single support (S.S.) and less in double support (D.S.) than did the subjects in the other groups. No significant between-trial differences were found for any of the variables in any of the three groups. These results indicate a high degree of stride-to-stride consistency even among subjects in both the institutionalized and high risk groups. Although stride rates and therefore stride times were similar for all groups, the independent subjects spent a greater proportion

Table 1

Summary of Basic Stride Characteristics of Three Functional Groups of Older Adults

Group	Variable	Stride 1	Stride 2
Community independent	*HV (M/S)	.96 (.17)	.99 (.23)
(N = 21)	*SL (M)	.55 (.09)	.58 (.12)
	SR (ST/S)	1.73 (.17)	1.73 (.20)
	SS (% of TOS)	67	68
	DS (% of TOS)	33	32
High risk of institutionalization	HV (M/S)	.76 (.26)	.77 (.27)
(N = 16)	SL (M)	.43 (.09)	.46 (.10)
	SR (ST/S)	1.71 (.30)	1.64 (.33)
	SS (% of TOS)	61	61
	DS (% of TOS)	39	39
Institutionalized	HV (M/S)	.71 (.19)	.72 (.22)
(N = 17)	SL (M)	.43 (.08)	.45 (.10)
	SR (ST/S)	1.57 (.31)	1.57 (.23)
	SS (% of TOS)	63	62
	DS (% of TOS)	37	38

*Statistically significant differences between groups ($p < .05$)

Table 2
Correlation Coefficients for Both Between-Variable and Within-Variable Relationships (N = 54)

Relationship	Variables	R
Between-variable relationships	HV vs SL	.91*
	HV vs SR	.75*
	SL vs SR	.43*
Within-variable relationships	HV (St. 1) vs HV (St. 2)	.88*
	SL (St. 1) vs SL (St. 2)	.79*
	SR (St. 1) vs SR (St. 2)	.90*

*Statistically significant at $p < .05$.

of each stride in single support. It appears that this is indicative of a normal adult walking pattern and that the other less capable groups had developed walking patterns with somewhat longer double support periods in order to enhance dynamic balance.

In order to assess both between-variable and within-variable relationships, Pearson Product-Moment correlation coefficients were developed (see Table 2). Strong statistically significant ($p < .05$) positive relationships were found between walking velocity and both length and rate of striding. In fact, 83% ($.91^2$) of the variance in velocity is common to stride length. There was also a statistically significant relationship between stride length and rate indicating that those subjects with longer strides tended to have higher stride rates.

Stride-to-stride consistency was again apparent in the within-variable (between-stride) correlation analysis. Statistically significant relationships were found for each of stride length, stride rate, and walking velocity measured in the two strides selected for analysis. These results reinforced findings from the analyses of variance that even the high risk and institutionalized subjects did not exhibit significant variability from stride to stride.

In addition to the stride characteristics previously described, analysis of several postural variables was also undertaken. These included hip (H), knee (K), and ankle (A) angles at each of takeoff (TO), mid-swing (SW), touchdown (TD) and mid-support (SU) as well as range of motion of the arm around the shoulder (DA) and forearm around the elbow (DF) during the stride and vertical deviation of the center of gravity (DCG). Analysis of variance was used to identify between-group differences in each of these variables. Summary statistics of the postural variables are included in Table 3.

Results of the postural analysis indicate that to a great extent the walking patterns of the three groups of older adults are similar. There are, however, two key differences. The community independent group exhibited significantly greater knee flexion at both the point of takeoff and at mid-swing. Although further analysis was not undertaken, it is possible to speculate that these stride characteristics of the independent group were integral in a more efficient walking pattern which resulted in longer strides and higher velocities. Subjective assessment of both the film and actual experimental performances indicated that many subjects in both the high risk and institutionalized groups walked with a rigid, shuffling style gait rather than the fluid, energy efficient walking pattern associated with normal adult locomotion.

Table 3

Summary of Postural Variables of Three Functional Groups of Older Adults

Variable	Independent (N = 21)	Group Mean Data High Risk (N = 16)	Institutionalized (N = 17)
HTO (radians)	3.20 (.17)	3.22 (.08)	3.15 (.26)
KTO*	2.49 (.08)	2.73 (.13)	2.69 (.10)
ATO	1.98 (.19)	1.96 (.10)	1.98 (.17)
HSW	2.87 (.12)	2.93 (.14)	2.94 (.24)
KSW*	2.33 (.21)	2.51 (.23)	2.68 (.24)
ASW	1.86 (.17)	1.91 (.19)	1.82 (.11)
HTD	2.71 (.11)	2.68 (.11)	2.79 (.21)
KTD	3.01 (.14)	2.79 (.15)	2.85 (.12)
ATD	1.92 (.06)	1.88 (.08)	1.95 (.14)
HSU	3.05 (.11)	3.03 (.10)	3.06 (.19)
KSU	2.87 (.12)	2.83 (.12)	2.84 (.16)
ASU	1.67 (.18)	1.64 (.10)	1.64 (.12)
DA	.38 (.08)	.39 (.17)	.62 (.27)
DF	.64 (.07)	.51 (.25)	.82 (.48)
DCG (M)	.06	.07	.08

*Statistically significant between-group differences at $p < 05$.

Analysis of arm and forearm motions during the stride revealed no significant between-group differences. It is of interest, however, to note that with respect to these movements, the high risk and institutionalized groups were much more erratic than the independent group. Some subjects walked with almost no arm actions while others exhibited a wide range of motion at both the shoulder and elbow.

Summary and Conclusions

In summary, it has been shown that selected biomechanical features can be used to differentiate between functional groupings of older adults. Although more detailed investigation is warranted, it appears that gait analysis may eventually provide a basic model from which intervention techniques can be developed in an attempt to rehabilitate high risk and possibly institutionalized individuals and return them to a state of community independence. With specific reference to the variables measured in this study, it is appropriate to draw several conclusions. First, variables in the walking stride can be used to differentiate between three functional groups of older adults. Second, community independent older adults walk with postural characteristics similar to other older adult groups but with higher walking velocities, longer stride lengths and greater knee flexion at the takeoff and mid-swing phases of the stride. Finally, although variance differs between groups, members of all three groups of older adults exhibit stride-to-stride consistency in all of the kinematic and postural variables measured in this study.

Acknowledgments

Supported in part by grant No. 92-91 from the Gerontology Research Council of Ontario, J.H. Duthie, principal investigator.

Supported in part by grant No. 9507 from NSERC through the University of Windsor Research Board, G.W. Marino, principal investigator.

References

BLENKNER, M. (1967). Environmental change and the aging individual. *Gerontologist*, **7**, 101-105.

COOPER, J.M., Adrian, M., & Glassow, R.B. (1982). *Kinesiology*. St. Louis: C.V. Mosby Co.

FALETTI, M.V. (1982). *Human factors analysis of functional ability in daily living* Presented at The Annual Meeting of the Gerontological Society of America. November, 1982.

FREKANY, G.A., & Leslie, D.K. (1975). Effects of an exercise program on selected flexibility measurements of senior citizens. *Gerontologist*, **15**, 182-183.

HAYER, W.J., Kafer, R.A., Simpson, S.C., & Hayer, F.W. (1976). Reinstatement of verbal behavior in elderly mental patients using operant procedures. *Gerontologist*, **14**, 149-152.

SIDNEY, K., & Shepard, R. (1977). Activity patterns of elderly men and women. *Journal of Gerontology*, **32**, 25-32.

Human Locomotion via Optimal Programming

Y. Morel, P. Bourassa, and B. Marcos
Université de Sherbrooke, Sherbrooke, Quebec, Canada

In the last decade, three methods frequently have been used in locomotion studies: namely, the direct dynamic problem approach (e.g., Boccardi & Pedotti, 1981; Onyshko & Winter, 1980), the inverse dynamic problem approach, and the optimization methods with minimizing principles (Beckett & Chang, 1968; Chow & Jacobson, 1971; Hatze, 1977; Vukobratovic, 1978).

Mathematical Model

The model for this study, Figure 1, consists of one segment for the upper body and a set of lower extremities consisting of the thigh and the shank. Motion takes place in the sagittal plane with rotation around roller type joints. External work comes from the muscular couple C_1 for the thigh and C_2 for the shank and the ankle reactions in the deploy and stance phase. Hip motion is prescribed over the stance, deploy, and swing phase of the stride cycle. Thus the trunk motion may be uncoupled and treated separately. Cyclic and symmetrical motion is assumed for each leg. With q_1 and q_2 as the generalized angular coordinates for the thigh and shank, the Lagrange equations become

$$\frac{d}{dt}(\partial L/\partial \dot{q}i) - \partial L/\partial qi = Qi \qquad (1)$$

where L is the Lagrangian and Qi the generalized external force. A state variable description using $\dot{q}_1 = q_3$, $\dot{q}_2 = q_4$, yields a system of four differential equations in q_1, $C_1(t)$, $C_2(t)$

$$\{\dot{q}_i\} = \{F(q)\} + \{C(t)\} + \{G(q,R_a,C_a,t)\} \qquad (2)$$

where {F} is the RHS contribution of the kinetic and potential energy in the Lagrange equations for the variables q_1 and q_2, {C(t)} is the generalized force vector for the muscular effort, and {G} the generalized force and moment at the ankle, considered zero in the swing phase.

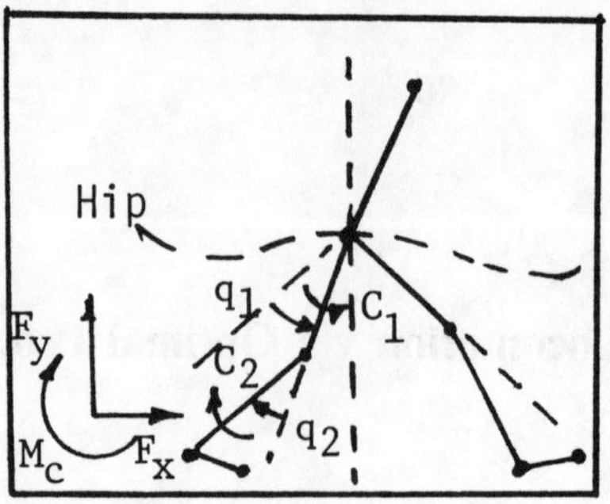

Figure 1—Segment model.

Stance Phase

Beginning at the end of the swing phase, this motion lasts for about 50% of the stride cycle. Ankle motion is prescribed so that two nonlinear constraints linking the hip to ankle projections in the X and Y axis must be satisfied at all times.

$$g_i(q_1,q_2) = 0, \ i = 1,2 \qquad (t_0 < t < t_1) \tag{3}$$

A Newton-Raphson algorithm applied to Equation 3 yields the angles q_1 and q_2. Differentials of first and second order of g_i are also zero and provide linear relationships for the angular velocities and accelerations q_3, q_4, $\dot{q}_3$, and $\dot{q}_4$. No optimization is necessary in this phase, and the procedure is almost straightforward. A simulation of this motion, Morel (1983), is given together with corresponding net moments given in Figure 2. The ankle moment was found to be predominant in this instance.

Deploy Phase

This phase which goes from the end of the stance phase t_1 and lasts until t_2 when the ground reaction force is zero is very important from the energetic point of view even

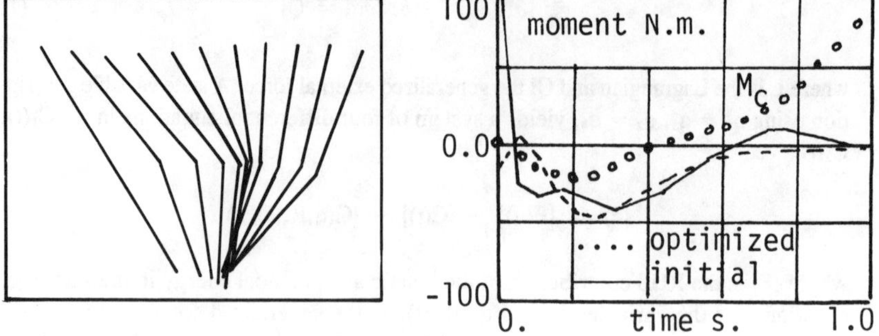

Figure 2—Stance phase sequence and net moment.

if it lasts for only 10% of the stride period. During that phase, the ankle is assumed to rotate around the ball of the foot at a radius d and at an unspecified angle $\alpha(t)$ with respect to the ground level. Ankle force and moment Ra and C_a are calculated from the ground reactions at the ball of the foot. This deploy phase problem is one in which the motion (Equation 2) must be satisfied simultaneously with the constraint (Equation 3) for the hip-to-ground projections and a minimized function of muscular activity in the form $(C_1^2 + C_2^2)$.

$$J = \min[\int_{t_1}^{t_2}(r_1C_1^2(t) + r_2C_2^2(t)]dt + J_3(t_2) \quad (4)$$

where $J_3(t_2)$ is a final state configuration aimed for at the end of the deploy phase.

The system consisting of Equations 2, 3, and 4 yields a solution for the four generalized coordinates q_i as well as for C_1, C_2, and α.

The final state functional J_3 is of the form

$$J_3 = \sum_{1}^{3} \gamma_i\phi_i(t_2) \quad (5)$$

where

$$\begin{aligned} \phi_1(t_2) &= [q_3(t_2) - q_3^*(t_2)]^2 \\ \phi_2(t_2) &= [q_4(t_2) - q_4^*(t_2)]^2 \quad (6) \\ \phi_3(t_2) &= [\alpha(t_2) - \alpha^*(t_2)]^2 \end{aligned}$$

with γi being appropriate weighting factors for each ϕi and q_i^* and α^* being specified, aimed-for values of the time variables $q_3(t)$, $q_4(t)$, and $\alpha(t)$.

Foot angle α is eliminated between g_1 and g_2, leading to one constraint equation $(s[q_i,t] = 0)$ for all t in the range (t_1, t_2). Introducing $\dot{q}_5(t) = s^2$, the constraint will tend to be satisfied if $J_2 = q_5(t_2)$ is minimized.

Finding optimal values of the muscular couples $C_1(t)$ and $C_2(t)$ in one integration is not a likely outcome. Instead, an iterative procedure is followed in which the generalized working reactions at the ankle are brought progressively through weighting factors from zero to full scale. At the end of each iteration, the initial control vector C_1, C_2 is corrected to take into account the increment in the weight factor of the ankle reactions. The sequence of operation is as follows. With

$$\begin{aligned} &\{C\}^t = \{o,o,C_1,C_2^o\},\ \{w\}^t = \{o,o,w_1,w_2^o\},\ \{q\}^t = \{q_1,q_2,q_3,q_4\} \\ &\{F\{^t = \{q_3,q_4,F_1(q_1,q_2),F_2(q_1,q_2)\},\ \{G\}^t = \{o,o,G_1,G_2\} \quad (7) \end{aligned}$$

where G_1 and G_2 are functions of q_1, q_2, and ankle reactions. The iterative process brings a solution to Equation 2 with an initially small weight vector {w}. This vector is increased iteratively until it becomes (0,0,1,1). After the ith iteration, Equation 2 becomes

$$\{q\}_i^o = \{F_i\}_i^o + \{C\}_i^o + \{w\}_i^t\{G\}_i^o \quad (8)$$

where $\{\ \}_i^o$ stands for an optimized vector at iteration i. Increasing $\{w\}_i$ by $\{dw_i\}$, let

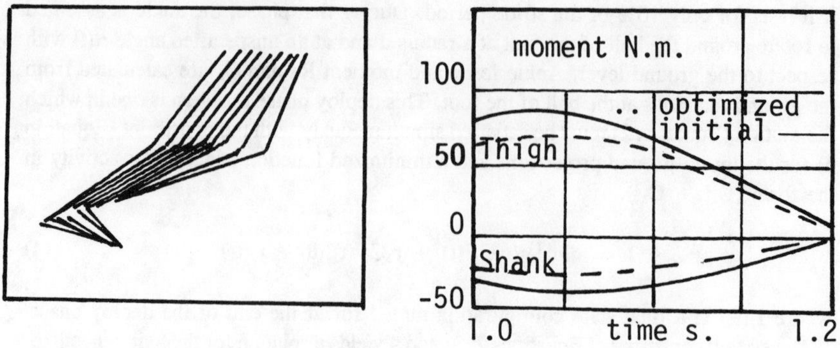

Figure 3—Deploy phase sequence and net moments.

$$\{C\}_{i+1} = \{C\}_i^o - \{dw_i\}_i^t\{G\}_i^o \tag{9}$$

then the integration of the system

$$\{q\}_{i+1} = \{F\}_{i+1} + \{C\}_{i+1} + \{w_i + dw_i\}^t\{G_i\}_{i+1} \tag{10}$$

leads to a new optimized vector. Typical simulation results for this phase are presented in Figure 3.

The iterative method so described is powerful but nevertheless is quite involved and other methods may prove to be much more efficient in this difficult phase. The model could be upgraded with an extra member from the ankle to the ground. A new ankle control vector element would thus appear together with two extra state variables for the member. There would meanwhile be no working reactions at the ankle joint, neither under the ball of the foot, and two constraint equations, g_i, would be available. With the minimization of the functional J a faster integrable system of nine equations would be available for this phase.

Swing Phase

This phase is characterized by the absence of ground reactions on the swinging leg and thus presents a lower level of difficulties. The swinging leg must clear the ground level at all times and must proceed in the forward direction. A constraint inequality of the form

$$g_3(q_1,q_2)\ 0$$

will thus provide feasible trajectories. Figure 4 depicts the leg motion in this phase and Figure 4 gives the corresponding optimal moments $C_1(t), C_2(t)$.

Conclusion

A human locomotion computer program was developed using a powerful method based on optimal programming. In the deploy phase, the minimization process was mean-

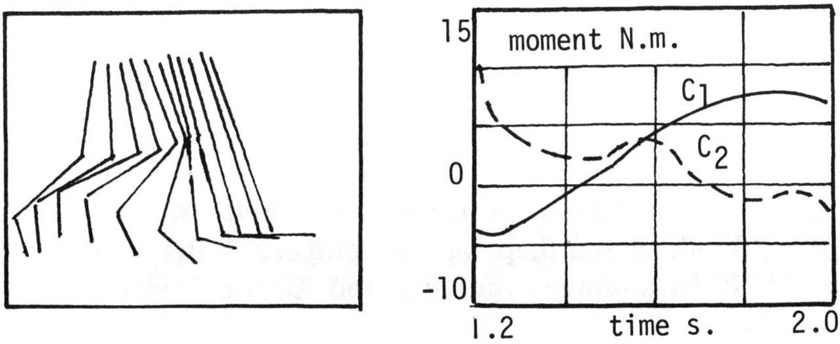

Figure 4—Swing phase sequence and net moment.

while hampered with a predominant contribution from the work of the external reactions at the ankle joint. In the swing phase, the constraint equation happened to be more simple and the convergence to an optimal solution easier. In the stance phase, apart from sensitivity to experimental data, the program development was almost straightforward.

Acknowledgment

This research was made possible with the help of Quebec FCAC grant (EQ-1783) and Canadian NSERC (A-2713).

References

BECKETT, R., & Chang, K. (1968). On the evaluation of the kinematics of gait by minimum energy. *Journal of Biomechanics*, **1**, 147-159.

BOCCARDI, S., & Pedotti, A. (1981). Evaluation of muscular moments at the lower limb joints by an on-line processing of kinematic data and ground reaction. *Journal of Biomechanics*, **14**, 35-45.

CHOW, C.K., & Jacobson, D.H. (1971). Studies of human locomotion via optimal programming. *Mathematical Biosciences*, **10**, 239-306.

HATZE, H. (1977). A complete set of control equations for the human musculoskeletal system. *Journal of Biomechanics*, **10**, 799-805.

MOREL, Y. (1983). Modelisation de la locomotion humaine par une méthode d'optimisation. Unpublished master's thesis, University of Sherbrooke, Quebec, Canada.

ONYSHKO, O., & Winter, D.A. (1980). A mathematical model for the dynamics of human locomotion. *Journal of Biomechanics*, **13**, 361-368.

VUKOBRATOVIC, M. (1978). Dynamics of the active articulated mechanisms and synthesis of artificial motion. *Mechanism and Machine Theory*, **13**, 1-56.

The Moment of the Calf Muscles in Walking and Stepping: A Comparison Between EMG-to-Force Processing and Kinetic Analysis

A.L. Hof
University of Groningen, Groningen, The Netherlands

C.N.A. Pronk and J.A. van Best
Erasmus University, Rotterdam, The Netherlands

In EMG-to-force processing (EFP) the force of a muscle (or the corresponding moment with respect to a joint axis) is determined from the electromyogram, which is related to the muscle activation, by means of a model of the muscle's contractile and elastic properties. The viability of this method has been successfully demonstrated for several types of contractions: isotonic, tilting, eccentric-concentric, on varous ergometers (Hof & van den Berg, 1981), but a similar demonstration for walking had not yet been done.

In kinetic analysis the total joint moment is deduced from the ground reaction forces and the positions and accelerations of the body segments by means of Newtonian equations. These two fundamentally different methods, EMG-to-force processing and kinetic analysis, are compared in this paper. The kinetic analysis was performed with the Integrated Gait Analysis System (IGAS) at Erasmus University, Rotterdam (Pronk, van Best, & van Eijndhoven, 1983). For the same subject, recordings were made of the ankle moment (IGAS) and the calf muscle moment around the ankle (EFP). If the moments due to other muscles can be neglected both curves should be identical.

Methods

Simultaneous recordings by means of EPF and IGAS were made on five subjects (see Table 1). They walked at average, slow, and fast speed and stepped up and down a 25-cm-high bench, three times each. The EMG of the tibialis anterior, antagonist of the calf muscles (triceps surae), was monitored throughout.

The EFP method and its calibration procedure have been described elsewhere (Hof & van den Berg, 1981). After the calibration the subject made a series of tilting contractions on the torque plate, which served as a reference. The raw EFP input signals

Table 1

Subject Data

Subject No.	Sex	Age (years)	Weight (kg)	Stature (m)	Leg Length (m)	Reference Moment (Nm)	Walking Speed v (m/s)	Step Length (m)
46	M	21	74.7	1.86	0.94	98	1.24	0.70
47	F	22	58.3	1.68	0.86	66	1.21	0.58
48	M	21	71.0	1.79	0.92	88	1.44	0.85
49	M	22	79.4	1.83	0.94	100	1.49	0.78
50	F	22	61.1	1.60	0.78	70	1.15	0.65

Note. The reference moment is defined as the moment in quiet standing on one leg, with the heel off the floor. The rightmost columns give speed and step length for the walking registrations of Figure 1.

(the EMGs of soleus and gastrocnemius and the goniometer signal for the ankle angle) were stored on an FM tape recorder (no. 1) and the processor output, the calf muscle moment M_m, on another one (no. 2) for later A/D conversion and computer analysis.

In the IGAS system the ground reaction force and the center of pressure were measured with a Kistler 9281A forceplate, inconspicuously mounted in a 10-m walkway. The positions of the body segments were recorded by a Selspot system, consisting of 14 LED body markers attached to the subject and with two cameras positioned at 5-m distances left and right of the walkway. Position and force data were sampled at 312 Hz, PCM modulated, stored on a tape recorder (no. 3) and processed off-line on a PDP 11/34 computer. The data processing was done with a system based on ENOCH (Gustafsson & Lanshammar, 1977; Pronk et al., 1983), which includes corrections for lens distortion and floor reflections and a calibration of the camera positions. A calibration was performed before each experiment. The moment of the ankle was calculated by means of Newtonian equations from the ground reaction force and the movement of the foot. Ground reaction force data were low-pass filtered up to 25 Hz.

The main source of error in the ankle moment M_a is the uncertainty in the position of the ankle axis. (Inertial and gravitational forces are very small for the foot in stance.) We estimated this error to be less than 1.5 cm. Together with a ground reaction force up to 1.5 times body weight, this results in a maximal IGAS error of 20 Nm.

IGAS yields the total ankle moment M_a, while EFP gives the moment due to the calf muscles M_m. A comparison is only possible as long as the moment due to other muscles can be neglected. In this study we assumed the moment from the calf muscle synergists (tibialis posterior and peronei) to be small. The moment due to the antagonist muscles cannot be neglected. We assumed that it was significant as long as there was a distinct EMG signal from the tibialis anterior and that it continued until 160 ms after cessation of this EMG. The latter time delay was estimated from previous calf muscle measurements. Only in the intermediate periods was a comparison possible between M_m and M_a.

A measure for this error is the coefficient of variation defined as follows:

$$\epsilon = \left[\int_{t_1}^{t_2} (M_m - M_a)^2\, dt \Big/ \int_{t_1}^{t_2} M_a^2 dt\right]^{1/2}$$

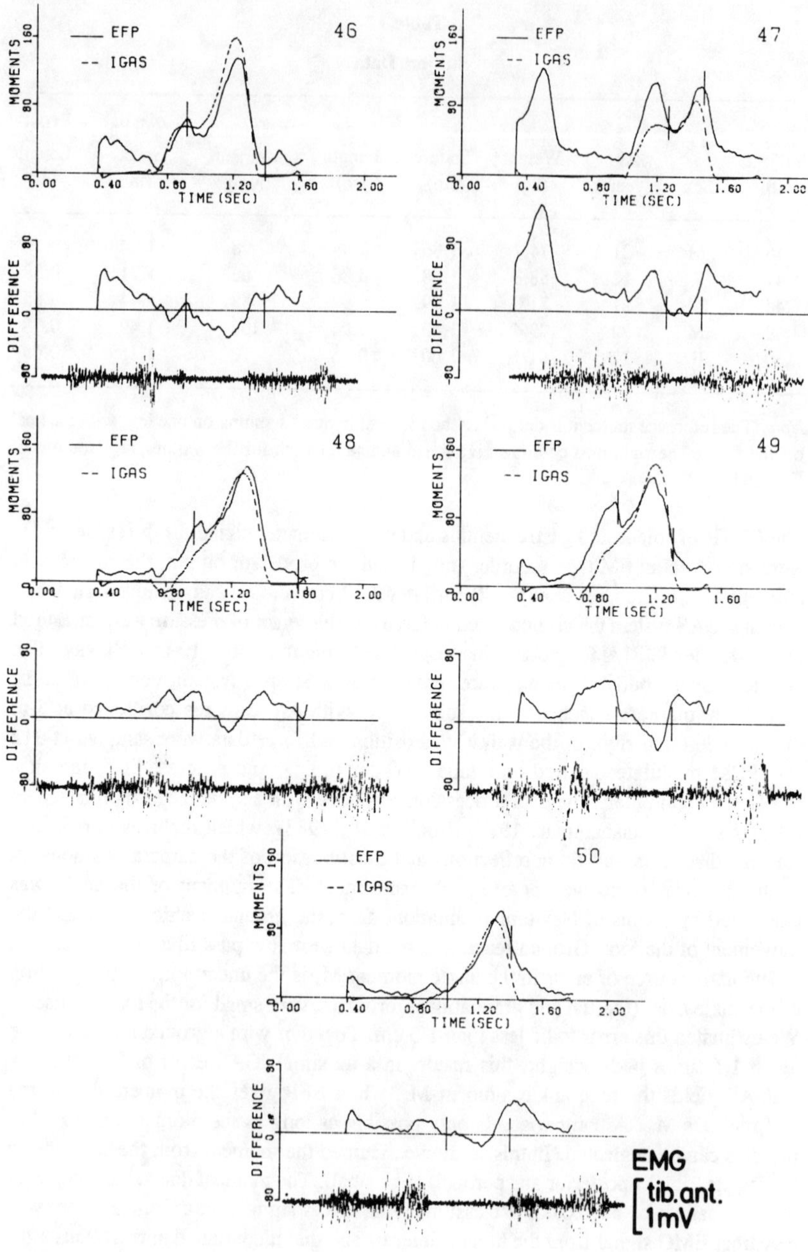

Figure 1—EFP calf muscle moment M_m (—), IGAS ankle moment M_a (----) and the difference $M_m - M_a$ for all subjects in normal walking. Below each figure is an original registration of the tibialis anterior EMG. The interval where the tibialis moment is supposed to be zero, and where EFP and IGAS curves can be compared, is between the vertical strokes. For speed and step length, see Table 1.

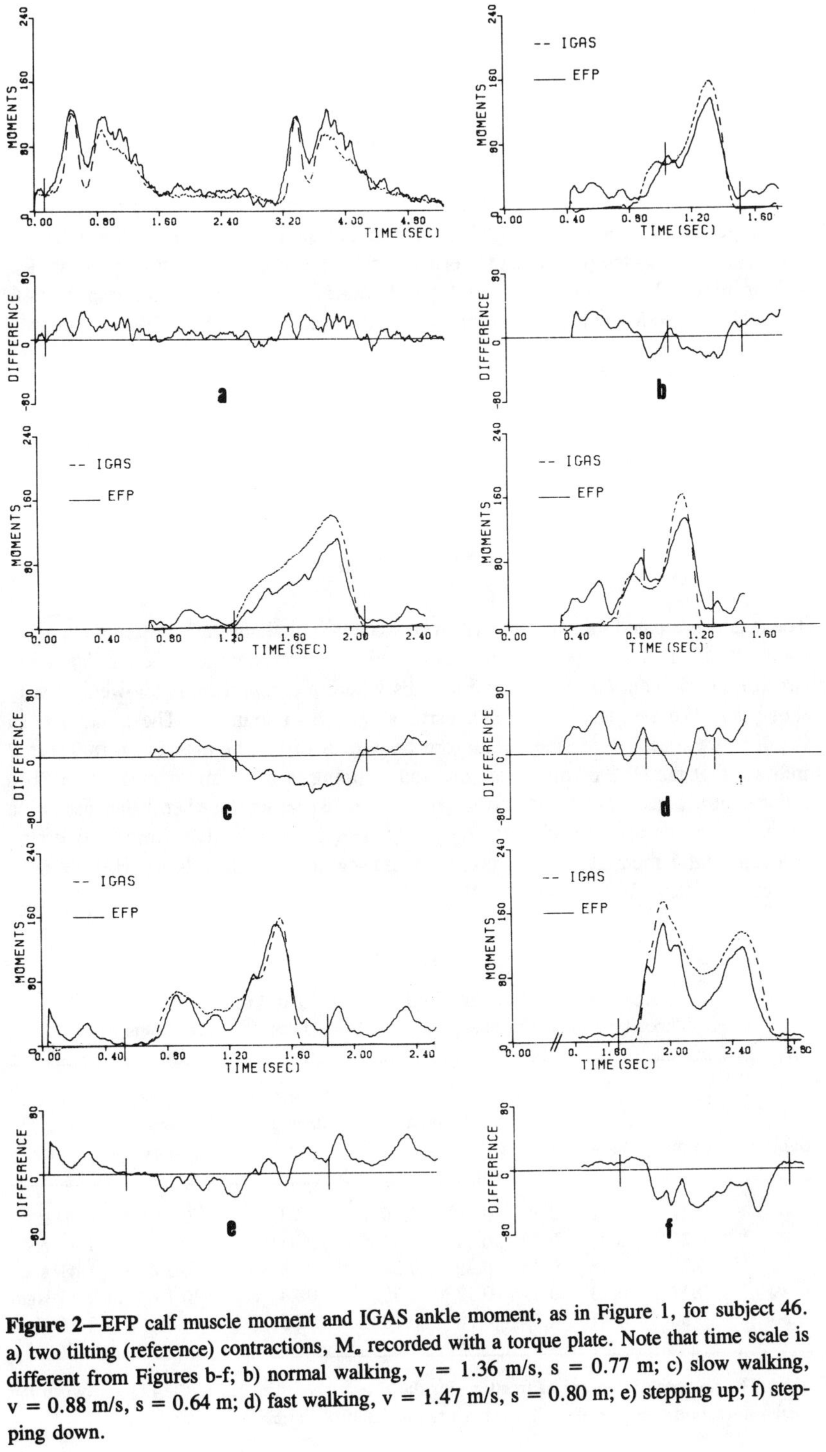

Figure 2—EFP calf muscle moment and IGAS ankle moment, as in Figure 1, for subject 46. a) two tilting (reference) contractions, M_a recorded with a torque plate. Note that time scale is different from Figures b-f; b) normal walking, v = 1.36 m/s, s = 0.77 m; c) slow walking, v = 0.88 m/s, s = 0.64 m; d) fast walking, v = 1.47 m/s, s = 0.80 m; e) stepping up; f) stepping down.

where t_1 and t_2 enclose the interval of measurement as defined above. (The factor $1/\epsilon$ is known as the signal to noise ratio.)

Results

Figure 1 shows data for one random recording at normal walking speed for each subject, together with the tibialis EMG. It is seen that the interval where the tibialis moment can be neglected (between the vertical strokes) can be quite short (e.g., in subject 47). In Figure 2 recordings of every type of contraction used is given for subject 46. This subject had a low tibialis activity throughout and thus EFP and IGAS values can best be compared.

In Table 2 the coefficient of variation ϵ has been given for all subjects and all types of contraction. A statistical analysis shows that ϵ in walking or stepping is not significantly greater than in the reference tilting contractions for three of the five subjects (F-test for equality of variance).

Discussion

The results presented show a comparison between two fundamentally different methods, each with its own sources of error. Taking this into consideration, the similarity between both moment curves, in the intervals where a comparison is allowed, is very satisfying. We estimated the IGAS error at 20 Nm maximum. The great part of the difference $M_m - M_a$ must therefore be due to EFP. The data given in Table 2 indicate that the EFP error in walking and stepping is of a similar magnitude as in tilting contractions. Results on these have been reported earlier, together with a discussion of the sources of error (Hof & van den Berg, 1981). Thus EMG-to-force processing is a dependable method for determining the muscle moment in walking (Hof, Geelen, & van den Berg, 1983).

Table 2

Coefficient of Variation, Time Weighted Average of Three (Walking and Stepping) and Two (Tilting) Contractions

	Walking			Stepping		Walking + Stepping	Tilting	
Subject	Normal	Slow	Fast	Up	Down	Average	(Reference)	F-ratio
46	0.21	0.32	0.20	0.19	0.24	0.24	0.25	0.90 n.s.
47	0.27	0.32	0.50	0.51	0.41	0.41	0.34	1.42 n.s.
48	0.24	0.28	0.23	0.25	0.24	0.24	0.20	1.41 n.s.
49	0.17	0.23	0.39	0.13	0.30	0.24	0.15	2.63 sign.
50	0.30	0.21	0.29	0.50	0.28	0.35	0.19	3.46 sign.

Note. The number of degrees of freedom for the F-test was estimated as the ratio between the total contraction time and the effective averaging time of the EFP.

When there is tibialis EMG activity, the difference $M_m - M_a$ might be interpreted as the moment of the antagonist muscle group. In slow walking the tibialis moment was always low, but at moderate and fast speed it increased in some of the subjects (no. 47, 48 and 49) up to 60 Nm around heel-strike. Subject 46, on the other hand, showed no evidence of substantial tibialis moment, except during swing. Interpreting $M_m - M_a$ as tibialis moment is hazardous, because the errors of both methods are cumulative. It seems, nevertheless, that we have obtained, for the first time, quantitative data on antagonist moments exerted in walking.

References

GUSTAFSSON, L., & Lanshammar, M. (1977). Unpublished doctoral dissertation, Uppsala University, Sweden.

HOF, A.L., & van den Berg, Jw. (1981). EMG to force processing, I-IV. *Journal of Biomechanics*, **14**, 747-792.

HOF, A.L., Geelen, B.A., & van den Berg. Jw. (1984). Calf muscle moment, work and efficiency in level walking; role of series elasticity. *Journal of Biomechanics*, **16**, 523-537.

PRONK, C.N.A., van Best, J.A., & van Eijndhoven, J.H.M. (1984). Data processing and software organization for an Integrated Gait Analysis System. In D. Winter, R. Norman, R. Wells, K. Hayes, & A. Patla (Eds.), *Biomechanics IX-B* (pp. 269-273). Champaign: Human Kinetics Publishers.

The Role of the Ankle Plantar Flexors in Level Walking

M. Fujita, N. Matsusaka, T. Norimatsu, and R. Suzuki
Nagasaki University School of Medicine, Nagasaki, Japan

T. Chiba
Ohmura Municipal Hospital, Nagasaki, Japan

There is some controversy concerning the role of the ankle plantar flexors in normal walking. Sutherland (1980) reported that the ankle plantar flexors do not propel the body forward although, paradoxically, maximum step length is not possible without the essential stabilizing effect of the plantar flexors. Simon (1978) concluded that the plantar flexors do not actively push or propel the body forward, but control the forward propulsive momentum, making it possible for the body to move farther forward from its base of support. While some authors agreed that there is a push-off or thrust upward through the leg occurring after heel-off, the entire push-off concept was questioned by others.

The purpose of this paper was to analyze the role of the ankle plantar flexors, especially during the second half of the stance phase.

Materials and Methods

Five normal male adult volunteers between 20 and 35 years of age were used. A flow chart of the methods used is shown in Figure 1. Each subject was asked to walk more than 10 times with bare feet in each of the walking styles on a flat table, on which a force plate was fixed. Electromyograms (EMGs), ground reaction forces (GRF), the foot switch (FS), and angular changes of the ankle and metatarsophalangeal (MP) joints were recorded simultaneously in various types of level walking. These types include normal, slow, fast, and longer stride styles of walking. The muscles examined were gastrocnemius (TS), peroneus longus (PL), tibialis posterior (TP), flexor hallucis longus (FHL), and flexor digitorum longus (FDL).

Results

Figure 2 shows simultaneous records of MEGs, GRF, FS and Ankle and MP angles. The Plantar flexors were shown to be active only during the midstance phase. They were not active at the moment of the peak of the backward component (B.C.) of the

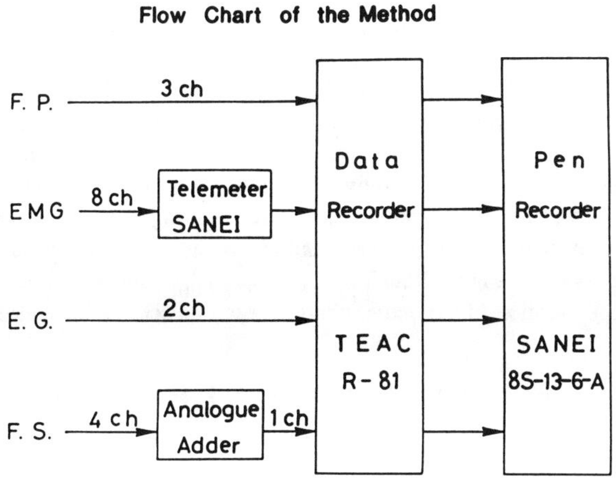

Figure 1—Force plate (FP), electromyograms (EMG), electrogoniometers (EG), and the foot switch (FS) were recorded simultaneously.

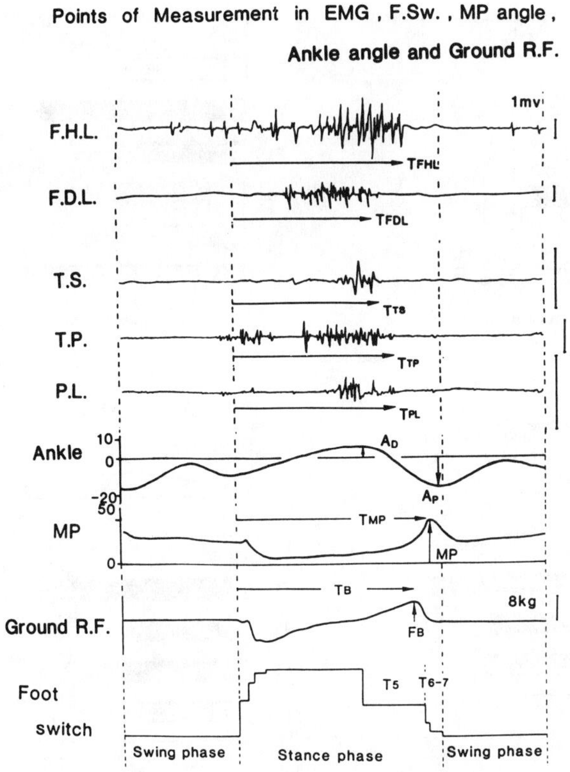

Figure 2—Simultaneous records of GRF, EMGs, FS, and ankle and MP angle. For details, see text.

ground reaction force. The peak of the B.C. appears prior to that of the peak of plantar flexion of the ankle and the extension of the first MP joint.

From the above data it would seem that the plantar flexors do not participate in the action of push-off in normal walking. During the last 10% of the stance phase the opposite foot is already in contact with the ground and the plantar flexors would not be required as a propulsive force. Moreover, we aimed to study the function of the plantar flexors in slow, fast, and longer stride styles of walking. To evaluate these findings more exactly key points were measured and compared between normal walking and other walking styles. In Figure 2 the T_{TS}, T_{PL}, T_{TP}, T_{FHL}, and T_{FDL} show the percentage of the time duration of the active phase of each muscle in the stance phase. A_D,

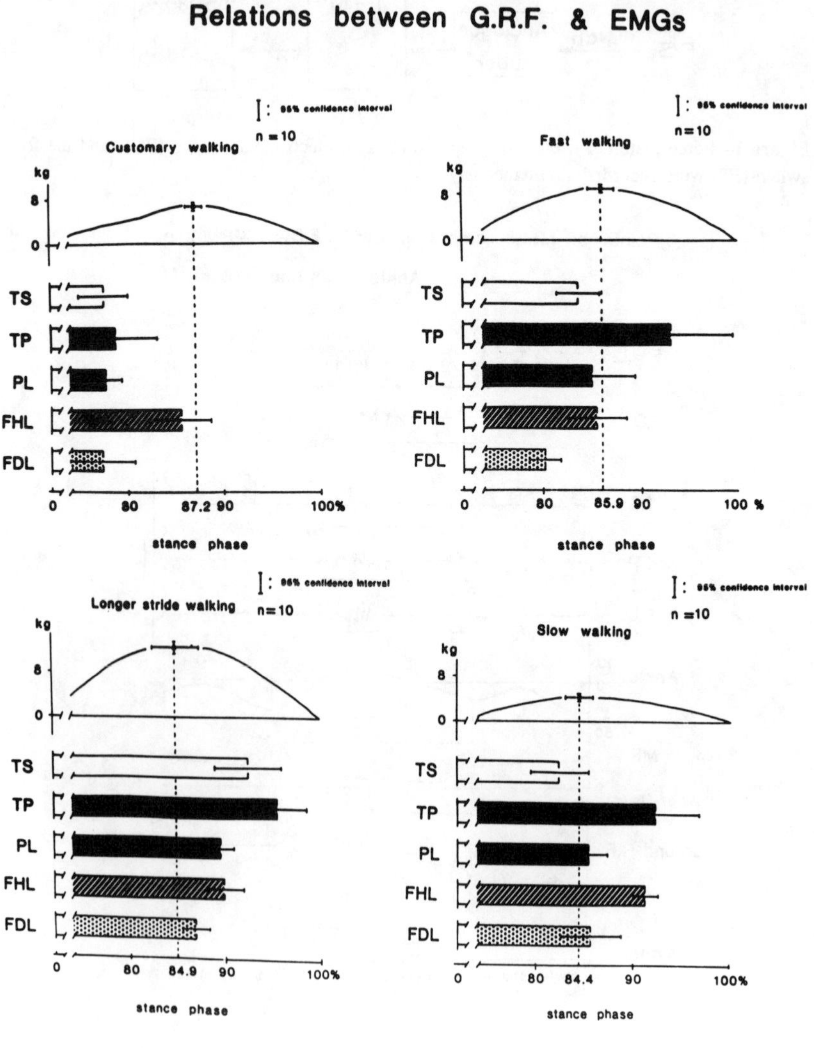

Figure 3—Relations between GRF and EMGs in each of the walking styles.

A_p, and MP are the peak of the angular changes of the ankle and the first MP joint. T_B is the percentage of the peak of the B.C. F_B is the peak of the B.C. T_{MP} is the percentage of the peak of dorsiflexion. T_5 and T_{6-7}show the percentage of floor contact time of the forefoot in the stance phase. As the parameters of the other walking styles were increased compared to those of customary walking, we suggest that the plantar flexors are required as propulsion forces. We paid particular attention to the relation beween GRF & EMGs, and when the active phase of plantar flexors is prolonged beyond the peak of the B.C. we suggested that the plantar flexors were required as propulsion forces. Figure 3 shows relations between GRF & EMGs in each of the

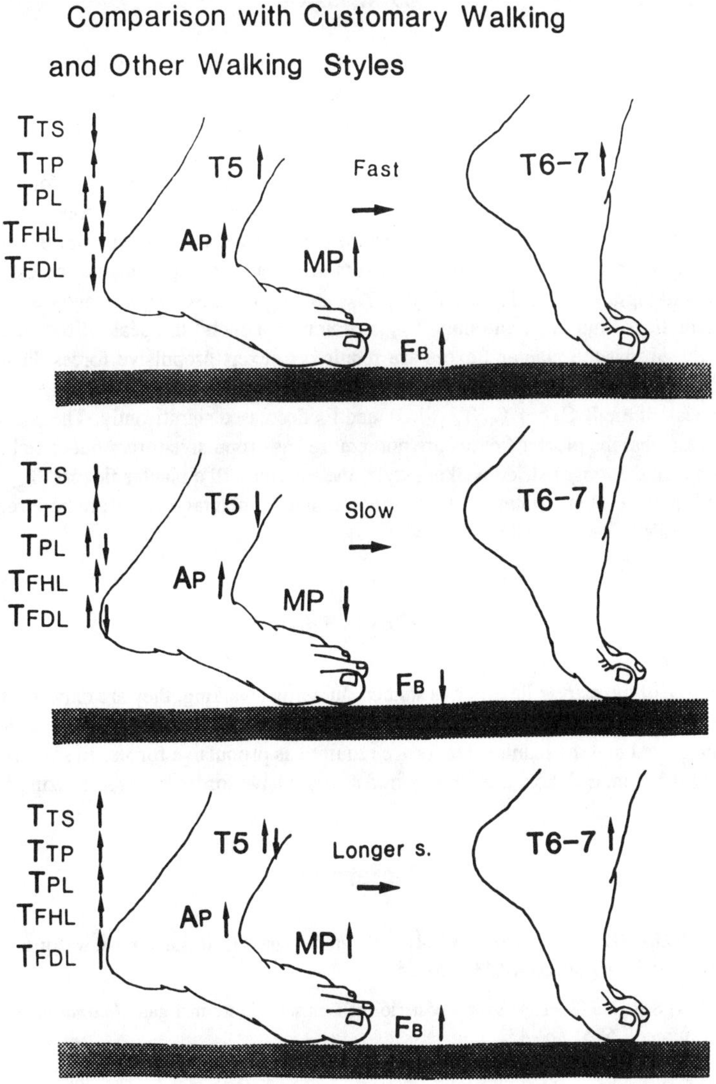

Figure 4—Comparison of normal walking with other styles of walking.

walking styles. In the fast walking style compared with normal walking, we notice that the active phase of each muscle is prolonged towards the toe-off period and T_{TP} is prolonged beyond the peak of the B.C. of the GRF. In the longer stride walking style compared with normal walking, we notice that the active phases of each muscle are prolonged beyond the peak of the B.C. In the slow walking style compared with normal walking, we find that the active phase of each muscle except T_{TS} is extended toward the toe-off period significantly (see Figure 4).

Discussion

In spite of many authors' studies, it has not yet been resolved whether plantar flexors contribute to push-off or roll-off in the second half of the stance phase. In our study we paid particular attention to this phase in which the function of the plantar flexors is most important. We would like to emphasize that it is important to research multilaterally the causal relation of many parameters. Many parameters, for example EMGs, FS, GRFs, and ankle and MP angle, were recorded simultaneously. As each of the parameters was increased in other walking styles compared with those in normal walking we suggest that the plantar flexors are required as propulsive forces. In the fast walking style, T_5, $T_{6\text{-}7}$, MP, A_D, A_P, and F_B increased significantly and in the plantar flexor muscles, and only T_{TP} is prolonged beyond the peak of the B.C. It is suggested that the plantar flexors are required to exert propulsive forces during the end of the stance phase. In the slow walking style, T_{TP} and T_{FHL} are prolonged beyond the peak of the B.C. but T_5, $T_{6\text{-}7}$, MP, and F_B decreased significantly. The above data suggests that the plantar flexors are not required as propulsive forces but as stabilizing forces. In the longer stride walking style, the activity of the plantar flexors, $T_{6\text{-}7}$, MP, and $F_B$increased significantly. It is supposed that the plantar flexors are here required as propulsion.

Conclusion

The role of the plantar flexors was studied. In normal walking, they are only accessories and act during roll-off to maintain floor contact. In fast and longer stride walking it is suggested that the plantar flexors are required as propulsive forces. In slow walking styles the plantar flexors are not required as propulsive forces but as stabilizing forces.

References

SUTHERLAND, D.H. (1980). The role of the ankle plantar flexors in normal walking. *Journal of Bone and Joint Surgery*, **62A**, 354-363.

SIMON, S.R. (1978). Role of the posterior calf muscles in normal gait. *Journal of Bone and Joint Surgery*, **60A**, 464-472.

Day-to-Day Repeatability of EMG Patterns in Gait

J.F. Yang

University of Waterloo, Ontario, Canada

Practice results in increased repeatability of a motor skill. Walking has thus been assumed to be consistent stride-to-stride and day-to-day. Researchers in gait have always reported single-day scores on a subject, usually single-stride scores within a day. This implies that a subject's gait is repeatable both within and between days, such that any one stride is a good representation of a subject's "true walking pattern." Although this assumption is likely valid for temporal and kinematic measures, it remains to be substantiated for kinetic and electromyogram (EMG) measures.

Previous studies (Arsenault & Winter, 1980; Elliott & Blanksby, 1976; Hershler & Milner, 1978) suggest that EMG patterns are reasonably repeatable within a day, such that an average of five strides adequately represents a subject's EMG patterns on 1 day. It remains to be determined whether this relationship holds for between-day comparisons.

The purpose of this study was to determine the repeatability of EMG patterns in gait, between days, for the medial hamstrings, lateral hamstrings, medial gastrocnemius, lateral gastrocnemius, and soleus muscles.

Methods

Bipolar silver/silver chloride surface electrodes 0.5 cm in diameter were applied to the muscle belly of the five muscle groups on the right lower extremity, after appropriate skin preparation, in 6 young adult subjects (height 1.61 to 1.89 m, mass 54.3 to 97.5 kg). The electrode sites were marked with silver nitrate to prevent positioning differences between days. The myoelectric and footswitch signals were telemetered via a 6-channel biotelemetry system. The received myoelectric signals were full-wave rectified and filtered at 3 Hz, to give a linear envelope (LE) signal, then A/D converted at 250 Hz.

Each subject was instructed to walk at his or her comfortable cadence; no other instructions were given. All subjects practiced walking on the 10 m walkway for a minimum of 10 min prior to data collection each day. The initiation and termination of data collection were controlled by two photoelectric cells, spaced approximately 5 m apart in the middle of the walkway, such that only strides taking place between the photocells were recorded. A total of 32 s of data were recorded for each subject on each day. This procedure was subsequently repeated on 2 other days, spaced 1 day and 1 week apart from the 1st day.

Eleven artifact-free strides from each subject on each day were selected for averaging. Each stride was time normalized such that heel-contact (HC) to HC was considered 100% stride time. Each stride was then sampled at 5% stride intervals, such that twenty EMG amplitude measures were selected from each stride representing 5% to 100% stride time. In addition, an average cadence and average walking velocity were calculated for each subject, each day.

The day-to-day difference between ensemble average EMG's for each subject was quantified by the root mean square difference (RMS diff) between the two curves:

$$\text{RMS diff} = \sqrt{\frac{1}{N}\sum_{i=1}^{N}(X_i - X_i')^2} \quad (1)$$

where X_i is the LE EMG level at point i in the stride time on one day, X_i' is the LE EMG level at point i in the stride time on another day, and N is the number of points sampled in the stride. This RMS difference was further normalized to the RMS level of the EMG signal on the first day, to provide an index reflecting the differences between two days as a proportion of the signal level on day one:

$$\text{RMS index} = \frac{\text{RMS diff}}{\sqrt{\frac{1}{N}\sum_{i=1}^{N} X_i^2}} \times 100\% \quad (2)$$

Results

The temporal results are shown in Table 1. It is evident that subjects showed highly repeatable walking velocities and cadences, day to day.

The EMG results were much more variable. The RMS differences and RMS indices are shown in Table 2. The average differences represent averages across subjects, with the range showing the best and worst cases. From the large range of values, it is evident that some subjects were highly consistent in their EMG patterns day to day, while others were extremely variable. The different responses between subjects are shown graphically in Figure 1 for the lateral hamstrings and in Figure 2 for the medial gastrocnemius.

Discussion

The interpretation of the mechanical consequences of differences in EMG amplitude and phase during dynamic movements requires the consideration of many factors. The

Table 1

Temporal Results

Subject Number	1	2	3	4	5	6
Cadence	110	108	114	112	98	112
(steps/min)	111	107	111	113	96	113
	113	107	113	117	99	108
Average velocity	1.23	1.24	1.47	1.09	1.25	1.20
(m/s)	1.26	1.22	1.44	1.11	1.17	1.24
	1.28	1.22	1.49	1.16	1.24	1.19

Table 2

Root Mean Square Differences Between Days

Muscle Group	RMS Difference (μV)		RMS Index (%)	
	Average	Range	Average	Range
Med ham	21	5 - 32	32	16 - 45
Lat ham	40	9 - 166	42	27 - 81
Med gas	43	12 - 103	29	10 - 74
Lat gas	46	9 - 99	51	14 - 90
Soleus	33	14 - 61	20	11 - 33

size of the muscle in question, the muscle length at the time, the velocity and type of contraction, are but a few of the important factors. Thus a difference in EMG amplitude during a time in the gait cycle when the muscle is contracting eccentrically may represent a much larger difference in the muscle force generated than the same EMG difference when the muscle is contracting concentrically. Similarly, EMG differences observed during a rapid contraction in the gait cycle will represent lower force differences than those observed during slower contractions, all else being equal.

Taking all the above factors into consideration, it is clear from the results that major differences were observed in some subjects. While some subjects showed highly repeatable EMG patterns in gait, with RMS differences as low as 5 μV, others exhibited large EMG differences day-to-day. In some cases, amplitude differences were accompanied by a complete alteration of the phasic EMG pattern. Yet this highly variable EMG pattern still resulted in the production of extremely consistent temporal patterns.

The results suggest that consistent temporal walking patterns can be achieved by a number of possible muscle activation patterns within a person. Mechanically, different muscle activation patterns result in different muscle forces applied to the segments. Since the final kinematic pattern of a segment is dependent on the net effect of all forces, it is likely that a trade-off occurs between the muscles involved, such that consistent temporal and kinematic patterns result.

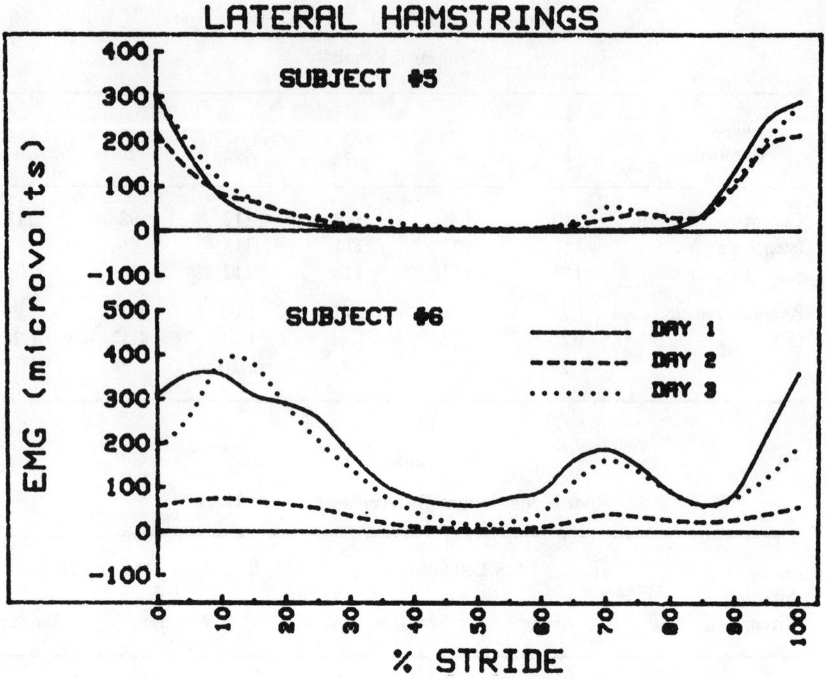

Figure 1—Day-to-day differences of the linear envelope EMG from the lateral hamstrings.

Evidence for muscle substitution between synergists has been reported under fatiguing conditions in the quadriceps muscle group (Viitasalo, 1983). Whether the same type of muscle substitution occurs in submaximal, nonfatiguing contractions, such as in gait, is not known. Results from net joint moment patterns of the lower extremity suggest that a more global type of trade-off occurs in walking. In stance, the hip, knee, and ankle in combination generate an extensor moment to prevent the lower limb from collapsing (Winter, 1980). The exact proportion generated at each joint, however, is variable from trial to trial within a person. Yet this variability in joint moments was accompanied by highly repeatable kinematic patterns (Winter, 1982). Thus it is likely that muscle trade-off occurs in gait, so long as some global constraint, such as the total support moment, is met.

The results of this study raise questions with regard to the interpretation of EMG data obtained from a single stride on a single day. The EMG pattern observed in a subject on a single day is likely only one of many different "normal" patterns available to that subject. There does not appear to exist one "true pattern" of EMG activity in gait for some subjects.

The results suggest that in gait, the study of any one muscle group, or any one parameter in isolation, could produce very limited results. It appears that the study of interrelationships between muscle groups could be more fruitful. Perhaps a "normal" pattern of interrelationships between muscles exists, to meet a more global constraint.

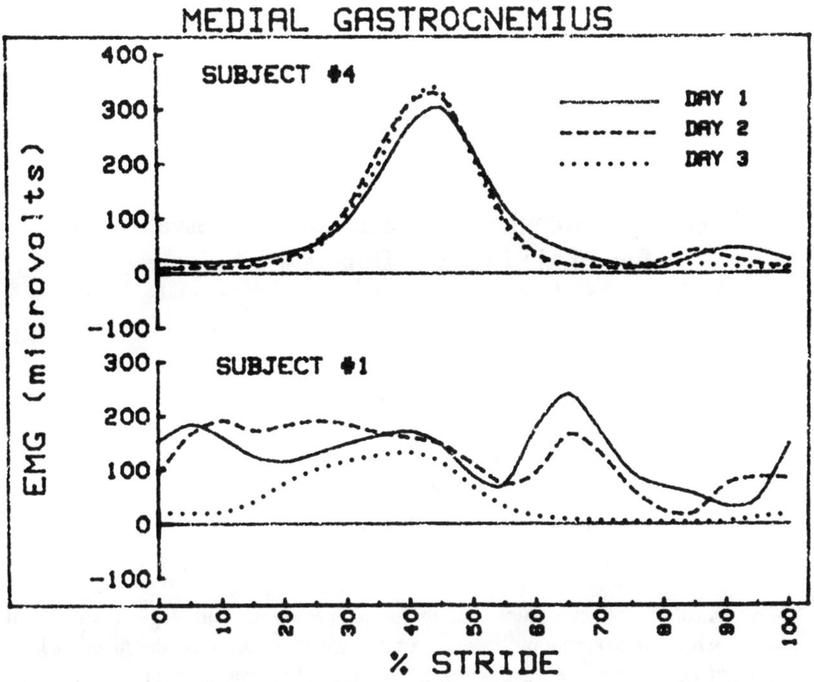

Figure 2—Day-to-day differences of the linear envelope EMG from the medial gastrocnemius.

Acknowledgments

ARSENAULT, B., & Winter, D.A. (1980). The repeatability of EMG in gait. *Proceedings of the Special Conference of the Canadian Society for Biomechanics: Locomotion I.* London, Ontario.

ELLIOTT, B.C., & Blanksby, B.A. (1976). Reliability of averaged integrated electromyograms during running. *Journal of Human Movement Studies*, **2**, 28-35.

HERSHLER, C., & Milner, M. (1978). An optimality criterion for processing EMG signals relating to human locomotion. *IEEE Transactions on Biomedical Engineering*, **25**(5), 413-420.

VIITASALO, J.T. (1983). Function of the knee extensor muscles during fatigue. In H. Matsui & K. Kobayashi (Eds.), *Biomechanics VIII-A* (pp. 271-277). Champaign: Human Kinetics Publishers.

WINTER, D.A. (1980). Overall principle of lower limb support during stance phase of gait. *Journal of Biomechanics*, **13**, 923-927.

WINTER, D.A. (1982). Motor patterns in normal gait—Are we robots? *Proceedings of the 2nd Biannual Conference of the Canadian Society for Biomechanics, Human Locomotion II.* Kingston, Ontario, Canada.

Electrophysiological and Kinesiological Evaluation for Ankle Muscles During Gait Cycle

M. Kumamoto
Kyoto University, Kyoto, Japan

O. Kameyama
Kansai Medical University, Kyoto, Japan

Conventional clinical evaluations for the postoperative and pathological walking of patients who have severe foot disorders are not generally based on the functional activities of leg muscles, but mainly based on external appearances and walking performances. Recently, EMG biofeedback treatments have been applied to the severe foot disorders and resulted in remarkable improvements of their gait. We have thought that the functional muscular activities of these patients were much more useful than the conventional clinical evaluations of them when we had to decide whether an operative treatment was needed or an EMG biofeedback treatment without an operation was applicable. However, we have confronted the fact that there was no standardized EMG gait pattern for Japanese people. A certain number of papers dealing with EMG patterns of Japanese gait have been reported, but the number of the subjects examined was not more than 20 and seemed too small to be referred to as a standard gait pattern.

In the present experiments, therefore, first we examined EMG recordings from leg muscles of healthy adult Japanese during a gait cycle, and attempted to determine what is an essential EMG pattern of Japanese gait. From the essential EMG patterns and their variations observed in the normal subjects, an evaluation based on the electrophysiological and kinesiological viewpoints (EK evaluation) was proposed. In this paper, we proposed the EK evaluation for the ankle muscles alone as a preliminary report. Second, the EK evaluation was applied to the severe foot disorder patients and compared with the conventional clinical evaluations.

Methods

Subjects employed in the normal gait experiments were 147 healthy adults of both sexes, ranging in age from 19 to 35 years, and in the pathological gait experiments for the EK evaluation there were 20 foot disorder, 13 congenital club foot, and 7 spastic foot-drop patients.

The congenital club foot patients (8 males, 12 feet and 5 females, 9 feet), who had the operative treatments within 1 year of age, were older than 7 years of age when the discharge pattern of the leg muscles during a gait cycle became an adult pattern (Okamoto & Kumamoto, 1972).

The subjects with spastic footdrop (5 males, 10 feet, and 2 females, 4 feet) were probably caused by cerebral palsy and they had no surgical treatment.

Muscles tested were Tibialis anterior (Ta), Gastrocnemius lateral head (Gl), Soleus (S), Peroneus longus (Pl), Tibialis posterior (Tp), of the ankle muscles and Vastus medialis (Vm), Rectus femoris (Rf), Semimembranosus (Sm) and Gluteus maximus (Gm) of the knee and hip muscles. EMGs were recorded with conventional surface electrodes and indwelling fine wire electrodes described by Basmajian (1978).

Foot switches were attached to the forepart of the sole and the heel to be able to record the following contact moments: Heel-strike—the moment when the heel first strikes the ground; Toe-on—the metatarsal heads contact with the ground; Heel-off—the heel is lifted up; and Toe-off—the foot is completely off the ground.

Angular changes of the knee and ankle joints, utilizing electrogoniometers, and motion pictures of side and front views, were simultaneously recorded with the EMGs.

Results

Normal Gait

The EMG pattern of the ankle muscles during a normal gait cycle were conclusively summarized (see Figure 1): (a) the EMG patterns of the Ta and the Gl represented

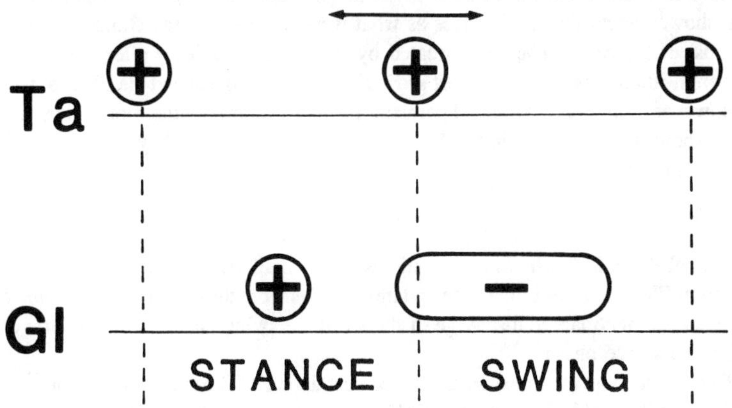

Figure 1—Diagram of the essential EMG pattern of the ankle muscles during a gait cycle shown in the healthy adult Japanese (147 subjects) without exception.

Ta: Tibialis anterior.
Gl: Gastrocnemius lateral head.
+: Electrical discharge existed.
-: No discharge observed.
Arrow: Variation in discharge phase.

the activities of the ankle muscles during a gait cycle. (b) The essential EMG patterns of the Ta and the Gl were as follows: (1) The burst of the Ta started before the toe-off and showed peak activity around the toe-off, performing dorsiflexion of the ankle where the ankle was plantarflexed for pushing off; (2) The sustained discharge of the Ta appeared prior to the heel-strike and continued through the early part of the stance phase so that the Ta could develop enough muscular tension to absorb the shock of heel-strike; (3) The Gl showed its main discharge in the latter half of the stance phase, functioning for pushing off; (4) The Gl showed almost complete electrical silence from the toe-off to the mid-swing phase, preventing footdrop. (c) Some variations in the EMG patterns of the Ta and Gl were observed in 70% to 30% of the recordings, but did not result in an abnormal gait.

From the results obtained in the normal gait experiments the EK evaluation was proposed and classified into four grades, Excellent, Good, Fair, and Poor, on the basis of the essential EMG patterns and their variations in the Ta and the Gl as shown in Table 1.

Pathological Gaits

Congenital Club Foot Patients. In post-operative walking, the EMG patterns of the knee and hip muscles were almost the same as the EMG patterns obtained in the present normal gait experiments.

The EK evaluation was applied to the EMG patterns of the Ta and the Gl during the postoperative walking of the patients, and the following results were obtained: Excellent, 3 cases (3 legs); Good, 7 cases (8 legs); and Poor, 1 case (2 legs).

The postoperative walking of the same patients was evaluated by conventional clinical evaluation methods such as those of Fredenhagen, Bost, and Abrams. The EK evaluation showed about the same grades as what were obtained by the clinical evaluations in most of the cases. The cases treated by the posterior release after the corrective cast application obtained the highest grades by both the clinical and the EK evaluations.

However, some cases who had pain and instabilities around the foot during the gait cycle showed poor grades in the EK evaluation, even though they had good evaluations clinically.

Cerebral Palsy Patients. Six of 7 patients, who could walk alone, showed somewhat different EMG patterns of the knee and hip muscles from those of the normal subjects—for instance, the marked discharge of the Rf in the swing phase or only a little of the Rf in the stance phase.

Two of the most severe patients showed sustained EMG patterns, apparently abnormal. As to the ankle muscles, the EMG patterns were classified into 6 Poors and 1 Fair for the EK evaluations.

The EMG biofeedback trainings (Bft) were applied to most of the CP patients referring to their EMG patterns. The Bft induced drastic changes in the EMG patterns and resulted in marvelous improvements of their walking forms in four cases of 6 patients where remarkable remissions in both EK and clinical evaluations were observed.

Table 1

	Essential Pattern	Phase	Variation	
Excellent	+	+	+	Kinesiologically reasonable muscular activities, general in normal gait
Good	+	+	±	Sustained small discharge in Ta or Gl seldom in normal
Fair	±	± ~ –	–	No essential discharge pattern— not in normal gait
Poor	–	–	–	Abnormal muscular activities
	(+) Each discharge clearly exists (+) Exists, but small (–) No essential pattern	(+) Variations distributed within normal range (±) Shift in phase of Gl (–) Shift in phase I of Ta	(+) Variations distributed within normal range (±) Slight Gl or sustained Ta in swing phase (–) Gl in swing phase	

Discussion

Our proposal of the electrophysiological and the kinesiological evaluation of the EMG patterns during a gait cycle is to examine the electrical discharge patterns of the representative antagonistic ankle muscles in terms of coordinating functions instead of the patterns of the individual muscular activities.

In the activities of the bifunctional antagonistic muscles of the thigh (the Sm, the Bf, and the Rf) their EMG patterns were reflecting the dynamic features where the upper body load was compensated for by the knee joint or the hip joint at the heel strike during a gait cycle (Kumamoto et al., 1981). Even in agonistic muscles of the Pectoralis major and the Posterior deltoid, their EMG patterns have been able to indicate the direction of horizontal resistant force during the arm extension movements (Okamoto, Takagi, & Kumamoto, 1966).

The CP patients gained almost poor EK evaluations for the ankle muscles, and they also showed abnormal EMG patterns in the thigh muscles. In the postoperative club foot patients, the EK evaluation on the ankle muscle activities was almost identical to the conventional clinical evaluations in most cases.

However, even though there was no abnormal pattern in the knee and hip muscles, the EK evaluations of the ankle muscle activities could reveal the functional disabilities which the conventional evaluations could not point out.

Therefore, the postoperative walking of the congenital club foot patients should be evaluated from the electrophysiological and kinesiological aspects in addition to the conventional clinical viewpoints as far as the function of gait is concerned.

References

OKAMOTO, T., & Kumamoto, M. (1972). Elecromyographic study of the learning process of walking in infants. *Electromyography,* **12,** 149-158.

BASMAJIAN, J.V. (1978). *Muscles alive: Their functions revealed by electromyography* (4th ed.). Baltimore: Williams & Wilkins.

KUMAMOTO, M., Oka, H., Kameyama, O., Okamoto, T., Yoshizawa, M., & Horn, L. (1981). Possible existence of antagonistic inhibition in double-joint leg muscles during normal gait cycle. In *Biomechanics VII-B* (pp. 157-162). A. Morecki, K. Fidelus, K. Kedzior, & A. Wit (Eds.), Baltimore: University Park Press.

OKAMOTO, T., Takagi, K., & Kumamoto, M. (1966). Electromyographical study of the extension of the upper extremity. *Japanese Journal of Physical Fitness,* **15,** 37-42.

Intertrial Variability for Selected Running Gait Parameters

H. Kinoshita
Hyogo University of Teacher Education, Japan

B.T. Bates and P. DeVita
University of Oregon, Oregon, USA

Coordinated movement is achieved through a self-regulating system involving numerous interactions between internal and external processes. Consequently, it is highly unlikely that a movement pattern will be performed identically from trial to trial. Descriptive parameters will fall within a bandwidth unique to the activity, the performer, and the conditions surrounding the performance. Even when conditions are carefully controlled and the activity is well learned, variability will be present. The understanding and evaluation of this variability is an important aspect of the study of human movement.

This study was undertaken to gain a better understanding of the variability associated with an individual's performance on selected kinematic and kinetic parameters describing the running gait pattern. Specifically, the purpose of the study was to examine the characteristics of the trial to trial variability of both ground reaction forces (GRF) and cinematographic data describing lower extremity function.

Methods

Three young males, free of lower extremity injuries, who were regularly running more than 65 km a week, volunteered as subjects. The data collection system consisted of a force platform interfaced to a computer, a high speed 8 mm motion picture camera, and a photoelectric timing system. The set-up and operation of the system have been described in detail elsewhere (Bates, James, Ostering, & Sawhill, 1980). Following an adequate number of warm-up and practice trials, subjects performed 15 successful trials in a barefoot condition. Running speed was monitored and controlled between 4.45 and 4.69 ms^{-1}. All subjects were heel-strikers and care was taken to ensure that same pattern under the barefoot condition tested. Force data (1000 Hz) and simultaneous rear view film data (150 frames/s) were obtained for each trial.

Film data were digitized and processed using a cubic spline smoothing technique to minimize errors as well as to generate 100 equally spaced estimates over the support period. Average pronation/supination angles (PA/SA) were then calculated for each 10% interval. In addition, two discrete values, the maximum angle of pronation (MP) and relative time to maximum pronation (TMP) wre obtained.

Ground reaction force data were normalized by dividing by body mass, and average values were computed for successive 10% intervals. Discrete parameters including the two maximum values (Z1, Z2) for the vertical component and the relative times to the maximum forces (TZ1, TZ2) were identified and evaluated. In addition, the absolute impulses for the three force components were determined.

Descriptive data along with intertrial variability were generated for the discrete parameters and each 10% interval of the support period for each subject. In addition, the variability data were examined to determine the number of trials necessary to obtain stable mean data.

Results and Discussion

The absolute and relative standard deviations for the discrete parameters are given in Table 1.

Figure 1 shows the 15 trial variability data for the vertical GRF component for the three subjects. The other components were similar in shape although magnitude differences obviously existed. Intertrial variability was considerably greater around the first maximum vertical force. This tendency can be observed from the absolute discrete data (see Table 1) which is further magnified by the relative standard deviations. Also evident from these data are a relative increase in absolute impulse variability for the X (15.05%) and Y (5.34%) components compared to Z (2.52%), suggesting that the mediolateral GRF component is the most variable.

The film data showed moderate variability values for the first 10% followed by a decrease and then a steady increase toward take-off (see Figure 2). The large variability at the end of the support period was assumed to be mostly the result of measurement error associated with perspective error due to leg mechanics at takeoff. The discrete parameters (MP, TMP) exhibited considerable relative variability.

Table 1

Mean Absolute and Relative Standard Deviation Values

Parameter	Abs. SD	Rel. SD (%)
Rel. time to 1st max vertical force (TZ1)	0.71%	11.79
Rel. time to 2nd max vertical force (TZ2)	2.20%	4.94
Rel. time to max pronation (TMP)	10.64%	24.04
1st max vertical force (Z1)	2.45 N/Kg	9.72
2nd max vertrical force (Z2)	0.68 N/Kg	2.24
Max pronation angle (MP)	1.410 deg	14.65
Absolute vertical imulse (ZI)	0.097 N.s/Kg	2.52
Absolute anteroposterior impulse (YI)	0.032 N.s/Kg	5.34
Absolute mediolateral impulse (XI)	0.0025 N.s/Kg	15.05

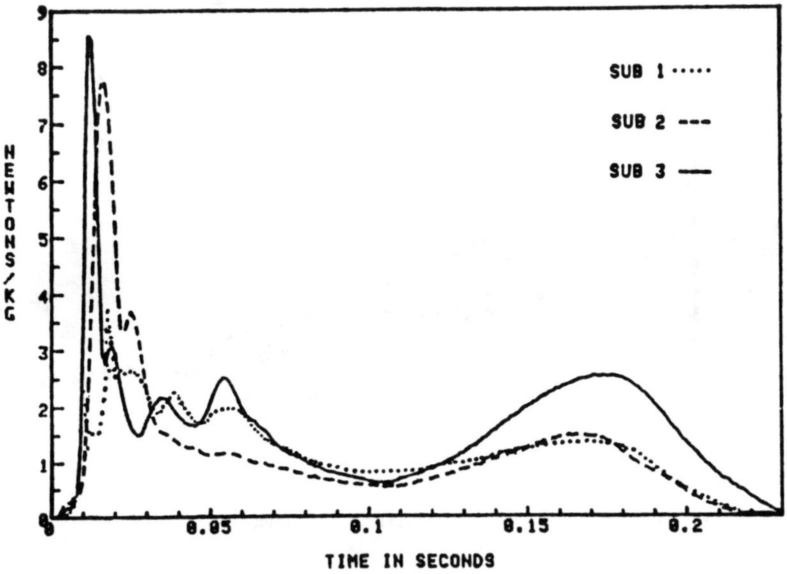

Figure 1—Standard deviation-time curves for the vertical force component.

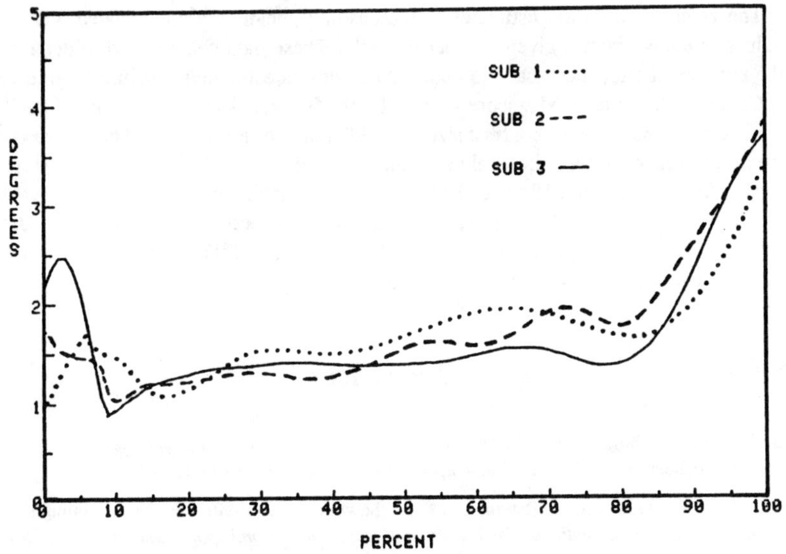

Figure 2—Standard deviation-time curves for film data.

The minimum number of trials necessary to obtain stable GRF and film data results can be estimated by examining the within-subject variability as a function of number of trials. Fifteen trial mean and standard deviation values were used as estimated popula-

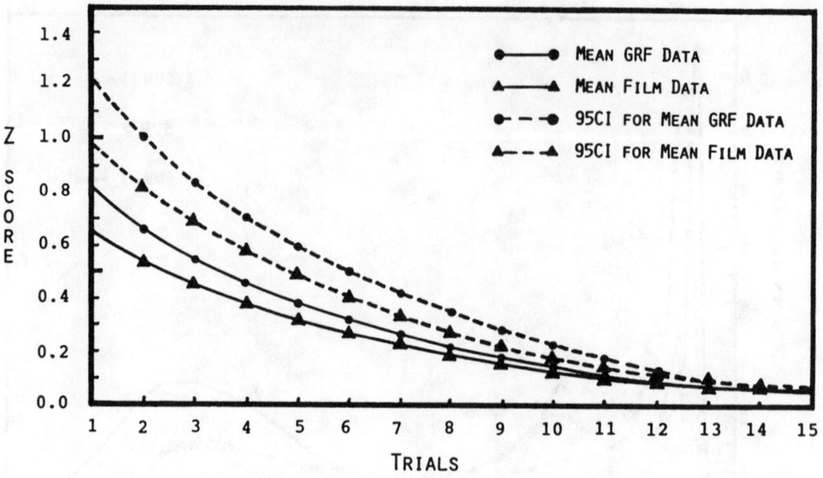

Figure 3—Relationship between absolute mean errors and number of trials for film and force data.

tion parameters to compute z-scores for successive cumulative trial means. In addition, the upper limit of the 95% confidence interval (95CI) was also computed for these data. The average results from all 10% intervals for both force and film data are presented in Figure 3.

The z-score values are estimates of the absolute mean errors associated with data values obtained from a given number of trials. These data can be used to determine the number of trials necessary to obtain data with an estimated absolute mean error.

Cohen (1969) suggested z-score values of .20, .50, and .80 as representing "small," "medium," and "large" contrast sizes. Based upon the results of this study, detecting mean differences of these general magnitudes would require 9, 3, and 1 trials, respectively, for film data and 10, 4, and 1 trials for force data. Corresponding 95CI values would be 10, 5, and 3, and 11, 7, and 4 for film and force data, respectively. Bates (1980) previously suggested 8 and 10 trials for mean and 95CI values for a z-score of .25, which are in agreement with these results.

References

BATES, B.T. (1980). Functional evaluation of footwear. *Proceedings of Biomechanics Symposium, Indiana University* (pp. 22-32). Indianapolis: Indiana State Board of Health.

BATES, B.T., James, S.L., Osternig, L.R., & Sawhill, J.A. (1980). Design of running shoes. *Proceedings of the International Conference on Medical Devices and Sports Equipment* (pp. 75-79). New York: American Society of Mechanics Engineers.

COHEN, J. (1969). *Statistical power analysis for behavioral sciences.* New York: Academic Press.

Electromyographic Patterns in Normal Pediatric Gait

R. Shiavi, B. McFadyen, and N. Green
Vanderbilt University
Nashville, Tennessee, USA

The study of the electromyographic (EMG) gait patterns of children with locomotor abnormalities is undertaken as a component of disability assessment. Adult normative data is used for comparison since only two studies concerning EMG patterns in normal pediatric gait have been reported (Griffin, Wheelhouse, Shiavi, & Bass, 1977; Sutherland, Olsen, Cooper, & Woo, 1980). These studies presented the on-off patterns with respect to age but did not consider walking speed. In order to provide a normative data base a large scale study of normal pediatric gait has been commenced. This paper presents the current results of this ongoing study.

Methods

EMG and foot-contact patterns were measured from normal children using Biosentry and B&L Engineering telemetry systems, respectively. Beckman Ag-AgCl surface electrodes on the right leg were used for all muscle measurements except the tibialis posterior in which bipolar wire electrodes were inserted. The foot contact patterns are measured and plotted in binary steps. Figure 1 shows an example with onset and offset of large toe (T), heel (H), first metatarsal (1), and fifth metatarsal (5) denoted. All measures are normalized against the percentage of stride time. The measurements were acquired in real-time using a PDP 11/23 computer system when the children were in the middle 5 meters of a 12-meter walkway. The children were requested to walk at self-selected speeds ranging from very slow to fast. The average linear envelope for each child's muscle was calculated using interpolation techniques (Shiavi & Green, 1983).

Results and Discussion

Twelve children have been studied whose ages range from 5 through 11 and whose walking speeds range from 0.8 to 1.9 m/s. The speeds and stride parameters agree with those found by Sutherland et al. (1980) and Fole, Quanbury, & Steinke (1979).

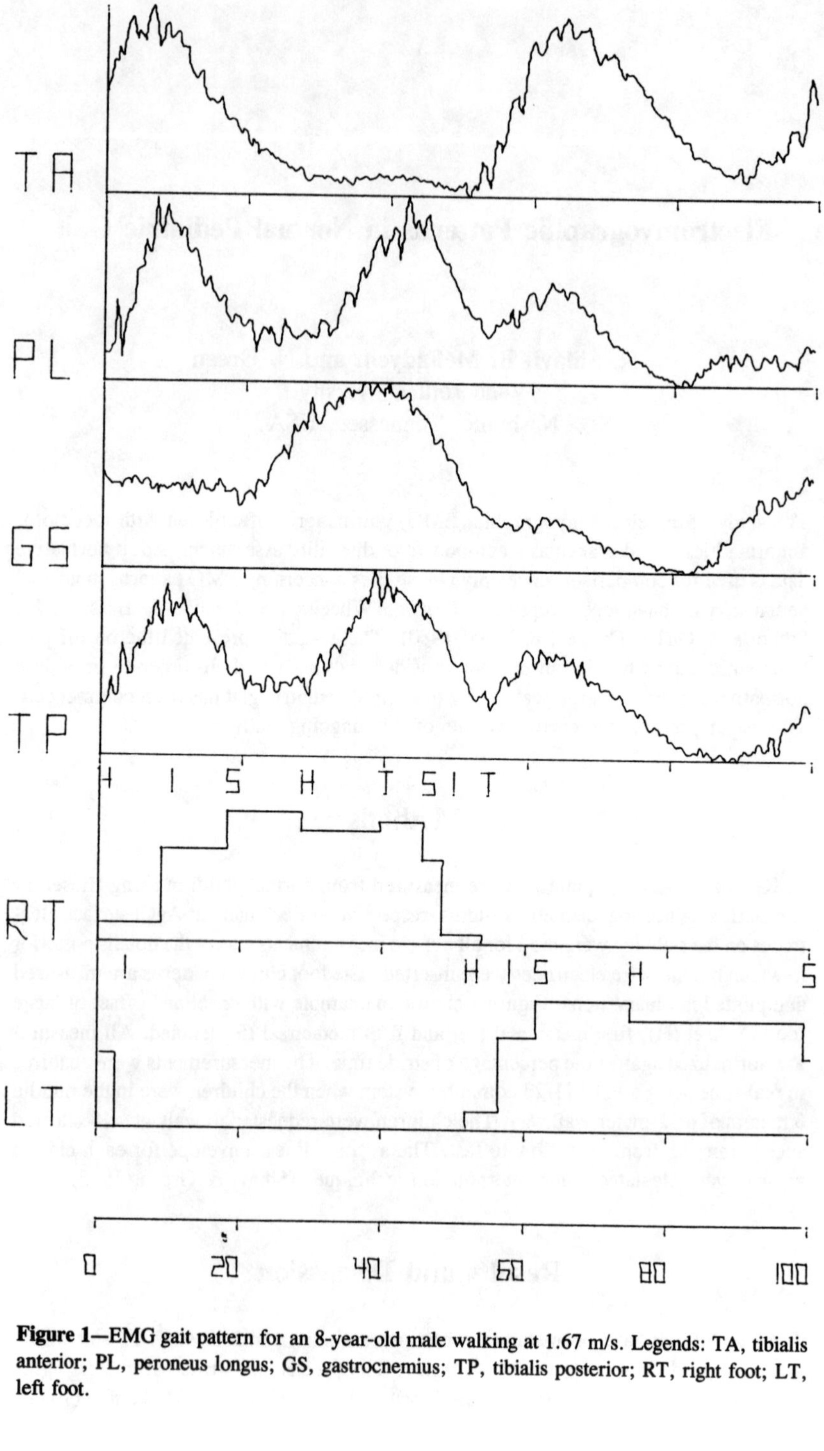

Figure 1—EMG gait pattern for an 8-year-old male walking at 1.67 m/s. Legends: TA, tibialis anterior; PL, peroneus longus; GS, gastrocnemius; TP, tibialis posterior; RT, right foot; LT, left foot.

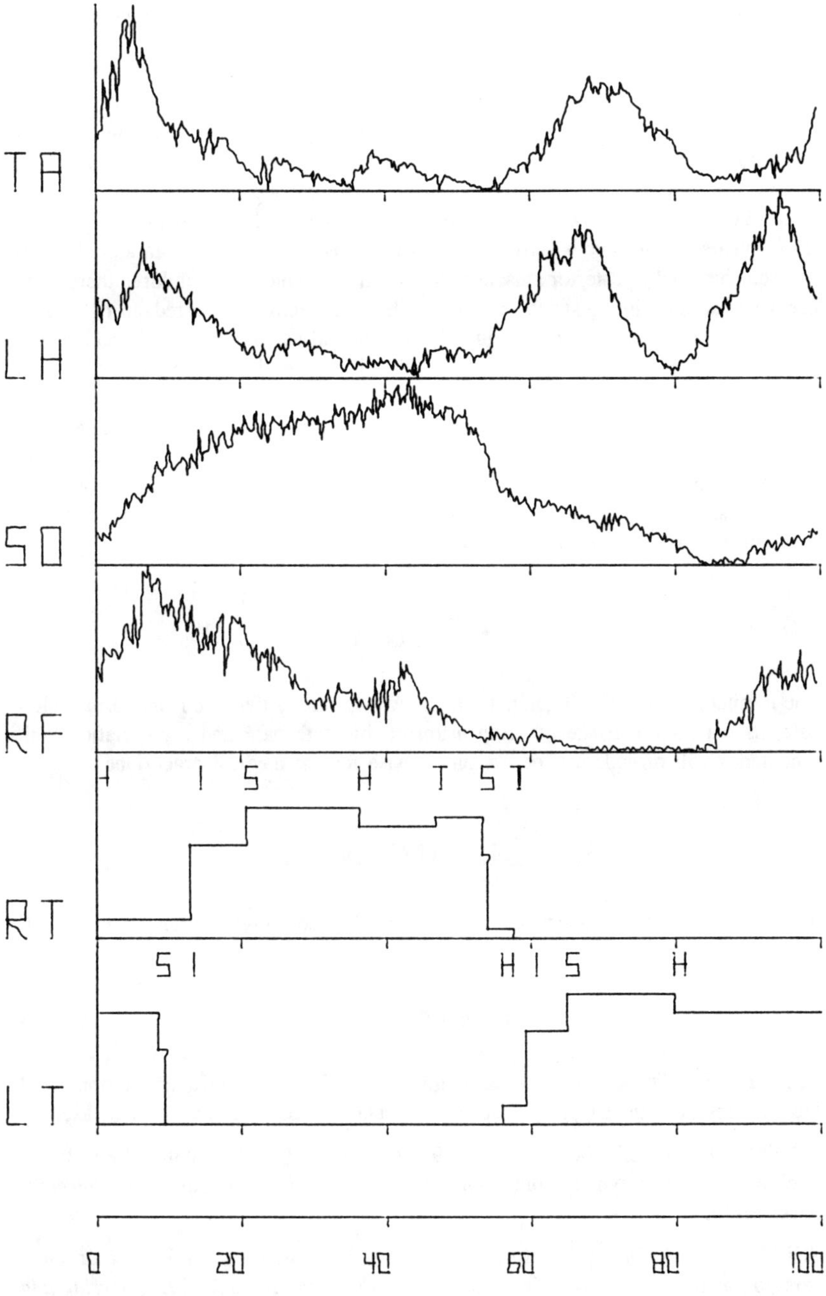

Figure 2—EMG gait pattern for an 8-year-old male walking at 0.65 m/s. Legends: TA, tibialis anterior; LH, lateral hamstring; SO, soleus; RF, rectus femoris; RT, right foot; LT, left foot.

The foot-contact patterns are the same as adult patterns, showing an onset and offset ordering of heel, fifth metatarsal, first metatarsal, large toe.

The tibialis anterior muscle had the classical bimodel pattern. Quiescence occurred only during midstance. The peroneus longus muscle consistently had one phase of activity commencing in late swing and continuing through initial midstance and a second phase during unloading. The classical midstance activity of the gastrocnemius and soleus muscles was found only in older children. The predominent pattern type observed commenced in late swing and continued through midstance. In children who were younger than 8 years old the soleus activity onset was almost a step rather than a gradual increase. The tibialis posterior muscle patterns were variable. In all children there were periods of activity during swing and stance. The stance activity occurred during loading and/or terminal midstance (see Figure 1). The lateral hamstring muscle had age and speed dependencies. For children 8 years old and younger the slow speed patterns had activity only during swing. As speed increased the duration of activity increased such that at fast speed the activity continued through midstance. Older children showed only the latter pattern at all speeds. The rectus femoris muscle had speed dependency. At the slower speeds the activity commenced in midswing and continued through initial midstance. As speed increased another phase of activity occurred during the stance-to-swing transition period (see Figure 2).

Conclusion

The normal pediatric EMG gait patterns show a definite difference from normal adult patterns but have consistency. Expansion of this data base and a correlation with kinematics will provide a good normative base for the medical practitioner.

Acknowledgment

This work has been supported in part by the United Cerebral Palsy Foundation.

References

FOLEY, C.D., Quanbury, A.O., & Steinke, T. (1979). Kinematics and normal child locomotion—A statistical study based on TV data. *Journal of Biomechanics,* **12,** 1-8.

GRIFFIN, P., Wheelhouse, W., Shiavi, R., & Bass, W. (1977). Habitual toe-walkers: A clinical and electromyographic gait analysis. *Journal of Bone and Joint Surgery,* **59A,** 97-101.

SHIAVI, R., & Green, N. (in press). Ensemble averaging of locomotor electromyographic patterns using interpolation. *Medical and Biological Engineering and Computers.*

SUTHERLAND, D., Olsen, R., Cooper, L., & Woo, S. (1980). The development of mature gait. *Journal of Bone and Joint Surgery,* **62A,** 236-253.

VI. PATHOLOGICAL GAIT

The Clinical Gait Lab: Form and Function

A.O. Quanbury
Rehabilitation Centre for Children,
Winnipeg, Manitoba, Canada

It is interesting to note the major trends and objectives in the study of human walking. The really early work at the turn of the century resulted from a genuine interest in, and a fascination for, what was observed in the everyday world. The common activity of walking caught the interest of the researchers of that era (Braune & Fisher, 1895; Marey, 1885), and with painstaking detail and laborious effort the first quantitatively descriptive data on walking were obtained. There appeared to be no driving practical application for many of these studies and the lack of suitable measuring and analyzing equipment would have prevented any large scale applications.

After World War II the interest in gait analysis was revived in the United States, this time for the purpose of designing better lower limb prosthesis for the many war veterans. The purpose was not so much to permit individual patient assessments but rather to obtain the necessary descriptive data on below knee, and above knee amputees to allow improvements in prosthesis designs that would benefit all amputees. As a consequence, these gait labs—most specifically the one at the University of California (Bresler and Berry, 1951; Bresler, Redcliffe, & Berry, 1951)—were in a research, rather than a clinical, setting and many of the techniques used for data collection were not suitable for clinical applications. Data collection and analysis was still laborious and time consuming because of the lack of modern electronic technology. However, the studies provided valuable normal and base line data that is still referred to today.

In the late 1960s and early 1970s research-oriented gait labs developed on a more widespread basis (Milner, Basmajian, & Quanbury, 1971; Paul, 1971; Sutherland & Hagy, 1972). Modern technology was promising automated motion analysis techniques (Winter, Greenlaw, & Hobson, 1972) and for the first time it appeared realistic to use the labs on a routine basis for a variety of applications, including rehabilitation medicine (Letts, Winter, & Quanbury, 1975; Perry, Hoffer, Giovan, Antonelli, & Greenberg, 1974). It is now some 10 to 15 years later and there are indeed a number of gait labs in clinical settings whose main purpose is patient service. However, research activities are still a significant part of the activities of many of these labs and there are still many questions to be answered: Has the age of the clinical gait lab really arrived, or is it just the beginning? Are the labs being utilized effectively for clinical applications? Are there barriers to their effective use in the form of funding, suitably trained personnel, acceptance by medical and paramedical personnel, and a true appreciation of what capabilities and limitations exist? These are questions that pertain not to the scientific or instrumentation aspects but to the application and orientation aspects of clinical gait laboratories. However, to help answer these questions the biomechanist and biomedical engineer must present to the clinical team members realistic capabilities and potentials of the gait lab as a clinical tool and help in the proper application and/or adaptation of this system for effective clinical use.

Clinical Research and Individual Patient Service

These two areas of rehabilitation medicine application of gait labs, although mutually supportive of each other, may have different requirements because the specific objectives are different. Because they are so complementary they nearly always exist together. Clinical research applications tend to look at a whole population group rather than an individual patient. The emphasis is to gain more specific knowledge of a gait pathology that may help in the improved treatment of the patient group as a whole. The design of a new prosthetic knee joint, based on a biomechanical gait analysis, or the improvement of a therapy program for stroke patients based on a gait assessment of stroke patients are examples of applied clinical research. Individual patient service, on the other hand, attempts to look at the specific patient and provide information that can be applied to the decision about treatment procedures for that patient.

Since patient service is fully in the area of the clinician or therapist who often lacks a biomechanical background there is an appeal to ''package'' the gait study results in a way more understandable to them. Data processing to provide indices of asymmetry of temporal or force platform data (Dewar & Judge, 1980) or compound angle curves whose shape is highly sensitive to gait changes (Hershler & Milner, 1980) is frequently encountered. These techniques have provided some sensitive indicators of gait dysfunction much in the way that oral temperature is a sensitive indicator of some systemic dysfunction (Schwartz, Heath, & Wright, 1933). It can be used as a measure of progress and response to treatment. Care must be taken in carrying this analogy too far. The permanently disabled person will be seeking a new optimal pattern of walking based on his altered capabilities (Winter, 1976). His best walking pattern will not, and probably should not, be the same as a normal person's. The patient with a traumatic, but transient injury may, however, return to normal, and monitoring of his return to normal may be meaningful.

These two application areas (clinical research and patient service) are sometimes approached from different professional viewpoints which can add to their differences. It is important to have a multidisciplinary team working together so that the full capabilities of the lab are applied to real clinically encountered problems. This team approach helps eliminate some of the criticism that exists regarding different lab facilities and techniques of analysis. The purely clinical person may feel the involvement and expense of doing a complete biomechanical assessment is unwarranted and the output is not understandable. The researcher may not understand the day to day problems of the clinician and the need to have quantitative data, even of a simple descriptive nature. Difficulties also arise when unreasonable expectations are placed on particular systems or the output from them. To attempt to make a biomechanical diagnosis from purely descriptive data can lead to erroneous conclusions (Winter, 1977).

Description vs. Analysis

There are many different forms of gait labs resulting in many different parameter sets collected and processed or analyzed: Cine systems, electrogoniometer systems, polarized light goniometer systems, video systems, infra red detection systems, foot switches, instrumented walkways, force platforms, and force shoes are all used. In the final analysis, however, they all fall into one of two major classifications: methods of obtaining descriptive data of gait patterns or methods and procedures to provide biomechanical analysis of the gait pattern. The data capable of providing analytical information may also be looked at in a descriptive light, of course.

The debate over these two gait lab functions still continues. Is there a clinical role for the gait lab that provides only descriptive data about the walking pattern? There are certainly a number of labs whose output falls into this category. Should all gait studies provide a biomechanical assessment that calculates muscle moments, joint forces, and energy level changes? That would certainly make clinical gait studies more involved and require considerable education of the clinical team members. To provide an answer, without conjecture, to the question of why a particular gait pattern is the way it is observed to be does require a biomechanical assessment. This still does not provide a medical diagnosis. This decision remains with the clinician and is based on input from a wide range of assessments and tests.

Monitoring a patient's change can be accomplished with descriptive data: kinetic, temporal, and floor reaction force patterns. Quantitative, rather than qualitative, descriptive data can be entered in the patient's chart for future reference. The equipment cost, labor cost, and patient inconvenience may be difficult to justify for this application. Care must be taken not to infer possibly erroneous biomechanical diagnosis from this data. Deviation of a patient's descriptive data from normal baseline data must be used with care also. As mentioned earlier, a permanently altered physical and/or neural system will have a different optimal pattern than normal.

The Future

Many gait pathologies involve movements in the frontal plane, or about the long axes of the bones. Single sagittal plane analysis, the most commonly used, is unsuitable for analyses in these other planes of motion. Full 3-D systems found primarily in research settings will now be needed in the clinical labs. Many gait disorders also result in inconsistent and highly variable walking patterns. To ensure that the data collected are representative, a number of consecutive strides must be analyzed. Stride-to-stride variation may in itself be useful information. For a full biomechanical analysis this also means a number of separate force platforms or a special force shoe with force transducers. Full acceptance by the medical community and a true appreciation by them of the clinical potential of the gait lab is still a limiting factor in some centres. At those centres where acceptance has been made we see the most positive application of the lab for clinical use. This is proof of the necessity of gaining this acceptance.

Summary

Technology has advanced to the point where routine clinical use of the gait lab is feasible. Kinesiology education programs and the introduction of more biomechanics teaching in undergraduate thereapy programs have produced personnel suitably trained for operating a clinical gait laboratory. The clinical application of biomechanical assessments of gait has been demonstrated repeatedly. The most effective use of the lab clinically can only come when it is accepted by the medical community that it is attempting to serve.

References

BRAUNE, C.U., & Fisher, O. (1895). *Der gang des menschen I-Teil: Verouche am un unbelastetan menschen*. Leipzig.

BRESLER, B., & Berry, F. (1951). *Energy levels during normal level walking*. Prosthetic Devices Research Project, University of California, Berkeley.

BRESLER, B., Radcliffe, C.W., & Berry, F.R. (1951). *Energy and power in the legs of above knee amputees during normal level walking*. Lower Extremity Amputee Research Project, University of California, Berkeley.

DEWAR, M.E., & Judge, G. (1980). Temporal asymmetry as a gait quality indicator. *Medical and Biological Engineering and Computers,* **18,** 689-693.

HERSHLER, C., & Milner, M. (1980). Angle-angle diagrams in the assessment of locomotion. *American Journal of Physical Medicine,* **59,** 109-125.

LETTS, R.M., Winter, D.A., & Quanbury, A.O. (1975). Locomotion studies, an aid in the clinical assessment of childhood gait. *Canadian Medical Association Journal,* **112,** 1091-1094.

MAREY, E.J. (1985). Locomotion de l'homme; image stereoscopic des trajectories que decrit dans l'espace un point du tronc pendant la marche, la course et les autres allures. *Comp. Rend. Acad. Sc. Paris,* 100.

MILNER, M., Basmajian, J.V., & Quanbury, A.O. (1971). Multi-factorial analysis of walking by electromyography and computer *American Journal of Physical Medicine,* **50,** 235-258.

PAUL, J.P. (1971, March). Load actions on the human femur in walking and some resultant stresses. *Exp. Mech.,* pp. 121-125.

PERRY, J., Hoffer, M.M., Giovan, P., Antonelli, D., & Greenberg, R. (1974). Gait analysis of the triceps surae in cerebral palsy. *Journal of Bone and Joint Surgery,* **56A,** 511-520.

SCHWARTZ, R.P., Heath, A.L., & Wright, J.N. (1933). Electrobasographic method of recording gait. *Archives of Surgery,* **27,** 926-934.

SUTHERLAND, D.H., & Hagy, J.L. (1972). Measurement of gait movements from motion picture film. *Journal of Bone and Joint Surgery,* **54A,** 787-797.

WINTER, D.A., Greenlaw, R.K., & Hobson, D.A. (1972). Television-computer analysis of kinematics of human gait. *Computers and Biomedical Research,* **5,** 498-504.

WINTER, D.A. (1976). The locomotion laboratory as a clinical assessment system *Medical Progress through Technology,* **4,** 95-106.

WINTER, D.A. (1977). Clinical Gait Laboratories—Measurement, Analysis or Assessment. *Proceedings of the Canadian Clinical Engineering Congress, Montreal.*

Kinematic and Kinetic Analysis of Normal and Pathological Gait

K. Mechelse, R. Pompe, J.A. van Best, C.N.A. Pronk, and J.H.M. van Eijndhoven
Erasmus University
Rotterdam, The Netherlands

Many investigators have studied pathological gait at the level of foot-floor contact and kinematic parameters (positions, angles and speed). However, only few data are available from kinetic analysis.

In this study the effect of local dysfunctions on the gait pattern of the lower part of the locomotor system (hip, knee, ankle) was investigated.

Material and Method

Gait patterns of 6 ambulatory patients with Becker muscular dystrophy (BE), 6 ambulatory patients with a hereditary motor and sensory neuropathy of Dyck type I (CMT, Charcot-Marie-Tooth type) and 10 normal controls (NC) matched for age, height, and weight were analyzed. The diagnosis was established by clinical examination, clinical neurophysiological data, serum creatine phosphokinase levels, muscle biopsy, and family history. A rough estimate of the motor ability was made (see Table 1).

The measurement system consisted of two Selspot cameras, placed on both sides of a 10-m walkway, and 14 light emitting diodes (LED's) attached to the subject. Ground reaction forces were measured with a Kistler force plate. Footswitches consisting of copper strips were attached under the heels and toes to the subject's own flexible-soled shoes. The footswitches were closed when the strips were in contact with the conducting surface of the walkway. Body segment dimensions and body weight were measured in order to calculate the mass distribution according to a nine-segment mechanical body model. Forces and moments at the joints were calculated by means of an integrated gait analysis system (IGAS) (Pronk, van Best, & van Eijndhoven, 1983). Each subject was asked to walk the length of the walkway at a comfortable pace as many times as necessary to make contact with his right foot with the force plate at least three times, and also three times with his left foot. A 2.4 m^2 area was seen by the cameras which permitted the registration of one complete stride.

Table 1

Assessment of Motor Ability

No.	Initials	Age Years	Sex	Wheel-chair	Walking Distance	Standing	Climbing Step	Gowers	Rank
BE 4	E.K.	15	M	No	>800 m	>20 min.	Yes	No	1
BE 3	M.K.	12	M	No	>800 m	>20 min.	Yes	Yes ?	2
BE 5	H.T.	12	M	No	>800 m	>20 min.	Yes	Yes	3
BE 6	R.S.	7	M	No	>800 m	>20 min.	difficult	Yes	4
BE 2	W.B.	40	M	No	<800 m	>20 min.	with help	Yes	5
BE 1	B.K.	30	M	Yes	<800 m	>20 min.	No	Yes	6
No.	**Initials**	**Age Years**	**Sex**	**Wheel-chair**	**Walking Distance**	**Standing**	**Climbing Step**	**Gowers**	**Rank**
CMT 3	F.v.O.	45	M	No	>800 m	>20 min.[*]	Yes	No	1
CMT 4	R.v.O.	16	F	No	>800 m	>20 min.[*]	Yes	No	2
CMT 6	C.B.	25	M	No	>800 m	±20 min.	Yes	No	3
CMT 2	C.C.B.	61	M	No	>800 m	<20 min.[*]	Yes	No	4
CMT 1	H.N.-B.	30	F	No	>800 m	<20 min.[*]	Yes	No	5
CMT 5	A.v.O.	12	F	No	<800 m	<20 min.	Yes	No	6

[*]Complaints distinguished otherwise identical subjects

Results

Kinematic Data.

Walking speed was the only linear variable which showed significant differences between the NC, CMT, and BE groups (mean ± SD: 1.31 ± 0.14 m/s; 1.16 ± 0.08 m/s; 1.09 ± 0.18 m/s). The stride length was decreased (1.42 ± 0.15 m; 1.38 ± 0.07 m; 1.30 ± 0.11 m) and the stride duration (1.08 ± 0.08 s; 1.14 ± 0.09 s; 1.19 ± 0.26 s) increased in the CMT and the BE group but the differences were not significant. Swing stance (0.64 ± 0.05; 0.63 ± 0.05; 0.63 ± 0.04; 0.64 ± 0.06) and step length/cadence (s/N) ratio (0.63 ± 0.09; 0.65 ± 0.05; 0.62 ± 0.18) and a symmetry factor (step length L/step length R: 1.00 ± 0.05; 1.00 ± 0.04; 1.02 ± 0.05) did not differ significantly. Neither did the cadence (112 ± 8; 106 ± 7; 106 ± 22).

The relation of stride duration, swing time, and stance time to walking speed was studied by comparing our data with the results of a study on normal young adults (see Figure 1). All data of our normal controls except those of a boy of 7 were within the normal range of this study. His stride and stance times were too short. Borderline values for swing time were found for CMT patient number 5 (increased swing time). The stride duration and the swing time were too long in BE 2. The stride duration and the stance time were too short in BE 6 (a 7-year-old boy) and BE 3 (a 12-year-old boy).

The maximum range of hip and knee "rotation" in flexion-extension did not differ significantly in the three groups (0.80 ± 0.08 rad; 0.81 ± 0.06 rad; 0.84 ± 0.12 rad and 1.17 ± 0.11 rad, 1.12 ± 0.09 rad; 1.21 ± 0.16 rad) (see Figure 2). The range

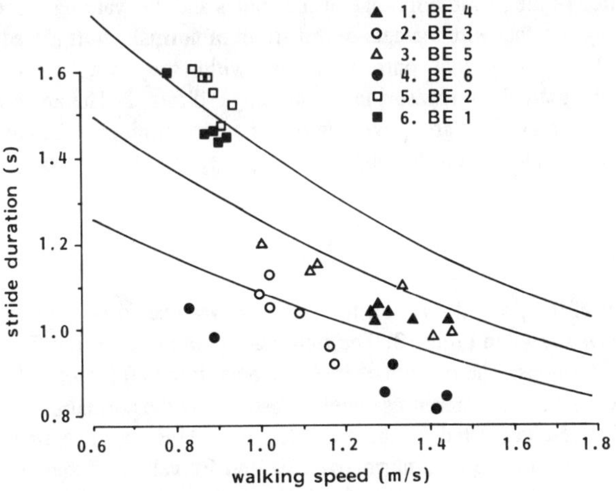

Figure 1—Relation of stride duration and walking speed in the Becker group. The relations of gait parameters of four healthy male subjects with the speed of locomotion (0.6-2.5 m/s) were fitted with second order regression analysis. The deviations of the measured values from these fitted curves were also fitted with a second order regression analysis. A 95% confidence interval was estimated by adding and subtracting 3x the latter relation to the first relation (Halbertsma, personal communication). In the figure, the fitted curves of stride duration and walking speed and the data of the Becker group are shown.

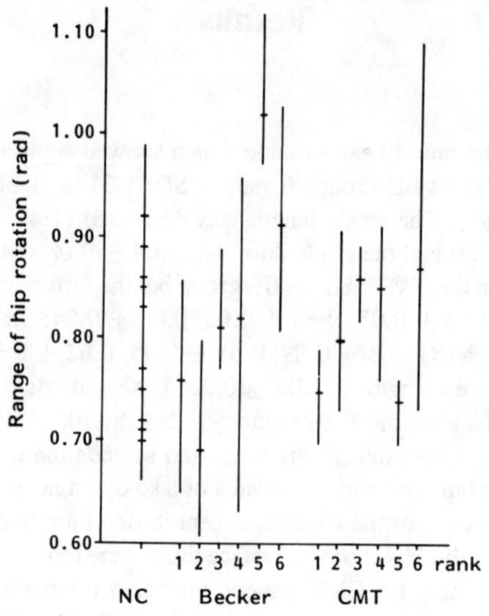

Figure 2—Maximum range of hip rotation in flexion-extension: the means and range of all normals and patients are shown.

of ankle plantar and dorsiflexion was lower in the CMT group and higher in the BE group (0.71 ± 0.19 rad; 0.59 ± 0.17 rad; 0.83 ± 0.24 rad), but the differences were not significant.

The relation of the maximum range of the angles and the walking speed was studied by comparing our data with the data of the group of normal adults already mentioned. The hip angles of our normal controls were not within the normal range of this group. However, the methods of measuring the hip angle differed. The ankle angles of our normal controls and CMT group were within the normal range. The values of the BE patients 1, 2, 3, and 6 were too high.

Kinetic Data:

The relation of the ground reaction force (Fz, z, vertical force) to walking speed in the 3 groups is shown in Figure 3. The figure does not indicate significant differences between the 3 groups. The relation of Fy (horizontal force-aft force) and walking speed did not show significant differences either. Subsequently the normalized dynamic ranges of Fz and Fy were corrected for the influence of walking speed. A first order regression analysis was done on the normalized Fz and Fy values of the NC group. This, equation has been used to calculate the forces at 1 m/s^{-1}. A comparison of the corrected Fz and Fy values of the three groups showed that almost all values of the CMT group were within the ranges of the normal controls, and most Fz and Fy values of the BE group were larger.

The ankle moment in the CMT group was significantly lower and in the BE group significantly higher than in the normals (9.27 ± 0.71; 8.37 ± 0.48; 9.38 ± 0.92). The hip moment of the BE group was significantly lower than the hip moment of the normals (4.22 ± 0.79; 4.04 ± 1.46; 3.22 ± 1.1). Both maximum negative hip mo-

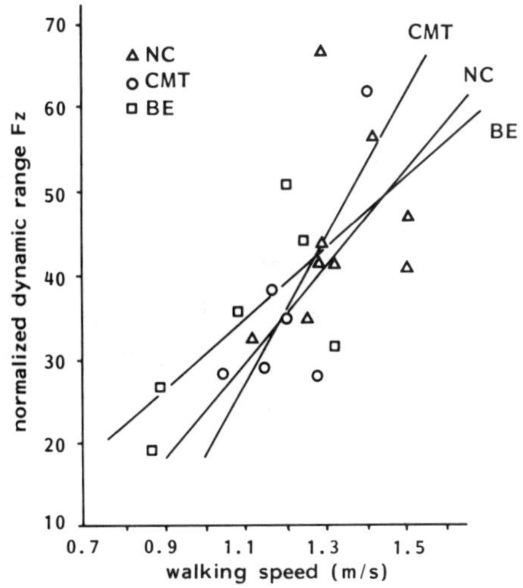

Figure 3—Relation of Fz and walking speed in the three groups. The lines were obtained by linear regression. NC △ Fz = 57.8 v − 34.5; r = 0.50; CMT ○ Fz = 86.8 v − 68.8; r = 0.78; BE □ Fz = 42.7 v − 12.5; r = 0.70.

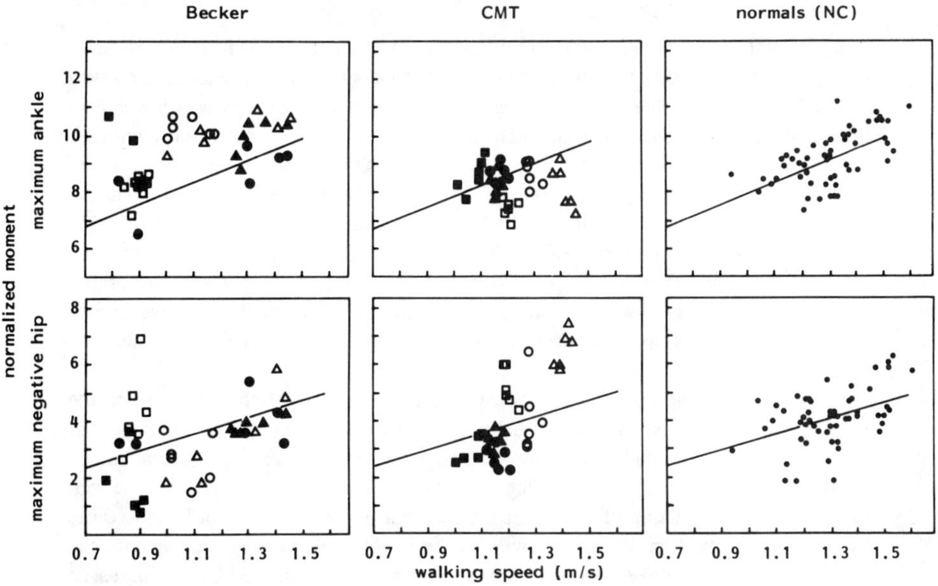

Figure 4—Normalized moments versus walking speed. ▲ 1 CMT 3; • 4 CMT 2; ▲ 1 BE 4; • 4 BE 6; ○ 2 CMT 4; □ 5 CMT 1; ○ 2 BE 3; □ 5 BE 2; △ 3 CMT 6; ■ 6 CMT 5; △ 3 BE 5; ■ 6 BE 1.

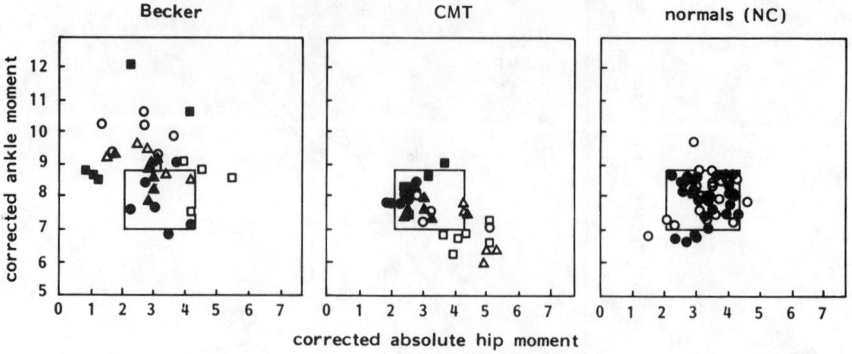

Figure 5—Corrected ankle moment versus corrected negative hip moment. Symbols as in Figure 4.

ment and the positive ankle moment during heel-off, normalized for body length and body weight, showed a linear regression with walking speed in normals (see Figure 4). Thereafter the moments of hip and ankle were according to the regression equation, corrected at a walking speed of 1 m/s^{-1}. The normalized moments show a greater ankle moment together with a smaller hip moment in the BE patients, while the reverse was found in the CMT patients (see Figure 5).

Discussion

The gait patterns of patients with weakness of the lower leg (CMT) or hip muscles (BE) and of normal controls were compared. The analysis of the kinematic data showed little of importance. The mean walking speed of the patient groups was lower than the mean walking speed of the normals. The decreased stride duration and increased stride length in the patient groups were, however, in accordance with the decreased walking speed except for one normal control and four Becker patients. In two patients and one normal control the stride duration was too small. All three were children and the effect is probably due to their height. In two Becker patients the stride duration was too long. Both were seriously handicapped and they had the lowest motor ability score. The range of hip and ankle angles differed in the three groups. However, when the walking speed is taken into account the ankle angle of the Becker patients only is too large.

The analysis of the kinetic data showed some interesting differences between the three groups with respect to the normalized hip and ankle moments during push-off. In the CMT group a smaller ankle moment together with a larger hip moment can be found. The weakness of the lower leg muscles could explain the smaller ankle moment, but the explanation of the larger hip moment is not obvious. An increased hip flexion to clear the foot during the swing phase is not likely because the patients had no foot drop. Perhaps the decreased ankle moment is compensated by the increased hip moment. In the Becker patients a smaller hip moment together with a larger ankle moment can be found. A smaller hip moment can be related to the weakness of the hip muscles. According to Sutherland and associates (1981), knee buckling in these patients during stance, due to a weak quadriceps, is opposed by a more forceful action of the ankle plantar flexors. This effect might explain the observed larger ankle moment. It would be of interest, however, to know whether the increase actually compensates the decrease of the hip moment. The occurrence of the compensating mechanism in other pathological gait patterns has been mentioned before and we think that our results support these findings (Strickland & Andriacchi, 1982; Sutherland et al., 1981).

Acknowledgment

Supported by the Princess Beatrix fund and the Foundation of Children Rehabilitation, "Adriaan Stichting," Rotterdam.

References

PRONK, C.N.A., van Best, J.A., & van Eijndhoven, J.H.M. (1983). Data processing and software organization with an Integrated Gait Analysis System. In D. Winter, R. Norman, R. Wells, K. Hayes, & A. Patla (Eds.), *Biomechanics IX-B* (pp. 269-273). Champaign: Human Kinetics Publishers.

STRICKLAND, A.B., & Andriacchi, T.P. (1982). Joint angular impulse as a gait parameter. *Journal of Biomechanics*, **15**(4), 335-336.

SUTHERLAND, D.H., Olshen, R., Cooper, L., Wyatt, M., Leach, J., Mubarakand, S., & Schultz, P. (1981). The pathomechanics of gait in Duchenne muscular dystrophy. *Developmental Medicine and Child Neurology*, **23**, 3-22.

Use of Normalized Profiles in the Assessment of the Kinetics of Pathological Gait

D.A. Winter
University of Waterloo,
Ontario, Canada

All medical diagnoses are characterized by a differential comparison of each patient with that of the normal population. Unfortunately, clinical gait assessments have, to date, lacked such a comparison, and, in fact, there have been very few kinetic analysis of pathological gait. At present, individual muscle forces cannot be reliably partitioned, but the net effect of all muscle forces at each joint can be analyzed using standard link segment analyses. The resultant moment of force profiles over the stride period are powerful diagnostic patterns with which to assess abnormal motor activity. Thus it would be highly desirable to base a diagnosis on the patient's moment of force pattern superimposed on those of normals. In this way a clinician or therapist could pinpoint the joint and muscle groups responsible for the pathological gait pattern. The purpose of this paper is to report the moment of force profiles for normal gait at natural and slow cadences and to demonstrate their use in the assessment of pathological case studies.

Methods

Moment of force patterns were derived from force plate and synchronized cinematographic data. Subjects or patients were recorded with a cine camera mounted on a tracking cart while they walked along a 10-m walkway. The camera rate was 50 Hz, and the anatomical coordinates were digitized with a Numonics Digitizer interfaced with a microprocessor. After correction for distortion and parallax, the absolute coordinates were digitally filtered with a 6 Hz zero lag low pass filter (Winter, Sidwall, & Hobson, 1974). Standard link segment analyses yielded the moment of force curves at the ankle, knee, and hip plus a recently defined support moment (Winter, 1980) which is the algebraic sum of moments at the three joints. As such it represents the total extensor or flexor pattern of the lower limb, and has been found to be extensor for almost all of stance and flexor for most of swing.

Subjects were classified according to their walking speed: natural or slow cadence (defined as 20 steps/min less than natural cadence). The stride period for each subject's record was normalized to 100% and each subject's moment was divided by his or her body mass. Then for each cadence group an ensemble average curve was calculated at each 2% interval of the stride period. The standard deviation at each of these intervals was also plotted; Figure 1 demonstrates such averaged profiles for the natural cadence group of 16 subjects.

Two case studies are presented. One is a knee joint replacement and a second is a hip arthroplasty patient. These comparative profiles appear in Figures 2 and 3.

Results and Discussion

Figure 1 is the plot of the normalized moments of force for the natural cadence group of subjects. The joint moments/body mass show the ankle to be quite consistent across

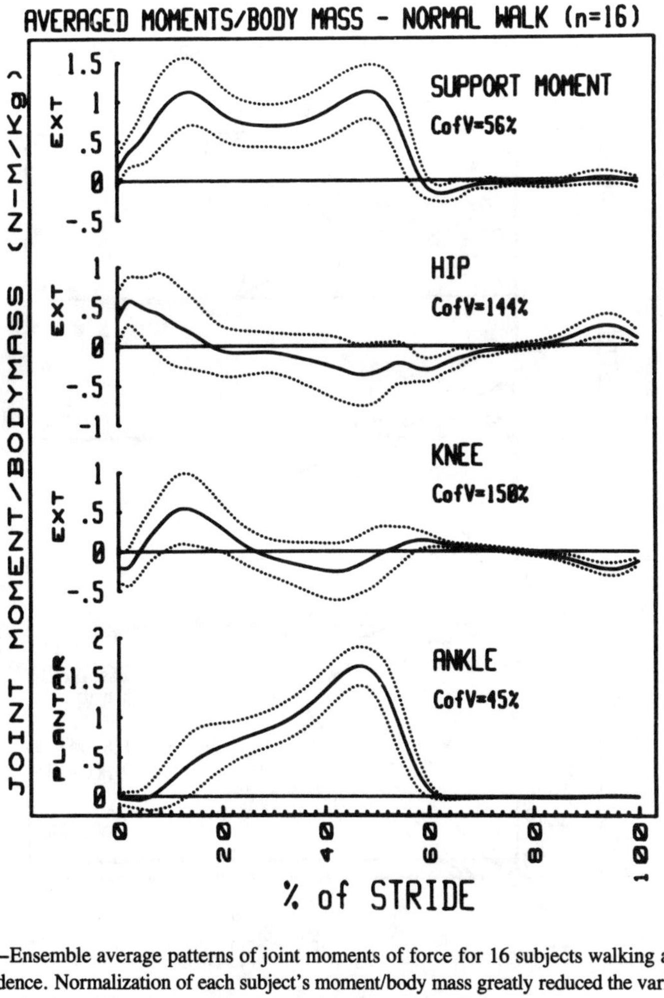

Figure 1—Ensemble average patterns of joint moments of force for 16 subjects walking at their natural cadence. Normalization of each subject's moment/body mass greatly reduced the variability as seen by the dotted line (one standard deviation either side of the mean).

the population group as indicated by the relatively low coefficient of variation (45%). The support moment pattern for these subjects is also fairly consistent (C of V = 56%). However, the variability at the hip and knee is quite high and is indicative of the tremendous flexibility of the motor patterns at these joints. Much of this flexibility is due to the anatomy of the double joint muscles, and some due to adaptions of the neuromuscular control system. Such variability is important in the diagnosis of patient patterns. First, it means that a patient will have to deviate considerably from the average pattern in order to be considered "abnormal." Secondly, the flexibility exhibited by the normal population is a forewarning that the patient population may also be as flexible.

Case Studies

Figure 2 shows the profiles of a female knee replacement patient who walked with a stiff legged stance (maximum flexion of 3°). Her cadence was 102 which suggests

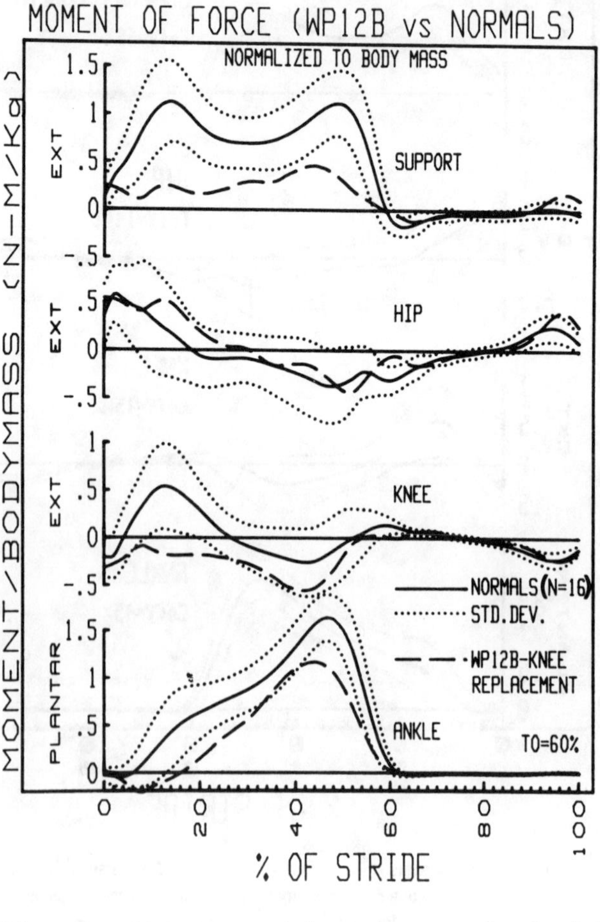

Figure 2—Joint moment of force patterns of a knee replacement patient plotted along with the patterns of natural walking of normals. See text for details.

a comparison with the kinetic profiles of the natural cadence group (cadence = 103 ± 6). Thus we see the moment of force patterns superimposed on the profiles of that normal population. Diagnostically we note that the ankle moment of force is slightly lower than normal but has a correct pattern. The knee motor pattern was quite abnormal and showed a distinct flexor pattern during entire stance. The hip pattern was also atypical and showed a compensation of the opposite direction: higher than normal extensor moments during stance. This total pattern change at these two joints could be accomplished by above-normal activity of the hamstring muscles and appears to be an attempt to prevent knee collapse using the hip extensors in lieu of the knee extensors. Such compensation is considered quite reasonable, as it would reduce the articulating forces at the knee and therefore reduce pain and danger of loosening the prosthesis. The support moment is less than normal over the total stance period; with such a stiff left during weight bearing, she did not need a major extensor activity to stabilize her knee joint.

Figure 3 shows a hip arthroplasty patient whose cadence was 98, indicating a comparison with the natural cadence group of normals. The ankle moment was a correct

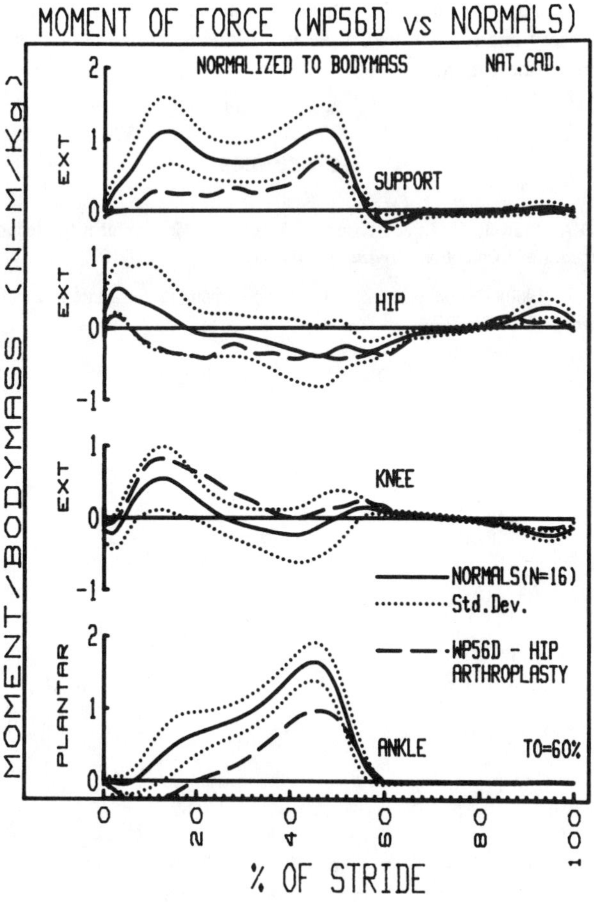

Figure 3—Joint moment of force patterns of a hip arthoplasty patient plotted along with the patterns of natural walking of normals. See text for details.

pattern but slightly less than normal. The hip pattern was flexor during all of stance (instead of a normal extensor pattern early in stance and flexor during later stance). The knee moment was within the normal range but was extensor for all of stance. What we see here is the reverse type of compensation from what was seen for the knee replacement patient. Especially during the critical weight acceptance phase the hip extensors were not active, presumably to prevent loading of the damaged hip joint. Compensation to control knee flexion during weight bearing was achieved by overactive knee extensors. The total extensor pattern of the lower limb, as indicated by the support moment, had a normal extensor pattern during stance but was of lower magnitude.

Conclusions

The assessment of pathological gait, as demonstrated by these comparisons of kinetic profiles, is greatly enhanced. Such techniques allow the clinician to quickly determine quantitatively what abnormal motor patterns exist and also see what form of compensation the patient may be using. With this information the rehabilitation team can assess previous surgery and therapy and make more intelligent decisions regarding future management of the patient.

References

WINTER, D.A., Sidwall, H.G., & Hobson, D.A., (1974). Measurement and reduction of noise in the kinematics of locomotion. *Journal of Biomechanics*, **7**, 157-159.

WINTER, D.A. (1980). Overall principle of lower limb support during stance phase of gait. *Journal of Biomechanics*, **13**, 923-927.

Gait Patterns and Muscle Function Following Van Nes Rotational Osteotomy

M. Glynn, A. Naumann, J. Mazliah, M. Milner, H. Galway, R. Gillespie, and I. Torode
Hospital for Sick Children,
Toronto, Canada

The condition of partial or complete absence of the femur, termed Proximal Femoral Focal Deficiency (PFFD), requires prosthetic management, since there is no possibility of leg equalization. There are three surgical procedures which may be performed to assist in prosthetic fitting. These are arthrodesis of the knee, amputation of the foot, and the Van Nes rotation plasty. The Van Nes osteotomy involves rotating the leg through 180°, so that the ankle of the short leg can work the knee joint of a prosthesis (Van Nes, 1950). This operation is highly controversial. Protagonists of the procedure state that it allows a controlled, smooth walking pattern (Kostuik, Gillespie, Hall, & Hubbard, 1975). Others feel that the theoretical advantage of this procedure is not clinically apparent (Amstutz, 1969). A detailed literature search has revealed no objective analysis of the function of the rotated foot. Such analysis could contribute to the resolution of this controversy.

A retrospective study of gait patterns and muscle function following Van Nes osteotomy was therefore undertaken to determine (a) the function of the rotated ankle by comparing the maximum voluntary isometric torques developed by the plantarflexors of the rotated and normal feet, (b) whether the timing of the electromyographic (EMG) patterns of the muscles of the rotated limb have altered from their normal pattern to control the prosthetic knee, and (c) whether these muscles demonstrate an ability to control moments about the prosthetic knee.

Clinical Material

Ten subjects took part in the study. Their ages ranged from 4 to 26 years (mean 15.8 years). Only 1 subject was 6 months following osteotomy at the time of the study, the other 9 having had the operation at least 1 year prior to the study. Six subjects had a marked malfunction of their hip abductors with a strongly positive Trendelenburg sign. The other four had clinically stable but abnormal hips.

Methods

Monitoring of EMG signals from the foot flexors and extensors was accomplished by means of surface electrodes. The signals were preamplified, band-pass filtered, rectified and averaged (33 mS time-constant). These signals, together with foot-floor contact patterns and ground-reaction forces (Kistler Model 9261A) were sampled at 100 Hz over a 10-second period during a comfortable walking speed and fast walking. The dynamics of the positions of body segments were determined by using two video cameras, a screen splitter, tape recorder and digitizer. A PDP11/34 minicomputer was used to acquire, analyze, and present the information. Torques developed about the rotated ankle during maximum voluntary isometric contractions to plantarflex the feet were mesured using a mechanical system in conjunction with the Kistler force platform (Mazliah, Naumann, White, Milner, & Carroll, 1983).

Results

The subjects could be grouped into two categories based on clinical and x-ray considerations. These were

(1) Ineffective hip abductor mechanism (Positive Trendelenburg sign). Of the six subjects in this category, two developed an extension moment, one a neutral moment, and three a flexion moment about the prosthetic knee during early to mid stance.

(2) Effective hip abductor mechanism (negative Trendelenburg sign). One of these subjects developed a neutral/flexion/extension pattern of moments around the prosthetic knee during stance. The other three developed a normal extension/flexion/extension/flexion pattern during this period.

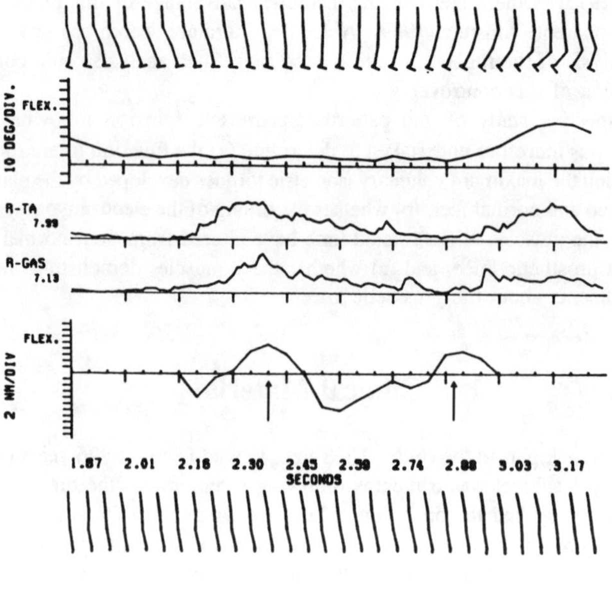

Figure 1—Prosthetic side.

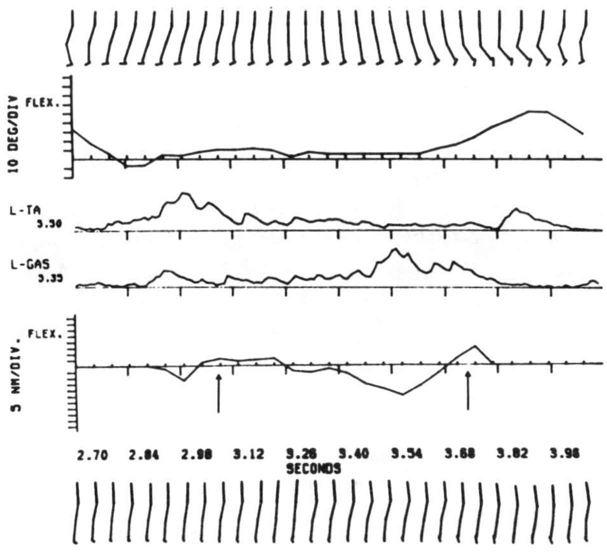

Figure 2—Normal side.

"Stick-diagrams" in sagittal (top) and frontal (bottom) planer, knee angle, tibialis anterior and gastrocnemius muscle activities, and knee flexion-extension moments vs. time—subject wearing high-heeled shoes.

Eight subjects flexed their prosthetic knee during the first 30% stance. In 4 of these, prosthetic knee flexion during this period was greater than 10°. In all subjects phasic activity of the gastrocnemius resembled that of the normal quadriceps at different speeds of walking.

A comparison between the moments developed about the rotated and normal ankles by the gastrocnemius showed ratios ranging from 0.83:1 to 8.50:1 with 1 subject showing a 52:1 ratio. Results from this latter subject are depicted in Figures 1 and 2. Each figure shows (from top bottom) the sagittal projection of joint markers, the prosthetic knee angle, tibialis anterior and gastrocnemius muscle activities, the knee flexion-extension moment, and the frontal projection of joint markers vs. time. The time between the two arrows depicts the single-limb support phase. These patterns were maintained with minor variations at fast speeds of walking. The subject depicted in Figures 1 and 2 was wearing shoes with a 2-inch heel.

Discussion

Clear differences exists between above-knee amputee walkers and those with a Van Nes prosthesis. The above-knee amputee relies on hip extensors to stabilize the prosthetic knee during stance by locking it into extension (Murray, Drought, Gardner, & Mollinger, 1980). The Van Nes subjects with a near normal hip demonstrate the ability to flex the prosthetic knee during stance. A contraindication for this procedure has been an unstable hip. However, we have demonstrated in such subjects that the gastrocnemius acts at a time when knee flexion moments occur and act to control the prosthesis at a time when control in above-knee prostheses is achieved by the use of hip extensors. Since the Van Nes subjects do not have normal hip control, we conclude that the presence of an unstable hip is an indication that this procedure be implemented. Measurements of the reflex arc of the gastrocnemius were made and appear to be normal. This indicates that these subjects, if they stumble, have the ability to reflexively activate muscles which can produce sufficient torques to stabilize the subjects and prevent their falling, and that this can occur at different walking speeds. This is achieved using a relatively inexpensive prosthesis with a simple hinge.

Significance

This study demonstrates the usefulness of the gait laboratory in resolving clinical controversies. The study described here also has implications for patients with osteosarcoma. Recently the Van Nas procedure has been used in the management of these tumors when located at the distal end of the femur (Kotz, & Salzer, 1982).

Acknowledgments

The authors would like to thank the Conn Smythe Research Foundation for supporting this work.

References

AMSTUTZ, H.C. (1969). The morphology, natural history and treatment of proximal focal femoral deficiency. In G.T. Atken (Ed), *Proximal femoral focal deficiency: a congenital anomaly.* Washington, D.C.: (pp. 50-76). National Academy of Sciences.

KOSTUIK, J., Gillespie, R., Hall, J., & Hubbard, S. (1975). Van Nes rotational osteotomy for treatment of proximal femoral focal deficiency and congenital short femur. *Journal of Bone and Joint Surgery,* **57A**(8), 1039-1046.

KOTZ, R., & Salzer, M. (1982). Rotation plasty for childhood osteosarcoma of the distal part of the femur. *Journal of Bone and Joint Surgery,* **64A**7, 959-969.

MAZLIAH, J., Naumann, S., White, C., Milner, M., & Carroll, N. (1983). Electrostimulation as a means of decreasing knee flexion contractures in children with spina bifida. *Proceedings of the 6th Annual Conference on Rehabilitation Engineering*, pp. 63-65.

MURRAY, M.P., Drought, A.B., Gardner, G.M., & Mollinger, L.A. (1980). Gait patterns of above-knee amputees using constant-friction knee components. Bulletin of Prosthetics Research, **17**,(2), 35-45.

VAN NES, C.P. (1950). Rotation-plasty for congenital defects of the femur. *Journal of Bone and Joint Surgery,* **32B,** 12-16.

Vectograms for Decision of Priorities in Bilateral Knee Joint Replacement

G.C. Santambrogio and R. Rodano
Polytechnic of Milan, Italy

Preliminary studies made by means of the Vector Diagram Technique (VDT) demonstrated that normal locomotion presents both inter- and intra-individual characteristics (Boccardi, Chiesa, & Pedotti, 1977; Pedotti, 1977). Further analyses performed by using the VDT on pathological subjects tackled the problem to verify and quantify the presence of abnormalities typical of a given impairment through a comparison between normal and pathological characteristics coming out from inter-individual similarities. Clinical applications of these findings were to evaluate the efficiency of gait in pathological subjects (Pedotti & Santambrogio, 1980), the efficacy of specific therapeutic treatments (Cova, Pedotti, Pozzolini, Rodano, & Santambrogio, 1980; Pedotti & Santambrogio, 1981) and the recovery after operation (Gualtieri, Luna, Nannini, Pedotti, & Rodano, 1981).

Particularly when serious and complex disease occurs, such as rheumatoid arthritis, the VDT analysis can provide very useful integrative information about the functional condition of the affected joints and can then offer helpful suggestions in order to plan the priorities of surgical interventions for multijoint replacements. The clinical decision as to which lower limb joint is to be replaced is usually based on the x-ray records and the evaluation of articular mobility in passive condition. Nevertheless, this information is not always sufficient to appraise which joint is more compromised so that, in such cases, personal considerations on pain and course of the disease guide the surgeon's decision.

The scope of this study is to present some recent results about the use of the VDT as an objective trend for decision of priorities in bilateral knee joint replacement.

Method

The VDT is a particular experimental procedure to obtain in vectorial form the evolution of the ground reaction force during the whole stance phase of a stride. Subjects to be tested by VDT were asked to walk along a pathway 15 m long at a fixed cadence imposed by a metronome. About two-thirds down the pathway, where steady-state ambulation was reached, a force plate was placed flush with the floor for the measure of the ground reaction forces; then the signals coming from the platform were processed as described by Pedotti (1977).

Normal vectograms taken at different cadences were collected from subjects presenting no clinical anomalies in order to identify interindividual similarities of normal ambulation.

Selected patients with monoarticular rheumatoid arthritis localized at the kne joint were assessed to determine the presence of recurrent characteristics typical of the disease, not distorted by contralateral or multijoint lesions. The vectograms of patients with both the knee joints equivalently affected by rheumatoid arthritis were examined and their vectograms were compared with the normal state and with those from monoarticular disease. The aim was to verify vectogram patterns, still identify the disease, and, moreover, discriminate the seriousness of the lesions.

Results

The vectogram patterns of normal walking have recurrent characteristics, some of which can be summarized as follows:

1. the envelope presents two maxima, at the impact and push-off phases respectively, and a central minimum;
2. the application point of the vector moves forward monotonically and presents a plateau in the last part of the stance phase;
3. the vector inclination changes continuously from backward to forward, making deceleration and acceleration phases quite clear and distinct from one another;
4. at a given cadence the results from the right and left leg are very similar and point out the symmetry of gait.

More detailed information regarding normal vectograms is outside the primary scope of this paper but is available in the references listed below. Figure 1 shows four of the vectograms obtained by testing 21 patients with monoarticular disease localized at the knee joint. The results illustrated refer to the affected side and were taken just a few days before prosthesis implantation. Significant and recurrent changes of the vectogram patterns with respect to the normal ones, and typical similarities among different subjects affected by the same impairment suggest the following considerations:

1. monoarticular rheumatoid arthritis localized at the knee joint provides vectograms with typical inter-individual similarities;
2. the envelope of these typical vectogram patterns does not present the initial maximum but a characteristic slope, due to a slower increase of the vector amplitude, takes place;
3. the application point moves rapidly towards the forefoot, where the vector reaches its maximal amplitude, pointing out a wider plateau during the last part of the stance phase;
4. the vector inclination still changes from backward to forward but decelaration and acceleration phases get partially entangled at the forefoot;
5. the vectograms taken from the affected and healthy leg point out the asymmetry of pathological gait (see Figure 2A and 2B);
6. after knee joint replacement, both the operated and the normal leg provide vectograms very similar to those obtained from normal subjects, thus indicating a full functional recovery (see Figure 2C and 2D).

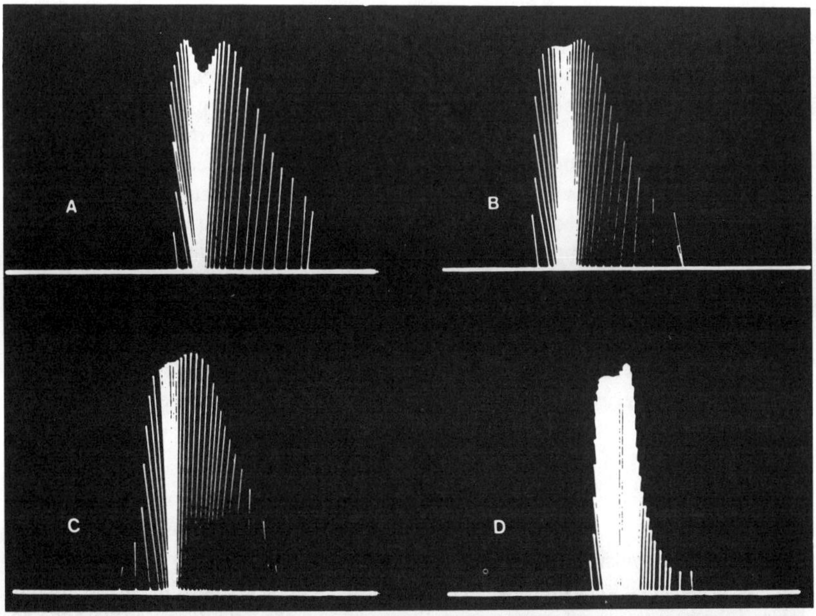

Figure 1—Vectograms from the affected leg of 4 different patients with monoarticular rheumatoid arthritis. All the subjects walked from right to left. The sample frequency is 50 Hz.

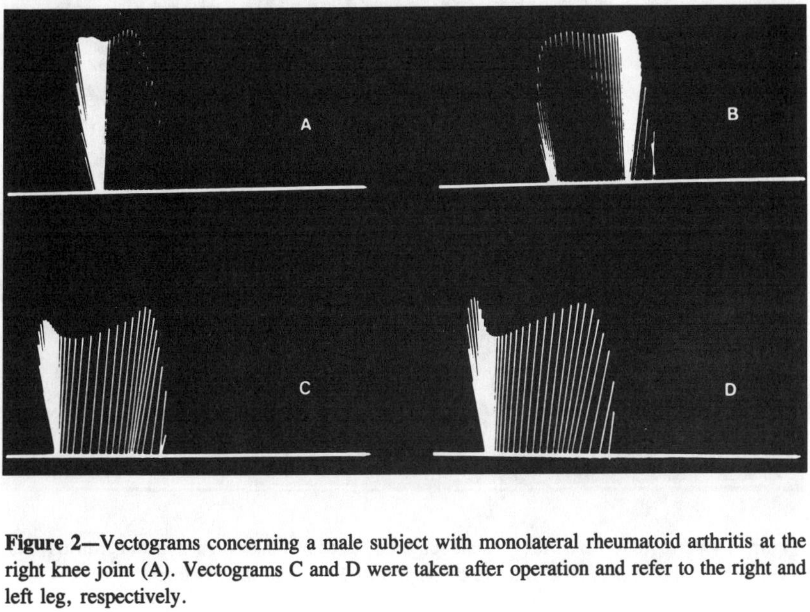

Figure 2—Vectograms concerning a male subject with monolateral rheumatoid arthritis at the right knee joint (A). Vectograms C and D were taken after operation and refer to the right and left leg, respectively.

The results collected from 19 patients with both the knee joints equally affected by serious rheumatoid arthritis indicate that the presence of typical vectogram patterns like those illustrated in Figure 1 identifies which joint is more compromised. In Figure 3, four vectograms concerning a bilateral case tested before (A and B) and after (C and D) prosthesis implantation at the right knee joint are reported.

On the basis of the vectograms discussed above, the following remarks can be drawn:

1. these characteristics are essentially the same observed in monoarticular impairment but concern repeatedly one leg only, even though both are equally affected;
2. the contralateral leg presents vectograms pointing out intra-individual characteristics which depend on the subject considered and the seriousness of the disease;
3. after surgical intervention the operated leg provides a quite normal vectogram pattern while the untreated one presents all the characteristics previously observed about the other leg prior to replacement;
4. these characteristics always involve the leg more compromised.

Conclusions

This study made by using the VDT upon selected patients with monoarticular rheumatoid arthritis at the knee joint indicates that typical vectograms are peculiar to the affected leg and the characteristics arising from these patterns are strictly related with the actual functional efficiency of the knee joint. When both the knee joints are involved, the vectogram previously observed from monoarticular impairment still appears but it involves one leg only. Prosthesis implantation removes this typical pattern from the

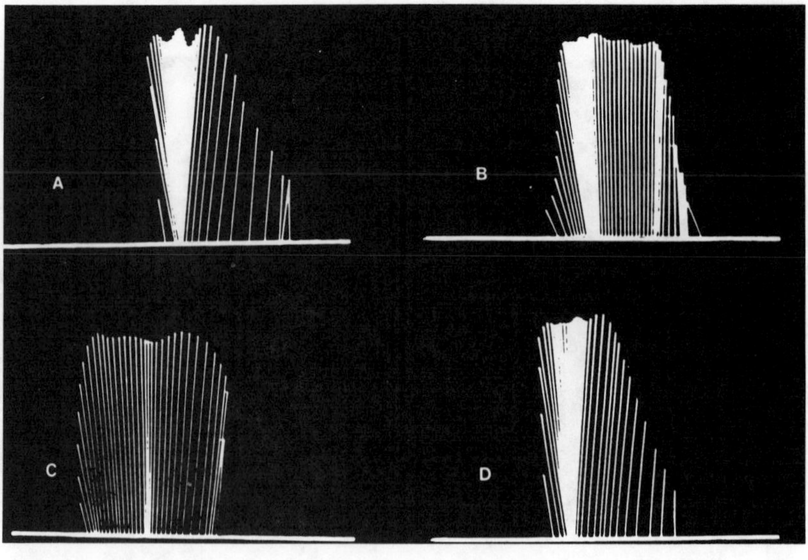

Figure 3—Vectograms concerning a male subject affected by bilateral rheumatoid arthritis at the knee joints. Diagrams A & C refer to the right leg, B & D to the left leg.

treated to the untreated leg because of the new level of motor efficiency typical of each leg. Therefore these considerations can be usefully taken into account in bilateral disease in order to verify which joint is more compromised and then to decide objectively which is to be operated first.

References

BOCCARDI, S., Chiesa, G. & Pedotti, A. (1977). A new procedure for the evaluation of normal and abnormal gait. *American Journal of Physical Medicine,* **56,** 163-182.

COVA, P., Pedotti, A., Pozzolini, M., Rodano, R., & Santambrogio, G.C. (1980). Procedure to use in orthopaedics for the analysis of the gait biomechanics in patients with various impairments. *Acta Orthopedica Belgica,* **46,** 545-557.

GUALTIERI, G., Luna, E., Nannini, G., Pedotti, A., & Rodano, R. (1981). *Monitoring gait of patients with total knee prosthesis by Vector Diagram Technique* (pp. 407-413). ISAM-Gent. Academic Press.

PEDOTTI, A. (1977). Simple equipment used in clinical practice for evaluation of locomotion. *IEEE Trans. BME,* **24,** 456-461.

PEDOTTI, A., & Santambrgio, G.C. (1980). Automatic analysis of gait efficiency in pathological subjects. *Second Mediterranean Conference on Medical and Biological Engineering,* 175-176.

PEDOTTI, A., & Santambrogio, G.C. (1981). Gait analysis in patients with hip prosthesis. In A. Morecki, K. Fidelus, K. Kedzior, & A. Wit (Eds.), *Biomechanics VII-A* (pp. 339-346). Baltimore: University Park Press.

Comparative Analysis of Some Pathologies of the Foot by the Vector Diagrams

P. Cova and R. Viganò
Instituto Orthopedico Gaetano Pini, Milan, Italy

A. Pedotti and R. Rodano
Centro di Bioingegneria—Fondazione Pro Juventute, Milan, Italy

Usually diagnosis of foot diseases is performed through an examination of the foot itself and through an inspection of the load distribution in static condition, that is, when the subject is in a standing position. In most cases the evaluation is done by visually inspecting the foot, by using photographic and x-ray records, and by using a podoscope to view the contact areas of the foot with the ground. However, a meaningful functional evaluation should also take into account the behavior of the foot during walking and consider the dynamic modifications introduced by the disorders.

The development of an efficient procedure to obtain quantitative data on dynamic characteristics has attracted wide attention (Betts, Franks, Duckworth, & Burke, 1980; Cavanagh & Mychiyoshi, 1980; Simkin, McQueen, & Stokes, 1979). However, most of the procedures available so far are either quite complicated to perform or require extensive data processing to obtain the requisite information. In the present paper we describe the results obtained with a vectogram technique which is easy to perform and also provides immediate quantitative information on dynamic characteristics of gait.

Methods

Sixty subjects suffering from diseases localized in the feet were analyzed with the vectogram technique which has been used previously for gait investigations (Boccardi, Chiesa, & Pedotti, 1977) and essentially consists of online processing of the signals provided by a piezoelectric force plate. The vectograms represent the spatial and temporal sequence of the ground reaction force during the stance phase of walking and constitute a meaningful synthesis of the whole movement indicating the dynamic variations of the load on the foot.

The main advantages of this procedure can be summarized as follows:

(1) the tests are easy to perform, even for subjects with large impairments of gait;

(2) it takes only a little time to perform these tests and quantitative results are obtained immediately;

(3) a large number of vectograms can be generated easily with high reliability;
(4) a clinical interpretation of the quantitative measurements obtained is possible.

The typical pattern of vectograms of normal gait in the advancing plane are well known (Cova, Pedotti, & Pozzolini, 1980) and may be summarized as follows: symmetry of gait, half-butterfly shape, closeness of vectors at the push-off phase, and progressive variation of the amplitude and inclination of the vectors during contact.

Results

Two broad classifications of pathology of the foot can be identified, depending on the nature of the load distribution. The first classification is the flat foot while the other classifcation includes pes cavus, pes equinus, and combinations of the two.

The vectograms of two subjects with bilateral flat feet are shown in Figure 1. Comparison of these diagrams with that from the normal subjects shows an evident alteration in the shape. The fast displacement of the vectors from the heel to the toe, which is typical of normal subjects, is replaced by a slower displacement of the vectors in the same direction.

There is also a concentration of vectors around the maximum vector during the impact phase, which is not the case with normal subjects. The concentration of vectors in the push-off phase, typical of normal subjects, is completely absent in the vectogram at the top of Figure 1 while it is spread over a greater length in the bottom of the vectogram of Figure 1. The changes in force patterns can be associated with the physiological characteristics of the flat foot; the larger surface area of contact induces a more uniform pressure distribution which in turn results in a more uniform time course of the point of application. Other variables such as contact duration, length of support-base, and

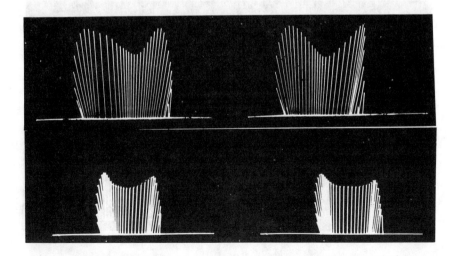

Figure 1—Four vectograms taken in the sagittal plane, direction of advance from right to left. The vectograms on the top are for a subject with more pronounced flat foot than the subject whose vectograms are shown at the bottom. The cadence of walk is 108 steps/min (5.5 km/hr). The vectors are sampled at 20-ms intervals.

the values of maximum and minimum forces are similar to those found in normal subjects.

Figure 2 represents the vectograms obtained from 2 subjects with bilateral pes cavus. All the diagrams are characterized by two concentrations of vectors localized to the heel and toe regions of the foot; the major loads are exerted in these regions. Between the areas of these two concentrations relatively few vectors are present, suggesting a quick displacement of the loads from the heel to the toe.

The apparent discrepancy between the time course of the contact shown by the movie records (quick initial contact of the toe, followed by the heel impact) and the time course of the vectograms is easily explained by the manner in which the point of application of force is computed. The point of application represents the location of the net weighted-average of the forces exerted by the ground on the various zones of the foot at any given instant. Thus, if equal force is exerted at the heel and the toe at a given instant, the point of application will appear to be in the middle of the foot, although there may be no physical contact between the ground and the middle region of the foot. Thus, the central vectors in this particular case do not signify a real physical contact of the central region with the ground, but are simply the results of our computational procedure.

The initial toe contact does not appear in the vectogram, because the reaction force generated during this contact is small and below the threshold level of force recording, and also because the duration of this contact is very small.

The vectograms of subjects with bilateral pes equinus are shown in Figure 3. In both cases, the pathology is more evident in the left side. The length of the support-base is approximately only 50% of the foot length, the impact phase is very short and there is a quick displacement of the loads from the central part of the foot to the toe. Also, there is only a single concentration of vectors which covers about 80% of the duration of the stance phase. The impact phase is characterized by maximal vectors which are

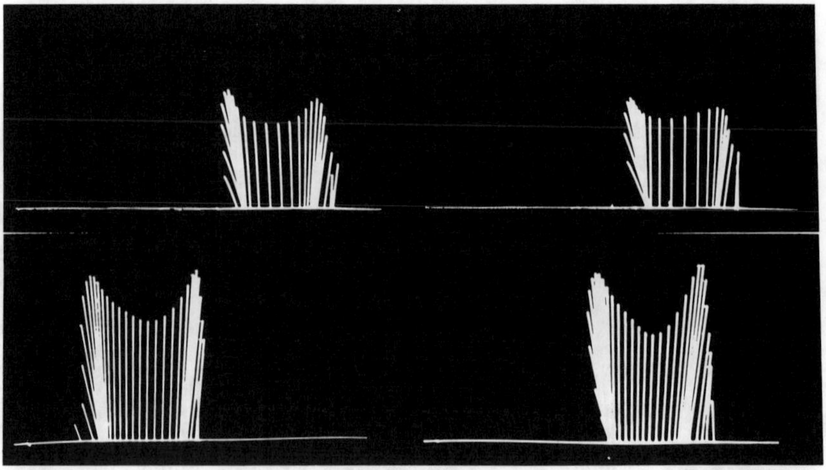

Figure 2—Four vectograms from 2 subjects with bilateral pes cavus. At the top are the vectograms of a subject with overload of all the metatarsal heads. At the bottom are the vectograms of a subject with overload of the first metatarsal heads. Direction of advance, cadence, and interval time between each vector are the same as in Figure 1.

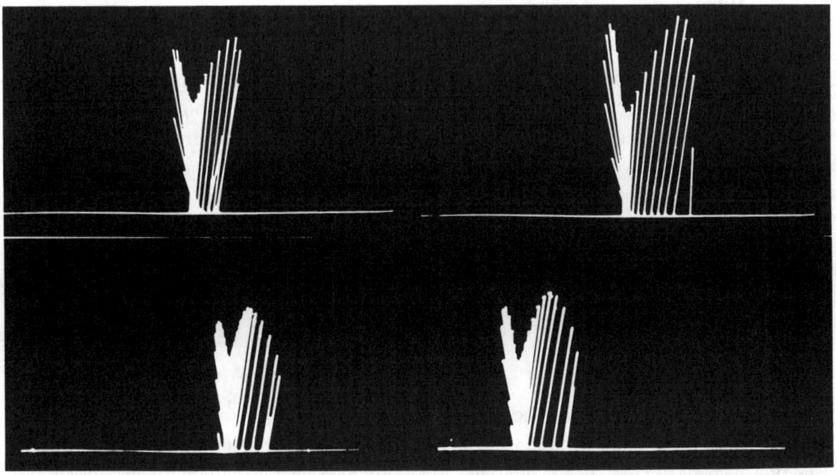

Figure 3—Four vectograms of 2 subjects, with bilateral pes equinus. For both subjects, the right foot, whose vectograms are shown on the left, is more pathological. Direction of advance, cadence, and interval time between each vector are the same as in Figure 1.

larger than those during the push-off phase. The foot with the worse pathology shows a first contact which is made by the toe region of the foot and has vectors of lower magnitude and denser concentration.

During the impact phase the foot appears to have a large negative inclination while making contact with the ground, possibly because of the rigidity associated with the ankle; during the push-off phase the subject is apparently not able to make the usual

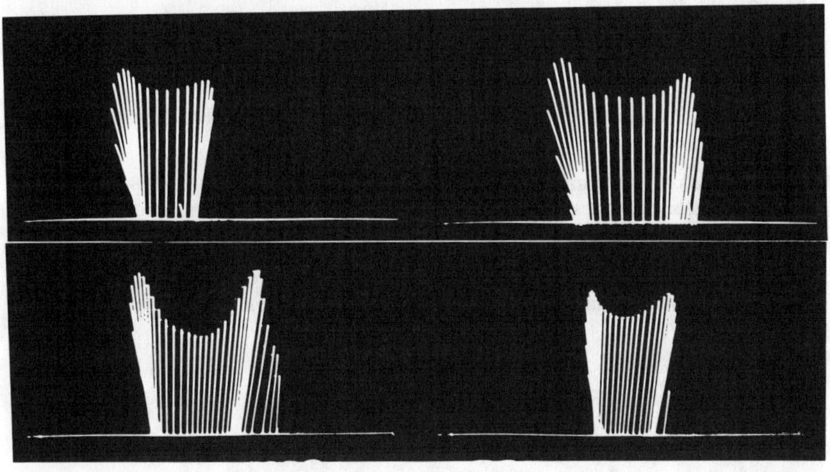

Figure 4—Four vectograms of 2 subjects with monolateral pes cavus-equinus. For the subject whose vectograms are shown at the top, the pes cavus-equinus is in the left side, while for the subject at the bottom it is in the right side. Both subjects have a contralateral pes cavus. Direction of advance, cadence, and interval time between each vector are the same as in Figure 1.

plantar flexion and is thus constrained to decrease the values of the forces and increase the duration of this phase.

The vectograms shown in Figure 4 are for 2 subjects with monolateral pes cavus-equinus. For both subjects, the second foot is cavus. Interestingly enough, the pes cavus-equinus has a vectogram that is intermediate between the vectograms for pes cavus and pes equinus. Two vector concentrations are present, there is a quick displacement of vectors in the central region, there is a reduced support base, and vectors appear in the toe region during the initial contact. The other vectograms confirm the previously stated characteristics for the pes cavus.

Conclusions

The vectogram technique has been applied to over 60 subjects in order to analyze the effects of two opposite pathologies of the foot, the flat foot and pathologies resulting in overloads of the toe. The pathologies affect the load distribution on the feet, as shown by the vectograms, in significant and characteristic patterns, even when visual inspection indicates that the gait is quite normal.

The vectogram along with other devices able to detect the load distribution under the foot may be particularly useful if the alteration in load distribution can be identified in the early stages of the pathological development.

References

BETTS, R.P., Franks, C.I., Duckworth, T., & Burke, J. (1980). Static and dynamic foot pressure measurements in clinical orthopaedics. *Medical and Biological Engineering and Computers*, **18**, 674-689.

BOCCARDI, S., Chiesa, G., & Pedotti, A. (1977). New procedure for evaluation of normal and abnormal gait. *American Journal of Physical Medicine*, **56**, 163-182.

CAVANAGH, P.R., & Mychiyoshi, A. (1980). A technique for the display of pressure distributions beneath the foot. *Journal of Biomechanics*, **13**, 69-75.

COVA, P., Pedotti, A., Pozzolini, M., Rodano, R., & Santambrogio, G.C. (1980). Procedure to use in orthopaedics for the analysis of the gait biomechanics in patients with various impairments. *Acta Orthopaedica Belgica*, **46**, 545-557.

SIMKIN, A., McQueen, A.K., & Stokes, I.A.F. (1979). Foot-ground pressure measurement: "Foot print" device. *Annual Report of the Orthopaedic Engineering Centre* (pp. 16-21).

The Influence of Pathological and Reconstructed Anatomy at the Hip Joint on Loads in the Lower Limb Joints

I.G. Kelly and D.L. Hamblen
The University of Glasgow, Glasgow, Scotland

T.R.M. Brown and J.P. Paul
University of Strathlyde, Glasgow, Scotland

Joint replacement at the hip for arthritic condition is one of the major advances in clinical practice in recent years. It offers relief of pain and some restoration of function to most patients with only a small percentage of risk or complications in the end results. Such large numbers of patients have received joint replacement that attention is increasingly being given to matters secondary to the major surgical procedure. There is currently considerable interest and activity in the field of international standards for the determination of the relevant test procedure to assess the strength of the femoral component of a total hip replacement. The validation of this procedure requires information on the magnitudes, directions, and frequencies of occurrence of the loads developed at the hip joint. While there is some work available for patients walking in a straight line, on a level surface, and at uniform speed under laboratory conditions, there is no large volume of data as yet relating to other locomotive activities. As further data become available relating to the performance of normal individuals and also to that of patients who have received joint replacements, it becomes possible to look in closer detail at the mechanics of the situation in respect of the position in which the components are placed and the effect of the load actions at the joint in question. Similarly, surgical treatment at one joint affects the loads transmitted at the other joints in the locomotor system and, for the patient, this may be beneficial or adverse.

The series of tests reported here relates to patients who exhibited a painful hip joint and no complaints of trouble at the other joints in the leg. For this trial, those patients were selected whose walking ability before surgery allowed them to undertake tests in a locomotion laboratory. They were fitted with either the standard Charnley total hip joint replacement or with the CAD Muller total hip replacement. The decision relating to choice of prosthesis was clinical in that the more difficult cases were dealt with by the Charnley procedure. Patients were submitted to a full clinical examination involving calibrated x-ray measurements of the legs before and after surgery to obtain the relative positions of the joints and the joint replacements at the level of the knee and the hip. The equipment utilized included two Kistler force platforms, each providing three components of force and three components of moment relative to the reference axes. The data from each force platform was sampled by a PDP11/34 computer at 0.02-s time intervals. The patients wore brief shorts to allow full viewing of legs and surface markers were applied to the skin in regions of bony prominences. Sufficient markers were applied to each limb segment to allow the recording in three dimensions of linear and angular movements. It was felt particularly important to measure in as exact a fashion as possible the rotation of the limb segments so that the actual axes of the joints could be defined at every instant in time, since one of the characteristics of degenerative joint disease is restricted or fixed angular position of the limb segments. The successive positions of the markers in space were monitored by the Strathlyde computer/television system (Jarrett, Andrews, & Paul, 1976). This system is interfaced to the PDP11/34 computer and gives real time acquisition of the coordinates of the marker points in three dimensions. The force data from the Kistler platforms are acquired during the "flyback" period of the television camera scan. Displacement data are acquired at 50 frames/s and the force and displacement data can be played back to the laboratory a few seconds after each test run to confirm that satisfactory records have been obtained. The patients were assessed prior to surgery and at intervals of 6 and 12 months thereafter.

The knowledge of the forces and the true position of the joint axes and joint centers at each instant in the analysis allowed the calculation of the resultant forces and moments transmitted between the body segments in three dimensions This data was then further analyzed in the manner described by Paul (1967), Morrison (1967), and Paul and Poulson (1974) to obtain the values of the forces transmitted at the hip and knee joint on the affected and contralateral sides.

The anatomical measurements were made for the hip joint positions on the affected (operated) side and the contralateral side. These measurements were taken from x-rays and expressed relative to the positions of the anterior superior iliac spines. An axis system was set up with dimension z along the line connecting the iliac spines and dimension y perpendicular to it. The coordinates of each hip joint relative to the adjacent iliac spine were measured as y and z coordinates. To eliminate the effect of the overall dimensions of each patient, the data on the affected side were normalized to the corresponding measurements for the nonaffected side and these were compared on a fractional basis. Figure 1 illustrates this procedure. A mirror image of the nonaffected side was imposed on the affected side and the coordinates of the effective center of rotation of one side relative to the other were measured. These are plotted in Figure 1, in which the mediolateral offset and the superior/inferior displacement are respectively given by:

$$\text{Mediolateral offset} = [z\ (\text{affected})/z\ (\text{healthy})] - 1.0$$
$$\text{Vertical offset} = [y\ (\text{healthy})/y\ (\text{affected})] - 1.0$$

Figure 1 shows these data with the symbol 'm' for those patients receiving the Muller device and 'c' for those using the Charnley device. Figure 1 shows the situation for the hips prior to surgery and it is seen that the degenerated joints have their effective center displaced in the superior direction with a slight tendency for the patients treated with the Charnley device to have more medial migration also.

Figure 2 shows the corresponding diagram for the relationships at the hip following joint replacement surgery. Due to the pelvic reaming undertaken routinely with the Charnley procedure, these patients have a hip joint placed considerably more medially, relative to the unaffected hip, than do patients who have the Muller procedure. The

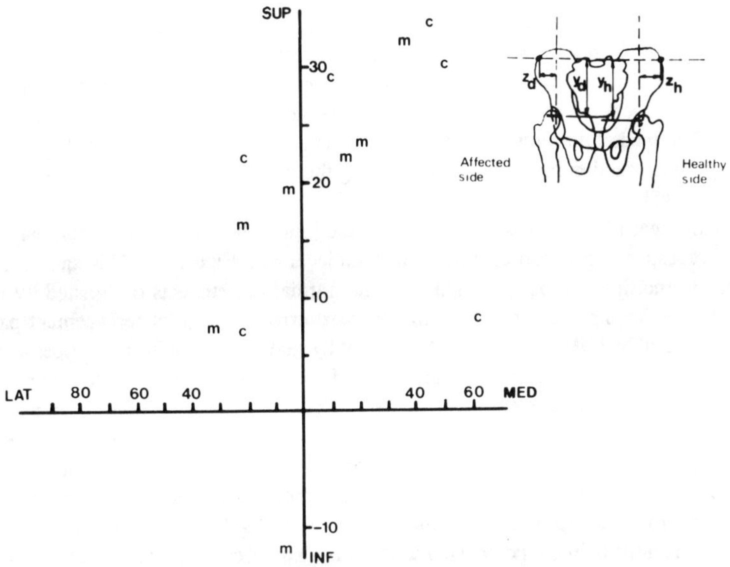

Figure 1—Displacement position of hip joint center due to degenerative disease. 'm' and 'c' represent patients treated with Muller and Charnley prostheses, respectively.

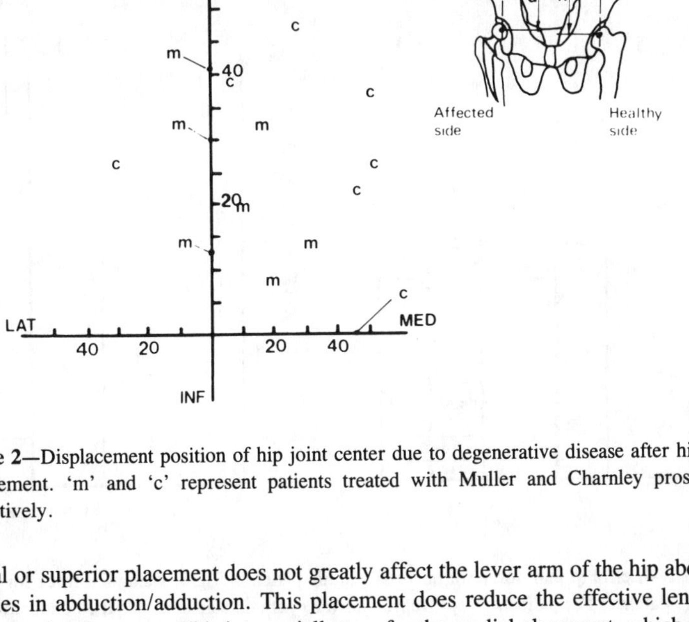

Figure 2—Displacement position of hip joint center due to degenerative disease after hip joint replacement. 'm' and 'c' represent patients treated with Muller and Charnley prostheses, respectively.

medial or superior placement does not greatly affect the lever arm of the hip abductor muscles in abduction/adduction. This placement does reduce the effective length of these muscles, however. This is especially true for the medial placement, which makes the line of action of these muscles more nearly vertical. Both procedures shorten the muscle, which should cause a reduction in the patient's ability to generate abduction force, through the length/tension relation. A limping gait would thus be likely in patients with the more extreme joint replacement, at least in the early stages of rehabilitation. The medial placement is advantageous in that it provides a shorter lever arm for the ground reaction force, thereby reducing the muscle activity necessary to keep the pelvis level.

With regard to the loads transmitted by the joints, the patients were compared with the corresponding group of normal test subjects (see Figure 3). This shows the hip force in multiples of body weight with the normal test subjects designated by N and the bar indicating plus or minus two standard errors. Hip joint replacement patients are indicated by HJR, 'h' referring to the healthy side, 'op' referring to the operated side.

While the level of statistical significance of the data for the joint replacement patient relative to the normal is not high, it is seen that a highly significant difference exists between the operated and nonoperated sides for the patients, with the nonoperated side carrying some 20% more load than the operated side. This may be due to the altered mechanics of the new joint, or to the patients being accustomed to an asymmetrical gait pattern. The knee joint loads on the nonoperated side of the patients are generally higher than those in the normals, and there is a significant difference between the left and right sides of the patients following operation at one hip joint. This shows that hip joint replacement at the left hip involves enhanced loading at the right knee, and vice versa. When the data of Figure 3 are correlated with the data of Figure 2, it is found that the most adverse joint loading is found in those patients in whom the geometry of the replaced joint deviates most from the normal. This conclusion has not been fully validated and further tests are being conducted to explore it.

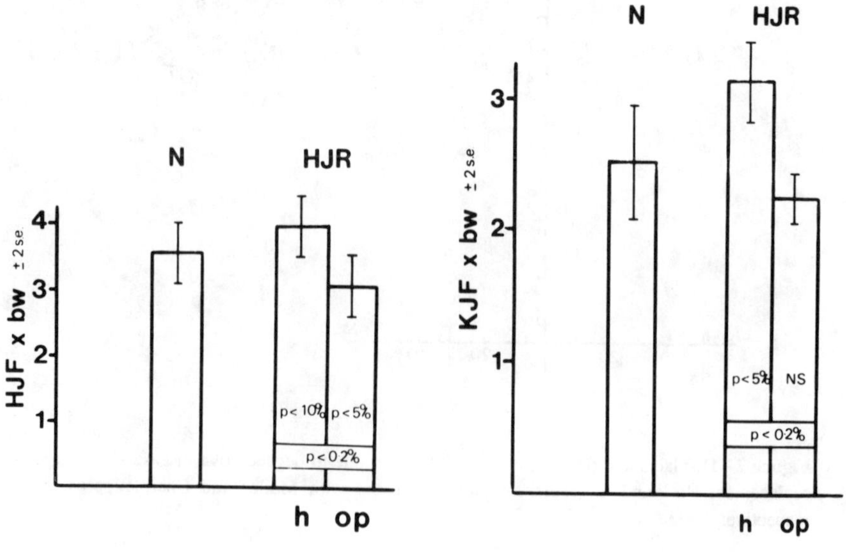

Figure 3—Left: hip joint force for normals and hip patients. Right: knee joint force for normals and hip patients. For patients, "op" represents the operated side and "h" the nonoperated side.

Acknowledgment

This work was supported by the Medical Research Council, the Scottish Home and Health Department, and the McFeat Bequest.

References

JARRETT, M.O., Andrews, B.J., & Paul, J.P. (1976). A television computer system for the analysis of human locomotion. *IERE Conference Proceedings No. 34.*

MORRISON, J.B. (1968). Bioengineering analysis of force actions transmitted by the knee joint. *Biomedical Engineering,* **3,** 614-170.

PAUL, J.P. (1967). Forces transmitted by joints in the human body. Proc. Symposium: Lubrication and Wear in Living and Artificial Human Joints. *Proceedings of the Institute of Mechanical Engineering,* VI81.3J:8-15.

PAUL, J.P.., & J. Poulson. (1964). The analysis of forces transmitted by joints in the human body. *Proceedings of the 5th International Conference on Experimental Stress Analysis,* (pp. 3.34-3.42). Udine, Italy.

The Use of Angle Diagrams for Quantitative Evaluation of Joint Function During Locomotion

P. Arnell
University of Manchester, United Kingdom

F. Johnson and J. Oborne
University of Nottingham, United Kingdom

Several authors have recommended angle diagrams as pictorial summaries of sagittal plane movements of lower limb joints during locomotion (Grieve, 1968; Mitchelson, 1975; Öberg & Lamoreux, 1979). These Lissajous figures are constructed by plotting angular changes of either two adjacent limb segments to form an angle diagram of the associated joint or two joints to form an angle-angle diagram. Use of angle diagrams is largely qualitative, relying on recognition of normal and abnormal configurations. Comparison of normal and pathological plots, at best, yields subjective impressions of differences in the functional joint movements they describe.

Kolstad, Wigren, and Öberg (1982) and Minford, Minns, and Brown (1983) have subjected angle plots to limited quantitative analysis, and have extracted from them dynamic joint ranges for comparative purposes. Only Hershler and Milner (1980) have made any comprehensive proposals of potentially useful numeric parameters to be derived from angle diagrams.

Our study was carried out to investigate the parameters that may be derived from angle plots, to define and interpret their relationships to biomechanical variables and to assess the clinical relevance of these relationships.

Method

A polarized light goniometer was used to obtain angular measurements. Limb-mounted photovoltaic cells received an amount of polarized light from a DC light source proportional to their sagittal plane displacement relative to absolute vertical. They generated a voltage analogue of their angular position scaled at 10 mV/° and sampled at 133 Hz.

A Crane polarized light goniometer, footswitches, and infrared light cells were interfaced to a DEC LSI-11 computer. Software was developed to acquire, store, and display data, to select a segment of data, and to construct and analyze angle diagrams.

Gait records were obtained from 17 controls with normal lower limb joints (9 females and 8 males aged 20 to 70 years) and from 12 patients with radiographic evidence of lower limb joint disease (9 females and 3 males aged 24 to 80 years with rheumatoid arthritis, osteoarthritis, or ankylosing spondylitis).

Transducers were mounted on the pelvis, thigh, calf, and foot of one limb per subject using skeletal references, and calibrated to the sagittal angulation of their limb segment. A gravity-protractor was used for measurement of thigh and calf angles and calibration error was estimated to be that of an experienced user of a manual goniometer, that is, ± 2.5° (Johnson, 1982). Pelvic angulation was set at 0° for the subject's normal stance posture and the foot transducer reflected the gradient of the footwear. Footswitches attached to lightweight insoles, constructed from footprint impressions to ensure accurate placement under the heel and metatarsal heads, were inserted in the shoes. Subjects were given time to become accustomed to walking with these electronic attachments.

All subjects were instructed to walk at their own comfortable walking speed during recording for a maximum of 10 traverses of the 9 m walkway. Infrared light cells at 2.25 m and 6.75 m timed the traverse of the middle 4.5 m length. Simultaneous flexion-extension angle/time records of the four lower limb segments and foot contact signals were obtained (see Figure 1).

In subsequent data analysis, synchronized complete cycles of angular data from the four limb segments were selected using foot contact records. Figure 1 shows the moveable cursors set to select the angular displacement records of the left limb segments between two successive left heelstrikes. Two such cycles were selected from each of a subject's runs to construct an average of 20 each of hip, knee, and ankle angle diagrams and knee-angle/hip-angle diagrams per subject (see Figure 2). Each of the four sets of diagrams was analyzed to provide the mean values of absolute area ($|A|$), square root absolute area ($\sqrt{|A|}$), perimeter (P), the ratio $\sqrt{|A|}/P$ and dynamic joint range (J). The average velocity of each traverse of the walkway was calculated using START (2.25 m) and STOP (6.75 m) timing signals.

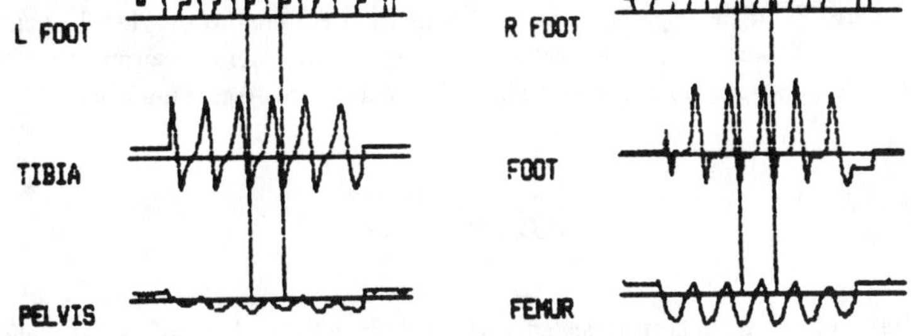

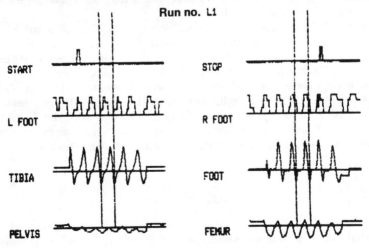

Figure 1—Example of left lower limb flexion-extension angle/time records, foot contact signals, and timing signals in a normal subject.

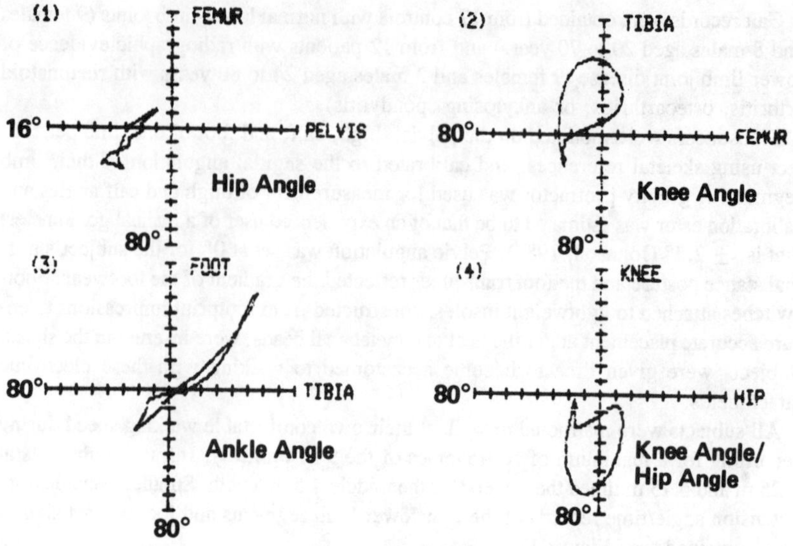

Figure 2—Example of the 4 angle diagrams of the left lower limb in a normal subject.

Results

Preliminary trials with the goniometer established that transducer output error of up to ± 5° is introduced when the light source and transducer are not coplanar. At a 3 m distance from the projector, the maximum field of view for accurate transducer output was confirmed to be 0.5 m from the projector line of sight. This output error was eliminated by moving the projector on a trolley during recording to maintain the subject within a 0.5 m range of the light source.

A summary of the ranges of mean values of derivatives $|A|$, $\sqrt{|A|}$, P, J, and $\sqrt{|A|}/P$ from the four types of angle diagrams and the range of mean average velocities of both the control and patient groups is given in Table 1. Pooled within-subject SD indicates the spread of derivative values for a single individual of either group.

Relationships between the derivative values were investigated and the correlation between $\sqrt{|A|}$ and J proved significant ($p < .001$). A plot of all mean values of $\sqrt{|A|}$ against P for the hip, knee, and ankle joints was found to provide 3 separate, joint-specific clusters of normal values (see Figure 3). Useful discrimination was demonstrated between normal and pathological data points (see Figure 4).

Discussion

The ability of all subjects to repeatedly select their individual, comfortable walking speed was confirmed (see Table 1), ensuring that angle diagrams were not altered by speed variation. Soft tissue movement is known to influence angular output of transducers but was assumed to be a constant error in all subjects.

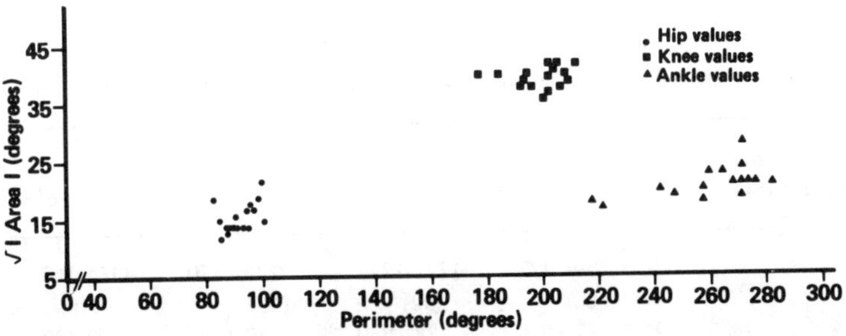

Figure 3—Mean values of $\sqrt{|A|}$ and P from hip, knee, and ankle angle diagrams of normal subjects.

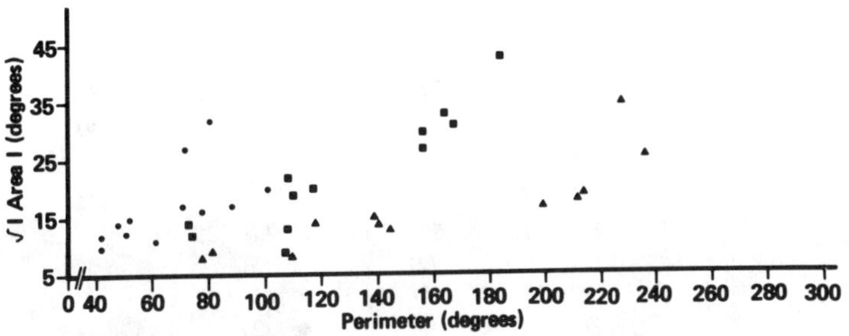

Figure 4—Mean values of $\sqrt{|A|}$ and P from hip, knee, and ankle angle diagrams of patients with lower limb arthropathies.

Unlike the other three types of plots, hip angle diagrams displayed wide qualitative within-subject variation in configuration, with consequent high pooled within-subject SD of |A| and $\sqrt{|A|}$ derivatives. This finding appears to reflect the special difficulties of excluding movements in the other orthogonal planes from hip flexion-extension measurements.

Our findings, which suggest $\sqrt{|A|}$ as an analogue for sagittal movement about the joint axis, concur with a similar proposal by Hershler and Milner (1980). It is possible that a larger sample will permit selection of boundary values for $\sqrt{|A|}$ for each joint which discriminate between normal and abnormal joint ranges.

Hip and ankle angle diagrams were seen to contain crossover patterns. The algorithm used to estimate areas attached positive or negative signs to successive self-contained loops. The significance of these sign differences merits further investigation. By transforming area into absolute area values, we may be discarding useful information about frequency differences between adjacent limb segments' movements and interseg-ment energy transfer.

Our data suggest a new relationship for P, that is, a function of the amount of forward pendular movement of the joint itself, with the mean value of P progressively increasing from proximal to distal joints.

Area and perimeter analysis challenge the use of angle diagrams for subjective impressions of function, as apparently dissimilar figures can yield approximately equal area and equal perimeter values.

Table 1

Range of Mean Values of Angle Diagram Derivatives and Average Velocities and the Pooled Within-Subject SD

Range of Mean Values (SD)		Hip	Knee	Ankle	Knee/Hip
$\|A\|$ (°)²	*N*	139- 472 (80)	1291-1782 (92)	284- 770 (43)	1115-1637 (85)
	P	97-1016 (73)	88-1830 (72)	69-1194 (51)	73-1238 (66)
$\sqrt{\|A\|}$ (°)	*N*	12- 22 (2)	36- 42 (1)	17- 28 (1)	33- 40 (1)
	P	10- 32 (2)	9- 43 (2)	8- 35 (1)	8- 35 (1)
P (°)	*N*	82- 100 (4)	177- 212 (7)	217- 282 (16)	178- 215 (9)
	P	42- 101 (4)	73- 184 (7)	78- 236 (10)	68- 168 (7)
J (°)	*N*	28- 40 (2)	59- 71 (2)	17- 39 (3)	
	P	11- 35 (2)	4- 64 (3)	5- 37 (2)	
$\sqrt{\|A\|}$/P	*N*	0.14-0.23 (0.03)	0.18-0.22 (0.01)	0.07-0.10 (0.01)	0.16-0.22 (0.01)
	P	0.18-0.39 (0.03)	0.09-0.23 (0.01)	0.08-0.15 (0.01)	0.12-0.23 (0.01)
Average Velocity (m/s)	*N*	1.11-1.62 (0.05)			
	P	0.20-1.18 (0.04)			

N = 17 controls with normal joints
P = 12 patients with joint pathology

Our results suggest $\sqrt{|A|}$ to be a measure of the phase difference between two limb segments. In pathological joints in which little or no movement occurs, this phase difference, and consequently the area value, can approach zero as the angle diagram configuration approaches a single line. In such pathological cases, P/$\sqrt{|A|}$ (Hershler & Milner, 1980) approaches infinity and no longer stands as an appropriate ratio. As P in these cases will always have a finite, non-zero value, we have selected the ratio $\sqrt{|A|}$/P. Normal hip angle, knee angle, and knee-angle/hip-angle diagrams all have different, characteristic shapes, yet their $\sqrt{|A|}$/P values are similar (see Table 1). Our data oppose the Hershler and Milner proposal that P/$\sqrt{|A|}$, and therefore its reciprocal, is a descriptor of specific geometric shape. As this index does not differentiate between hip and knee joints in normal subjects, it is unlikely to be a useful discriminator of pathology.

Our plots of mean values of $\sqrt{|A|}$ against P discriminate between the 2 subject groups. The small sample size precludes the use of a sophisticated statistical technique such as discriminant analysis. This would permit investigation of boundary points, for the three joints, between the control and patient groups. Further investigation along these lines is in progress.

Conclusions

Our proposals for the biomechanical relationships of $\sqrt{|A|}$ and P are based on both normal and pathological gait data and therefore suggest a clinical potential for these numeric derivatives.

Our plot of mean values of $\sqrt{|A|}$ against P may offer a clinical method of quantifying change in joint function during locomotion. It is suggested that shift of pathological data points relative to the joint-specific clusters of normal values over time may prove to be a sensitive indicator of such change. Measurement of these position changes may provide, therefore, the means of quantifying change in joint function and thus of assessing the outcome of surgical or conservative forms of treatment.

Acknowledgments

We would like to thank the NWRHA and SHA for their help in establishing the Hope Hospital Gait Analysis Laboratory. We are grateful for the statistical advice from R. Barr, Department of Extramural Studies, and O.V. Smith, ARC Epidemiology Unit, University of Manchester, U.K.

References

GRIEVE, D.W. (1968). Gait patterns and the speed of walking. *Biomedical Engineering*, **3**, 119-122.

HERSHLER, C., & Milner, M. (1980). Angle-angle diagrams in the assessment of locomotion. *American Journal of Physical Medicine*, **59**, 109-125.

JOHNSON, F. (1982). The knee. *Clinics in Rheumatic Diseases*, **8**, 677-702.

KOLSTAD, K., Wigren, A., & Öberg, K. (1982). Gait analysis with an angle diagram technique. *Acta Orthopaedica*, **53**, 733-743.

MINFORD, A.M.B., Minns, R.A., & Brown, J.K. (1983). Asymmetry of gait in normal children demonstrated by polarized light goniometry. *Child Care, Health and Development*, **9**, 97-108.

MITCHELSON, D.L. (1975). Recording of movement without photography. In D.W. Grieve, D.I. Miller, D. Mitchelson, J.P. Paul, & A.J. Smith (Eds.), *Techniques for the analysis of human movement* (pp. 33-65). London: Lepus Books.

ÖBERG, K., & Lamoreux, L.W. (1979). Gait assessment of total joint replacement patients by means of step parameters and hip-knee angle diagrams. In R.M. Kenedi, J.P. Paul, & J. Hughes (Eds.), *Disability* (pp. 274-284). London: MacMillan Press.

The Application of Walking Analysis to Clinical Examination

T. Norimatsu, S.M. Yoh, M. Fujita, Y. Nagatani, N. Matsusaka, R. Suzuki, Y. Okumura, and T. Okajima
Nagasaki University, Nagasaki, Japan

Recently gait studies have progressed remarkably, and recent advances in the treatment of patients with locomotor disturbances have emphasized the need for a practical method to evaluate the progress of therapy. In practice, performing gait analysis with patients is not practical because of pain and other physical problems. To overcome this problem we have undertaken a step by step approach, using serial photography to study two aspects of gait analysis. For the first step, we concentrated on two areas. That is we studied two problems as follows:

(1) How much deviation is there in the walking cycle of the same subject?

(2) How many frames are needed to analyze human gait exactly?

We reported these results at the Eighth Congress of Biomechanics in 1981 at Nagoya.

For the second step, we asked, How many walking cycles are needed for the purpose of clinical treatment?

Figure 1 shows the typical mean pattern of angular changes.

The first trace shows 1/24 s time duration, the second trace 1/36 s, and the third trace 1/60 s with standard deviation. These results mean than there is no noticeable deviation and 1/24 is sufficient to analyze the human gait.

We reported this result at the Fifth ISEK in 1982 in Yugoslavia.

Figure 2 shows one of the typical examples of this result. In this example the trace of five walking cycles is situated within the standard error of 15 walking cycles and there is almost no difference in pattern compared with the trace of 7 walking cycles.

In spite of the severe standard error the results show that 5 walking cycles is sufficient to analyze the human gait.

Materials and Methods

From the results mentioned above, we recorded 1 normal subject and 2 patients for five walking cycles at 1/24s.

The normal subject—Subject A

The O.A. of the ankle joint patient—Subject B

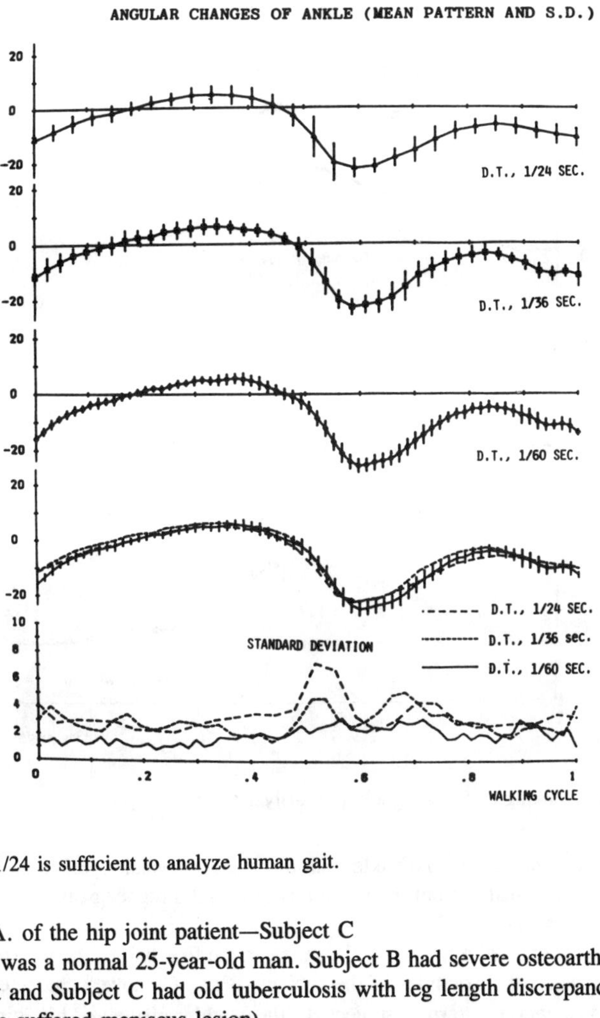

Figure 1—1/24 is sufficient to analyze human gait.

The O.A. of the hip joint patient—Subject C

Subject A was a normal 25-year-old man. Subject B had severe osteoarthritis of the ankle joint and Subject C had old tuberculosis with leg length discrepancy (the opposite knee suffered meniscus lesion).

Angular changes of body rotation, shoulder joint, elbow joint, hip joint, pelvis, pelvic tilt, knee joint, and M.P. joint of the foot of the 3 subjects were measured from serial photographs (shown in Figure 3).

To calculate the standard deviation of the different walking cycles, Cramer's rule was applied. The mean pattern and normalization of the angular changes of five walking cycles were calculated and pictured by a PC-8000 computer.

Results

In the normal subject, the normalization pattern of angular changes of the shoulder girdle, shoulder, and elbow joint were of the convex type, but the pelvis and hip joint were of the concave type.

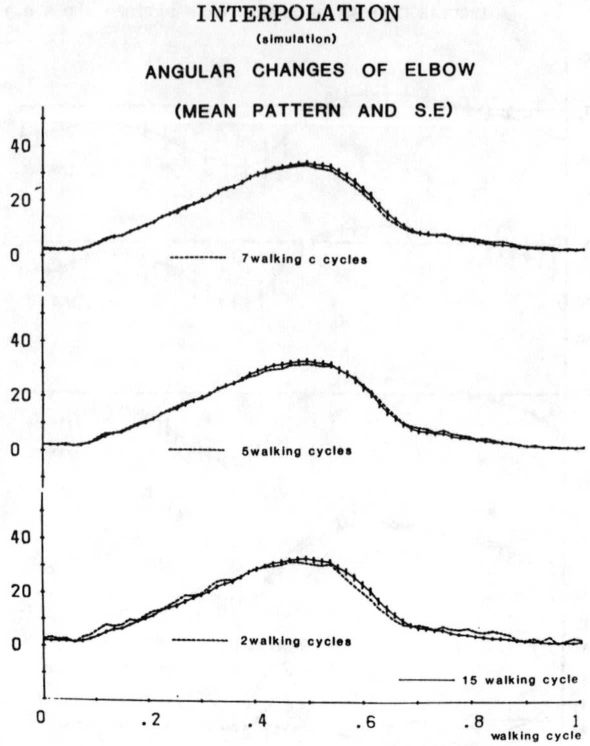

Figure 2—Five walking cycles is sufficient to analyze the human gait.

In relation to other joints, the ankle joint and metatarsophalangeal joint of the foot showed a two-peak pattern, but the ankle joint showed a higher peak in the first half of the walking cycles.

The knee joint showed a smooth lower curve in the first half of the walking cycles. On the other hand, the examples of the subjects B and C showed characteristically.

In the normalization pattern of subject B, the shoulder girdle and hip joint were of a concave type and only the shoulder joint was of a convex type. The shoulder girdle pattern of subject B was completely different from that of a normal pattern.

In the normalization pattern of subject C, the elbow joint was quite different compared with those of the other two subjects. The pattern of the knee joint was similar to that of subject B.

Summary and Conclusion

It is difficult to evaluate the human gait only by the mean pattern of angular changes (shown in Figures 1 & 2), but the normalization pattern of the angular changes made it easy as follows:

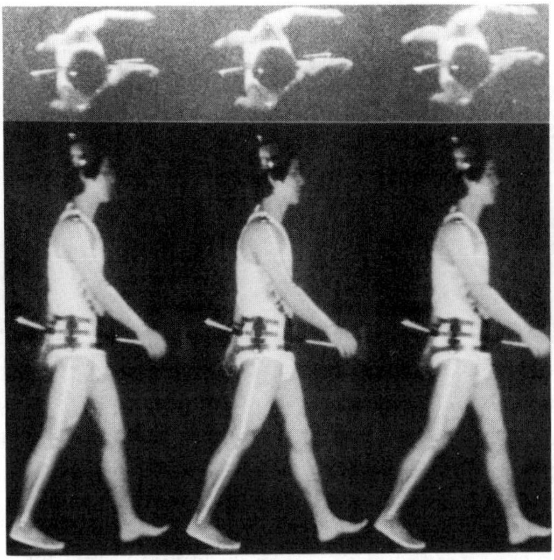

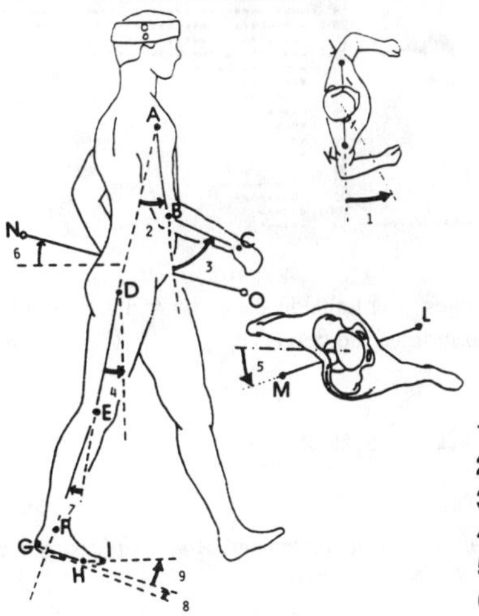

1—shoulder girdle
2—shoulder
3—elbow
4—hip
5—pelvis
6—pelvic tilt
7—knee
8—ankle
9—M.P. joint of the foot

Figure 3—Method of measuring the angle (arrows indicate plus)

1. The pattern of angular changes was examined easily.
2. Differences in the phase were also examined easily.
3. Correlations were recognized as patterns.
4. Normalization of angular changes was useful for clinical evaluation.

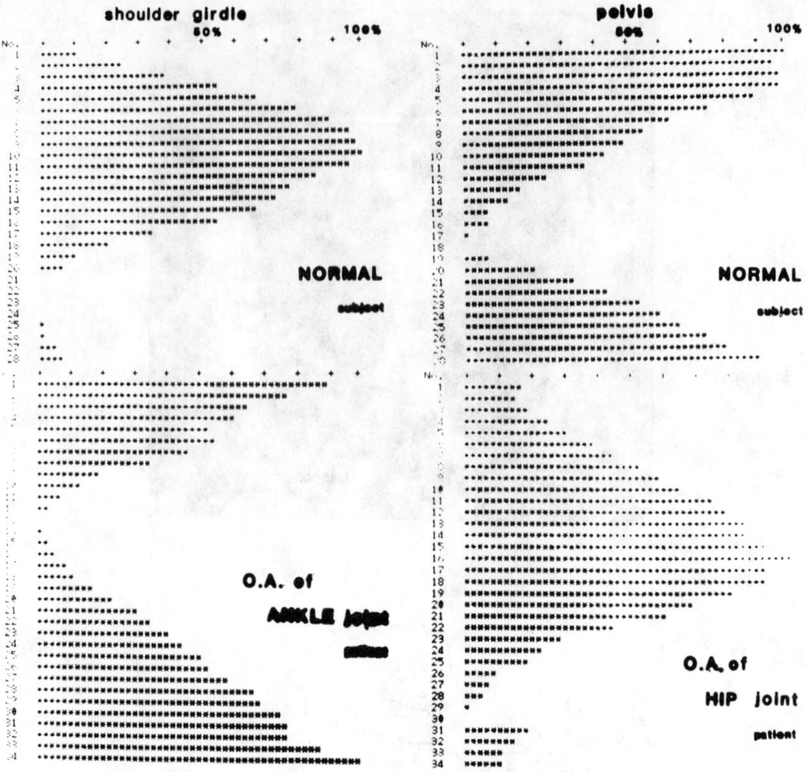

Figure 4—Normalization of angular changes. The left side shows normalization of the shoulder girdle and the right side shows normalization of the pelvis.

References

MURRAY, M.P. (1967). Gait as a total pattern of movement. *American Journal of Physical Medicine,* **46,** 290-333.

MURRAY, M.P., Drought, A.B., & Kory, R.C. (1964). Walking patterns of normal men. *Journal of Bone and Joint Surgery,* **46A,** 335-360.

MURRAY, M.P., Kory, R.C., Clarkson, B.H., & Sepic, S.B. (1966). A comparison of free and fast speed walking patterns of normal men. *American Journal of Physical Medicine,* **45,** 8-24.

MURRAY, M.P., & Clarkson, B.H. (1966). The vertical pathways of the foot during level walking: I. Range of variability in normal men. *Journal of the American Physical Therapy Association,* **46,** 585-589.

MURRAY, M.P., & Clarkson, B.H. (1966). The vertical pathways of the foot during level walking. II. Clinical examples of distorted pathways. *Journal of the American Physical Thereapy Association,* **46,** 590-599.

Comparisons of Spatial and Temporal Patterns in Walking

R.L. Craik, C. Leiper, B.A. Cozzens,
T. Roaper, and M. Hanlon
Moss Rehabilitation Hospital, Philadelphia, Pennsylvania, USA

Evaluation of locomotor performance has gained increased acceptance as a clinical tool. However, the lack of normative age base by which to develop realistic age-appropriate goals and the lack of a data base which defines the effect of walking velocity on spatial and temporal footfall patterns, prevent realistic comparisons between disabled and healthy populations. These data bases would enhance the ability of the clinician to determine whether a gait deviation is age-dependent, velocity-induced or related to pathology. The purpose of this study was to determine whether evaluation of footfall patterns indicated ambulatory dysfunction when performance of a disabled individual was compared to nondisabled individuals walking at similar velocities.

Methods

Thirty healthy adult subjects ranging in age from 21 to 85 years and free from orthopedic, neurological, or systemic complaint were selected to provide a healthy reference data base. Sixteen hemiparetic subjects, 9 left and 8 right, with a mean time since onset of 82 days, were selected from the inpatient population of a rehabilitation hospital. Criterion for selection was the ability to walk with a step length long enough to clear the previous ipsilateral foot print. A second group of 16 subjects with a below-knee amputation were selected from the same hospital population. These subjects had undergone amputation an average of 63 days prior to testing and had not previously walked using a prosthesis. Characteristics of all groups are given in Table 1.

The instrument used was a 3.8 m long walkway consisting of pressure sensitive switches. This walkway consists of left and right sections arranged with 256 pressure switches on each side. The switches are oriented perpendicular to the direction of the subject's progression. The walkway is joined by a flat ribbon connector to a control box which contains digital circuitry and a microprocessor. As the subject progresses over the substrate, time and location of each switch closure and release is recorded by a microprocessor. Temporal and spatial resolution of the walkway is within ± 0.015 s and ± 0.0075 m, respectively, with a switch closure threshold less than 2.5 kg.

Table 1

Characteristics of Sample

		Healthy	Hemiparetic	Amputee
Sex	M	15	6	10
	F	15	10	6
Age (y)	x	49	60	60
	(SD)	(20)	(13)	(14)
Height (m)	x	1.71	1.66	1.72
	(SD)	(0.10)	0.08)	(0.10)
Leg Length (m)	x	0.93	0.88	0.99
	(SD)	(0.07)	(0.05)	(0.06)
Side Involved	L		8	5
	R		8	11
Onset to Test (days)	x		82	61
	(SD)		(68)	(39)
	(median)		(51)	

Street clothes and shoes with heels not higher than 5 cm were worn by all subjects. Subjects traversed a 3.8 m walkway several times. Instructions to each healthy volunteer covered a range of four speeds and included the command, ''Walk as slowly as possible.'' Both groups of disabled subjects were instructed to walk at a comfortable velocity. Data collection continued until a minimum of five steady state steps were collected for each limb. Parallel bars surrounded the walkway for those subjects who required upper extremity assistance for ambulation. Data were collected on two occasions for the amputee subjects, on the initial day of walking with a prosthesis and again at discharge from the hospital.

Performance variables were the average walking velocity, step times, step lengths, stance times, swing times, and double support times for left and right legs. Step length was normalized by dividing the mean value obtained at each speed by each subject's leg length. Average performance was computed for each subject for each of the performance variables. The temporal variables were normalized to the gait cycle (foot-off to subsequent foot-off on the ipsilateral side). A one way analysis of variance was used to determine whether there was a significant difference ($p<.05$) in walking performance among the four groups. A Tukey test was used to define the significant differences between the groups. Symmetry in performance between the two limbs was examined using a repeated measures design. The resolution of the instrumentation required that any temporal asymmetry less than 0.03 s and any spatial asymmetry less than 0.015 m be ignored, that is, considered symmetrical.

Results

The mean walking velocity at the very slow speed for the healthy sample was 0.36 m/s with a standard deviation of ± 0.16. The hemiparetic sample walked at a similar

Table 2

Means and Standard Errors for the Temporal Variables

		Healthy	Hemi-paretic	Amputee Initial	Amputee Discharge
Stance	x	71	78	81	74
	(SE)	(1)	(2)	(1)	(1)
Swing	x	29	22	19	26
	(SE)	(1)	(2)	(1)	(1)
Double	x	20	26	23	18
Support	(SE)	(1)	(2)	(1)	(1)

speed of 0.27 ± 0.16 m/s. The amputees walked significantly slower than the healthy sample on the initial day of walking, 0.22 ± 0.08 m/s ($p<.05$). At discharge, the amputees' walking velocity had improved to 0.42 ± 0.17 m/s, which was significantly faster than initial performance and faster than the velocity demonstrated by the hemiparetic sample ($p<.05$). There were no differences in the performance of the involved limb for the temporal variables compared to the normal subjects. However, stance time was longer and swing time was shorter for the uninvolved limb in both the amputees and the hemiparetics compared to the healthy sample ($p<.05$) (see Table 2).

Swing time for the uninvolved limb is identical to the period of single support for the involved limb. The shortened swing time and protracted stance time for the uninvolved limb in the hemiparetics and amputees reflect an inability or unwillingness to trust the involved limb during its period of single support (see Figure 1). Thus, the performance dysfunction is manifested as a compensation by the sound limb, since performance of the involved limb was similar to that of the healthy sample. In addition to these deviations, subjects with a hemiparesis also showed a protracted double support period compared to the healthy sample ($p<.05$). By discharge, these values in the amputee sample were not significantly different from those of the healthy sample.

Conclusions

The results of this study support the assumption that objective evaluation of footfall patterns distinguishes healthy from pathological gait. Moreover, the differences in walking demonstrated by the amputees at the onset and termination of treatment suggest that the parameters are sensitive indicators of change in performance. When the normative data base includes walking velocities consistent with those of the patient, the data provide more specific indications of gait deficiencies and suggest areas for further intervention. Although the sample is small, the results of this study also suggest that footfall patterns may distinguish between types of locomotor deficits. Ongoing studies include an expansion in the number of subjects in the normative data base so that age and sex can be matched for the patient, and attempts to correlate severity of impairment with locomotor performance.

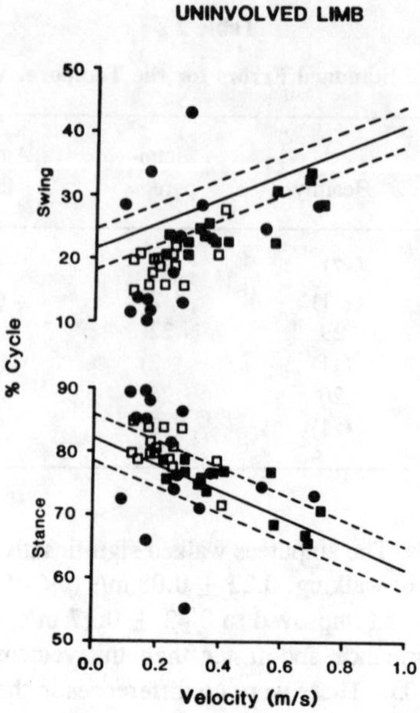

Figure 1—The % of the gait cycle spent in swing and stance. The solid and dashed lines represent the mean and standard deviation for the healthy sample. ● hemiparesis, □ amputee (initial), ■ amputee (discharge).

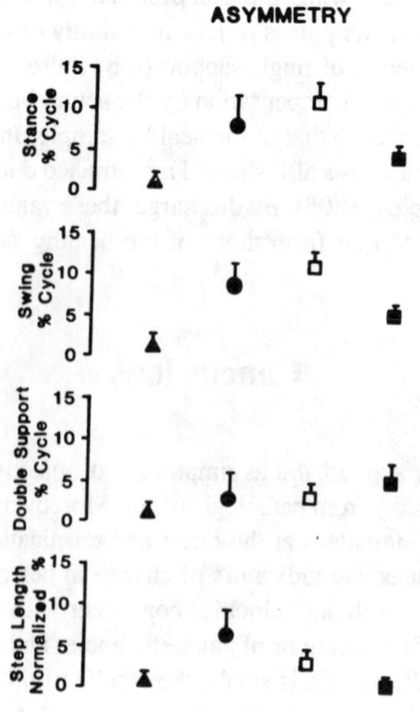

Figure 2—The differences between the limbs. Each difference is represented in % cycle for the temporal variables and % leg length for step length. ▲ healthy, • hemiparesis, □amputee (initial), ■ amputee (discharge).

Acknowledgment

This work was supported in part by NIHR grant #23-P-55518/5.

Electromyographic Features of Lower Limb Muscles for Neurogenic and Myogenic Patients During Standing Posture and Locomotion

K. Sakamoto, T. Usui, M. Muro, and A. Nagata
The University of Electro-Communications, Tokyo, Japan

It is generally difficult to discriminate between the normal person and the dystrophy patient and still more difficult to detect the progression of the deterioration of the nervous and the muscular function. Various procedures such as (a) biochemical analysis (Caruso & Buchthal, 1965), (b) muscle biopsy (Engel, 1967), (c) contraction experiments (Desmedt, 1968), and (d) electromyography (EMG) (Mechler, Csenker, Fekete, & Dioszeghy, 1982; Nagata, Muro, & Murakami, 1982) have been attempted. In these procedures, the following effective parameters have been reported. The diminished absolutely refractory period of muscle fiber bundles produced by electrical stimulation is a useful index in identifying disease (Caruso & Buchthal, 1965). The motor nerve conduction velocity (Mechler et al., 1982) and the disorder of muscle contraction (Desmedt, 1968) can also antedate anomaly. The surface EMG in isometric and dynamic contraction (Nagata et al., 1982) is able to distinguish the neurogenic and the myogenic patients from the normal person. However, EMG study on standing and walking representing the activity of daily life and the study based on the classification of the functional stages have not been done. In this paper, in order to evaluate quantitatively neuro-muscular function for patients, the surface EMGs of lower limb muscles during standing and walking for various diseases and for all functional stages are analyzed.

Methods

Subjects were 10 normal healthy males aged 20.0 ± 3. 0 years and 11 neuromuscular male patients aged 44.0 ± 11.3 years. The latter group included patients with progressive muscular dystrophy (PMD), myotonic dystrophy (MD), and neurogenic muscular atrophy (NMA) (see Table 1). Functional stages for standing and walking in neurogenic and myogenic patients are classified from observation of the gait of patients and from the basis of transfer movement in activities of daily living (ADL). The criterion is defined in Table 2. The muscular function (surface EMG) of these subjects was examined during the maintenance of erect posture within 1 min and during straight

Table 1

Disease and Function Stage

Disease	Function Stage**	Number of subject
A. Myogenic		
(1) PMD (progressive muscular dystrophy)		
(a) Limb-Girdle (LG)	2, 4, 5	5
(b) Facioscapulohumeral (FSH)	3, 4	
(2) MD (myotonic dystrophy)	1, 2, 3, 3	4
B. Neurogenic		
(1) NMA (neurogenic muscular atrophy)		
(a) Kugelberg-Welander (KW)	3	
(b) Charcot-Marie-Tooth (CMT)	4	2
C. Normal	1 for all	10

**See Table 2.

Table 2

Function Classification in Neurogenic and Myogenic Disease

Function Stage	Physical Activity
1	Walk without assistance
2	Walk with the aid of railing
3	Walk without assistance but not able to rise from chair
4	Walk in long-leg braces but require assistance for balance
5	Stand in long-leg braces but not able to walk with assistance

walking for a distance of 6 m. Surface EMG's from the rectus femoris (R.F.), tibialis anterior (T.A.), gastrocnemius (G.) and soleus (S) were measured by bipolar disk electrodes (ϕ = 6 mm), the distance between the electrodes being 2.5 mm.

The integrated EMG (IEMG) was measured at 5 interval second, and the mean value of IEMG was calculated (μV/s). The power spectrum was calculated by fast Fourier transform of the EMG data with the sampling time of 1 ms, and the mean frequency of the power spectrum (Mean Power Frequency MPF) was recorded.

Results and Discussion

The typical EMGs for the neurogenic and the myogenic patient during standing and walking are shown in Figure 1, along with EMGs of normal subjects. The EMGs of

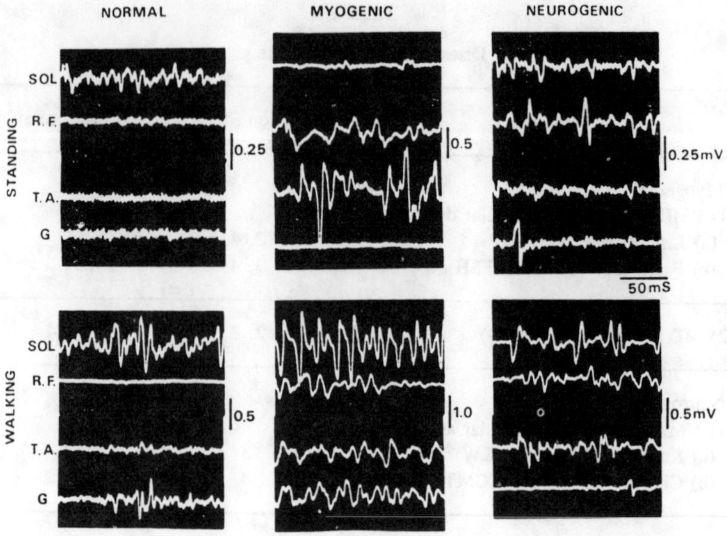

Figure 1—Surface EMGs of normal, myogenic, and neurogenic subjects.

four lower limb muscles of the patients show differences in both amplitude and frequency, as compared with EMGs of normals. Especially for PMD patients (see Table 1) during standing posture, the slow and large amplitude of EMG signals from the rectus femoris and tibialis anterior were seen. During walking, slow and larger amplitudes of EMG signals for soleus and gastrocnemius were observed. During standing, PMD patients tended to lean backwards, while during walking they tended to lean forward. Therefore, the muscles of lower limbs move more actively to hold normal posture.

IEMG values confirm the above observation (see Figures 2-5). The pattern represen-

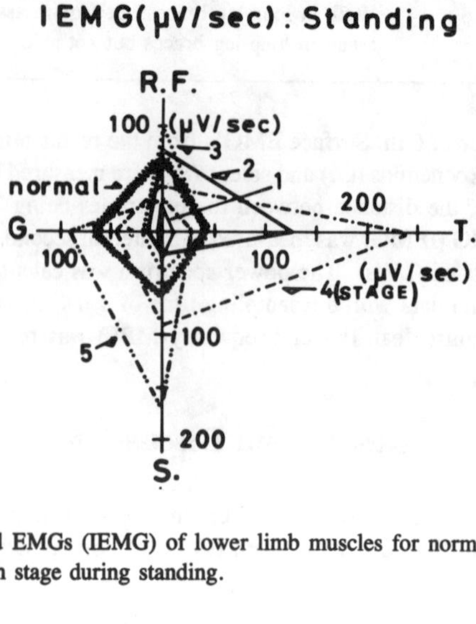

Figure 2—Integrated EMGs (IEMG) of lower limb muscles for normal subjects and patients classified by function stage during standing.

tation is useful for taking the features of IEMG during functional stages and in disease as discussed below.

In the standing posture, IEMG of the tibialis anterior in functional stage 4 and the value of the soleus and gastrocnemius in functional stage 5 show very large values (see Figure 2). This represents high activity of muscles for the adjustments of the ankle joint. In the PMD patients, similar results were obtained (see Figure 3). Since PMD includes patients with various functional stages (see Table 1), the results of Figure 3 indicate the characteristic of the disease. Moreover, PMD gives a larger value of IEMG of the rectus femoris. This means that PMD patients have to control with a lot of effort not only the ankle joint but also the knee joint during standing.

In locomotion, functional stage 2 shows the largest value of IEMG and the higher functional stages 3 and 4 show smaller values when compared with functional stage

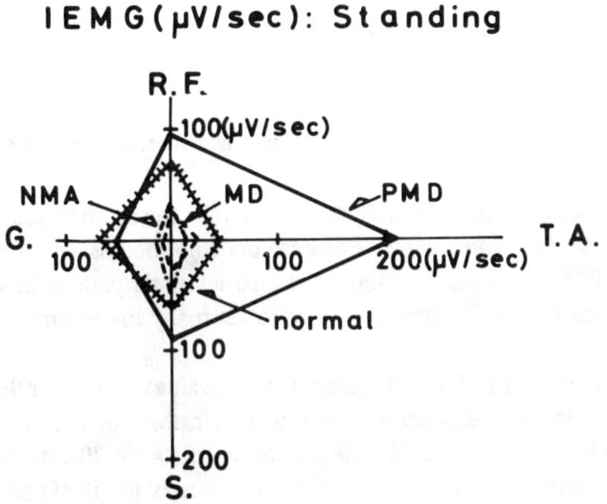

Figure 3—IEMG of lower limb muscles for disease in standing.

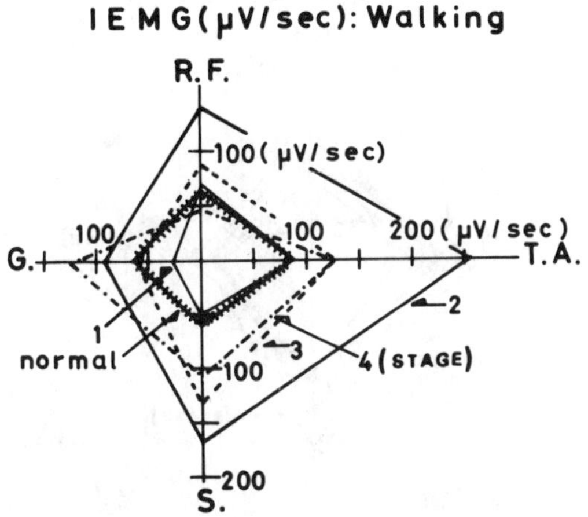

Figure 4—IEMG of lower limb muscles in normal subjects and patients by function stage during walking.

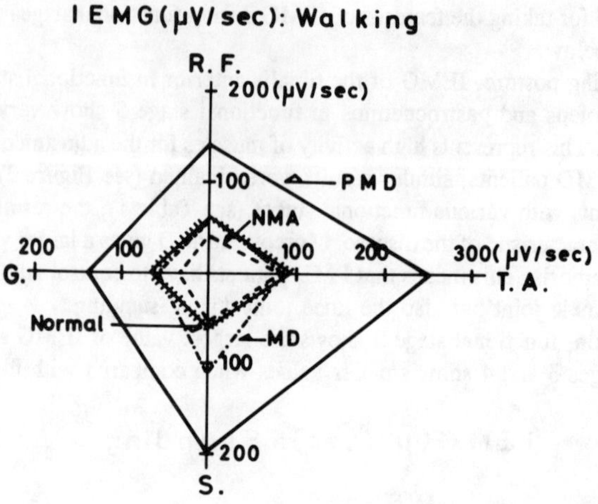

Figure 5—IEMG of lower limb muscles in normal and disease conditions during walking.

2. But they show larger values than those obtained for the normals (see Figure 4). The results suggest that the progression of disease does not necessarily give larger values of IEMG. The PMD patients also show the characteristic pattern in walking as well as in standing (see Figure 5); they also have to set many lower limb muscles to work for walking.

As for the frequency of EMG during standing, the value of MPF for the rectus femoris and tibialis anterior for PMD patients is lower than that for the normals, while the value for the soleus is higher than that for the normals by 20 Hz. During walking, the values of the MPF for the soleus and gastrocnemius for PMD patients are lower than those for the normals. Thus, EMGs for the most active muscle show both slower

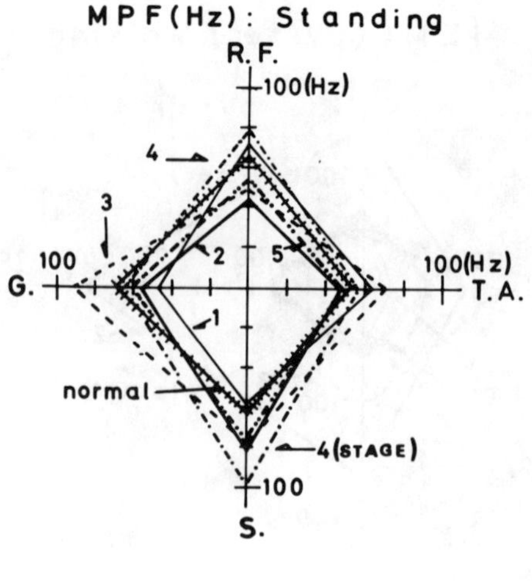

Figure 6—Mean power frequency (MPF) of lower limb muscles in normal subjects and patients by function stage in standing.

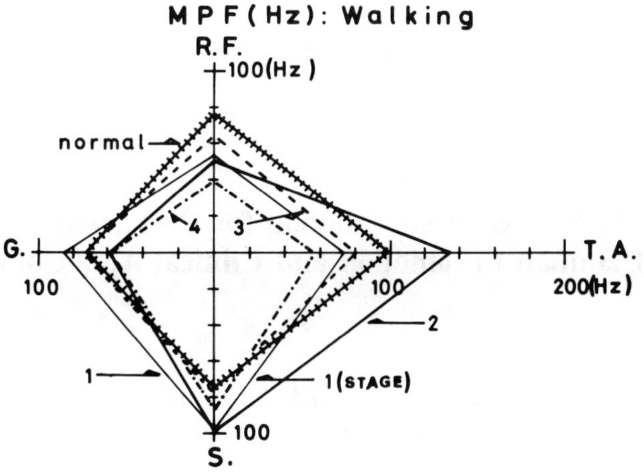

Figure 7—MPF of lower limb muscles in normal and disease conditions by function stage during walking.

and larger amplitude waves for PMD patients (see Figures 1, 4, and 5). In general, the MPF values for the soleus in walking for all the subjects are higher than those in standing by 10 to 20 Hz. This tendency is seen clearly for PMD patients and for functional stages 1 and 2. The progression of the disease (i.e., increase of functional stage in Table 2) shows higher values of MPF in standing, but the opposite result is obtained in walking as shown in Figures 6 and 7. In the EMG signals from rectus femoris, all the patients show a lower value of MPF in walking than that of the normal subjects. As for the tibialis anterior and gastrocnemius, a clear distinction between patients and normals was not observed.

Thus these representations are an effective way of evaluating the degree of muscular function for various diseases.

References

CARUSO, G., & Buchthal, F. (1965). Refractory period of muscle and electromyographic findings in relatives of patients with muscular dystrophy. *Brain,* **88**, 29-50.

DESMEDT, J.E., & Emeryk, B. (1968). Disorder of muscle contraction processes in sex-linked (Duchenne) muscular dystrophy, with correlative electromyographic study of myopathic involvement in small hand muscles. *American Journal of Medicine,* **45**, 853-873.

ENGEL, W.K. (1967). Muscle biopsies in neuromuscular diseases. *Pediatric Clinics of North America,* **14**, 963-995.

MECHLER, F., Csenker, E., Fekete, I., & Dioszeghy, P. (1982). Electrophysiological studies in myotonic dystrophy. *Electromyography and Clinical Neurophysiology,* **22**, 349-356.

NAGATA, A., Muro, M., & Murakami, K. (1982). Power spectrum analysis of surface EMG on isometric and dynamic voluntary contractions for neuromuscular disease. *Japanese Journal of Electroencephalography and Electromyography,* **10,** 111-120.

New Heuristic Method for Prediction of Human Locomotion and Clinical Application

Y.E. Toshev and G.Y. Brankov
Bulgarian Academy of Sciences, Sofia, Bulgaria

When confronted with a nonlinear engineering problem the usual approach is to avoid the nonlinear aspect using linearization. Such approach is very common in the field of the interdisciplinary sciences including biomechanics. However, it is important to note that the prediction of the behavior of living systems is a very delicate problem which cannot be solved with classical mathematical methods.

The Volterra/Wiener representation for nonlinear systems makes it possible to obtain the optimal input/output relation for cases when both the input and the output values of the parameters can be measured in discrete-time form. As a means of the general theory only the input/output relation in the case of two input parameters $\overline{X}_1$ and $\overline{X}_2$ and one output parameter $\overline{Y}$ is presented:

$$\overline{Y} = A_o + A_1\overline{X}_1 + A_2\overline{X}_2 + A_3\overline{X}_1\overline{X}_2 + A_4\overline{X}_1^2 + A_5\overline{X}_2^2 \qquad (1)$$

Obviously it is necessary to carry out a minimum of six experiments in order to determine the six coefficients in Equation 1. However, when four input parameters are used the number of coefficients increases to 70 and the total number is 200,000. It is clear that it is not possible to carry out such a large number of experiments. In addition, years of computing time are required.

The New Heuristic Method

The proposed heuristic method is very similar to the mass animal selection which involves the following steps:

1. Selection of one male and one female animal group (analogous to establishing hypotheses).
2. Breeding and receiving the first propagation (analogous to obtaining the first generation of solution).
3. Selection of the "best" newborn animals (analogous to selecting the "best" solutions).

4. Next breeding and new propagation (analogous to obtaining the second generation of solutions).

5. New selections, etc., up to the moment when the properties of the newborn begin to degenerate (analogous to continuing the computing until the successive solutions become less valid).

The most important question is: What is the meaning of "best" solutions? But can the question, What is the meaning of "best" animals? be easily answered? To find and to use successfully a criterion for the "best" solution (or for "best" animal) is a question of long experience and deep understanding of the specific problem.

Suppose there are "n" input parameters $X_1, X_2, X_3, \ldots X_n$, one output parameter Y and convenient measuring devices. Usually we do more than the minimum number of experiments and every measurement is repeated 2 or 3 times. That means that for every specific set of measured input parameters a value for Y and its standard deviation σ will be obtained. The normalization of the experimental data is a well known procedure. So the "global" description of our system is:

$$Y = F(X_1, X_2, X_3, \ldots X_n) \tag{2}$$

Equation 2 is replaced with a series of "partial" descriptions:

$$\begin{aligned} Y_1 &= f(X_1, X_2) \\ Y_2 &= f(X_1, X_3) \\ &\vdots \\ Y_m &= f(X_{n-1}, X_n) \end{aligned} \tag{3}$$

where m is equal to the number of all possible combinations of the input parameter pairs. In other words in the places of $\bar{X}_1$ and $\bar{X}_2$ from Equation 1 all possible pairs of the input parameters are systematically substituted. For every pair the values Y_1, $Y_2, \ldots Y_m$ are calculated and compared with the experimental value Y. Some solutions are selected using the criterion:

$$\chi^2_{xx} = \sum_{i=1}^{P} \frac{(Y_{calc\ xxi} - Y_i)^2}{\sigma_i} \tag{4}$$

where χ^2_{xx} = the criterion for every possible pair XX, $Y_{calc\ xxi}$ = the calculated value for every pair XX using Equation 1, Y_i = the experimental output values for every experiment, i = number of experiments, p$\geqslant$ = number of the input parameters, σ_i = standard deviations of the experimental values Y_i.

After the selection using Equation 4 new series of "partial" descriptions are selected for the second generation of solutions:

$$\begin{aligned} Z_1 &= f(Y_1, Y_2) \\ Z_2 &= f(Y_1, Y_3) \\ &\vdots \\ Z_k &= f(Y_{\ell-1}, Y_{\ell}) \end{aligned} \tag{5}$$

$$
\begin{aligned}
Z_1 &= f(Y_1, Y_2) \\
Z_2 &= f(Y_1, Y_3) \\
&\;\;\vdots \\
Z_k &= f(Y_{1-1}, Y_e)
\end{aligned} \tag{5}
$$

where 1 = the number of selected solutions from the first generation, k = the number of all possible combinations of the pairs of selected solutions.

Using Equation 4 all the procedures are repeated to select the solutions for the next generation. Remember that if the pair X_4X_8 is selected from the first generation it means that the value of $Y_{calc}\ X_4X_8$ is selected. In other words, the value of $\bar{Y}$ from Equation 1 is selected when $\bar{X}_1$ and $\bar{X}_2$ are replaced with the experimental values of the input parameters X_4 and X_8. For example, in the second generation $\bar{X}_1$ and $\bar{X}_2$ from Equation 1 will be replaced by $Y_{calc}X_4X_8$ and $Y_{calc}X_3X_7$.

From generation to generation the criterion χ^2 will decrease $\chi^2 = 0$ in the case of coincidence of both predicted and experimental values. There will be one moment when χ^2 will begin to increase, at which the procedure is stopped.

Results

Using this method 23 normal subjects were investigated. The goal was to predict the human ground reaction during walking (both left and right feet) on the basis of selected body parameters which are easy to measure:

X_1 = body height
X_2 = ground-iliospinale
X_3 = ground-tibiale
X_4 = tibiale-spherion
X_5 = distance cristarum
X_6 = conugata externa
X_7 = hip circumference
X_8 = knee circumference
X_9 = body weight

Using the original force-plates (Toshev & Baev, 1976) the experimental ground reactions are obtained and the first 10 harmonics are calculated. Using the heuristic method both amplitudes and phases of the first 10 harmonics are predicted step-by-step. Using the inverse Fourier transformation it is easy to obtain the ground reactions as functions of the time.

Four generations of solutions were sufficient to obtain a relatively small value for χ^2 criterion for all 10 harmonics. The step-by-step values of χ^2 for the first harmonica of the vertical component of the right foot ground reactions are shown in Table 1.

This method can be applied in a clinical setting as noted in the following outline:

1. Measurement of 9 body parameters (weight, lengths, circumferences) is conducted.
2. These data are then put into the computer and the predicted ground reactions are immediately obtained (either graphically or numerically).
3. Measurement of the patient's ground reactions and data storage are conducted.
4. Finally the experimental and the predicted data are compared and a diagnostic result is proposed.

Table 1

Chi-Square Value for the First Five Harmonics

Number of the Generation	χ^2 for the Generation
1	230.00
2	115.00
3	1.23
4	0.28*
5	1.16

*The "best."

Acknowledgment

The authors would like to thank Professor Richard C. Nelson, Director of the Biomechanics Laboratory at Penn State University (U.S.A.) for his kind support.

Reference

TOSHEV, Y.E., & Baev, P.P. (1976). The design and construction of a new device for measuring the vertical and the horizontal components of the support reactions during human locomotion. *Biomechanics,* **3**, Sofia. (In Bulgarian)

Ambulatory Consistency of the Visually Impaired

J. Hamill
Southern Illinois University, Carbondale, IL, USA

K.M. Knutzen
Western Washington University, Bellingham, WA, USA

B.T. Bates
University of Oregon, Eugene, OR, USA

The ability to move efficiently through one's environment is fundamental. The development of this skill is of primary importance to all individuals. While the average person utilizes all of his/her senses for gathering information, the visually impaired individual must move through his/her environment in the absence of visual feedback. However, those who are visually impaired find it necessary, and indeed, advantageous to be mobile and independent in travel. The degree of mobility demonstrated by a visually impaired individual has been shown to be enhanced by orientation and mobility training and/or the use of mobility devices (Dawson, 1981). Prior to this training, the gait of a visually impaired individual is hesitant and somewhat less efficient. A measure of efficient locomotion may be gained from the consistency of the gait pattern as evidenced by the symmetry of the lower limbs. Fisher and Gullickson (1978) suggested that consistency in the locomotor pattern should be maintained in order to minimize the metabolic cost of locomotion. Recognizing that there may be unique compensations during ambulation by visually impaired individuals, a study was undertaken to determine the consistency of the kinetics of the support phase of the walking gait of several visually impaired individuals who had orientation and mobility training.

Methods

Subjects

Ten healthy individuals, 5 visually impaired and 5 sighted, served as subjects for the study. The mean age for the visually impaired group was 28.8 years (± 7.4) and for the normal group was 30.2 years (± 7.3). Subjects were required to complete a questionnaire and to sign an informed consent form in accordance with University policy prior to participation in the study. Two of the visually impaired subjects were totally blind and three had some degree of light perception. Each visually impaired subject was a proficient independent traveller who utilized a mobility device and had extensive mobility and orientation training.

Apparatus and Procedures

The experimental set-up consisted of a Kistler Multicomponent Measuring Platform (Type 9261A), centered in a large testing area, interfaced via a Trans Era analog-to-digital converter to a Tektronix 4052 graphics system. Data sampling was accomplished at 1000 Hz. Eight raw voltage signlas were amplified and electronically processed to provide the three orthogonal ground reaction force components. Walking speed (1.70 m/s) was monitored by a photoelectric timing system over a criterion distance (5 m).

The subjects participated in the data collection session in a random order and without reference to group classification. Upon arrival at the experimental area, each subject was fitted with a standard pair of running shoes to eliminate subject-shoe interactions. A nylon guide wire, parallel to the direction of movement, was extended over a 10 m distance to serve as a tracking device. Contact with the line was accepted although the visually impaired subjects were encouraged to make minimal use of the track. Subjects were allowed adequate practice to assure consistent walking speed and a "normal" footfall pattern on the force platform. Each subject was required to complete ten successful trials with both the right and left limbs. A successful trial was one in which the subject contacted the force platform in a normal stride pattern at the designated pace.

Data Analysis

The first step in data evaluation was to normalize the force-time data by dividing by body mass to aid in making between-subject comparisons. Individual trial data were then evaluated to obtain a set of 20 variables consisting of 11 vertical, 5 antero-posterior and 4 medio-lateral values previously identified as representative descriptors of ground reaction force data (Bates, Osternig, Sawhill, & Hamill, 1981). Ten trial mean values for each variable were computed and absolute differences between the corresponding right and left limb values for each subject were calculated. Ratios of the right to left limb values for each variable were also evaluated.

Results

Mean right/left values, ratios and the mean absolute differences for each variable for each group are presented in Table 1. The differences between the mean right/left values for the sighted group were less than 6% for 19 of the 20 variables. The remaining value describing a medio-lateral force component variable (algebraic impulse 0 to 100% of support) exhibited a difference of 12%. The visually impaired group exhibited mean differences of less than 8% for 18 of the 20 variables. Two variables describing the medio-lateral components of total excursions for 30 to 60% of the support phase and algebraic impulse over the total support phase revealed differences of 11% and 13% respectively between the limbs.

Table 1

Mean Values, R/L Ratios and Absolute Differences for Ground Reaction Force Variables

	Visually Impaired				Sighted			
	Right	Left	x Diff.	Ratio	Right	Left	x Diff.	Ratio
Vertical								
Rel. time to first max. force	22.00	22.00	2.20	1.00	22.93	23.58	1.95	0.97
1st max. force	11.30	11.24	0.42	1.01	10.60	10.79	0.43	0.98
Rel. time to min. force	25.54	24.69	0.25	0.99	24.85	24.73	0.17	1.00
Min. force	10.97	10.96	0.10	1.00	10.47	10.72	0.38	0.98
Rel. time to 2nd max. force	72.08	72.61	0.46	0.99	74.43	73.89	0.69	1.01
2nd max force	11.52	11.42	0.41	1.01	11.09	11.08	0.27	1.00
Average vertical force	8.16	8.13	0.20	1.00	7.78	7.87	0.21	0.99
Impulse to 1st max. force	0.120	0.130	0.04	0.92	0.121	0.128	0.01	0.95
Impulse to 1st min. force	0.150	0.150	0.01	1.00	0.139	0.138	0.01	1.01
Total impulse	6.10	6.09	0.15	1.00	6.06	6.14	0.29	0.99
Support time	0.75	0.75	0.01	1.00	0.78	0.79	0.01	0.98
Antero-Posterior								
Rel. time to max. braking	14.97	15.05	1.27	0.99	15.71	16.72	1.35	0.94
Rel. time to transition	47.30	48.74	1.65	1.01	52.64	52.11	1.62	1.01
Rel. time to max. propelling	82.42	83.17	0.51	0.99	83.14	83.56	0.61	0.99
Braking impulse	0.380	0.380	0.01	1.00	0.336	0.333	0.04	1.01
Propelling impulse	0.400	0.410	0.01	0.98	0.351	0.344	0.02	1.02
Medio-Lateral								
Algebraic impulse (0-30%)	-0.001	-0.001	0.000	0.92	-0.001	-0.001	0.000	1.00
Total excursions (30-60%)	0.697	0.630	0.045	1.11	0.604	0.609	0.089	0.99
Algebraic impulse (0-100%)	0.019	0.017	0.006	1.14	0.019	0.017	0.005	1.12
Total excursions (0-100%)	0.303	0.283	0.034	1.07	0.258	0.251	0.024	1.03

Force: newtons/kg of body mass; Time: percent; Impulse: newton sec.

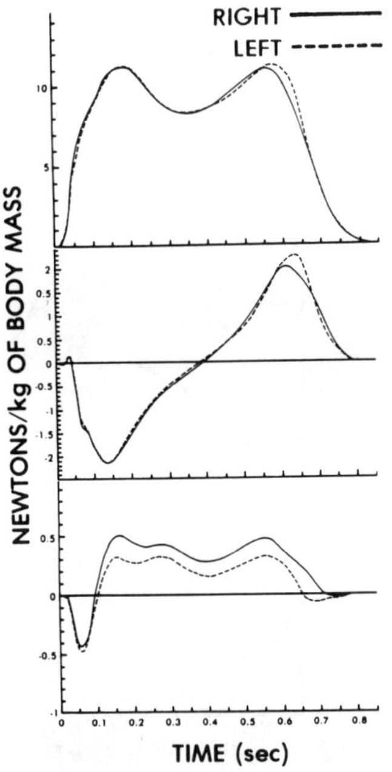

Figure 1—Mean (5 subjects; 10 trials/subject) right and left vertical, antero-posterior, and medio-lateral ground reaction force components of the visually impaired group.

Figures 1 and 2 contain graphic representations of the mean force-time curves for each group (5 subjects/group). These curves were obtained using a point-by-point averaging process based upon a predetermined point, the first maximum vertical force.

Discussion

Ground reaction forces are the source kinetics of ambulation and have often been used as primary descriptors of the support phase of human gait. Human gait is a repetitive activity consisting of distinct cycles with each cycle made up of 2 steps. According to Fisher and Gullickson (1978) the human body appears to have developed kinetically to decrease energy expenditure during ambulation and the metabolic cost is minimized when there is symmetry between the kinematic, kinetic, and temporal variables of the lower limbs. Assuming consistency in the gait pattern, the ratio of right limb to left limb values should approach unity. DuChatinier, Molen, & Rozendal (1970) proposed that this was attainable only in normal gait.

The similarity of the right/left ratios for the variables describing the components of the ground reaction force indicated that the visually impaired subjects had achieved a consistent kinetic ambulatory pattern. Variations between corresponding parameters for the visually impaired group were matched by similar variations for the sighted individuals. Comparisons between the mean absolute differences for the right and left limbs indicated that the extent of variability within the two groups was about equal.

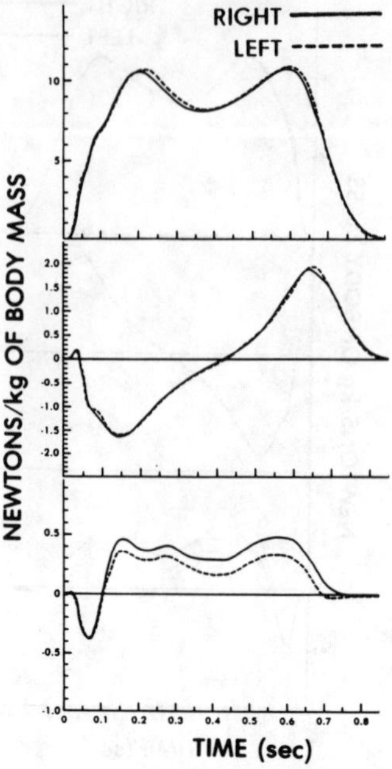

Figure 2—Mean (5 subjects; 10 trials/subject) right and left vertical, antero-posterior, and medio-lateral ground reaction force components of the sighted group.

While there were recognizable differences between the groups for selected variables, basic symmetry within each group was maintained. It may be concluded that the consistency of the gait pattern of the visually impaired is comparable to the consistency of normal gait patterns.

References

BATES, B.T., Osternig, L.R., Sawhill, J.A., & Hamill, J. (1981). Identification of critical variables describing ground reaction forces during running. In H. Matsui & K. Kobayashi (Eds.), *Biomechanics VIII-B* (pp. 633-640). Champaign: Human Kinetics Publishers.

DAWSON, M.L. (1981). A biomechanical analysis of gait patterns of the visually impaired. *American Corrective Therapy Journal*, **35**(3), 66-71.

DUCHATINIER, K., Molen, N.H., & Rozendal, R.H. (1970). Step length, step frequency and temporal factors of stride in normal human walking. *Anatomy,* **73,** 214-227.

FSHER, S.V., & Gullickson, G. (1978). Energy cost of ambulation in health and disability. *Archives of Physical Medicine and Rehabilitation,* **59,** 124-133.

Kinematic Analysis of the Walking Gait of Sighted and Congenitally Blind Children: Ages 6 to 10 Years

H.E. MacGowan
University of Minnesota, Minneapolis, Minnesota, USA

Analysis of walking has been of interest to researchers in several areas, such as physical medicine (Aptekar, Ford, & Bleck, 1976; Edelstein, 1965; Lamoreaux, 1971), child development (Ames, 1937; Gessell, 1928; McGraw, 1936; Shirley, 1931), developmental medicine (Burnett & Johnson, 1974a, 1974b; Scrutton, 1969; Sutherland, Olshen, Cooper, & Woo, 1980) and physical education (Finley & Karpovich, 1964; Gollnick & Karpovich, 1964. Professionals in physical medicine have focused on the analysis of both normal and pathological gaits (Andriacchi, Ogle, & Galante, 1977; Murray, 1967; Murray, Drought, & Kory, 1964), while researchers in child development (Ames, 1937, Gessell, 1928; Frailberg, 1968; McGraw, 1969) have focused upon the sequential development of the walk. In physical education, the limited research identified (Finley & Karpovich, 1964; Gollnick, Tipton, & Karpovich, 1964; Karpovich & Wilklow, 1959) has studied the walking pattern of able-bodied adults or children. There is limited research in child development (Frailberg, 1968; Frailberg, Fraiberg, Fraiberg & Gibson, 1966; Freedman & Cannady, 1971) or physical medicine (Sutherland et al., 1981), and none identified in physical education which has looked at the walking gait of children with a disability.

As mastery of walking is a developmental process, it is reasonable to expect that any phenomena affecting the developmental process could affect mastery of independent walking. Such may be the case with children who are born with a physical or mental disability; for example, a visual impairment, mental retardation, physical malformation, or cerebral palsy, in particular, children who are born blind and have not viewed the walking pattern. A result may be that these children do not master the adult walk pattern by school age.

Problem

The problem of the study was to provide a quantitative description of selected parameters of the walking gait of congenitally blind and sighted children between the ages of 6 and 10.

Methods

Parameters Studied

The 35 parameters studied included six linear parameters—stride length, ratio of the stride length to leg length and to total body height, vertical displacement of the center of gravity, trunk and total body velocity. The nine temporal parameters included total cycle time, the time of total support, double support, single support, and the swing phase, as well as the percentage of total cycle time spent in total support, double support, single support, and the swing phase plus stride frequency. Finally, the positional parameters studied were the angle of the hip, knee, and ankle during early, mid- and late swing, heel contact, midstance and toe-off, plus the trunk angle.

Subjects

A total of 12 girls and 6 boys (nine matched pairs), with a mean age of 9 years, were filmed walking a distance of 10 to 15 m. The congenitally blind and sighted children were matched according to sex, age, height, and weight. The correlation for these factors was, excluding sex, .81, .78, and .90, respectively.

Subject Preparation and Filming

Self-adhesive markers were placed on the subjects' estimated joint centers of the right shoulder, elbow, wrist, hip, knee, and ankle, as well as on top of the head. The subjects walked barefoot and were filmed on three out of six trials.

Filming Procedure

Filming was conducted in an environment familiar to the subject, such as the classroom or cafeteria area in the child's school. The filming was done using a 16mm Photosonic 1PL high speed camera with a 50mm Schneider lens and a camera speed of 50 frames/s.

The camera lens was set 1 m above the ground and 8.9 m from, and perpendicular to, the movement plane. The film plane was parallel to the movement plane and the single camera filmed movement in the sagittal plane only.

Film Analysis

Two of the three trials filmed were digitized and the average of the two trials taken as each subject's results. The film was digitized using rear screen projection with the joint centers being represented by the centers of the black and white markers. Every alternate frame was digitized and digitizing began with the fourth frame prior to right heel touchdown and was completed six frames following completion of one walk cycle. Digitizing reliability for the investigator was .996.

The right side of the subject only was digitized, as the left hip, shoulder and portions of the limbs were not visible throughout the whole cycle. As each trial was digitized, the data were transferred to the University of Minnesota Computer Center for further analysis. The computer data yielded both kinetic and kinematic information; however, only the kinematic information was used in this study. During the digitizing, the walk cycle was divided into six phases so that specific points in time could be readily identified. These phases were (a) right-heel touch down, (b) double support, (c) right single support, (d) double support, (e) left single support, and (f) double support.

Results

There were a total of 11, of the 35 parameters studied, which were statistically significant (t (8) = 2.306, p<.05), and these are listed in Table 1.

Discussion

The mean stride length of both sighted and congenitally blind children is shorter than that reported in the literature for children (Grieve, 1969; Sutherland et al., 1980) and adults (Boenig, 1977; Murray, 1967; Murray et al., 1964). The significantly shorter stride length of the congenitally blind children could be due in part to the lack of visual feedback about the environment, accompanied by considerable apprehension while moving. This apprehension may result in a slower, more controlled movement in an attempt to avoid encounters with object that could cause injury. Another result of the slower movement is a slower total body velocity and a smaller stride frequency for the congenitally blind children.

Table 1

Mean Values for the Significant Linear, Temporal and Positional Parameters

Parameters	Sighted	Congenitally Blind
Linear:		
Stride length (m)	.82	.64
Trunk velocity (m/s)	1.00	.74
Total body velocity (m/s)	1.01	.75
Ratio stride/leg	1.17	.93
Ratio stride/height	.62	.49
Temporal:		
Total support time (s)	.46	.55
Double-support time (s)	.18	.23
% cycle in double support	21.33	25.55
Positional (in degrees):		
Ankle at early swing	109	102
Knee at midswing	122	134
Ankle at toe-off	118	106

Ratios of the stride length to leg length, or total body height, are calculated in an effort to provide a comparative value, in order that appropriate comparisons can be made between subjects and groups of subjects. The ratio of the leg to total body height for the subjects in this study was .9; therefore, it is possible to discuss the stride length to total body height ratio and assume similar results for the stride length to leg length ratio. The ratio of stride length to total body height for the sighted children (.62) falls within the reported range for adults of .5 to 1.1 (Folley, Quanbury, & Steinke, 1979; Smidt, 1971). The congenitally blind children's ratio (.49) falls only just outside (.01) this range. As total body height was one of the factors upon which subjects were matched and the congenitally blind children had a shorter stride length, it might be expected that this ratio be smaller.

The congenitally blind children have a slower total body velocity (.75m/s), and as might be expected, take longer to complete a cycle. The total cycle time for the congenitally blind children was .92 s and for the sighted .84 s, both values being smaller than the adult cycle time of 1 s (Murray, 1967; Murray et al., 1964).

Accompanying the longer total cycle time of the congenitally blind child is a longer support time (.55 s), in particular, double support (.23 s), compared to .46 s and .18 s, respectively, for the sighted children. Increases in total cycle time are due in the main to increases in support and double support times. This demonstrated that less time is spent in single support, or the swing phase. This is evidenced in this study as the congenitally blind children spent 34% of total cycle time in the swing as compared to 37% by the sighted children. The congenitally blind children also spent a higher percentage of time in double support (25.55%) than the sighted children (21.33%), demonstrating that less time was spent in the swing.

The angle of the congenitally blind children's right ankle (102°) during early swing was less than for the sighted children (109°), that is, they dorsiflexed the ankle more. During toe-off however, the angle of the right ankle for the congenitally blind children (106°) was also less than for the sighted children (118°) which resulted in the ankle being plantar flexed less. The right knee of the congenitally blind child during midswing was slightly more extended (134°) than that of the sighted child (122°).

There appears to be no scientific explanation for the differences in joint positions during the walk. Blind individuals have been observed to shuffle along. This may result in a more extended knee during the swing and the foot being plantar flexed more throughout the cycle. The subjects in this study did not shuffle along; however, some apprehension on the part of the congenitally blind subjects may have resulted in the different joint angles.

Conclusions

The congenitally blind children in this study walked differently from the sighted children. They walked more slowly than the sighted children which is evidenced in the slower velocities. Accompanying the slower velocities is a longer cycle time. More time was spent in total support and double support by the congenitally blind children. The right ankle of this group of children was dorsiflexed more during early swing and the right knee extended more during midswing. The congenitally blind children also plantar flexed their right ankle more during toe-off.

Acknowledgments

I wish to thank the University of Minnesota Computer Center and the Graduate School for financial assistance in the purchase and processing of film, and data analysis.

References

AMES, L. (1937). The sequential patterning of prone progression in the human infant. *Gen. Psych. Mono.*, **19**(4), 409-460.

ANDRIACCHI, T., Ogle, J., & Galante, J. (1977). Walking speed as a basis for normal and abnormal gait measurement. *Journal of Biomechanics*, **10**, 261-268.

APTEKAR, R., Ford, F., & Black, E. (1976). Light patterns as a measure of assessing and recording gait. 1: Methods and results in normal children. *Developmental Medicine and Child Neurology*, **18**, 31-36.

BOENIG, D. (1977). Evaluation of a clinical method of gait analysis. *Physical Therapy*, **57**(7), 795-798.

BURNETT, C., & Johnson, E. (1971). Development of the gait in childhood: Part 1. *Developmental Medicine and Child Neurology*, **13**, 196-206.

BURNETT, C. (1971). Development of the gait in childhood: part 2. *Developmental Medicine and Child Neurology*, **13**, 207-215.

EDELSTEIN, J. (1965). Biomechanics of normal ambulation. *Journal of the Canadian Physical Association*, **17**, 174-185.

FINLEY, F. & Karpovich, P. (1964). Electromyographic analysis of normal and pathological gait. *Research Quarterly*, **35**(3), 379-384.

FOLLEY, C., Quanbury, A., & Steinke, T. (1979). Kinematics of normal child locomotion: a statistical study based on T.V. data. *Journal of Biomechanics*, **12**, 1-8.

FRAIBERG, S. (1968). Parallel and divergent patterns in blind and sighted infants. *Psy. Study Child*, **23**, 264-300.

FRAIBERG, S., Fraiberg, L., & Gibson, R. (1966). The role of sound in the reach behavior of blind infants. *Psy. Study Child*, **21**, 327-357.

FREEMAN, D., & Cannady, C. (1971). Delayed emergence of prone locomotion. *Journal of Nervous Disorders*, **153**(2), 108-117.

GESSELL, A. (1928). *Infancy and human growth*. New York: The MacMillan Co.

GOLLNICK, P., & Karpovich, P. (196). Electrogoniometric study of locomotion and some athletic movements. *Research Quarterly*, **35**(3), 357-367.

GOLLNICK, P., Tipton, C., & Karpovich, P. (1964). Electrogoniometric study of walking on high heels. *Research Quarterly*, **35**(3), 370-377.

GRIEVE, D. (1969). The assessment of gait. *Physiology*, **55**, 452-460.

KARPOVICH, P., & Wilkow, L. (1959). A goniometric study of human foot in standing and walking. *U.S.A.F. Medical Journal*, **10**(8), 885-903.

LAMOREAUX, L. (1973). Kinematic measurements in the study of human locomotion. *Bulletin of Prosthetics Research*, **25**, 1-24.

MCGRAW, M. (1969). *The neuromuscular maturation of the human infant.* New York: Hafner Publishing Company.

MURRAY, M. (1967). Gait as a total pattern of movement. *American Journal of Physical Medicine,* **46**(1), 290-303.

MURRAY, M., Draught, A., & Kory, M. (1964). Walking patterns of normal men. *Journal of Bone and Joint Surgery,* **46A,** 335-359.

SCRUTTON, D. (1969). Footprint sequences of normal children under five years old. *Development Medicine and Child Neurology,* **11,** 44-53.

SHIRLEY, M. (1931). *The first two years: A study of twenty-five babies.* Minneapolis: University of Minnesota.

SMIDT, G. (1971). Hip motion and related factors in walking. *Physical Therapy,* **51**(1), 9-21.

SUTHERLAND, D., Olshen, R., Cooper, L., & Woo, S. (1980). The development of mature gait. *Journal of Bone and Joint Surgery,* **62A,** (3), 336-353.

Gait Performance and Dynamic Characteristics of Weight-Bearing on the Patellar Tendon in PTB Prostheses

J. Mizrahi
Technion, Israel Institute of Technology, Haifa, Israel
and
Loewenstein Rehabilitation Hospital, Raanana, Israel

R. Seliktar
Technion, Israel Institute of Technology, Haifa, Israel
and
Drexel University, Philadelphia, Pennsylvania, USA

A. Bahar
Beilinson Hospital, Petah Tiqva, Israel

Z. Susak
Loewenstein Rehabilitation Hospital, Raanana, Israel

T. Najenson
Loewenstein Rehabilitation Hospital, Raanana, Israel

The patellar tendon bearing (PTB) prosthesis is a well accepted prosthesis for below-knee amputees. In this prosthesis preferred areas are being used to transfer the load beween socket and stump. Especially convenient is the patellar tendon (PT) area, which can withstand relatively high loads without developing pain or suffering from damage. Other areas include the sides of the tibia, as well as the distal and posterior parts of the stump. In the routine process of socket manufacturing, no standard procedure exists by which the force transmission between the socket and stump can be accurately controlled and optimized.

Load transmission between the socket and the stump was qualitatively investigated by Taft (1969) and Faulkner and Pritham (1973). Quantitatively, this matter was studied by Pearson, Holmgren, March, and Oberg (1973) and Lebiedowski and Kostewicz (1977), through measuring the pressures at the interface. The load borne at the distal end of the stump was studied by Katz, Susak, Seliktar, and Najenson (1979).

The purpose of this study was to determine the relation between geometry of the PT insertion in the socket and the dynamic load transmitted through the PT. The latter was measured by a load cell. Gait analysis was used to evaluate the optimal configuration of the PT insertion.

Methods

Patients

Two right below-knee male amputees volunteered and took part in the experiments. In both cases, amputation was for traumatic reasons. The subjects were in an excellent state of health and had adapted well to their prostheses (modular, Otto Bock), while achieving good walking ability. The patients were 32 and 17 years of age, height 1.78 and 1.72 m and mass 101 and 70 kg, and were denoted A and B respectively. The time between amputation and tests was 8 months for patient A and 18 months for patient B. Both patients were supplied with new prostheses, which they started using approximately two weeks prior to the beginning of the tests to allow themselves to adapt to these protheses, to make necessary modifications, and thus to ensure satisfactory fitting.

Instrumentation

A two-component load cell was designed and constructed to monitor the vertical and horizontal components of the force transmitted through the PT. The load cell was mounted on a specially constructed frame which was attached proximally to the previously cut PT insertion of the socket and anchored distally to the modular part of the prosthesis. Horizontal and vertical translations of the mounting frame were carried out to displace the PT insertion relative to the socket and thus to control the geometry variation during investigation. The load cell and mounting frame as attached on the prosthesis are shown in Figure 1.

Gait was monitored on a 10 m walkway including at halfway two Kistler piezoelectric force platforms, from which the foot-ground forces were recorded. Simultaneously, the PT forces on the prosthesis as sensed by the load cell, were monitored. Computer processing of the data included partial and total time-integration of the forces measured, analysis of the peak forces, timings within each of the force curves, time symmetry and impulse symmetry between the feet.

Testing Procedure

The dynamic foot-ground and PT forces were monitored on the walkway for every geometrical configuration of the PT insertion. The initial configuration obtained in the "standard" manufacturing procedure of the socket was taken as the reference zero position, about which translations of the PT insertion inwards and outwards, and upwards and downwards were carried out at predetermined increments. For each configuration, three walking tests were performed. Acceptability of each walking test was established by implementing the gait consistency criterion (Seliktar, Yekutiel, & Bar, 1979).

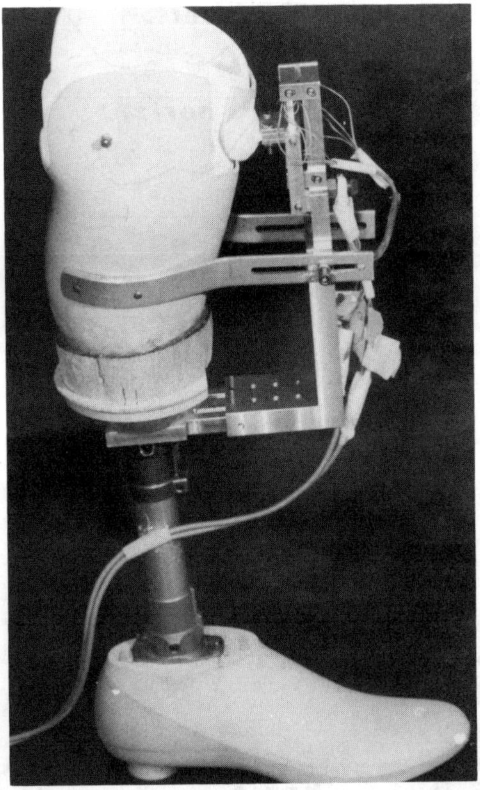

Figure 1—Load cell and mounting frame as attached on the prosthesis.

Optimization of the PT Insertion

The procedure described allows one to easily obtain the relation between geometry of the PT insertion and the load transmitted by the PT. It is, however, of interest to establish a procedure by which the geometry of the PT insertion can be optimized. For that purpose, gait performance criteria were determined as follows:

For each patient, characteristic features of the footground forces were established. These features normally expressed deviations from the typical course of normal force traces. The next step was to quantify the deviations and to relate them to the variation in geometry of the PT insertion. The best geometric configuration was that which minimized the characteristic abnormalities for each patient.

Results and Discussion

A typical pattern of the dynamic forces recorded during a walking test is shown in Figure 2, in which the foot-ground as well as the PT forces are displayed. These curves

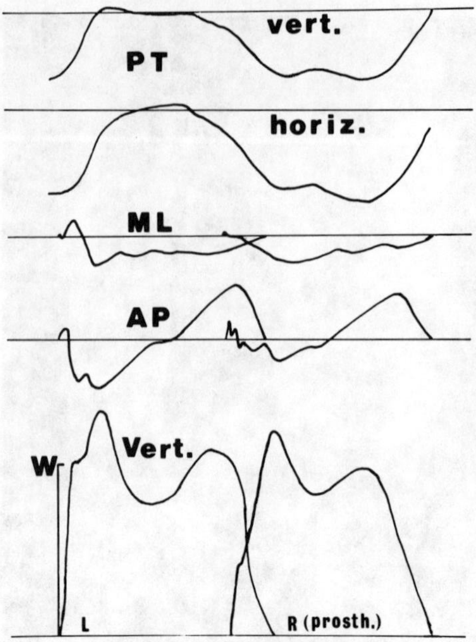

Figure 2—Pattern of forces during walking (patient A). PT = patellar tendon, ML = medio-lateral, AP = antero-posterior, Vert = vertical, W = body weight.

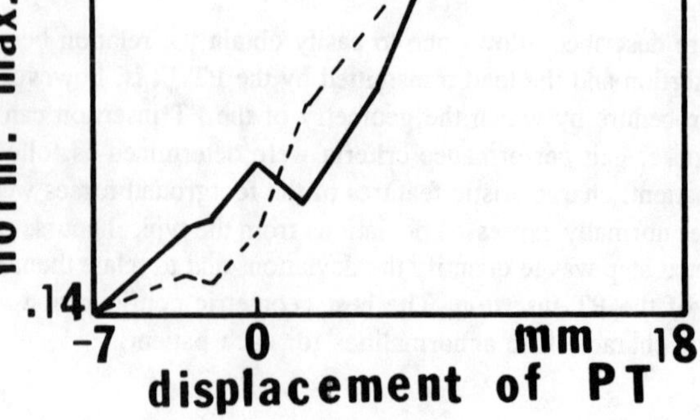

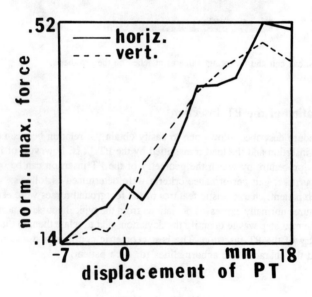

Figure 3—Relation between positioning of PT insertion and the force transmitted by PT (patient A).

correspond to Patient A and the PT insertion in this case is translated inwards by 15 mm relative to the zero reference position (see testing procedure).

The relation between position of the PT insertion and the force transmitted there is illustrated in Figure 3, which corresponds to Patient A. Both vertical and horizontal components were of comparable magnitudes and increased monotonously with displacing the PT insertion inwards. Flattening of both curves occurred near the maximum 18 mm inwards displacement, in which a considerable discomfort was experienced by the patient.

The foot-ground reaction curves of Patient A consistently displayed disturbances in the braking phase of the amputated leg with the prosthesis. It was assumed that these disturbances were the cause of discomfort to the patient and of difficulties in his gait. It was therefore important to minimize their effect through the correct positioning of the PT insertion. A number of criteria based on comparative evaluation of the braking phase were attempted.:

(1) braking phase versus pushing phase in the amputated leg (symmetry within the functioning of the leg)

(2) braking phase in the amputated leg versus the braking phase in the sound leg (braking symmetry between the legs).

The specific parameters on which the above criteria were implemented were: peak forces, impulses of the forces involved, and timings within the force curves.

Figure 4 is an example which shows the variations for Patient A of both braking and pushing impulses on the prosthetic leg versus the horizontal position of the PT insertion. These two curves cross one another at two points, of which the higher

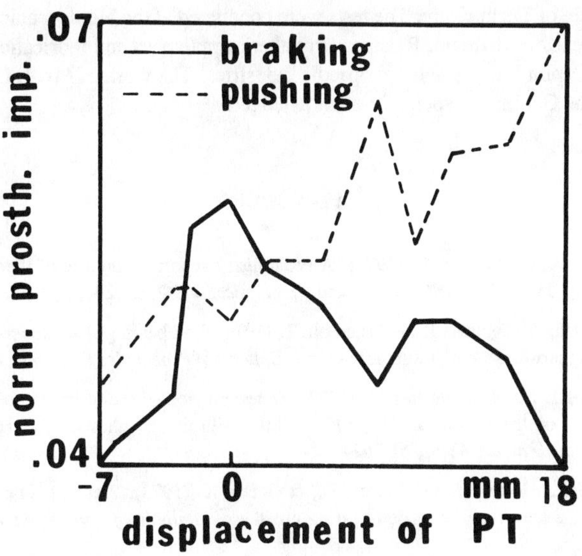

Figure 4—Variation of braking and pushing impulses versus horizontal position of the PT insertion (patient A).

(+2 mm) was selected to indicate correct positioning of the PT insertion, being preferred by the subject tested. A similar positioning was also indicated when plotting the other above mentioned possibilities.

In Patient B abnormalities in the foot-ground forces were reflected by the reduced peak magnitudes of the vertical forces on the amputated leg. This was also noticed, though to a lesser extent, by the decreased antero-posterior forces on the amputated leg. An optimization procedure similar to the one described for Patient A was made for Patient B but with the appropriate criteria.

The procedure undertaken often suggested more than one optimal geometric configuration, which satisfied the required criteria. Furthermore, not all criteria led to the same solution. In Patient A, for example, the optimal positioning reached by the impulse criterion was +2 mm inwards, whereas that determined by the timing criterion was +8 mm inwards. In such cases the actual solution was that which corresponded with the subjective preference of the patients. For both patients the criteria based on the impulses of the force curves and on the impulsing symmetry between the legs were found both successful and satisfactory from the patients' subjective feeling. Optimal positioning of the PT insertion was found at +3 mm inwards for Patient A and nearly the reference zero position for Patient B. These positionings should be taken in fitting the final socket for each patient.

Acknowledgments

This research was supported by the Fund for Promotion of Research at the Technion, Israel Institute of Technology. The tests were conducted at the biomechanics laboratory of the Loewenstein Hospital, Raanana. Prosthetic appliances and fabrication were contributed by Gapim Ltd., Israel Orthopaedic Industries. The work of Mr. I. Onna, Senior Prosthetist of Gapim is especially acknowledged.

References

FAULKNER, V., & Pritham, C. (1973). A preliminary report on the use of thermography as a diagnostic aid in prothetics. *Orthotics and Prosthetics,* **27**(4), 27-29.

KATZ, K. Susak, Z., Seliktar, R., & Najenson, T. (1979). End-bearing characteristics of patellar-tendon-bearing prostheses—a preliminary report. *Bulletin of Prosthetics Research,* **10**-32, 55-68.

LEBIEDOWSKI, N., & Kostewicz, J. (1977). Determination of the order of pressure exerted by static forces on the skin of the lower limb stump with the prosthesis. *Chirurgia Narzadow Ruchu Ortopedia Polska,* **42**(6), 615-624.

PEARSON, J.R., Holmgren, G., March, L., & Oberg, K. (1973). Pressures in critical regions of the below-the-knee patellar-tendon-bearing prostheses. *Bulletin of Prosthetics Research,* **10**, 10-19, 52-76.

SELIKTAR, R., Yekutiel, M., & Bar, A. (1979). Gait consistency test based on the impulse-momentum theorem. *Prosthetics & Orthotics International,* **3,** 91-98.

TAFT, C.B. (1969). Radiographic evaluation of stump socket fit. *Artificial Limbs,* **13**(2), 36-40.

A Biomechanical Study of Normal and Amputee Gait

R. Lewallen, A.O. Quanbury, K. Ross, R.M. Letts
Rehabilitation Centre for Children,
Winnipeg, Manitoba, Canada

Through the use of kinetic and kinematic data and the inverse dynamic relationship a powerful new tool has been developed for the study of pathological gait. This tool allows one to determine the joint moments in the sagittal plane of the ankle, knee, and hip joint. Initial studies have been reported using this tool to determine joint moments in a few normals and a few gait abnormalities (Groh and Baumann 1975; Winter, 1980; 1981). This study is an effort through the use of a gait laboratory to determine the gait cycle, joint angles, and joint moments in the sagittal plane at the ankle, knee, and hip in normal subjects and secondly to compare this normal data with above and below knee amputees.

Methods

Seventeen normal subjects were studied to obtain a control population with which to compare 6 above-knee (AK) and 6 below-knee (BK) amputees. The amputees were young and healthy and had required amputation for trauma in ten cases. The seventeen normal subjects consisted of 8 children and 9 adults. This allowed for a range of body weights ranging from 18 to 85 kg and body heights from 108 cm to 178 cm. Each subject with few exceptions underwent a minimum of two gait studies on each leg. High speed cine film and a standard marking system was used to obtain kinematic data with simultaneous recording used for the gait study have been described previously (Bresler & Frankel, 1950; Winter, 1980; 1981; Winter, Greenlaw & Hobson, 1972; Winter, 1974; Winter, Sidwall, & Hobson, 1974).

The kinematic data were finally combined with the force platform data and a biomechanical model was used to determine individual joint members in the sagittal plane (Bresler & Frankel, 1950) as well as the overall support moment (Winter, 1980; 1981).

Results

Kinematic Analysis

The joint angles were determined for the normal group and compared to both amputee groups. This data show that ankle plantar flexion is increased 15° above normal on the remaining normal leg in both above and below knee amputees. A tendency for the hip to remain in flexion was noted in both the normal and amputated leg in the AK amputee. Otherwise, the range of motion of both types of amputees paralleled the normal population but was slightly decreased.

A comparison of the joint angles was made between the adult normal subjects and the pediatric normal subjects and they were essentially the same.

The stride length and walking speed were calculated from the digitized cine film data for each individual. The stride length and walking speed were highly variable and were dependent on the mood and height of the individual. The step length (which is one-half the stride length)-to-height ratio should be .46 (Murray, Drought, & Kory, 1964). The step length-to-height ratio is .45 in our normals compared to .37 and .40 in our AK and BK amputees, respectively.

The stance and swing phase of gait in the normal group was 61% ± 2% and 39% ± 2%, respectively.

The stance phase of gait was found to be increased in the above AK and BK amputees as compared to the normals by 12.4% and 4.0% in the AK and BK groups, respectively.

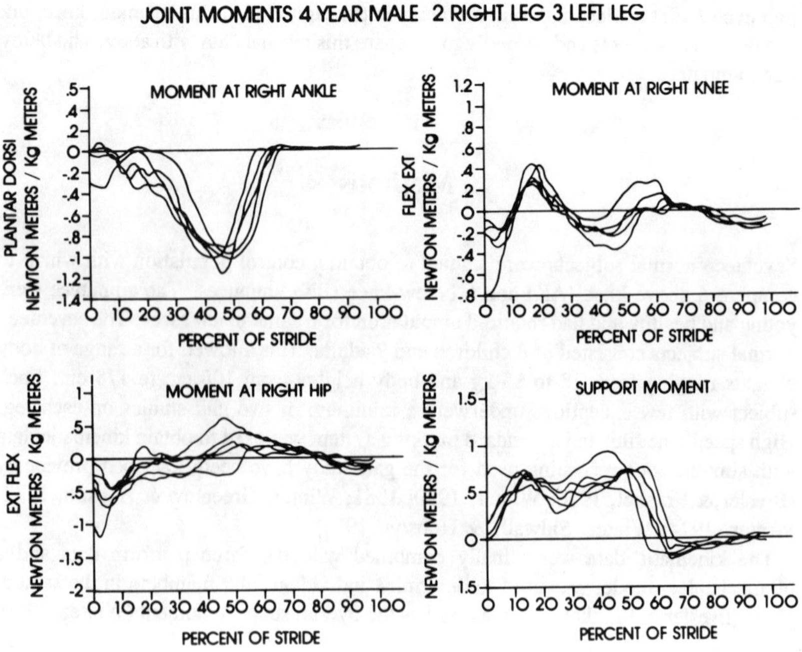

Figure 1—Normalized moment curve.

The force plate data revealed that there was no difference in the progression of the center of pressure in the three groups in the medial-lateral direction or the anterior-posterior direction. The force plate data were normalized by dividing them by the body weight of each individual to allow one to compare the forces easily. There was very little difference between children and adults, except that the first peak of the vertical force curve was larger than the second in children but smaller for adults. The AK and BK amputees showed markedly similar ground reaction forces on their amputated limbs. The posterior force exerted by the amputated limb on initial heel contact was reduced by 50% or more in the AK and BK groups. The normal limbs of the AK and BK amputees showed normal or slightly increased force plate forces as compared to normals. The intact limbs in the amputees differed mainly in the decreased second vertical force peak in the BK amputees as compared to the AK amputees.

The joint moments were studied carefully in the 11 normals who had a minimum of three gait studies. The moments were normalized against the body weight and height. The normalized plots are quite consistent, and are similar to those of a normal subject described by others (see Figure 1) (Winter, 1980; 1081).

The joint moments for the AK and BK amputees were compared to the normals. In comparing the AK amputees to the normals, the knee flexion moment and the hip extension moment of the sound limb was increased slightly above the normal range. The remaining sound limb knee extensor moment and the 2nd support moment peak was below the normal range. The rest of the joint moments were within normal range on the sound limb. The joint moments of the amputated leg were either low normal or below normal.

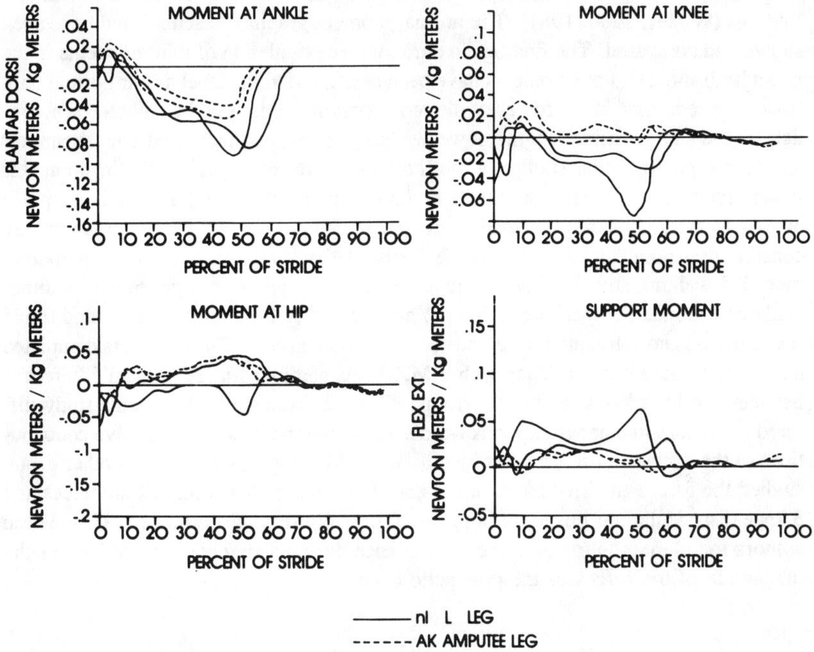

Figure 2—Joint moments for AK amputee.

In the BK amputees, the normal leg generated an increased ankle dorsi-flexion moment, whereas the rest of the joint moments were normal or below normal. On the amputated side, the joint moments were low normal or below normal.

Discussion

This study consisted of two parts. The first part was to gather data on a normal group of subjects who were allowed to walk at their normal pace on a force platform and walkway. The second part of the study consisted of a group of amputees who were allowed to walk at their normal cadence on a walkway and force plate. In this way, our normal group could be used as a comparison to the amputee population.

In this study, our normal pediatric population (youngest age 4) had very similar kinematic and kinetic components of gait as compared to our adult population. In fact, the youngest subject had a very consistent pattern of gait. This agrees with previous studies on the gait of children once the child is age 4 and older (Beck, Andriacchi, Kuo, Fermier, & Galante, 1981; Foley, Quanbury, & Steinke, 1979; Grieve & Gear, 1966).

The force plate data for children show an increased vertical force curve at initial weight acceptance as compared to adults. This increased vertical force curve may be a function of an increased walking speed of children and therefore an increased force at weight acceptance. Although each child was encouraged to walk at a normal rate, the child may have walked faster in laboratory conditions.

The joint moment curves were quite revealing. The moments at the ankle and the support moments were the most consistent in each individual. This is similar to Winter's findings (Winter, 1980, 1981). The normal moment graphs of each normal have been shown and compared. The findings in our amputee population of joint moments in the intact limb show that the moments have been decreased to the level of our normal controls. The reduction in the forces of the remaining intact limb in the amputee is secondary to the shorter stride length, slower walking speed, and increased double support and stance phase. Other studies in the literature have made similar findings but this is the first study to quantify the forces in the sagittal plane of the joints in the amputee as well as the kinetic and kinematic parameters. Studies on amputees from an energy consumption point of view (Waters & Perry, 1976) found that traumatic amputees modified walking speed to keep relative energy costs within normal limits. Another study determined the mechanical load in normal and AK amputee subjects and found that there was no substantial difference between the 2 groups. These authors completed a force plate and kinematic study with E.M.G. data and found no substantial differences between the joint loads in the two groups (Groh & Baumann, 1975). This study differed from ours in that assumptions were made about the action and relative contributions of the various muscle groups by utilizing E.M.G. and x-ray data. Another group studied the forces and kinematics at the knee joint in 5 children—2 AK amputees and 3 knee disarticulations during swing phase (Hoy, Whiting, & Zernicke, 1982). These authors found that the forces at the normal knee during swing phase were 2 times the magnitude of the forces of the prosthetic knee.

Conclusion

This study has used the instrumented gait laboratory to determine the kinematics and kinetics of normal gait and compared this data with the kinematics and kinetics of amputee gait. Data have been gathered on 17 normals and 12 amputees. No significant differences could be found between the children and adults in this study. Joint moment data were determined for the ankle, knee, and hip as well as the overall support moment. The ankle moment data and support moment data show less variability, and one can predict the approximate ankle and support moments of a normal individual using this data. In using this tool in the study of amputee gait, we found that the joint moments in the sagittal plane of the amputee's intact extremity are normal or below normal. The moments and support moments of the amputated extremity are lower than the intact limb confirming the notion that the amputated extremity behaves as a passive support. The intact limb does not develop increased forces in the joints because of a slower walking speed, decreased step-length and increased double support and stance phase of gait. The gait laboratory is a powerful tool in the study of prosthetic and amputee gait.

Acknowledgment

This work was funded in part by the War Amps of Canada. The authors wish to acknowledge the assistance of Mr. R. Borchert for developing the necessary computer programs and Mr. G. Dyck for his assistance in the literature review.

References

BECK, R.J., Andriacchi, T.P., Kuo, K.N., Fermier, R.W. & Galante, J.O. (1981). Changes in the gait patterns of growing children. *Journal of Bone and Joint Surgery,* **63A,** 1452-1456.

BRESLER, B., & Frankel, J.P. (1950). The forces and moments in the leg during level walking. *Transactions of the American Society of Mechanical Engineers,* **72,** 27-36.

FOLEY, C.D., Quanbury, A.O., & Steinke, T. (1979). Kinematics of normal child locomotion—A statistical study based on T.V. data. *Journal of Biomechanics,* **12,** 1-6.

GRIEVE, D.W., & Gear, R.J. (1966). The relationships between length of stride, step frequency, time of swing and speed of walking for children and adults. *Ergonomics,* **5,** 379-399.

GROH, H., Baumann, W. (1975). Joint and muscle forces acting in the leg during gait. In P.V. Komi (Ed.), *Biomechanics V-A,* pp. 328-333. Baltimore: University Park Press.

HOY, M.G., Whiting, W.C., & Zernicke, R.f. (1982). Stride kinematics and knee joint kinetics of child amputee gait. *Archives of Physical Medicine and Rehabilitation,* **63**, 74-82.

MURRAY, M.P., Drought, A.b. & Kory, R.C. (1964). Walking patterns of normal men. *Journal of Bone and Joint Surgery,* **46A,** 335-360.

WATERS, R.L., & Perry, J. (1976). Energy cost of walking of amputees, influence of level of amputation. *Journal of Bone and Joint Surgery,* **68A,** 42-46.

WINTER, D.A. (1980). Overall principle of lower limb support during stance phase of gait. *Journal of Biomechanics,* **13,** 923-927.

WINTER, D.A., Greenlaw, R.K., & Hobson, D.A. (1972). Television—computer analysis of kinematics of human gait. *Computers and Biochemical Research,* **5,** 498-504.

WINTER, D.A., Quanbury, A.O., Hobson, D.A., Sidwall, H.G., Reimer, G., Trenholm, B.C. Steinke, T., & Schlosser, H. (1974). Kinematics of normal locomotion—A statistical study based on T.V. data. *Journal of Biomechanics,* **7,** 479-486.

WINTER, D.A., Sidwall, H.G., & Hobson, D.S. (1974). Measurement and reduction of noise in kinematics of locomotion. *Journal of Biomechanics*, 7, 157-159.

Panel Discussion on
The Present and Future of Clinical Gait Labs

Clinical gait laboratories are a reality in our society today. They have a variety of characteristics and each is somewhat unique in terms of equipment, data collected, form of analysis used and data presentation. This is, to a large part, the result of inhouse development of equipment to meet specific needs and requirements and a continuation of original presentation techniques. As more labs are established with commercially available equipment, more homogeneity of techniques and analyses and presentation forms will emerge. Until then, there will likely be very little standardization among labs.

The present fully equipped clinical gait laboratory requires a capital equipment outlay of around $250,000 and has an annual operation budget of about $80,000/year. It is located either within a health care facility or adjacent to one. This is necessary to provide close association with clinical staff and for convenience of inpatients referred to the lab. Research is a necessary component of the clinical gait lab. To be most successful clinically, research projects should attempt to answer specific clinical problems. The outgrowth of the research project should be a clinically applicable and useful function of the lab—directed towards a specific problem area in pathological gait.

The clinical lab is invariably supported by an interdisciplinary team that may include physicians, engineers, technicians, therapists and other clinical representatives. There is not a clear consensus of who should direct the lab. Ideally, the job description, which includes both engineering and medical knowledge components, should decide the most appropriate team member to direct the lab. However, as a clinical service function within a hospital, the present adimnistrative structure often dictates a physician as the director. The physician can take the overall responsibility of patient treatment and provide the necessary medical link with the heads of all other hospital departments, who are traditionally physicians. The problem does not appear to be one of overcoming one's professional pride but rather the development of a true team spirit to provide the best service for the patient, keeping in mind the various constraints, including the administrative structure, that are often imposed on the team.

Sufficient clinical and medical input to the team can be difficult to obtain. Physicians' clinical practices often preclude their direct involvement in the lab's activities on an ongoing basis. Clinical input from fellows in surgery and physical medicine has been a solution to this problem in some centers. They become involved in the lab on a specific project-by-project basis for the term of their fellowship. A certain amount of their fellowship time can be devoted to gait lab activities and the continuity of the medical fellowship program ensures the continuity of direct medical involvement with the laboratory. The physician must support the lab activities by making the fellow's time available, referring his or her patients to the lab, and giving serious consideration to the output from the lab.

It is recognized that the fully equipped gait laboratory can provide a number of functions regarding the study of pathological gait. These include description, monitoring, analysis, evaluation, and assessment. At the present time the main clinical function of most, if not all, gait labs is gait description. The lab is able to provide a quantitative description of a patient's gait (or certain elements of it) on the particular day of the study. If this description can be taken as representative of the patient's gait at that point in time, then the data can be used to compare pre- and post-treatment conditions, to evaluate the effectiveness of a treatment procedure to produce change, or simply to monitor long term changes in a patient's walking pattern. These quantitative descriptions may be made from individual temporal or kinematic parameter values or curves or may be obtained from a composite of gait descriptors that includes a weighted sum of many gait parameters.

With the advancement of technology the data collection aspect of a gait study is becoming less and less time consuming. This has permitted the study of larger numbers of subjects with both normal and pathological gait. As a result, statistical procedures are becoming more realistic as a means of identifying patient types. The future trend appears to be the classification of patients based on the value of one or more measured gait parameters. Longitudinal studies of these patient groups will show the effectiveness of various treatment procedures—again, on a statistical basis. At the present time this method of using the gait lab to influence patient treatment appears more realistic than using individual biomechanical analysis to diagnose the condition and thereby determine the optimal treatment procedure.

Turnaround time and the availability of gait lab results are influenced by the nature of the center within which the lab functions. Turnaround times vary, among centers, from half a day to six weeks. It is recognized that credibility of results should not be sacrificed for quick turnaround and that certain clinical situations do exist where short turnaround times are necessary. Equally important is the requirement to present the data in a form understandable by the clinician. The education of the clinician in areas of biomechanics is seen as part of the solution to this problem. Various graphic presentation techniques, (for example, the superposition of kinematic data, force platform data, and EMG data) are already being used in an attempt to present a large number of parameters simultaneously and in an understandable manner.

The emphasis on the study of gait rather than other body motions and activities can be defended on the importance of walking and the frequency of complaints from patients regarding walking difficulties. However, it is likely that future laboratories may look at pathological movement analysis in general rather than just at walking.

The clinician gait laboratory is still a developing entity. There are many unanswered questions regarding the most significant parameters to accurately describe and assess pathological gait and continuing research is needed to address specific problem areas. For the next several years each lab will likely retain its own individual nature that is best suited to the staffing, funding, application, and environment of the medical community of which it is a part.

Panel Discussion

Present and Future of Clinical Gait Laboratories.

Arthur Quanbury, Rehab Centre for Children, Winnipeg, Manitoba, Canada

Daniel Antonelli, Rancho Los Amigos Hospital, Downey, California, USA
Lee Kirby, Nova Scotia Rehab Centre, Halifax, Nova Scotia
Keith Laughman, Mayo Clinic, Rochester, Minnesota, USA.
Morris Milner, Ontario Crippled Children's Centre, Toronto, Ontario, Canada
Sheldon Simon, Children's Hospital, Boston, Massachusetts, USA